DUXBURY

Preliminary Edition

Introductory Applied
Biostatistics

Ralph B. Dagostino, Sr.
Boston University

Lisa M. Sullivan
Boston University

Alexa S. Beiser
Boston University

THOMSON
™
BROOKS/COLE

Australia • Canada • Mexico • Singapore • Spain • United Kingdom • United States

THOMSON
BROOKS/COLE

Acquisitions Editor: Carolyn Crockett
Assistant Editor: Ann Day
Editorial Assistant: Rhonda Letts
Technology Project Manager: Burke Taft
Marketing Manager: Joseph Rogove
Marketing Assistant: Jessica Perry
Advertising Project Manager: Tami Strang
Project Manager, Editorial Production: Belinda
Krohmer
Print/Media Buyer: Jessica Reed
Permissions Editor: Sommy Ko
Cover Designer: Denise Davidson
Cover Image: PhotoDisc®, Getty Images™
Cover Printer: Webcom Limited
Printer: Webcom Limited

Printed in Canada.

1 2 3 4 5 6 7 07 06 05 04 03

ISBN 0-534-40689-0

For more information about our products, contact
us at:
Thomson Learning Academic Resource Center
1-800-423-0563
For permission to use material from this text,
contact us by:
Phone: 1-800-730-2214
Fax: 1-800-730-2215
Web: http://www.thomsonrights.com

Brooks/Cole—Thomson Learning
10 Davis Drive
Belmont, CA 94002
USA

Asia
Thomson Learning
5 Shenton Way #01-01
UIC Building
Singapore 068808

Australia/New Zealand
Thomson Learning
102 Dodds Street
Southbank, Victoria 3006
Australia

Canada
Nelson
1120 Birchmount Road
Toronto, Ontario M1K 5G4
Canada

Europe/Middle East/Africa
Thomson Learning
High Holborn House
50/51 Bedford Row
London WC1R 4LR
United Kingdom

Latin America
Thomson Learning
Seneca, 53
Colonia Polanco
11560 Mexico D.F.
Mexico

Spain/Portugal
Paraninfo
Calle/Magallanes, 25
28015 Madrid, Spain

INTRODUCTORY APPLIED BIOSTATISTICS

Ralph B. D'Agostino, Sr., Lisa M. Sullivan, Alexa S. Beiser

Contents

Introductory Applied Biostatistics

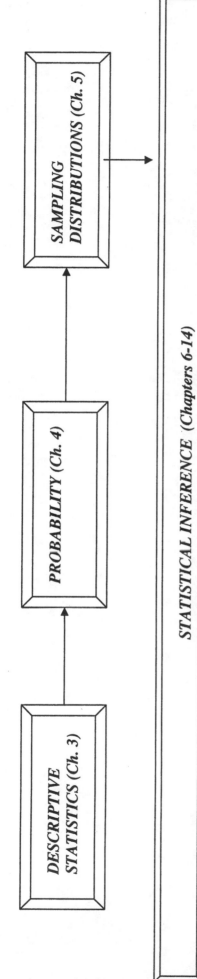

DESCRIPTIVE STATISTICS (Ch. 3) → PROBABILITY (Ch. 4) → SAMPLING DISTRIBUTIONS (Ch. 5) →

STATISTICAL INFERENCE (Chapters 6-14)

Outcome Variable	Grouping Variable(s)/Predictor(s)	Analysis	Chapter(s)
Continuous	-	Estimate μ, Compare μ to Known, Historical Value	6/13
Continuous	Dichotomous (2 Groups)	Compare Independent Means (Estimate/Test ($\mu_1 - \mu_2$)) or the Mean Difference(μ_d)	7/13
Continuous	Discrete (> 2 Groups)	Test the Equality of K Means using Analysis of Variance ($\mu_1 = \mu_2 = \ldots \mu_k$)	10/13
Continuous	Continuous	Estimate Correlation or Determine Regression Equation	11/13
Continuous	Several Continuous or Dichotomous	Multiple Linear Regression Analysis	11
Dichotomous	-	Estimate p, Compare p to Known, Historical Value	8
Dichotomous	Dichotomous (2 Groups)	Compare Independent Proportions (Estimate/Test ($p_1 - p_2$))	8/9
Dichotomous	Discrete (>2 Groups)	Test the Equality of k Proportions (Chi-Square Test)	8
Dichotomous	Several Continuous or Dichotomous	Multiple Logistic Regression Analysis	12
Discrete	Discrete	Compare Distributions Among k Populations (Ch-Square Test)	8
Time to Event	Several Continuous or Dichotomous	Survival Analysis	14

CHAPTER 1: Introduction

Biostatistics is the application of the principles of statistics to the medical or health field. The techniques we discuss here can be applied broadly; our emphasis in the applications is generally medically oriented.

Why is statistics or biostatistics important? Consider an example. Most people believe that it is beneficial in terms of cardiovascular health to exercise regularly and to eat a low fat diet. It is this underlying belief that often motivates us to exercise on a day when we otherwise might skip exercising, or to feel guilty when we consume a high fat meal or maybe two desserts. Should we be motivated by this belief or truly feel guilty? Maybe this idea of exercising and eating right is just a rumor that got out of control!

Why do we believe that exercising and eating healthy is associated with better cardiovascular health? Well, there are a number of research studies that have shown that people who exercise regularly and eat low fat diets are less likely to suffer from cardiovascular disease. An example of one such study is the Framingham Heart Study (Kannel WB: Habitual level of physical activity and risk of coronary heart disease: the Framingham Study. *Can.Med.Assoc.J.* 1967;96:811-812., Dawber TR, Kannel WB, Pearson G, Shurtleff D: Assessment of diet in the Framingham Study; methodology and preliminary observations. *Health News* 1961;38:4-6.). The Framingham Heart Study started in 1948 and involved over 5000 men and women. At the beginning, each participant in the study had a complete physical examination. The participants came back for repeat examinations every two years and the study is still going on 2002. Along with monitoring such things as blood pressure, cholesterol, exercise and nutrition, the study investigators also monitored whether participants had heart attacks or other cardiovascular disease over time. Much of what we understand today about cardiovascular disease was derived from this very important study.

The details of investigations based on important studies often get reported in the newspaper or recapped on television newscasts. Should we believe that just because there was an association between exercise and cardiovascular disease among the participants in the Framingham Heart Study (or some other study) that the same would hold for us? This idea of generalizing or inferring associations that are observed in a group of participants under study to the population at large is the crux of statistics.

Everyday we are inundated with information. Information comes from newspapers, magazines, book, television newscasts and the Internet. We need to distinguish important information from not-so-important information. We need to determine when the results of research studies apply to us and when they do not. It is unlikely that many of us will ever participate in a research study. However, we rely on well conducted research studies to inform us of important associations such as those between certain behaviors (e.g., exercise and diet) and better health. Statistics will provide us the tools to analyze and interpret studies.

Our primary goal in this book is to provide background for readers to apply and appropriately interpret statistical applications in the medical field. We have three specific aims:

- To provide an overview of *Statistical Vocabulary*,
- To describe *Statistical Methodology and Interpretation*, and
- To introduce *Statistical Computing Techniques*.

Throughout the book, we present vocabulary associated with each application along with explicit definitions of terms and concepts. We provide readers with the tools to implement appropriate statistical techniques along with sufficient detail to understand each concept in a broad sense.

We cover a variety of methodologic topics from descriptive statistics and probability theory to statistical inference. The majority of the book is dedicated to methods for statistical inference including one and two sample tests for means and proportions, analysis of variance techniques and correlation and regression. In each topic area, we discuss the methodology, including assumptions and statistical formulas, along with appropriate interpretation of results. We introduce and discuss applications through real examples, most of which are taken from our work in applied biostatistics. We have purposely selected examples involving relatively few subjects to illustrate computations while minimizing the actual computation time. All of the techniques described can be applied to larger problems in practice.

Statisticians almost always use the computer and one statistical computing package or another to implement statistical techniques. There are a number of statistical computing packages available and most can be implemented on any computing platform (e.g., personal computer, Macintosh, UNIX, mainframe). In this book we use SAS (SAS Institute Inc., SAS® User's Guide: Statistics, Version 8.02. Cary, NC: SAS Institute Inc., 1999-2001) to illustrate statistical computing techniques. In each chapter we present SAS programming code and output for each technique. Our objectives are to familiarize the reader with both the format and content of computer output and to provide appropriate interpretation of results. In the statistical computing sections we present programming techniques for interested readers. An introduction to the statistical computing package SAS is contained in Appendix A. This introduction includes an overview of the SAS system, techniques for creating and executing SAS programs and examples of various techniques for entering data. Throughout the statistical computing sections we include advanced statistical computing techniques in the sections marked "*" for more sophisticated readers. These sections can be omitted, as appropriate. The omission of these sections will not disrupt the progression of material.

Although statistical techniques are almost always implemented using statistical computing packages, we feel strongly that readers must master techniques by hand before moving to computer applications. We therefore focus on explicit computations throughout this book. Wherever possible, we illustrate techniques by hand and then perform the corresponding analysis using SAS.

Introduction

CHAPTER 2: Motivation

2.1 Introduction
2.2 Vocabulary
2.3 Population Parameters
2.4 Sampling and Sample Statistics
2.5 Statistical Inference

2.1 Introduction

To motivate the topics covered in detail in subsequent chapters, here we present a simple example to illustrate the major concepts in applied biostatistics. In an overview fashion, we present a number of vocabulary terms along with brief definitions. General formulas are indicated in shaded boxes. We leave it to the reader to continue into subsequent chapters for more explicit and involved discussions of each term and concept. We recommend that readers refer back to this example from time to time as a means of reminding themselves about the various concepts and their interrelationships.

In Section 2.2 we present some general vocabulary terms and the example. In Section 2.3 we discuss population parameters. In Section 2.4 we discuss sampling and sample statistics and in Section 2.5 we describe statistical inference.

2.2 Vocabulary

Statistical analysis is the analysis of characteristics of subjects of interest. *Subjects* are the units on which these characteristics are measured. In most medical applications, subjects are human beings. However, subjects may be cells, blood samples, or animals used in research experiments. The characteristics are measurable properties, such as age, systolic blood pressure, outcome of surgery (e.g., success, failure), total cholesterol level, and are called *variables*. It is important to define variables explicitly in each application along with appropriate measurement units. For example, age could be measured in years, weeks, days, hours, etc. An investigation involving human subjects would most likely measure the age of each subject in years (except, possibly an investigation of newborns in which case weeks or days may be more appropriate). An investigation involving blood samples might measure the age of each sample in days or hours.

In statistical applications we work with *data elements* or *data points*. Each data element is a representation of a particular measurable characteristic, or variable. Data elements may be subjects' ages measured in years (e.g., 51, 29, 36), systolic blood pressures measured in millimeters of mercury (mmHg) (e.g., 140, 160, 110), or classifications of disease stage (e.g., stage I, stage II, stage III).

In all statistical applications, it is important to define the *population* of interest explicitly. A population is simply the collection of <u>all</u> subjects of interest. For example, if we are interested in

American males with heart disease then the collection of <u>all</u> American males with heart disease would constitute the population. A *sample* is a subset of the population of interest. For example, American males aged 50 with heart disease would constitute one sample from the population of all American males with heart disease. One hundred randomly selected American males with heart disease would constitute another sample from the population of all American males with heart disease (assuming that there are many more than 100 American males with heart disease).

There are many different samples that can be selected from any given population. The number of distinct samples that can be taken from a given population depends on the numbers of subjects in both the population and in the sample. In subsequent chapters we will outline explicit formulas to determine the number of possible samples from a particular population. We will also present some of the more popular methods used to select subjects from a population into a sample.

The number of subjects in a population, or the *population size*, is denoted N. The number of subjects in a sample, or the *sample size*, is denoted n. Any descriptive measure based on a population is called a *population parameter,* or simply a *parameter*. Any descriptive measure based on a sample is called a *sample statistic*, or simply a *statistic.* N is an example of a parameter, while n is an example of a statistic.

Data elements may have been measured on each member of a population or on each member of a sample. In most applications, populations are very large. There are exceptions, however, and an analyst cannot tell simply by looking at the data set or by its size alone (i.e., the number of data elements), whether the collection of data elements comprise a population or a sample. Someone involved in the design of the study, such as the statistician or an investigator in the particular substantive field, must convey such information to the data analyst.

Data elements (or observations) deriving either from a population or a sample are denoted X, where X is a variable name, or placeholder, representing the characteristic of interest. We will illustrate the use of this notation through examples in Chapter 3.

Example 1. Suppose we have a population consisting of 5 individuals who are 65 years of age or older (i.e., N=5). In this example, our population is small and our interest lies solely in these 5 individuals. Suppose, we are interested in analyzing the number of visits to primary care physicians over a 3 year period in our population. We survey each member of our population and assess the number of visits to primary care physicians over the previous 3 years. We exclude emergency room visits and hospitalizations from our assessment. The characteristic of interest, or the variable under investigation, is the number of visits to primary care physicians in 3 years. The population data are shown below and displayed in Figure 2.1 in a dot plot where each observed data point is indicated by a dot. The horizontal scale represents the number of visits to primary care physicians in 3 years.

Subject Number	Number of Visits to Primary Care Physicians In 3 Years
1	2
2	4
3	6
4	10
5	18

Figure 2.1   Number of Primary Care Visits In 3 Years (N=5)

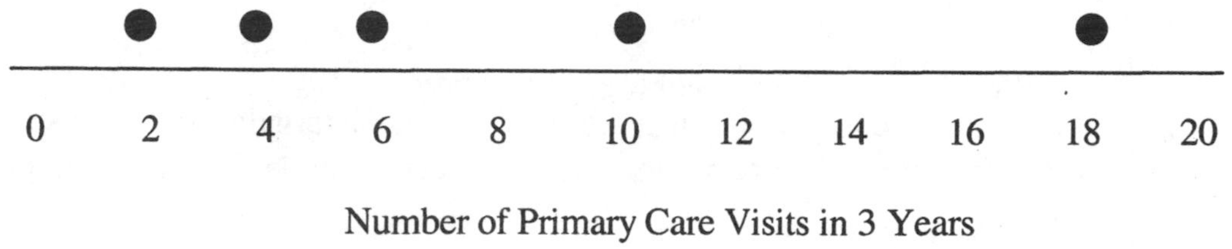

Number of Primary Care Visits in 3 Years

In this population, 3 of 5 subjects reported 6 or fewer visits to primary care physicians in 3 years. Two of the subjects reported many more visits (10 and 18). Recall that each subject is 65 years of age or older, which might explain the magnitude of reported numbers of visits. The two subjects reporting 10 and 18 visits to primary care physicians in 3 years may have chronic illnesses which require more frequent follow-up, or acute illnesses or conditions which required frequent attention over shorter periods of time. As will become apparent in the remainder of this book, interpreting data (even a single characteristic measured on what seems to be a relatively homogenous collection of subjects (individuals 65 years of age or older)) is often quite complicated.

Suppose that the population in Example 1 included N=500 individuals 65 years of age or older, or N=5,000 individuals 65 years of age or older. In both cases, it would be impossible to understand the population with respect to the reported numbers of visits to primary care physicians simply by inspecting the observed data elements. Even in the smaller population (N=500), there would be too many data elements to draw conclusions simply by inspection. It is generally necessary to summarize characteristics measured on a population to understand the characteristic under investigation. Several summary measures are described below.

2.3 Population Parameters

Any measure computed on a population is called a *population parameter*, or simply a parameter. There are many parameters; we will present only a few here. The first parameter is the *population size*, denoted N. In Example 1, N=5.

It is generally of interest to describe a population in terms of its average value on a particular characteristic. For Example 1, we wish to address the question, what is the typical number of visits that patients 65 years of age or older make to primary care physicians in 3 years? The *population mean*, denoted μ ("mu"), addresses this question and is computed by summing the values and dividing by the population size.

$$\mu = \frac{\Sigma X}{N} \qquad\qquad (2.1)$$

where: Σ (upper case "sigma") denotes summation,
X is a placeholder which represents the characteristic under consideration (e.g., number of primary care visits in 3 years), and N denotes the population size.

In Example 1, the population mean is 8:

$$\mu = \frac{\Sigma X}{N} = \frac{2+4+6+10+18}{5} = 8$$

The mean is a very useful parameter. Since the population in Example 1 is so small, it is not necessary to summarize the data elements to understand the population with respect to the number of primary care visits in 3 years. If the population size was N=500 (or N=5,000) the population mean would provide a very useful summary.

In Example 1, no two individuals reported the same number of visits to primary care physicians in 3 years. In addition to summarizing the population with respect to what a typical value looks like (i.e., $\mu = 8$), it is generally of interest to understand variability in the characteristic of interest. If we take the population mean 8 as representative of a typical number of primary care visits in 3 years, it is of interest to understand how close each individual in the population is to that typical value. In particular, are all of the reported numbers of visits close to 8 or are they widely spread above and below 8?

There are several measures of variability or dispersion, the first we discuss is the *population range*. The population range is defined as the difference between the largest, or maximum, score and the smallest, or minimum, score. In Example 1, the population range is 18 - 2 = 16. Some report the range as between 2 and 18, while others report the range as 16. Either is acceptable.

A more sophisticated and intuitive measure of dispersion is the *population variance*, denoted σ^2 ("sigma squared"). The population variance is based on "deviations from the mean" or distances between each observation and the population mean. The following table displays the data elements from Example 1 along with deviations from the mean (i.e., distances from $\mu=8$):

X	$(X - \mu)$
2	-6
4	-4
6	-2
10	2
18	10
	0

Notice that the sum of the deviations from the mean is zero, a property of the population mean. Recall, the goal is to generate an estimate of the dispersion in the population, in particular the dispersion in numbers of visits to primary care physicians in 3 years relative to the population mean, $\mu = 8$.   In the table above we computed the deviations from $\mu = 8$ for each subject in our population. Inspecting the deviations we see a fair amount of variation in the reported numbers of visits. The first subject reported 6 fewer visits than the mean, the second subject reported 4 fewer visits than the mean, the last subject reported 10 more visits than the mean number of visits. Since this population is small, we can evaluate the variability in the population by inspection. Again, imagine a population of size N=500 (or N=5,000) in which it would be impossible to inspect deviations from the mean and draw conclusions regarding variability.

There are several techniques which can be employed to summarize the magnitude of these deviations from the mean. Notice that it is not practical to take the mean deviation (i.e., sum the deviations and divide by N) since the sum of the deviations is always zero. The most popular method used to assess variation is the average squared deviation from the mean, or the *variance*. This measure proves to be the most straightforward mathematically. Another summary measure is the mean absolute deviation. However, this measure can be difficult mathematically especially in mathematical proofs which are beyond the scope of this discussion.

The following table displays the numbers of primary care visits in 3 years, denoted X, deviations from the mean, and squared deviations from the mean, respectively.

X	$(X - \mu)$	$(X - \mu)^2$
2	-6	36
4	-4	16
6	-2	4
10	2	4
18	10	100
	0	160

Motivation

The sum of the squared deviations from the mean is 160. The population variance is the average of the squared deviations from the mean, and is defined as follows:

$$\sigma^2 = \frac{\Sigma(X-\mu)^2}{N} \qquad (2.2)$$

In Example 1, the population variance is 32:

$$\sigma^2 = \frac{\Sigma(X-\mu)^2}{N} = \frac{160}{5} = 32$$

The population variance for Example 1 is interpreted as follows. The number of primary care visits in 3 years is, on average, 32 visits squared from the mean of 8. In computing the variance we squared each deviation from the mean to capture the magnitude of the deviations (since the negative deviations cancelled the positive deviations). To return to our original units, we take the square root, which produces the *population standard deviation*, denoted σ ("sigma"):

$$\sigma = \sqrt{\sigma^2} = \sqrt{\frac{\Sigma(X-\mu)^2}{N}} \qquad (2.3)$$

The standard deviation is $\sigma = \sqrt{32} = 5.7$. The population standard deviation in Example 1 is interpreted as follows. The number of primary care visits in 3 years is about 5.7 visits from the mean of 8.

The population standard deviation is the most widely used parameter to describe the dispersion in a population and is interpreted as the typical deviation from the mean. It is difficult to quantify large and small values of the standard deviation, as the magnitude of the standard deviation depends on the measurement scale of the characteristic under investigation. The same is true for the mean value; however, the interpretation of the mean is more intuitive.

The standard deviation is particularly useful in comparing populations with respect to a particular characteristic. For example, suppose we have two populations and wish to make comparisons with respect to numbers of primary care visits in 3 years. Suppose the two populations have equal mean numbers of primary care visits (i.e., $\mu_1 = 8$ and $\mu_2 = 8$). Suppose that the standard deviations in numbers of visits are 5.7 and 2.3, respectively (i.e., $\sigma_1 = 5.7$ and $\sigma_2 = 2.3$). The difference in the standard deviations indicates that the numbers of visits are more dispersed in population 1 as compared to population 2, while both populations have the same means. In

Motivation

population 2, the reported numbers of visits are more tightly clustered about the mean of 8 (i.e., the reported numbers of visits deviate by about 2.3 visits from the mean of 8 visits).

2.4 Sampling and Sample Statistics

In Section 2.3 we outlined a number of parameters which are used to describe a population. In most statistical applications, we do not have the entire population available, but instead have only a subset or sample of individuals selected from the population of interest.

In many situations it is impossible and/or impractical to analyze the entire population. In some cases the population may be so large that it is impossible to measure a specific characteristic or set of characteristics on each subject. In other cases it may be too costly to measure certain characteristics on each subject. For example, suppose we wish to analyze a particular characteristic that is measured by a laboratory test which costs more than $2,000 per subject. It may not be financially feasible to conduct tests on every member of the population. In some applications, it may be too time consuming to analyze the entire population. In such a case it may take so long to collect and analyze the data that the results are not useful. We therefore rely on samples or subsets of the population for analysis.

If the sample of subjects (i.e., the subset of the population) is representative of the population, then it is reasonable to think that what we observe in the sample would be similar to what we would observe in the population, if the entire population were observed or analyzed. This notion is the basis for statistical inference and we will provide a more complete justification in Chapters 4 and 5.

There are many different samples that can be drawn from any given population. The number of distinct samples that can be taken from a given population depends on the numbers of subjects in both the population and the sample. (Processes for selecting samples will be discussed in several subsequent chapters).

To illustrate the concept of sampling, suppose in Example 1 we enumerate all possible samples of size 3 from our population of 5 individuals. There are 10 samples of size $n = 3$ that do not contain any individual more than once (i.e. there are 10 simple random samples without replacement). The samples are given below where X_1 denotes the number of visits to primary care physicians reported by the first individual selected, X_2 denotes the number of visits reported by the second individual selected, and so on (See also Figure 2.2). For example, the first sample consists of subjects 1, 2, and 3 and the second sample consists of subjects 1, 2 and 4.

	Sample	
X_1	X_2	X_3
2	4	6
2	4	10
2	4	18
2	6	10
2	6	18
2	10	18
4	6	10
4	6	18
4	10	18
6	10	18

Notice how many distinct samples are possible from this very small population. Imagine how many samples of size n=10 (or n=50) are possible from a population of N=500 (or from a population of size N=5,000).

Figure 2.2   Simple Random Samples of Size n=3 from a Population with N=5

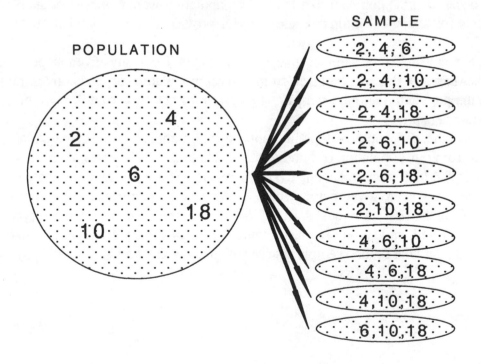

The samples shown above are exhaustive (all inclusive) under sampling without replacement. Each sample shown above is equally likely. That is, the chance that any one sample is selected from our population is 1/10, or 10%.

Measures computed on a population are called parameters, while measures computed on a sample are called *sample statistics* or simply *statistics*. (In Chapter 3 we present and interpret a variety of statistics.) For illustration purposes, suppose we compute the *sample mean*, denoted $\overline{X}$ ("X bar") for each sample shown above. As in the case of the population mean, the sample mean is computed by summing the values and dividing by the *sample size* (i.e., $\overline{X} = \Sigma X/3$). Suppose, we also compute the *sample range* (i.e., maximum-minimum) for each sample. The samples, their means and ranges are given below:

X_1	Sample X_2	X_3	Sample Mean $\overline{X} = \Sigma X/n$	Sample Range=Max-Min
2	4	6	4.0	4
2	4	10	5.3	8
2	4	18	8.0	16
2	6	10	6.0	8
2	6	18	8.7	16
2	10	18	10.0	16
4	6	10	6.7	6
4	6	18	9.3	14
4	10	18	10.7	14
6	10	18	11.3	12

Notice that the sample means and sample ranges vary, depending on the individuals selected into each sample. The enumeration of all possible sample means is called the *sampling distribution of the sample means*. The distribution of the sample means is shown in a dot plot in Figure 2.3. A sampling distribution is the listing of all values of a statistic (e.g., $\overline{X}$) based on all possible samples generated under a particular sampling strategy.

Figure 2.3 Sampling Distribution of the Sample Means (n=3)

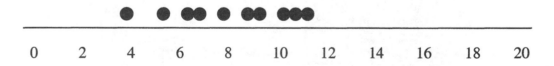

| 0 | 2 | 4 | 6 | 8 | 10 | 12 | 14 | 16 | 18 | 20 |

Mean Number of Primary Care Visits in 3 Years

Motivation

To reiterate, we generally never have an entire population available, instead we have a sample. In statistical applications we attempt to draw inferences about a population based on that sample. The theory behind statistical inference is based on the relationship between the sampling distribution of the sample statistic and the population parameter (described in detail in Chapter 5). Here we introduced a population in Example 1 and enumerated all possible samples of size n=3 without replacement. In the following section we introduce the concept of statistical inference.

2.5 Statistical Inference

Suppose the administrators of a particular health maintenance organization (HMO) wish to estimate the number of visits to primary care physicians that their enrollee's 65 years of age or older make in 3 years in order to allocate resources. Suppose that the HMO has N=5,489 members who are 65 years of age or older (assume that some sociodemographic characteristics such as age and gender are available on each enrollee of the HMO in a centralized database while the number of primary care visits is not). It would be very time consuming and costly to survey each of the N=5,489 enrollees with regard to the number of primary care visits they made in the past 3 years. Statistical inference techniques can be used to estimate the mean number of visits among all members of the HMO who are 65 years of age or older based on a subset or sample of such enrollees of the HMO.

In statistical inference (discussed extensively in Chapters 6-13), we have a single sample and wish to make inferences about unknown population parameters based on sample statistics. In Figure 2.4 we illustrate a common situation in which the population mean of a particular characteristic, μ, is unknown. To generate an estimate of this population mean we take a sample of subjects from the population and compute various statistics (e.g., the sample mean $\overline{X}$). We use this information from the sample to make inferences about the unknown population mean. Inferences about a population are valid if the sample is a random sample.

Figure 2.4 Statistical Inference

POPULATION

$\mu = ?$

SAMPLE

$n, \overline{X}$

Motivation

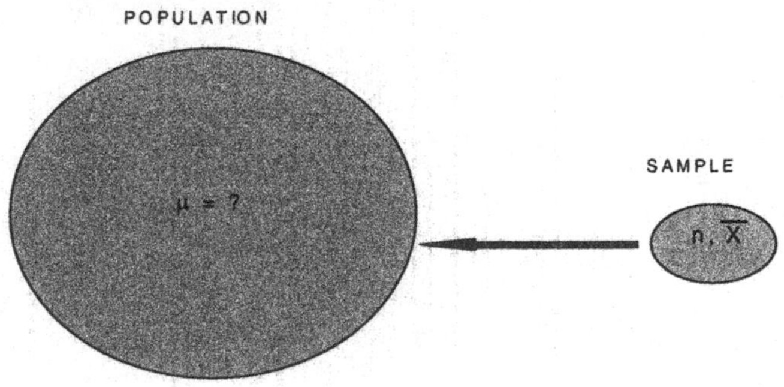

It is intuitive, but worth noting, that the larger the sample size (larger n), the "better" the inference about the population parameter. We will define explicit criteria for determining how "good" estimates are in Chapter 6. We will use our simple Example 1 to illustrate the notion of statistical inference and to raise issues which will be addressed in complete detail in subsequent chapters.

In Example 1, we know the population and we computed the population mean, $\mu = 8$. Notice that only one sample out of ten had a sample mean ($\overline{X}$) of 8. The sample means ranged from 4.0 to 11.3. In practice we will not know the population and we will have only a single sample from the population of interest. Suppose that by chance we happened to select the first sample (which is as likely as any other sample to be selected). The reported numbers of primary care visits among individuals in the first sample ($\overline{X} = 4.0$) are much lower than the numbers reported by other individuals in the population. If this were a real application we would not know the population mean (or anything else about the population) and we would underestimate the mean number of primary care visits in 3 years among enrollee's 65 years of age or older based on this first sample. If we happened by chance to select the last sample ($\overline{X} = 10.7$), we would overestimate the mean number of visits in 3 years among enrollee's 65 years of age or older. In practice we have one sample and, in fact, never know for sure if we are underestimating population parameters, overestimating population parameters or are right on target. We will, however, be able to quantify how much "error" is in our estimate of the population parameter. We must be willing to accept some error due to the fact that our sample is only a subset of the population. In practice, we will also generally have samples of much larger size (larger n). Example 1 is a simple (somewhat unrealistic) example, used only to illustrate concepts.

If we compute the mean of the sample means (i.e., sum the 10 sample means ($\overline{X}$'s) and divide by 10) we see that the mean of the sample means is equal to 8.0, the value of the population mean. Therefore, on the average, the sample mean $\overline{X}$ is equal to the population mean μ. Based on this property, we say that the sample mean is an *unbiased* estimator of the population mean.

We computed the population range as 16, and only 3 samples of ten have a range of 16. The other samples all have ranges less than 16. On the average, the sample range is not equal to the population range but is less than the population range. The sample range is a biased estimator of the population range.

In statistical inference, which will be discussed in the majority of the chapters in this book in detail, the first step in addressing a problem of estimating a population parameter is choosing a sample statistic with good properties, such as unbiasedness. Statistical inference is based on probability theory which will be reviewed in Chapter 4. We begin our discussions with techniques to summarize sample data in Chapter 3. Table 2.1 summarizes the notation and formulas for commonly used population parameters. The notation and formulas for sample statistics will be presented in Chapter 3.

Table 2.1 Summary of Notation: Population Parameters

PARAMETER	NOTATION/FORMULA	DESCRIPTION
Population Size	N	Number of subjects in population
Population Mean	$\mu = \dfrac{\Sigma X}{N}$	Typical or average value
Population Variance	$\sigma^2 = \dfrac{\Sigma (X - \mu)^2}{N}$	Average squared deviation from the mean
Population Standard Deviation	$\sigma = \sqrt{\dfrac{\Sigma (X - \mu)^2}{N}}$	Typical deviation from the mean

CHAPTER 3: Summarizing Data

3.1 Introduction

The first step in solving any problem is a clear understanding of the problem. In statistics, the basis for any analysis is a clear understanding of the data. Describing, summarizing and presenting data are central to all statistical applications. Before any statistical inferences about a population parameter (e.g., μ) are made based on a sample statistic (e.g., $\overline{X}$), the sample must be summarized appropriately. In Figure 3.1 we illustrate the concept of statistical inference. Specifically, we want to estimate the mean of a population based on an analysis of a sample. For example, suppose we want to estimate how frequently individuals in a population exercise. For this analysis, suppose we measure exercise as the number of hours of exercise per week. Suppose the population of interest includes all persons with adult onset (or Type II) diabetes. These individuals are supposed to exercise (along with following a special diet) as part of their treatment plan. It would be impossible to study all persons with adult onset diabetes, so instead we select a sample. We measure the number of hours of exercise per week among the members of our sample and use that information to make generalizations about exercise in the population. In statistical inference we take into account for the fact that we did not study the entire population, this is described in detail in Chapter 6. Before we get to statistical inference, however, we must understand how to appropriately summarize our sample. In this chapter we discuss both numerical and graphical descriptions, summaries and presentations of sample data.

Introductory Applied Biostatistics

DESCRIPTIVE STATISTICS (Ch. 3) → PROBABILITY (Ch. 4) → SAMPLING DISTRIBUTIONS (Ch. 5)

STATISTICAL INFERENCE (Chapters 6-14)

Outcome Variable	Grouping Variable(s)/ Predictor(s)	Analysis	Chapter(s)
Continuous	-	Estimate μ, Compare μ to Known, Historical Value	6/13
Continuous	Dichotomous (2 Groups)	Compare Independent Means (Estimate/Test $(\mu_1-\mu_2)$) or the Mean Difference(μ_d)	7/13
Continuous	Discrete (> 2 Groups)	Test the Equality of K Means using Analysis of Variance $(\mu_1=\mu_2=\ldots\mu_k)$	10/13
Continuous	Continuous	Estimate Correlation or Determine Regression Equation	11/13
Continuous	Several Continuous or Dichotomous	Multiple Linear Regression Analysis	11
Dichotomous	-	Estimate p, Compare p to Known, Historical Value	8
Dichotomous	Dichotomous (2 Groups)	Compare Independent Proportions (Estimate/Test (p_1-p_2))	8/9
Dichotomous	Discrete (>2 Groups)	Test the Equality of k Proportions (Chi-Square Test)	8
Dichotomous	Several Continuous or Dichotomous	Multiple Logistic Regression Analysis	12
Discrete	Discrete	Compare Distributions Among k Populations (Ch-Square Test)	8
Time to Event	Several Continuous or Dichotomous	Survival Analysis	14

Figure 3.1   **Statistical Inference**

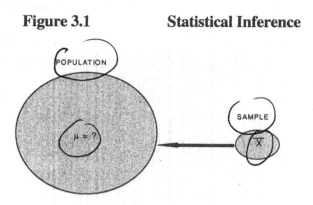

In Section 3.2 we provide background including vocabulary and notation. We also draw the distinction between continuous and discrete variables. In Section 3.3 we outline an array of descriptive statistics and graphical methods and use a number of examples to illustrate their applications. Numerical and graphical summaries for continuous variables are described in Sections 3.3.1 and 3.3.2, respectively. Numerical and graphical summaries for discrete variables are described in Sections 3.3.3 and 3.3.4, respectively. We illustrate computations and graphical displays both by hand and using SAS. In Section 3.4 we summarize key formulas. In Section 3.5 we provide SAS program code used to generate descriptive statistics and graphical displays.

3.2 Background

3.2.1 Vocabulary

In statistics we work with *data elements* or *data points*. Each data element is a representation of a particular measurable characteristic, or variable. For example, data elements may be individuals' ages measured in years (e.g., 51, 29, 36), or patients' systolic blood pressures measured in millimeters of mercury (mm Hg) (e.g., 140, 160, 110).

Data elements are measured on *subjects* or *units of measurement*. Though we think of subjects as human beings in most applications, it is not always the case. Subjects might be animals involved in a research experiment and we might be interested in the time it takes each animal to complete a specific task. In this case the time to complete the task is the variable of interest, and the observed times (in minutes) are the data elements. Subjects could be hospitals in a specific region of the country (e.g., Northeast) and we might record the number of orthopedic procedures performed in each hospital in one year. In this case the number of orthopedic procedures performed in one year is the variable of interest, and the observed number of procedures (measured in each hospital) are the data elements. The subjects or units of measurement are specific to each application and must be defined explicitly.

Subjects comprise either a *population* or a *sample*. A population is a collection of all subjects of interest. For example, if we are interested in American males with heart disease then the collection of all American males with heart disease would constitute the population. A sample is a subset of the population of interest. For example, one hundred randomly selected American males with heart disease would constitute one sample from the population of all American males with heart disease.

There are many different samples from any given population. The number of distinct samples that can be taken from a given population depends on the numbers of subjects in both the population and in the sample. In subsequent chapters we will outline explicit formulas to determine the number of possible samples from a particular population. We will also present some of the more popular methods used to select subjects from a population into a sample. For now, we assume that the subjects in any sample are selected at random from the population of interest and that the sample, as a whole, is representative of the population of interest.

3.2.2 Classification of Variables

Each variable under investigation can be classified as either *continuous* or *discrete*. Continuous (or measurement) variables assume, in theory, any value between the minimum and maximum value on a particular measurement scale. Variables such as age, height, weight, cholesterol level and systolic blood pressure are examples of continuous variables. Discrete variables take on a limited number of values, or categories. Discrete variables can be either *ordinal* or *categorical* variables. Ordinal variables take on a limited number of values or categories, but the categories are ordered. For example, symptom severity is an example of an ordinal variable with response options: Minimal, Moderate, Severe. Self-reported health status is another example of an ordinal variable with response options: Excellent, Very Good, Good, Fair, Poor. Categorical variables take on a limited number of categories and the categories are unordered. Gender is an example of a categorical variable with two response options: Male, Female. Marital status is another example of a categorical variable with the following response options: Married, Separated, Divorced, Widowed, Never Married. It is important to determine the nature of the variable under investigation as the most appropriate techniques used to summarize it depend on whether it is continuous or discrete.

3.2.2 Notation

Any descriptive measure based on a population is called a *population parameter,* or just a *parameter*. Any descriptive measure based on a sample is called a *sample statistic*, or just a

Introductory Applied Biostatistics

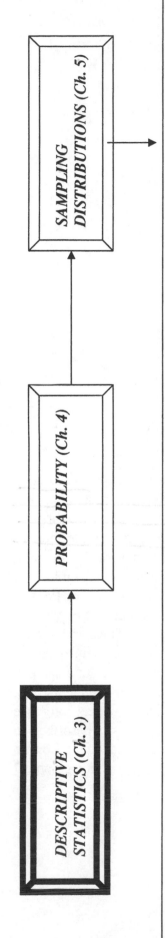

DESCRIPTIVE STATISTICS (Ch. 3) → PROBABILITY (Ch. 4) → SAMPLING DISTRIBUTIONS (Ch. 5) →

STATISTICAL INFERENCE (Chapters 6-14)

Outcome Variable	Grouping Variable(s)/Predictor(s)	Analysis	Chapter(s)
Continuous	-	Estimate μ, Compare μ to Known, Historical Value	6/13
Continuous	Dichotomous (2 Groups)	Compare Independent Means (Estimate/Test $(\mu_1-\mu_2)$) or the Mean Difference(μ_d)	7/13
Continuous	Discrete (> 2 Groups)	Test the Equality of K Means using Analysis of Variance $(\mu_1=\mu_2=\ldots\mu_k)$	10/13
Continuous	Continuous	Estimate Correlation or Determine Regression Equation	11/13
Continuous	Several Continuous or Dichotomous	Multiple Linear Regression Analysis	11
Dichotomous	-	Estimate p, Compare p to Known, Historical Value	8
Dichotomous	Dichotomous (2 Groups)	Compare Independent Proportions (Estimate/Test (p_1-p_2))	8/9
Dichotomous	Discrete (>2 Groups)	Test the Equality of k Proportions (Chi-Square Test)	8
Dichotomous	Several Continuous or Dichotomous	Multiple Logistic Regression Analysis	12
Discrete	Discrete	Compare Distributions Among k Populations (Ch-Square Test)	8
Time to Event	Several Continuous or Dichotomous	Survival Analysis	14

CHAPTER 3: Summarizing Data

3.1 Introduction

The first step in solving any problem is a clear understanding of the problem. In statistics, the basis for any analysis is a clear understanding of the data. Describing, summarizing and presenting data are central to all statistical applications. Before any statistical inferences about a population parameter (e.g., μ) are made based on a sample statistic (e.g., $\overline{X}$), the sample must be summarized appropriately. In Figure 3.1 we illustrate the concept of statistical inference. Specifically, we want to estimate the mean of a population based on an analysis of a sample. For example, suppose we want to estimate how frequently individuals in a population exercise. For this analysis, suppose we measure exercise as the number of hours of exercise per week. Suppose the population of interest includes all persons with adult onset (or Type II) diabetes. These individuals are supposed to exercise (along with following a special diet) as part of their treatment plan. It would be impossible to study all persons with adult onset diabetes, so instead we select a sample. We measure the number of hours of exercise per week among the members of our sample and use that information to make generalizations about exercise in the population. In statistical inference we take into account for the fact that we did not study the entire population, this is described in detail in Chapter 6. Before we get to statistical inference, however, we must understand how to appropriately summarize our sample. In this chapter we discuss both numerical and graphical descriptions, summaries and presentations of sample data.

statistic. The number of subjects in a population, or the *population size*, is denoted N. The number of subjects in a sample, or the *sample size*, is denoted n. μ is an example of a parameter, while $\overline{X}$ is an example of a statistic.

The collection of data elements, called the *data set*, may have been measured on each member of a population or on each member of a sample. An analyst cannot tell simply by looking at the data set or by its size alone (i.e., the number of data elements), whether the collection of data elements comprise a population or a sample. Someone involved in the design of the study, such as the statistician or an investigator in the particular substantive field, must convey such information to the data analyst.

Data elements (i.e., observations) deriving either from a population or a sample are denoted X_i, where X is a variable name, or placeholder, representing the characteristic of interest and the subscript i refers to the subject number on which the measurement was taken. For example, suppose we have a sample consisting of 4 individuals or subjects (i.e., n=4) and measure the age of each subject. Suppose the sample data set is as follows:

$$21 \qquad 25 \qquad 32 \qquad 26$$

The first subject is 21 years of age, the second subject is 25 years of age and so on. If we let X=age, we can denote each data element as follows:

$$X_1 = 21 \qquad X_2 = 25 \qquad X_3 = 32 \qquad X_4 = 26.$$

NOTE: In some applications the subscript, denoting the subject number, will be omitted for simplification.

This notation is employed to simplify computations and mathematical formulas which are illustrated in the next section.

3.3 Descriptive Statistics and Graphical Methods

In this section we present an array of descriptive statistics and graphical methods used to summarize sample data. In all cases we assume that the subjects comprising the sample were selected at random from the population of interest (i.e., the sample is representative of the population). We present a series of examples and in each example illustrate different descriptive statistics. General formulas are indicated in shaded boxes. In almost all cases we consider small samples (i.e., small n) for ease of illustration. In practice, these same descriptive statistics can be applied to larger samples (i.e., larger n). In Section 3.3.1 we present numerical summaries for continuous variables. In Section 3.3.2 we present graphical summaries for continuous variables. In Section 3.3.3 we present numerical summaries for discrete variables and in Section 3.3.4 we present graphical summaries for discrete variables. In Section 3.4 we summarize each of the descriptive statistics and provide rationale for their use in general. We encourage the reader to refer continuously to Section 3.4 as he or she reads through the examples in this section.

3.3.1 Numerical Summaries for Continuous Variables

We now use examples to present descriptive statistics to summarize continuous or measurement variables.

Example 3.1. Suppose we select a random sample of seven subjects from a population of patients 50 years of age with diagnosed coronary artery disease. We measure systolic blood pressures, in millimeters of mercury (mm Hg), on each subject. The sample data are given below:

$$121 \quad 110 \quad 114 \quad 100 \quad 160 \quad 130 \quad 130$$

Because the sample size is small (n=7) we can summarize the sample with respect to systolic blood pressure by inspection. The lowest blood pressure in the sample is 100 and the highest blood pressure is 160. Neither of these values would be considered clinically problematic. The remaining blood pressures are between 110 and 130 and seem quite reasonable given that the subjects were sampled from the population of patients 50 years of age with diagnosed coronary artery disease. Had the patients been sampled from the population of patients 30 years of age free of cardiovascular disease, these observed systolic blood pressures might be slightly higher than one might expect.

If the sample included not 7 patients, but instead 700, it would be impossible to summarize the sample with respect to systolic blood pressure by inspecting the values. In fact, the same would probably be true if the sample included 20 subjects. In most applications, it is necessary to use statistical techniques to summarize a sample. Several are described below.

To simplify the computations that follow, we represent each data element (i.e., each observed systolic blood pressure) using the variable X. Here X denotes systolic blood pressure, and the subscripts (i=1,2,...,7) denote the subject number in the sample:

$$X_1 = 121 \quad X_2 = 110 \quad X_3 = 114 \quad X_4 = 100 \quad X_5 = 160 \quad X_6 = 130 \quad X_7 = 130$$

It is generally of interest to summarize a continuous variable with respect to location. Location refers to the center of the data set, and addresses the question, What is a typical systolic blood pressure?

The subscripts can be dropped, because the subject number (i.e., subscript) has no impact on the computations that follow. To organize our calculations, we arrange the data elements in a column, as shown below. Notice that the data elements are ordered from smallest to largest (this is not necessary but is sometimes convenient) and that the data element 130 is listed twice, as subjects numbered 6 and 7 each have systolic blood pressures of 130.

X_i
100
110
114
121
130
130
160

The first descriptive statistic we consider is the *sample mean*, denoted $\overline{X}$ ("X bar"). The sample mean is one statistic which summarizes the average value of a sample. The sample mean gives a sense of what a typical value looks like. To compute the sample mean, we sum all of the observations and divide by the sample size. The sample mean of the systolic blood pressures is given below:

$$\overline{X} = (100 + 110 + 114 + 121 + 130 + 130 + 160) / 7.$$

In mathematics, the symbol Σ (upper case "sigma") denotes summation. The sample mean of the systolic blood pressures can be represented as follows:

$$\overline{X} = \Sigma X_i / 7,$$

where $\Sigma X_i = 100 + 110 + 114 + 121 + 130 + 130 + 160$. In general, the sample mean is denoted:

$$\overline{X} = \frac{\Sigma X_i}{n} = \frac{\Sigma X}{n} \tag{3.1}$$

NOTE: In the latter version of the formula, the subscript i is suppressed and the summation is understood to be over all subjects in the sample.

The mean systolic blood pressure is $\overline{X} = 865/7 = 123.6$. Reviewing the data elements, we see that some of the observed systolic blood pressures are above the mean of 123.6, while others are below the mean. The mean of 123.6 is interpreted as the average, or typical, systolic blood pressure in the sample. In journal articles and research reports, readers are generally not shown the actual data elements. Instead, summary statistics such as the sample size and sample mean are provided.

The sample mean is referred to as the balancing point, or pivot point, of the sample since the sum of the distances between observations below the mean and the sample mean are equal to the sum of the distances between observations above the mean and the sample mean (See dot plot in Figure 3.2).

Figure 3.2 Sample Mean as Balancing Point

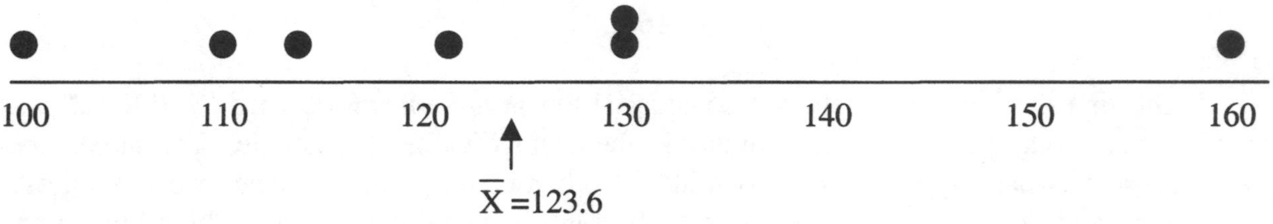

$\overline{X} = 123.6$

These distances or "deviations from the mean" are denoted $(X - \overline{X})$. The following table displays the data elements along with their respective deviations from the mean (i.e., distance from $\overline{X} = 123.6$):

X	$(X - \overline{X})$
100	-23.6
110	-13.6
114	-9.6
121	-2.6
130	6.4
130	6.4
160	36.4
865	-0.2*

* This sum is theoretically zero. The difference here is due to rounding.

Summarizing Data

The sample mean measures the *location*, or central tendency of the sample. The location is very important in interpreting sample data. However, two very different samples might produce the same sample mean. Consider a second sample of seven subjects from the population of patients 50 years of age with diagnosed coronary artery disease. Again we measure systolic blood pressures, in millimeters of mercury (mm Hg), on each subject. The sample data are given below:

$$120 \quad 121 \quad 122 \quad 124 \quad 125 \quad 126 \quad 127$$

The sample mean for this sample is: $\overline{X} = (120 + 121 + 122 + 124 + 125 + 126 + 127) / 7 = 865/7 = 123.6$. The sizes and the means are the same between samples, yet the samples are quite different. For a more complete understanding of the data, we also need a measure of the *dispersion*, or spread, in the sample. Measures of dispersion address whether the data elements are tightly clustered together or widely spread. Specifically, we are interested in whether the data elements are tightly clustered about the sample mean, or if the data elements are widely spread above and below the mean.

The goal is to generate an estimate of the dispersion in the sample, in particular the dispersion of the data elements about the sample mean. The deviations from the mean sum to zero, since the negative deviations "cancel out" the positive deviations (See Figure 3.2). Of real interest is the magnitude of these deviations. There are several techniques which can be employed to summarize the magnitude of the deviations from the mean. One method is the mean absolute deviation (MAD) which is simply the mean of the absolute values of the deviations from the mean:

$$MAD = \frac{\Sigma |X - \overline{X}|}{n}$$

This is not generally used for mathematical reasons which are beyond the scope of this book.

The more popular statistic, which proves to be the most straightforward mathematically, is based on __squared__ deviations from the mean, and is called the *sample variance*, defined as:

$$s^2 = \frac{\Sigma (X - \overline{X})^2}{n-1} \tag{3.2}$$

NOTE: The denominator in the sample variance is (n-1) and not n as was the case with the sample mean.

The following table is used to organize data for the computation of the sample variance. The table displays the data elements, deviations from the mean, and squared deviations from the mean, respectively.

X	$(X - \overline{X})$	$(X - \overline{X})^2$
100	-23.6	556.96
110	-13.6	184.96
114	-9.6	92.16
121	-2.6	6.76
130	6.4	40.96
130	6.4	40.96
160	36.4	1324.96
865	-0.2*	2247.72

*The sum is not exactly zero due to rounding.

For Example 3.1, the sample variance is:

$$s^2 = \frac{2247.72}{6} = 374.6$$

The sample variance is interpreted as the average squared deviation from the mean. Therefore, on average, systolic blood pressures in our sample are 374.6 units squared from the sample mean of 123.6. This information is important; however, in its present form it does not exactly achieve our original goal which was to compute a measure of the typical deviation from the mean in the sample. Recall that we summed the square of each deviation from the mean since their sum was zero. Because of this step, the sample variance does not address our original objective directly. To return to our original units, we compute what is called the *sample standard deviation*, denoted s, defined as the square root of the sample variance.

$$s = \sqrt{s^2} \tag{3.3}$$

The sample standard deviation of the systolic blood pressures is given below:

$$s = \sqrt{374.6} = 19.4$$

After taking the square root we have a statistic which can be interpreted as the typical deviation from the mean. In this sample, systolic blood pressures are about 19.4 units from the sample mean. It is often difficult to interpret the value of a standard deviation (e.g., Is 19.4 large, small or appropriate?). The standard deviation, however, is very useful for comparing samples. Recall the second sample of n=7 subjects we introduced selected from the same population. The second sample had the same size (n=7) and the same mean ($\overline{X}$=123.6), however, the standard deviation for the second sample is s=2.6. The standard deviation in the second sample is much smaller, because all of the observations are tightly clustered around the sample mean of 123.6. There is much more variability in the systolic blood pressures measured among patients with diagnosed coronary artery disease in the first sample (sample 1: s = 19.4) as compared to the second sample (s=2.6).

An alternative formulation is available for computing the sample variance, which is mathematically equivalent to the formulation provided in (3.2). This alternative formulation is called the *computational formula for the sample variance* and is given below. The formulation provided in (3.2) is called the *definitional formula*.

$$s^2 = \frac{\Sigma X^2 - (\Sigma X)^2/n}{n-1} \tag{3.4}$$

where ΣX^2 = the sum of the squared observations, and

$(\Sigma X)^2$ = the square of the sum of the observations.

The computational formula (3.4) can be easier to work with than the definitional formula given in (3.2), because the components in the computational formula are in most cases easier to compute (i.e., ΣX^2 and $(\Sigma X)^2$). We will now illustrate the use of the computational formula for the sample variance using data from Example 3.1. The following table displays each data element, along with each data element squared.

X	X^2
100	10,000
110	12,100
114	12,996
121	14,641
130	16,900
130	16,900
160	25,600
865	109,137

Using the computational formula (3.4):

$$s^2 = \frac{109{,}137 - (865)^2 / 7}{7 - 1}$$

$$s^2 = \frac{109{,}137 - 106{,}889.3}{6} = \frac{2247.7}{6} = 374.6$$

As noted, the computations can be somewhat easier with the computational formula as compared to the definitional formula. To implement the computational formula, we need only to compute the sum of the data elements and the sum of the squared data elements, as opposed to deviations and deviations squared for each data element. The reduced number of calculations with the computational formula reduces the chance of error in computations.

A standard data summary for a continuous variable in a sample consists of three statistics:

- sample size (n),

- sample mean ($\overline{X}$), and

- sample standard deviation (s).

These three statistics provide information on the number of subjects in the sample, the location and dispersion of the sample, respectively.

Summarizing Data

We purposely chose a small data set to illustrate these statistics. It is easy to imagine applications with much larger sample sizes in which it would be impossible to view the entire sample. In such cases the sample size, mean and standard deviation provide a very informative and useful summary. Publications and reports almost always include these statistics.

As a general guideline, descriptive statistics should include no more than one decimal place beyond that observed in the original data elements. For example, the systolic blood pressures are recorded as whole numbers. Therefore, descriptive statistics are presented to the nearest tenths place (i.e., one decimal place).

The standard summary for Example 3.1 is n=7, $\overline{X}$=123.6 and s=19.4. There are a number of other descriptive statistics beyond what we have called the standard summary statistics (i.e., n, $\overline{X}$, and s) that are also widely used for continuous variables. Other descriptive statistics include the median and quartiles.

The *sample median* is defined as the middle value. It is the value which has as many values above it as below it. The median is computed by arranging the data elements from smallest to largest, and successively counting from the right and left to arrive at the median, or middle, value. For example, if we arrange our 7 data elements from smallest to largest, and count in from the right and left, simultaneously, we arrive at the median or middle value after 3 steps:

```
Step 1:    100  110  114  121  130  130  160
Step 2:    100  110  114  121  130  130  160
Step 3:    100  110  114  121  130  130  160
                          ⇑
                        Median
```

Since the number of observations in this sample is odd (n=7), this procedure produces a single number, the median value, upon successively counting in from the right and left. In Example 3.2, we will illustrate the same procedure with an even number of data elements. The interpretation of the median in Example 3.1 is as follows. Half (50%) of the systolic blood pressures are greater than 121 and half (50%) are less than 121.

Both the mean and the median are statistics which measure the average or typical value of a particular characteristic. The median is particularly useful when there are extreme values (either very small or very large as compared to other values) in the sample. Suppose in Example 3.1 that the maximum value was not 160 but instead 260. The sample mean would be: $\overline{X}$ = (100 + 110 + 114 + 121 + 130 + 130 + 260) / 7 = 137.9 which does not look like a typical value (since 6 of 7 observations are below it). The sample mean is affected by extreme values. In this case, the value 260 inflates the mean value and it is no longer representative of a typical value. A better measure of location in this situation is the median, which is still 121 and is more representative of a typical value. In the absence of extreme values, the sample mean is considered a better measure of location since all observations contribute to the sample mean. As we work through more examples, it will become clear which measure of location (the mean or the median) is more appropriate in specific applications.

Because the sample size in Example 3.1 is small (n=7), the method of successively counting into the middle of the ordered data set to locate the median is easy to implement. When the sample size is larger, a more efficient method for computing the median involves 2 steps. In the first step, we compute the *position of the median in the ordered data set,* and in the second step we locate the median value. When the number of observations is odd, the <u>position</u> of the median is computed as follows:

$$\frac{n+1}{2} \tag{3.5}$$

For Example 3.1, the median is in the 4th position ((7+1)/2 = 4) in the ordered data set and in this example is equal to 121. The median represents the middle value, to further describe the sample, we now analyze the top and bottom halves.

The first and third *quartiles* are the values that separate the bottom and top 25% of the data elements, respectively. The first quartile of the sample, denoted Q_1, is the sample value that holds at least 25% of the data elements at or below it and at least 75% above or equal to it. The third quartile, denoted Q_3, holds at least 25% of the data elements at or above it and at least 75% below or equal to it. The median is also referred to as the second quartile, Q_2. The best way to determine the quartiles is to follow the two step procedure outlined above (i.e., first compute the positions of the quartiles in the ordered data set, and then locate the values). When the number of observations in a sample is odd, the positions of the quartiles are determined by the following formula.

$$\left[\frac{n+3}{4} \right] \tag{3.6}$$

where   [k] is the greatest integer less than k. For example, [2.1] = 2,
[2.9] = 2, [5.0] = 5, [10.8] = 10, and so on.

For Example 3.1,

$$\left[\frac{n+3}{4} \right] = \left[\frac{7+3}{4} \right] = \left[\frac{10}{4} \right] = [2.5] = 2$$

In Example 3.1, the quartiles are in the 2nd positions from the top and bottom of the ordered data set. The first quartile is $Q_1 = 110$ and the third quartile is $Q_3 = 130$. In Example 3.1,

Summarizing Data

approximately 25% of the systolic blood pressures are 110 or lower and approximately 25% of the systolic blood pressures are 130 or higher. Again, since the sample in Example 3.1 is so small (n=7) we do not need all of the statistics described to summarize and interpret these data. In larger samples, the quartiles are very informative statistics to understand the distribution of a particular characteristic.

The *mode* of the data set is defined as the most frequent value. In Example 3.1 the mode is 130, since it appears twice while the remaining values appear only once. A sample can have one mode or several modes. A sample with no repeated values has no mode

Other very informative descriptive statistics include the *minimum* and *maximum* values. In Example 3.1, the minimum is 100 and the maximum is 160. These values can be very useful, especially with regard to identifying *outliers*. Outliers are values which exceed the "normal" or expected range of values. For example, suppose ages are recorded on each of 20 individuals participating in an experimental study. Suppose the mean age for the sample is 83.5 with a standard deviation of 5.6. Suppose the minimum age is 70 and the highest 5 ages, in descending order, are 110, 90, 89, 89 and 87. Assuming that each age was recorded accurately, an age of 110 might be considered an outlier. It is not an incorrect value, just a value outside, in this case above, the normal range. Outlying values can be determined by an expert in the particular substantive area, or by using one of several statistical definitions (See Example 3.3). The statistical analyst need not do anything in particular with respect to outliers, only be aware of their existence and their impact on certain descriptive statistics (e.g., the sample mean).

Another descriptive statistic which addresses dispersion in a data set is the *range*. The range is defined as the maximum value minus the minimum value. In Example 3.1 the range = 160-100 or 60. Some investigators report the range as "100 to 160", while others report the range as 60. Both reports are appropriate. As noted, the range addresses dispersion in the sample. In Example 3.1 the observed systolic blood pressures cover 60 units. The range is based on only 2 values in the sample, the maximum and minimum. Though it is a very useful statistic, it can be somewhat misleading especially in the presence of outliers. For example, if the maximum value was 260 instead of 160 and all other data elements were unchanged, the range would be 260-100=160. This would suggest much more dispersion in the sample than the range of 60 (based on the data presented in Example 3.1) when only a single observation changed. We suggest that the range be interpreted with caution, and that the standard deviation be used to address dispersion in a sample.

Consider the following samples, call them samples A, B, C and D. The samples are all of the same size (n=11), have the same means (50) and ranges (100) , yet the standard deviations are different. How are the samples different?

| | Sample | | | |
Raw Data	A	B	C	D
	0	0	0	0
	50	10	20	0
	50	20	20	0
	50	30	20	0
	50	40	20	0
	50	50	50	50
	50	60	80	100
	50	70	80	100
	50	80	80	100
	50	90	80	100
	100	100	100	100

Summary Statistics

	A	B	C	D
n	11	11	11	11
$\overline{X}$	50	50	50	50
Range	100	100	100	100
s	22	33	35	50

Based on the standard deviations, the first sample has the least variation among observations while the last sample has the most. The range, in this example, does not discriminate among samples.

SAS Example 3.1. The following descriptive statistics were generated using SAS Proc Univariate (See interpretation below and Section 3.5 for more details) and the data in Example 3.1. An interpretation of the relevant components appears after the output.

SAS Output for Example 3.1

Summary Statistics

Summary Statistics

The UNIVARIATE Procedure
Variable: sbp (systolic blood pressure)

Moments

N	7	Sum Weights	7
Mean	123.571429	Sum Observations	865
Std Deviation	19.3550781	Variance	374.619048
Skewness	1.04211435	Kurtosis	1.63467176
Uncorrected SS	109137	Corrected SS	2247.71429
Coeff Variation	15.663069	Std Error Mean	7.31553189

Basic Statistical Measures

Location		Variability	
Mean	123.5714	Std Deviation	19.35508
Median	121.0000	Variance	374.61905
Mode	130.0000	Range	60.00000
		Interquartile Range	20.00000

Tests for Location: Mu0=0

Test	-Statistic-		-----p Value------	
Student's t	t	16.89165	Pr > $\lvert t \rvert$	<.0001
Sign	M	3.5	Pr >= $\lvert M \rvert$	0.0156
Signed Rank	S	14	Pr >= $\lvert S \rvert$	0.0156

Quantiles (Definition 5)

Quantile	Estimate
100% Max	160
99%	160
95%	160
90%	160
75% Q3	130
50% Median	121
25% Q1	110
10%	100
5%	100
1%	100
0% Min	100

Summarizing Data

```
              Extreme Observations
       ----Lowest----           ----Highest---
       Value        Obs         Value        Obs
         100          4           114          3
         110          2           121          1
         114          3           130          6
         121          1           130          7
         130          7           160          5
```

Interpretation of SAS Output for Example 3.1

The SAS Univariate Procedure is used to generate descriptive statistics on a continuous variable. SAS generates a number of descriptive statistics in the section labeled 'Moments'; we will highlight only a few.

The sample size is 7 (notice that SAS uses upper case 'N' as opposed to lower case 'n' to denote sample size), the sample mean is 123.6 and the sample standard deviation is 19.4. The sum of the observations (i.e., $\Sigma X_i = \Sigma X$) is 865, and the sample variance, s^2, is 374.6. The skewness of a sample indicates the degree of asymmetry in the sample distribution. Values close to 0 are indicative of symmetry. The kurtosis of a sample indicates the thickness in the tails of the distribution (i.e., the degree of clustering of observations at the extremes). Again, values close to 0 indicate lack of clustering in the tails. Estimates of skewness and kurtosis are somewhat unreliable in small samples and should be interpreted with caution. The normal distribution (discussed extensively in Chapter 4) has skewness=0 and kurtosis=0.

The uncorrected sum of squares, 'Uncorrected SS', is the sum of the observations squared (i.e., $\Sigma X^2 = 109,137$, which we used in the computational formula for the sample variance). The corrected sum of squares, 'Corrected SS', is the numerator of the sample variance, $\Sigma (X - \overline{X})^2 = 2247.7$. The coefficient of variation , 'Coeff Variation', is defined as the ratio of the sample standard deviation to the sample mean, expressed as a percentage (i.e., $CV = (s / \overline{X}) * 100$). The standard error of the mean 'Std Error Mean' is defined as $s/\sqrt{n}$.

The next part of the SAS output summarizes the most popular summary statistics for continuous variables in the section entitled 'Basic Statistical Measures.' Several measures of location are provided (the mean, median and mode) as are several measures of dispersion (standard deviation, range and interquartile range). The range is 160-100 or 60; the interquartile range, the difference between the first and third quartiles is 20. The most appropriate measures of location and dispersion depend on whether there are outliers in the dataset. If there are no outliers, the mean and standard deviation are the most appropriate measures of location and dispersion, respectively. If there are outliers, the median and interquartile range are the most appropriate measures of location and dispersion, respectively. The next part of the SAS output contains 'Tests for Location,' these will be discussed in detail in Chapter 6.

The next part of the SAS output displays the 'Quantiles' (or percentiles) of the variable, where the kth quantile is defined as the score that holds k% of the data below it. For example the maximum value is equivalent to the 100th quantile and equal to 160 in SAS Example 1. Similarly, the 75th quantile is equivalent to the third quartile (130), and so on. SAS also presents the 99th quantile which is equal to 160 in SAS Example 1. Since this is a small data set, these fine classifications are unnecessary and not meaingful.

SAS then prints the 'Extreme Observations' in the data set. In particular, the five smallest and five largest values are printed. Next to each value, in parentheses, is the observation number (i.e., the position of the observation in the data set). For example, the smallest value is 100, the fourth observation among the seven.

Example 3.2. Eight subjects are randomly selected from a population of patients with hypertension. Total serum cholesterol, in mg/100 ml, is measured on each subject and the sample data are given below.

197   212   211   184   260   233   245   219

In the following we summarize the cholesterol data using the statistics introduced in Example 3.1. Since total serum cholesterol is a continuous variable (as was systolic blood pressure), the same statistics will be computed. We will limit our discussion here to items and concepts that are distinct from those addressed in detail in Example 3.1. Again, a small sample size is used to illustrate the calculation of descriptive statistics while keeping actual computation time to a minimum. In practice, the same techniques can be applied to larger samples.

We will let the variable X denote total serum cholesterol. The table below displays the data elements, which have been ordered from smallest to largest, along with the value of each data element squared (which is used in the computation of the sample variance below):

X	X^2
184	33,856
197	38,809
211	44,521
212	44,944
219	47,961
233	54,289
245	60,025
260	67,600
1761	392,005

Summarizing Data

The number of patients, or sample size is n = 8. The mean cholesterol level in this sample is :

$$\overline{X} = \frac{\Sigma X}{n} = \frac{1761}{8} = 220.1 \,.$$

Notice that some of the total cholesterol levels are above the mean while others are below. This will always be true, as displayed in Figure 3.2 using Example 3.1. The sample mean represents a typical cholesterol level in this sample.

To address dispersion in the sample, we use the computational formula for the sample variance (3.4):

$$s^2 = \frac{\Sigma X^2 - (\Sigma X)^2/n}{n-1} = \frac{392,005 - (1761)^2/8}{7}$$

$$= \frac{392,005 - (3,101,121)/8}{7} = \frac{392,005 - 387,640.125}{7}$$

$$= \frac{4364.875}{7} = 623.6 \,.$$

Generally, the variance is not used to summarize dispersion. Instead, the sample standard deviation is computed:

$$s = \sqrt{623.6} = 25.0 \,.$$

The sample standard deviation represents how far each total cholesterol level is from the mean of 220.1. Again, by itself the standard deviation is often difficult to interpret. In particular, it is generally difficult to quantify what value of a standard deviation is considered large and what value is considered small. Individuals with substantive knowledge of the characteristic under investigation might have a feel for what is large and small with respect to the standard deviation.

Our standard summary of the cholesterol levels (a continuous variable) is: n = 8, $\overline{X}$ = 220.1, and s = 25.0.

The maximum cholesterol level is 260 and the minimum is 184. The range is 260-184 or 76. There is a substantial difference between the smallest and largest cholesterol levels, a difference of 76 units. Recall, the range is another measure of dispersion. In general, the range is less useful as a measure of dispersion than the sample standard deviation.

The median value, or the value that holds 50% of the cholesterol levels above it and 50% of the cholesterol levels below it, can be computed by arranging the data from smallest to largest, and successively counting from the right and left to arrive at the median, or middle, value:

Step 1: ~~184~~ 197 211 212 219 233 245 ~~260~~
Step 2: ~~184~~ ~~197~~ 211 212 219 233 ~~245~~ ~~260~~
Step 3: ~~184~~ ~~197~~ ~~211~~ 212 219 ~~233~~ ~~245~~ ~~260~~
⇑   ⇑

Two Middle Values

Summarizing Data

Because the number of observations in this sample is even (n=8), there are two middle values. *When the sample size is even, the median is defined as the mean of the two middle values :*

$$\text{Median} = \frac{212 + 219}{2} = 215.5$$

In Example 3.2, 50% of the cholesterol levels are above 215.5 and 50% of the cholesterol levels are below 215.5. We now have two statistics which represent a typical cholesterol level, the sample mean and the sample median. Although the convey different information, in general only one is necessary. Which is the best statistic to address location in this sample?

Reviewing the cholesterol levels in the sample, there do not appear to be outliers at either extreme in which case the mean is a better measure of location. An individual with clinical expertise would, however, be in a better position to make that assessment. In Example 3.3 we will present guidelines for assessing outliers based on statistical formulations.

In this example, as in Example 3.1, the sample size is small and it is easy to order the data elements from smallest to largest and to count into the middle from right and left to determine the median. In applications where the sample size is larger, it may be more efficient to use the two step method described in Example 3.1. First we compute the position(s) of the middle value(s) in the ordered data set and then locate those value(s). When the number of observations is even, there are two middle values, and their positions are computed as follows:

$$\frac{n}{2}, \text{ and } \left(\frac{n}{2}\right) + 1 \tag{3.7}$$

For Example 3.2,

$$\frac{n}{2} = \frac{8}{2} = 4^{\text{th}} \text{ position}$$

$$\left(\frac{n}{2}\right) + 1 = \left(\frac{8}{2}\right) + 1 = 5^{\text{th}} \text{ position}$$

The median is the mean of the observations in the 4th and 5th positions in the ordered data set (i.e. {212+219}/2 = 215.5).

To further describe the sample we now compute the quartiles. As noted in Example 3.1, the best way to determine the quartiles is to first compute the positions of the quartiles in the ordered data set and then to locate the values. When the number of observations in the sample is even, the positions of the quartiles are determined by the following formula.

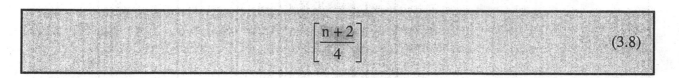

$$\left[\frac{n+2}{4}\right] \tag{3.8}$$

For Example 3.2,

$$\left[\frac{n+2}{4}\right] = \left[\frac{8+2}{4}\right] = \left[\frac{10}{4}\right] = [2.5] = 2 \,.$$

The quartiles are in the 2nd positions from the top and bottom of the ordered data set. The first quartile is $Q_1 = 197$ and the third quartile is $Q_3 = 245$.

SAS Example 3.2. The following descriptive statistics were generated using SAS. For Example 3.2 we produced the abbreviated summary statistics as opposed to the more extensive summary illustrated in Example 3.1. The following descriptive statistics were produced by SAS Proc Means (See interpretation below and Section 3.5 for more details) using the data in Example 3.2. A brief interpretation appears after the output.

SAS Output for Example 3.2

```
                        Summary Statistics

                       The MEANS Procedure
           Analysis Variable : chol total serum cholesterol

    N         Mean         Std Dev        Minimum         Maximum
   ------------------------------------------------------------------
    8    220.1250000     24.9710547    184.0000000     260.0000000
   ------------------------------------------------------------------
```

Interpretation of SAS Output for Example 3.2

The SAS Means Procedure is used to generate descriptive statistics on a continuous variable. There are many different statistics which can be requested. The statistics shown above are the default statistics. The sample size is 8 (notice that SAS uses upper case 'N' as opposed to lower case 'n' to denote sample size), the sample mean is 220.1 and the sample standard deviation is 25.0. The minimum and maximum cholesterol levels are 184 and 260. SAS displays summary statistics to 8 decimal places by default. Users should round appropriately to report summary statistics.

Example 3.3 A sample of 51 individuals are selected for participation in a study of cardiovascular risk factors. The following data represent the ages of enrolled individuals measured in years (continuous variable). Here age is measured in the usual way with a person being recorded as 65, for example, until the day he/she turns 66. The data are as follows:

60	62	63	64	64	65	65	65	65	65	65	66	66
66	66	66	67	67	67	68	68	68	70	70	70	71
71	72	72	73	73	73	73	73	75	75	75	75	76
76	77	77	77	77	77	79	82	83	85	85	87	

The number of subjects, or sample size, is n = 51. In this example we have a much larger sample size than in previous examples. Here it is not possible to interpret the age data simply by inspecting the values. Instead, we need summaries of location and dispersion.

The mean age in this sample is:

$$\overline{X} = \frac{\Sigma X}{n} = \frac{3637}{51} = 71.3 .$$

In order to assess dispersion in the sample, we will compute the sample standard deviation. As a first step, we compute the sample variance using the computational formula presented in Example 3.1:

$$s^2 = \frac{261,439 - (3637)^2/51}{(51-1)} = 41.4 .$$

The sample standard deviation is:

$$s = \sqrt{41.4} = 6.4 .$$

In general, participants' ages deviate from the mean of 71.3 by 6.4 years. Notice that the magnitude of the standard deviation of the ages is much smaller than the standard deviations we computed on the systolic blood pressures in Example 3.1 ($s_{SBP} = 19.4$) and on the cholesterol levels in Example 3.2 ($s_{CHOL} = 25.0$). The participants in the study described in Example 3.3 are very homogeneous with respect to age. It is possible that the study objectives were focused on individuals 60 years of age or older. Most, if not all, studies have very explicit inclusion and exclusion criteria. These criteria must be recognized in order to appropriately interpret summary statistics.

Summarizing Data

As noted earlier, in many publications and research reports, investigators do not present raw data (i.e., observations measured on each member of a sample); instead they present summary statistics. Suppose that we did not have the actual ages of each participant here, instead we had only the summary statistics: n=51, $\overline{X}$ =71.3 and s=6.4.

The mean and standard deviation are used to understand where the data are located and how they are spread. The *Empirical Rule*, given below, can be used to learn more about a particular characteristic based on the these commonly available statistics:

Empirical Rule (3.9)

Approximately 68% of the observations fall between $\overline{X}$ - s and $\overline{X}$ + s

Approximately 95% of the observations fall between $\overline{X}$ - 2s and $\overline{X}$ + 2s

Approximately all of the observations fall between $\overline{X}$ - 3s and $\overline{X}$ + 3s

---

Using the data in Example 3.3 the Empirical Rule (3.9) indicates that approximately 68% of the ages fall between 71.3-6.4=64.9 and 71.3+6.4=77.7, approximately 95% of the ages fall between 58.5 and 84.1, and almost all of the ages fall between 52.1 and 90.5. Because we have the actual observations here, we computed the percentages of 51 observations that actually fell into each range. The following table illustrates how closely the Empirical Rule approximates the distribution of ages in this sample.

Range	Empirical Rule Percent of Observations	Percent of Sample Data
64.9 - 77.7	Approximately 68%	78.2%
58.5 - 84.1	Approximately 95%	94.1%
52.1 - 90.5	Almost All	100%

The Empirical Rule suggested that approximately 68% of the ages would fall between 64.9 and 77.7. In Example 3.3, 78.2% of the ages actually fell between 64.9 and 77.7. Similarly, the Empirical Rule suggested that approximately 95% of the ages would fall between 58.5 and 84.1. In Example 3.3, 94.1% of the ages actually fell between 58.5 and 84.1. Finally, the Empirical Rule suggested that almost all of the ages would fall between 52.1 and 90.5 and in fact they do.

The computation of the mean and standard deviation are cumbersome with a sample of 51 observations. We now use SAS' Proc Univariate to generate descriptive statistics for the data in Example 3.3.

SAS Example 3.3. The following descriptive statistics were generated using SAS Proc Univariate and the data in Example 3.3. An interpretation of the relevant components appears after the output.

SAS Output for Example 3.3

```
                        Summary Statistics

                     The UNIVARIATE Procedure
                     Variable:  age   (age in years)

                              Moments
N                         51    Sum Weights                  51
Mean                71.3137255  Sum Observations           3637
Std Deviation        6.4358067  Variance              41.4196078
Skewness            0.57609039  Kurtosis              -0.2678579
Uncorrected SS          261439  Corrected SS          2070.98039
Coeff Variation     9.02463958  Std Error Mean        0.90119319


                   Basic Statistical Measures
          Location                      Variability
     Mean     71.31373     Std Deviation          6.43581
     Median   71.00000     Variance              41.41961
     Mode     65.00000     Range                 27.00000
                           Interquartile Range   10.00000


                   Tests for Location: Mu0=0
           Test          -Statistic-      -----p Value------
           Student's t   t  79.13256      Pr > |t|    <.0001
           Sign          M      25.5      Pr >= |M|   <.0001
           Signed Rank   S       663      Pr >= |S|   <.0001
```

Summarizing Data

```
Quantiles (Definition 5)
   Quantile       Estimate
   100% Max            87
   99%                 87
   95%                 85
   90%                 79
   75% Q3              76
   50% Median          71
   25% Q1              66
   10%                 65
   5%                  63
   1%                  60
   0% Min              60


          Extreme Observations
    ----Lowest----        ----Highest---
   Value      Obs        Value       Obs
      60        1           82        47
      62        2           83        48
      63        3           85        49
      64        5           85        50
      64        4           87        51
```

Interpretation of SAS Output for Example 3.3

The SAS Univariate Procedure generates a number of descriptive statistics in the section labeled 'Moments'; we will highlight only a few. The sample size is 51, the sample mean is 71.3 and the sample standard deviation is 6.4. The sum of the observations (i.e., ΣX) is 3637, and the sample variance, s^2, is 41.42. In the "Basic Statistical Measures' section, SAS also provides the median (71) and the mode (the mode is 65 which appears six times in the sample). The range in ages is 87-60 or 27, the difference between the first and third quartiles (called the *Interquartile Range* = $Q_3 - Q_1$) is 10 years.

The oldest subject in the sample is 87, 75% of the subjects are 76 years of age or younger, 50% of the subjects are 71 or younger, 25% of the subjects are at or below age 66. The middle 90% of the sample are between 63 and 85 years of age (5% are above 85 and 5% are below 63).

If we compute statistics by hand, we determine the position of the median using (3.5) because n is odd:

$$\frac{n+1}{2} = \frac{51+1}{2} = 26^{\text{th}} \text{ position}$$

The median is in the 26th position (in the ordered data set) and equal to 71.

The positions of the quartiles are determined by (3.6):

$$\left[\frac{n+3}{4}\right] = \left[\frac{51+3}{4}\right] = \left[\frac{54}{4}\right] = [13.5] = 13$$

The quartiles are in the 13th positions from the top and bottom of the ordered data set: $Q_1 = 66$ and $Q_3 = 76$.

 Outliers in a sample can be determined using a number of different definitions. We present two of the more popular definitions. The first is based on the Empirical Rule (3.5) and is as follows:

Observations outside the range : $(\overline{X} \pm 3s)$	(3.10)

For Example 3.3, this range is (52.1 to 90.5). There are no values outside of this range and therefore no outliers according to definition (3.10). A second definition is as follows:

Observations above $Q_3 + 1.5\,(IQR)$ or below $Q_1 - 1.5\,(IQR)$	(3.11)

more popular than 3.10

$$\text{where } IQR = \text{Interquartile Range} = Q_3 - Q_1.$$

For Example 3.3, the upper limit is $(76 + 1.5(10)) = 91$ and the lower limit is $(66 - 1.5(10)) = 51$. There are no values exceeding the upper limit or falling below the lower limit, as defined by (3.11). In Example 3.3 the most appropriate measure of location is the sample mean and the most appropriate measure of dispersion is the sample standard deviation. If there are outliers in a data set the most appropriate measure of location is the median and the most appropriate measure of dispersion is the interquartile deviation ($IQR/2$). If there are no outliers in the data, then the sample mean and standard deviation are the most appropriate measures of location and dispersion, respectively.

3.3.2 Graphical Summaries for Continuous Variables

 Graphical presentations can also be very useful for summarizing data. The choice of the most appropriate presentation for any application can be difficult. One must always remember that the primary goal for generating any graphical presentation of data is to provide a simple, complete and accurate representation of the data

41

One very informative graphical display for continuous variables is the *Box and Whisker plot*. The Box and Whisker plot incorporates the minimum and maximum, the median and the quartiles.

Example 3.1. The Box and Whisker plot for Example 3.1 is given in Figure 3.3. The minimum, first quartile (Q_1), median, third quartile (Q_3) and maximum are noted on the top line. These notations are not part of the Box and Whisker plot, they are only noted here for reference.

Figure 3.3   Box and Whisker Plot for Example 3.1

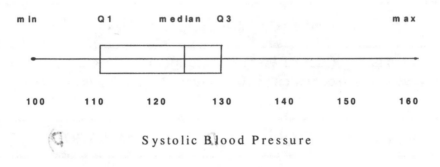

The lengths of the four sections of the Box and Whisker plot, separated by the first quartile, the median and the third quartile, reflect the clustering of observations within each section. The shorter the sections, relative to the others, the more observations are clustered. The Box and Whisker plot in Figure 3.3 indicates some clustering of observations between the median and third quartile (as indicated by the shorter section) and less clustering (more spread) among the observations between the third quartile and the maximum. With only n=7 observations in Example 3.1 we are probably making too much of these relationships among observations. If the Box and Whisker plot was based on a larger sample, we could make more general statements about the distribution of observations based on the plot.

Box and Whisker plots are extremely useful for comparing samples. For example, suppose we have a sample of patients free of coronary artery disease and measure their systolic blood pressures. Suppose we generate summary statistics (i.e., minimum, Q_1, median, Q_3, maximum) on this sample. Figure 3.4 displays the Box and Whisker plots for the sample of patients with diagnosed coronary artery disease (Example 3.1) and a sample of subjects free of coronary artery disease:

Summarizing Data

Figure 3.4   Using Box and Whisker Plots to Compare Samples

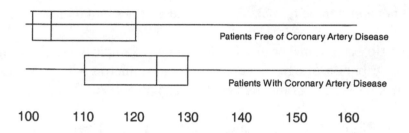

Systolic Blood Pressure

The ranges (i.e., the horizontal lines connecting minimum and maximum values) of observed systolic blood pressures are similar in the two samples. However, the majority of blood pressures in the sample of patients free of coronary artery disease are lower than those in the sample of patients with coronary artery disease. In fact, the middle 50% of the blood pressures (i.e., observations between Q_1 and Q_3) in the sample of patients free of coronary artery disease are between 102 and 120, while the middle 50% of the blood pressures in the sample of patients with coronary artery disease are between 110 and 130.

Example 3.2. The Box and Whisker plot for the cholesterol data is shown in Figure 3.5.

Figure 3.5   Box and Whisker Plot for Example 3.2

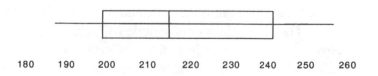

Total Serum Cholesterol

Summarizing Data

Based on the Box and Whisker plot, it appears that there is slight clustering at the lower end of the distribution as compared to the upper end (e.g., the lengths of the sections between the minimum and first quartile and between the first quartile and median are shorter than the length of the section between the median and the third quartile). Again, because the plot is based on such a small sample is it not appropriate to make much of these very slight indications.

Example 3.3 The Box and Whisker plot for the data in Example 3.3 is a very informative display of the data. The distribution of ages is approximately symmetric with a few values trailing off at the upper end. The middle 50% of the ages fall between 66 and 76.

Figure 3.6   Box and Whisker Plot for Example 3.3

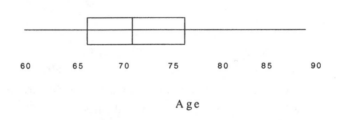

Age

Another useful graphical presentation for continuous variables is the *Stem and Leaf plot*. In the Stem and Leaf plot, each data element (e.g., each systolic blood pressure, cholesterol level, age) is split into two pieces: the stem and the leaf. The leading digits (e.g., hundreds place or tens place) denote the stem, while the trailing digits (e.g., tens and units or just the units place) denote the leaf. It is up to the analyst to determine the appropriate point at which to split the data elements. Best results are achieved when there are between 6 and 12 unique stems.

In Example 3.3, ages are recorded in years and consist of 2 digits. We let the first digit (the tens place) denote the stem and the second digit (the units place) denote the leaf. (In other applications there can be more than one digit in the stem and/or the leaf.) Ages of individuals in the sample ranged from 60 to 87. Therefore, the stems include 6, 7, and 8 (the first digit, or tens place, of each age). Since our objective is to have at least 6 stems, we will split each decade (60's, 70's, 80's) into two parts (low 60's, high 60's, low 70's, high 70's, low 80's and high 80's) to generate the Stem and Leaf plot.

To construct the Stem and Leaf plot, we list the stems vertically, from smallest to largest, and draw a vertical line. The line separates the stems from the leaves, which appear to the right of their respective stems. The leaves consist of the second digit (or units place) of each age. For example, the first observation is 60 (stem=6, leaf=0), the second observation is 62 (stem=6., leaf=2), and so on. Leaves are displayed on the same horizontal line as their respective stems, separated by spaces. The Stem and Leaf plot for the age data is shown in Figure 3.7.

Figure 3.7   Stem and Leaf Plot for Example 3.3

```
6 0 2 3 4 4
6 5 5 5 5 5 6 6 6 6 7 7 7 8 8 8
7 0 0 0 1 1 2 2 3 3 3 3 3
7 5 5 5 5 6 6 7 7 7 7 7 9
8 2 3
8 5 5 7
```

The Stem and Leaf plot is essentially a histogram with horizontal bars instead of vertical bars, but shows more detail in that the actual values of observations are displayed as opposed to only the number of observations in each category (or stem).

The 'plot' option in the SAS Proc Univariate statement generates some graphical displays of continuous data. More elaborate graphical presentations are available in other SAS procedures.

SAS Example 3.3.   The following graphical displays were generated using SAS Proc Univariate and the data in Example 3.3. An interpretation appears after the output.

SAS Output for Example 3.3

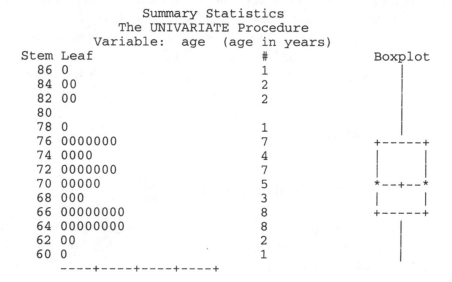

Summarizing Data

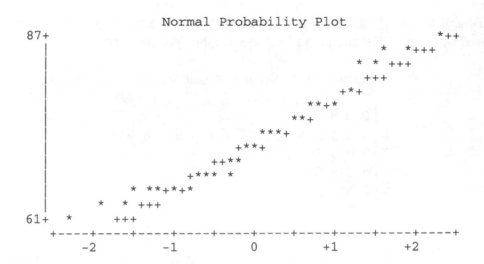

Normal Probability Plot

Interpretation of SAS Output for Example 3.3

The Stem and Leaf plot (labeled 'Stem Leaf') for the age data is displayed on the left. Notice that the scale is displayed from largest (top) to smallest (bottom). SAS uses an internal algorithm to determine how to split data elements into stems and leaves. In this example, SAS elected to use the tens and units place as the stems and the first decimal place as the leaves. SAS shows only the even ages and uses, for example, 62 for ages 62 and 63. Age 64 is used to display ages 64 and 65, and so on. Because each age was recorded to the nearest whole unit, the leaf for each value is 0.

To the right of the Stem and Leaf plot is a vertical Box and Whisker plot. The top and bottom horizontal lines indicate the third and first quartiles, respectively. The center horizontal line indicates the median. The "+" sign indicates the value of the sample mean. In this example, the sample mean is very close to the sample median (71.3 and 71, respectively). SAS will indicate extreme values (or outliers) using asterisks "*" or zeros "0", as follows. Any value which exceeds $Q_3 + 1.5(Q_3 - Q_1)$ or falls below $Q_1 - 1.5(Q_3 - Q_1)$ is denoted by "0". Any value which exceeds $Q_3 + 3(Q_3 - Q_1)$ or falls below $Q_1 - 3(Q_3 - Q_1)$ is denoted by "*". In this example there are no such values, thus there are no asterisks or zeros in the Box and Whisker plot.

3.3.3 Numerical Summaries for Discrete Variables

We now present descriptive statistics that are used to summarize discrete variables using examples. Recall, discrete variables are variables which can take on only a limited number of response options. These response options can be ordered (these variables are called ordinal variables) or unordered (these variables are called categorical variables). It is not critical to draw the distinction between ordinal and categorical variables. However, it is critical to draw the distinction between continuous and discrete variables because the numerical summaries (and graphical summaries presented in the next section) are quite different for continuous and discrete variables.

Example 3.4. Suppose we select a random sample of 50 subjects from the population of patients receiving care at a particular community health center. Each subject is given a self-administered survey containing a series of questions which address medical history and personal or background characteristics. One of the questions in the background characteristics section of the survey asks patients their current marital status. The response options are: Married, Separated, Divorced, Widowed, Never Married. Each respondent is instructed to select the one category (of five possible) that reflects his/her current marital status. Marital status is an example of a categorical variable because the response options are unordered. The sample data are summarized below:

Marital Status	Number of Individuals (f)
Married	24
Separated	5
Divorced	8
Widowed	2
Never Married	11
Total	**50**

The number of times each data element (or response option) is observed is called its *frequency*, denoted f. The table above which displays each response option and its frequency is called a *frequency distribution table*. A frequency distribution table is a useful summary for discrete data. Notice that the sum of the frequencies is equal to the sample size (i.e., $\Sigma f = n = 50$).

For these data, a frequency distribution table with one additional column, shown below, is a more useful summary.

Table 3.1   Frequency Distribution Table for Example 3.4

Marital Status	Number of Individuals (f)	Relative Frequency (f/n*100)
Married	24	48%
Separated	5	10%
Divorced	8	16%
Widowed	2	4%
Never Married	11	22%
Total	**50**	**100%**

This *frequency distribution table* displays each response option (first column), along with the frequency each is observed (second column). The *relative frequency* (third column) incorporates the sample size and is computed as follows: Relative Frequency = f/n, where n denotes the sample size (n=50). The relative frequency can be presented as a proportion (i.e., a decimal value between 0 and 1), or as a percent (i.e., a value between 0 and 100). Relative frequencies are often used to summarize a discrete variable. In this sample, 48% of the patients are currently married, 10% are separated, 16% are divorced, 4% are widowed, and 22% were never married.

SAS Example 3.4.   The following descriptive statistics were generated using SAS. For Example 3.4 we produced a frequency distribution table using SAS Proc Freq. A brief interpretation appears after the output.

SAS Output for Example 3.4

```
                    Frequency Distribution Table

                       The FREQ Procedure

                                      Cumulative    Cumulative
   marital      Frequency    Percent   Frequency      Percent
   ----------------------------------------------------------------
   Divorced          8        16.00          8        16.00
   Married          24        48.00         32        64.00
   Never Married    11        22.00         43        86.00
   Separated         5        10.00         48        96.00
   Widowed           2         4.00         50       100.00
```

Summarizing Data

Interpretation of SAS Output for Example 3.4

The SAS Freq Procedure is used to generate a frequency distribution table for a discrete (either categorical or ordinal) variable. The most useful summary of this variable is given by the relative frequencies, which appear in the column labeled 'Percent' above. SAS produces additional information in the frequency distribution table, namely the cumulative frequency and the cumulative percent. These summaries are most appropriate for ordinal variables and we discuss their interpretation in Example 3.5.

Example 3.5. Consider again the survey described in Example 3.4. Suppose that one of the questions in the medical history section of the survey asks patients to assess their current health status. The question in the survey reads "How would you rate your current health?" The response options are: Excellent, Very Good, Good, Fair, Poor. Each respondent is instructed to select the one category (of five possible) that best describes his/her current health. Health status is an example of an ordinal variable because the response options are ordered; Excellent is better than Very Good, Very Good is better than Good, and so on. The sample data are summarized below:

Health Status	Number of Individuals (f)
Excellent	19
Very Good	12
Good	9
Fair	6
Poor	4
Total	**50**

For these data, we compute additional descriptive statistics (i.e., additional columns), shown below:

Summarizing Data

Table 3.2   Frequency Distribution Table for Example 3.5

Health Status	Frequency (f)	Relative Frequency (f/n)	Cumulative Frequency	Cumulative Relative Frequency
Excellent	19	0.38	19	0.38
Very Good	12	0.24	31	0.62
Good	9	0.18	40	0.80
Fair	6	0.12	46	0.92
Poor	4	0.08	50	1.00
Total	**50**	**1.00**		

This frequency distribution table displays each response option (first column), along with the frequency each is observed (second column). The *relative frequency* (third column) incorporates the sample size and is computed as follows: Relative Frequency = f/n, where n denotes the sample size (n=50). Here we report the relative frequency as a proportion (i.e., a decimal value between 0 and 1). Relative frequencies are often used to summarize a discrete variable. In this example, 38% of the patients in the sample reported that their health was excellent. Twenty four percent reported that their health was very good, 18% reported that their health was good, 12% that their health was fair and 8% reported that their health was poor.

The *cumulative frequency* (fourth column) indicates the number of data elements at or better than each response option. For example, 19 patients reported that their health status was excellent. Thirty one patients reported that their health was very good or better (very good or excellent). Forty patients reported that their health was good or better (good or very good or excellent). Notice that the cumulative frequency for the last response option (poor) is equal to the sample size (i.e., n=50). The *cumulative relative frequency* is shown in the fifth column and is computed by taking the ratio of the cumulative frequency to the sample size. (Again, the cumulative relative frequency can be presented as a proportion or a percent.)

Ordered scales, such as the one used in this example to measure health status are very popular, particularly in the social sciences. Sometimes, numeric values are assigned to each of the response options to produce an ordered scale. These numeric values are then used to generate summary statistics. For example, suppose we assign the following values to the response options:

1 = Poor   2 = Fair   3 = Good   4 = Very Good   5 = Excellent

Summarizing Data

According to the above, higher values reflect better health status. There are many possible assignment strategies which might be employed in applications such as this. For example, we could have assigned values in reverse order such that lower scores reflect better health. The exact strategy influences the interpretation of summary statistics (described below).

Once the numerical values are assigned, we can then compute the mean and standard deviation using the same formulas used in Examples 3.1, 3.2 and 3.3. In applications in which numerical values are assigned to distinct response options, usually equally spaced assignments are made (e.g., 1, 2, 3, 4, 5; 0, 25, 50, 75, 100). Equally spaced numerical values imply that the distance between one response option and the next is the same across all possible responses. This may be true in many response scales, but not true in others. This issue as well as the exact numerical values assigned to each response option must be evaluated carefully when computing and interpreting numerical summary statistics such as the mean and standard deviation.

SAS Example 3.5. The following descriptive statistics were generated using SAS. For Example 3.5 we produced a frequency distribution table using SAS Proc Freq and generated descriptive statistics using the numerical assignments shown above. A brief interpretation appears after the output.

SAS Output for Example 3.5

Frequency Distribution Table
The FREQ Procedure
health status

health	Frequency	Percent	Cumulative Frequency	Cumulative Percent
Poor	4	8.00	4	8.00
Fair	6	12.00	10	20.00
Good	9	18.00	19	38.00
Very Good	12	24.00	31	62.00
Excellent	19	38.00	50	100.00

Summary Statistics
The MEANS Procedure
Analysis Variable : health health status

N	Mean	Std Dev	Minimum	Maximum
50	3.7200000	1.3099307	1.0000000	5.0000000

Summarizing Data

Interpretation of SAS Output for Example 3.5

The SAS Freq Procedure produced a frequency distribution table for the health status variable. The SAS Means Procedure is generally used to generate descriptive statistics on a continuous variable (e.g., Example 3.2). Here, we assigned numerical values to each response option (described above) and requested the default statistics. The sample size is 50, the sample mean is 3.7 and the sample standard deviation is 1.3. The minimum and maximum values are 1 and 5, respectively. All of these statistics must be interpreted with the specific numerical assignment strategy in mind. For example, the mean of 3.7 suggests that a typical health status assessment in this sample is Very Good (the response option assigned the numerical value 4).

Example 3.6 Recall the sample of 51 individuals selected for participation in a study of cardiovascular risk factors described in Example 3.3. Data were collected on each subject reflecting his/her age measured in years (a continuous variable). In Example 3.3 we summarized the sample using an array of descriptive statistics which are appropriate for continuous variables (e.g., n, $\overline{X}$, s, median, quartiles). Sometimes, continuous data are organized into categories as a means of summarizing. The range of the ages was 60 to 87. Suppose we collapse the ages into 5 year categories: 60-64, 65-69, 70-74, 75-79, 80-84, and 85-89. In this example, these classifications are intuitive. In other examples, it is up to the investigator to define the most meaningful categories. In all cases, the most useful presentations include between 6 and 12 distinct categories or classes. In general, with fewer than 6 categories information may be lost, while with more than 12 categories data may be too sparse (and therefore not useful as a summary).

Collecting the numbers of subjects in each class (See the raw data in Example 3.3) results in the following frequency distribution table:

Age Class	Numbers of Individuals (f)
60-64	5
65-69	17
70-74	12
75-79	12
80-84	2
85-89	3
Total	**51**

We now add the relative frequencies, cumulative frequencies and cumulative relative frequencies to the frequency distribution table:

Summarizing Data

Table 3.3   Frequency Distribution Table for Example 3.6

Age Class	Frequency (f)	Relative Frequency (f/n)	Cumulative Frequency	Cumulative Relative Frequency
60-64	5	0.10	5	0.10
65-69	17	0.33	22	0.43
70-74	12	0.24	34	0.67
75-79	12	0.24	46	0.91
80-84	2	0.04	48	0.95
85-89	3	0.06	51	1.01
Total	**51**	**1.00**		

The expanded table includes each response option (age class) and the frequency or number of subjects in each class. The relative frequency is computed by dividing the frequencies by the sample size (n=51). Here the relative frequencies are presented as proportions. The relative frequencies are the best summary of the age data: 10% of the subjects are between 60 and 64 inclusive, 33% are between 65 and 69, and so on. Almost half of the subjects are between 70-79 and 10% are 80 or older.

The cumulative frequency indicates the number of subjects in the particular age class or younger. For example, 5 subjects are in the 60 to 64 class, 22 are 69 or younger, 34 are 74 or younger. The cumulative relative frequency is shown in the fifth column and is computed by taking the ratio of the cumulative frequency to the sample size. In this example, 67% of the subjects are 74 or younger.

SAS Example 3.6.   The following frequency distribution tables was generated using SAS Proc Freq. A brief interpretation appears after the output.

```
                    Frequency Distribution Table
                        The FREQ Procedure

                                        Cumulative    Cumulative
       ageclass    Frequency    Percent  Frequency      Percent
       -----------------------------------------------------------
       60-64           5         9.80         5          9.80
       65-69          17        33.33        22         43.14
       70-74          12        23.53        34         66.67
       75-79          12        23.53        46         90.20
       80-84           2         3.92        48         94.12
       85-89           3         5.88        51        100.00
```

Interpretation of SAS Output for Example 3.5

The SAS Freq Procedure produced a frequency distribution table for the age class variable. The relative frequencies (labeled 'Percent') and the cumulative relative frequencies are probably the best summary for this ordinal variable.

3.3.4 Graphical Summaries for Discrete Variables

Graphical presentations can also be very useful for summarizing discrete data. A popular graphical display for a categorical variable is the bar chart, a popular graphical display for an ordinal variable is the histogram. Bar charts and histograms are based on either the frequency or the relative frequency of responses in each category. We will illustrate bar charts and histograms using Examples 3.4, 3.5 and 3.6.

Example 3.4. Marital status is an example of a categorical variable because the response options are unordered: Married, Separated, Divorced, Widowed, Never Married. The sample data are displayed below in a bar chart (produced in MS Word using the chart feature).

Figure 3.8   Bar Chart for Example 3.4 (Categorical Variable)

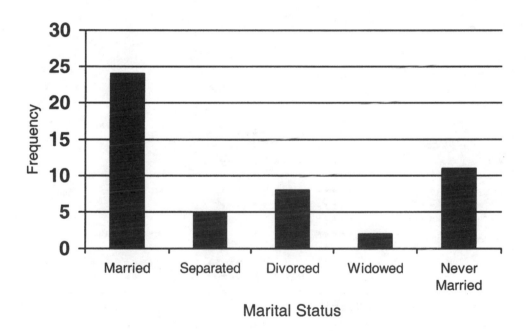

In the bar chart, each response option is listed along the horizontal axis. The vertical axis is scaled to accommodate either the frequencies (as in the example shown here) or relative frequencies. Rectangles are drawn above each response option to reflect the number of subjects (frequency) or proportion/percent of subjects (relative frequency) in each. In a bar chart, there are breaks between rectangles indicating distinct, unordered response options. In histograms

Summarizing Data

(illustrated in Examples 3.5 and 3.6), there are no breaks between rectangles suggesting order in the response options.

SAS Example 3.4.   The following bar chart was generated using SAS Proc Chart.

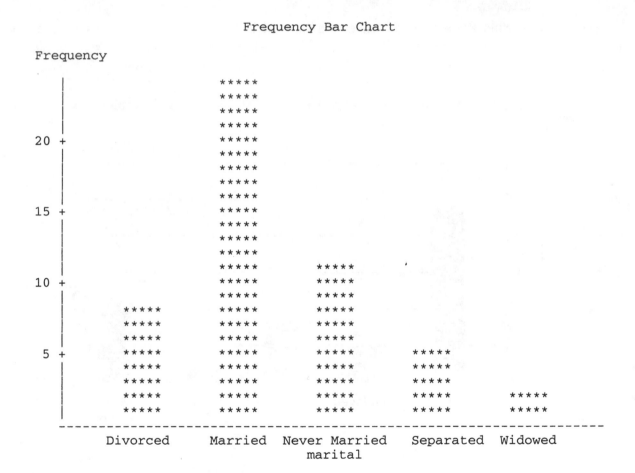

The SAS Chart Procedure is used to generate a bar chart for a categorical variable. The user can specify whether frequencies or relative frequencies are plotted on the vertical axis. Here we plotted frequencies.

Example 3.5. Health status is an example of an ordinal variable because the response options are ordered: Excellent, Very Good, Good, Fair, Poor. The sample data are displayed below in a relative frequency histogram.

Figure 3.9 Relative Frequency Histogram for Example 3.5 (Ordinal Variable)

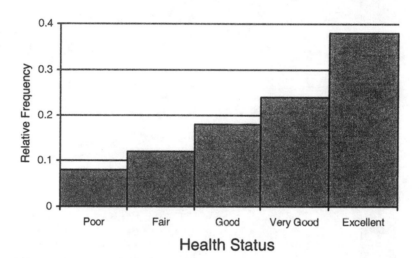

In this histogram, each response option is listed along the horizontal axis and the vertical axis is scaled to accommodate the relative frequencies. Rectangles are drawn above each response option to reflect the proportion of subjects in each. There are no breaks between rectangles suggesting the ordering of the response options. From the histogram it is clear that most subjects reported excellent health status, while fewer by comparison reported fair or poor health status.

SAS Example 3.5.   The following relative frequency histogram was generated using SAS Proc Chart.

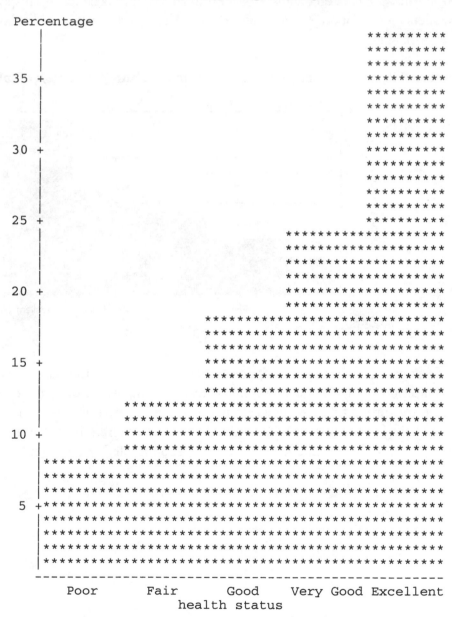

Summarizing Data

The SAS Chart Procedure can also be used to generate a histogram by specifying certain options (e.g., no breaks between response options). In this histogram the relative frequencies are expressed as percentages. The SAS code to generate the bar chart in Example 3.4 and the relative frequency histogram in this example are contained in Section 3.5

Example 3.6 The age class variable is an ordinal variable created from the continuous variable presented in Example 3.3. A frequency histogram for the age classes is shown below.

Figure 3.10   Frequency Histogram for Example 3.6 (Ordinal Variable)

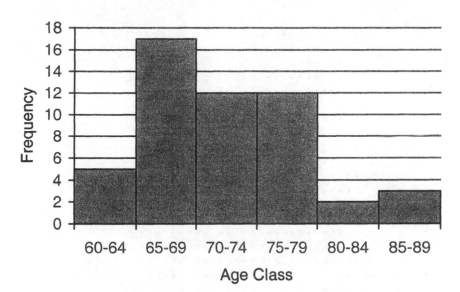

SAS Example 3.6.   The following frequency histogram was generated using SAS Proc Chart.

SAS Output for Example 3.6

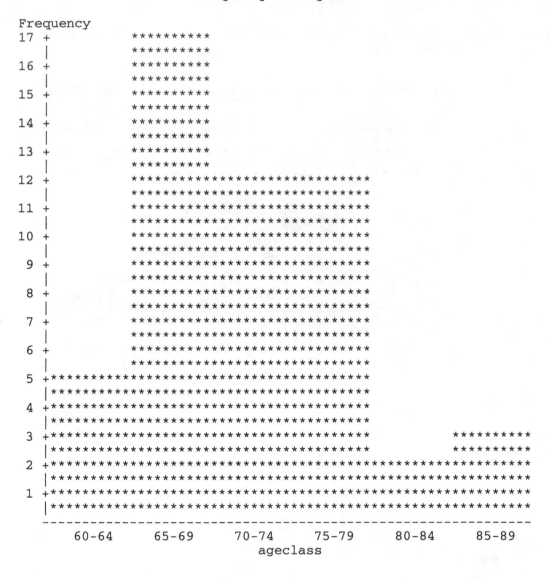

Frequency Histogram

Interpretation of SAS Output for Example 3.6

The SAS Chart Procedure was used to generate a frequency histogram. The SAS code used to generate the histogram is contained in Section 3.5. From the frequency histogram, we quickly see that the majority of the subjects in the sample are between 65 and 79 years of age.

3.4   Key Formulas

Mathematical formulas and short descriptions of each of the statistics illustrated in Section 3.3 are outlined below.

Numerical Methods: Location		
STATISTIC	NOTATION/FORMULA	DESCRIPTION
Minimum	Min	smallest value
Maximum	Max	largest value
Mean	$\overline{X} = \dfrac{\Sigma X}{n}$ *keep Σx for s^2*	average or typical value
Median	Order the data from smallest to largest Compute position(s) of middle value(s) n odd : (n+1)/2 n even : (n/2) and (n/2)+1, compute mean	middle value, holds 50% above it and 50% below it
Quartiles	Q_1 Q_3 Order data from smallest to largest Compute position(s) of quartiles n odd $\left[\dfrac{n+3}{4}\right]$ n even $\left[\dfrac{n+2}{4}\right]$	holds 25% of the data at or below it holds 25% of the data at or above it

Summarizing Data

Numerical Methods: Dispersion		
STATISTIC	NOTATION/FORMULA	DESCRIPTION
Range	Maximum - Minimum	
Variance	$s^2 = \dfrac{\Sigma(X - \overline{X})^2}{(n-1)}$ $= \dfrac{\Sigma X^2 - (\Sigma X)^2/n}{n-1}$	average squared deviation from the mean
mean $\ell\ell$ Standard Deviation	$s = \sqrt{s^2}$	typical deviation from the mean
Empirical Rule	$\overline{X} \pm s$ $\overline{X} \pm 2s$ $\overline{X} \pm 3s$	contains 68% of the distribution contains 95% of the distribution contains almost all of the distribution
Interquartile Range _(IQD)_	$Q_3 - Q_1$	difference between first and third quartile (middle 50%)
median $\ell\ell$ Interquartile Deviation	$\dfrac{Q_3 - Q_1}{2}$	mean difference between quartiles and median

Outliers 1). _Outside $\overline{X} \pm 3s$_

2). _Above $Q_3 + 1.5\,IQD$ or below $Q_1 - 1.5\,IQD$_

Numerical Methods: Distribution		
STATISTIC	NOTATION/FORMULA	DESCRIPTION
Frequency	f	number of subjects in each response category
Relative Frequency	f/n	proportion of subjects in each response category

Graphical Methods: Continuous Variables

Box and Whisker Plot

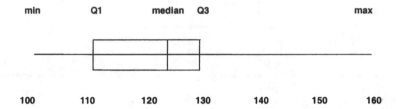

Stem and Leaf Plot

```
6 | 0 2 3 4 4
6 | 5 5 5 5 5 5 6 6 6 6 6 7 7 7 8 8 8
7 | 0 0 0 1 1 2 2 3 3 3 3 3
7 | 5 5 5 5 6 6 7 7 7 7 7 9
8 | 2 3
8 | 5 5 7
```

Summarizing Data

64

Graphical Methods: Discrete Variables

Histogram (Frequency or Relative Frequency Histogram)

Note the difference

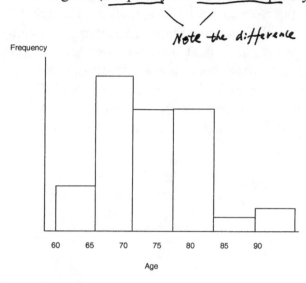

NOTE: Class limits, for classes of equal length, are displayed along the horizontal axis of the histogram. For ordinal variables no breaks are left between the rectangles. For categorical variables, there are breaks between the rectangles producing a bar chart (see below).

Relative Frequency Histogram

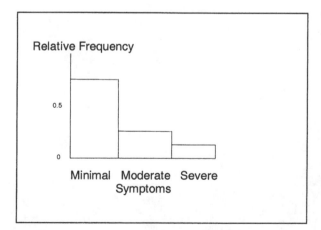

Bar Graph

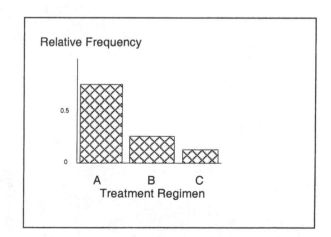

Summarizing Data

3.5   Statistical Computing

Following are the SAS programs which were used to generate the descriptive statistics and graphical summaries presented in Section 3.3 for Examples 3.1 through 3.6. The SAS procedures used and brief descriptions of their use are noted in the header to each example. Notes are provided to the right of the SAS programs (*in italics*) for orientation purposes and are not part of the programs. In addition, there are blank lines in the programs that follow which are solely to accommodate the notes. Blank lines and spaces can be used throughout SAS programs and are generally used to enhance readability. (More details are contained in Appendix A: Introduction to Statistical Computing Using SAS.) A summary of the SAS procedures used in the examples is provided in at the end of this section.

Continuous Variables

SAS EXAMPLE 3.1 Systolic Blood Pressure Data

Proc Univariate (Summary Statistics for Continuous Variable)

Seven subjects are randomly selected from a population of patients 50 years of age with diagnosed coronary artery disease. Systolic blood pressures, measured in millimeters of mercury (mm Hg), are taken on each subject and are listed below. Summarize the sample data using SAS.

121   110   114   100   160   130   130

Program Code

options ps=64 ls=80;	*Formats the output page to 64 lines in length, 80 columns in width*
data in;	*Beginning of Data Step*
input sbp;	*Inputs variable **sbp***
label sbp='systolic blood pressure';	*Attaches descriptive label to variable **sbp***
cards;	*Beginning of Raw Data section*
121	*actual observations*
110	
114	
100	
160	
130	
130	
run;	*End of Raw Data section*
title 'Summary Statistics';	*Title for output*
proc univariate;	*Procedure call. Proc Univariate generates summary statistics*
var sbp;	*Specification of variable **sbp***
run;	*End of Procedure section*

SAS EXAMPLE 3.2 Total Serum Cholesterol Data
Proc Means (Abbreviated Summary Statistics for Continuous Variable)

Eight subjects are randomly selected from a population of patients with hypertension. Total serum cholesterol, in mg/100 ml, is a continuous variable measured on each subject. The sample data are given below. Summarize the sample data using SAS.

197   212   211   184   260   233   245   219

Program Code

options ps=64 ls=80;	*Formats the output page to 64 lines in length, 80 columns in width*
data in;	*Beginning of Data Step*
input chol;	*Inputs variable **chol***
label chol='total serum cholesterol';	*Attaches descriptive label to variable **chol***
cards;	*Beginning of Raw Data section*
197	*actual observations*
212	
211	
184	
260	
233	
245	
219	
run;	*End of Raw Data section*
title 'Summary Statistics';	*Title for output*
proc means;	*Procedure call. Proc Means generates abbreviated summary statistics*
var chol;	*Specification of variable **chol***
run;	*End of Procedure section*

SAS EXAMPLE 3.3 Age Data (continuous variable)
Proc Univariate (Summary Statistics for Continuous Variable)

A sample of 51 individuals are selected for participation in a study of cardiovascular risk factors. The following data represent the ages of enrolled individuals, measured in years. Summarize the age data using SAS.

60	62	63	64	64	65	65	65	65	65	65
66	66	66	66	66	67	67	67	68	68	68
70	70	70	71	71	72	72	73	73	73	73
73	75	75	75	75	76	76	77	77	77	77
77	79	82	83	85	85	87				

Program Code

options ps=64 ls=80;	*Formats output page*
data in;	*Beginning of Data Step*
input age;	*Inputs variable **age***
label age='age in years';	*Attaches label to variable **age***
cards;	*Beginning of Raw Data section*
60	*actual observations (one observation per line,*
62	*51 lines total)*
63	
64	
64	
65	
.	
.	
.	
83	
85	
85	
87	
run;	*End of Raw Data section*
title 'Summary Statistics';	*Title for output of Proc Univariate*
proc univariate plot;	*Procedure call (Proc Univariate); the plot option generates several graphical displays of the data*
var age;	*Specification of variable **age***
run;	

Summarizing Data

Discrete Variables

SAS EXAMPLE 3.4 Marital Status Data
Proc Freq (Frequency Distribution Table for Discrete Variable)
Proc Chart (Bar Chart for Categorical Discrete Variable)

A sample of 50 subjects respond to a survey. The following data reflect marital status' of the subjects in the sample. Summarize the sample using SAS.

Marital Status	Number of Individuals (f)
Married	24
Separated	5
Divorced	8
Widowed	2
Never Married	11
Total	**50**

Program Code

```
options ps=64 ls=80;
data in;
  input marital $ 1-15 f;


cards;
Married          24
Separated        5
Divorced         8
Widowed          2
Never Married    11
run;
```

Formats output page
Beginning of Data Step
*Inputs alphanumeric ($) variable **marital** from columns 1-15 and numeric variable f*

Beginning of Raw Data section
actual observations:
*two entries per line (**marital** and f)*

Summarizing Data

```
title 'Frequency Distribution Table';
proc freq;
```
Procedure call (Proc Freq to generate frequency distribution table)

```
  tables marital;
```
*Specification of variable **marital***

```
  weight f;
```
weight f specifies variable containing frequencies (f)

```
title 'Frequency Bar Chart';
proc chart;
```
Procedure call (Proc Chart to generate frequency bar chart)

```
  vbar marital/freq=f discrete;
run;
```
*Specification of variable **marital**; vbar option requests a vertical histogram, freq=f specifies variable containing frequencies (f), discrete indicates that variable is discrete in nature as opposed to continuous*

SAS EXAMPLE 3.5 Health Status Data

Proc Format (Attaches Descriptions to Discrete Response Options)
Proc Freq (Frequency Distribution Table for Discrete Variable)
Proc Chart (Histogram for Ordinal Discrete Variable)
Proc Means (Abbreviated Summary Statistics)

The sample of 50 subjects described in SAS Example 3.4 also report their health status in the survey. The following data reflect the self-reported health status' of the subjects in the sample. Summarize the sample using SAS.

Health Status	Number of Individuals (f)
Excellent	19
Very Good	12
Good	9
Fair	6
Poor	4
Total	**50**

Summarizing Data

Program Code

options ps=64 ls=80;	*Formats output page*
proc format;	*Procedure Call (Proc Format to attach descriptive labels to response options)*
value hlth 5='Excellent'	*Specification of a name for format (hlth)*
4='Very Good'	*Description for response options*
3='Good'	
2='Fair'	
1='Poor';	
run;	

data in;	*Beginning of Data Step*
input health f;	*Inputs two numeric variables: health and f*
format health hlth.;	*Attaches description to each response option*
label health='health status';	*Attaches label to variable health*

cards;	*Beginning of Raw Data section*
5 19	*actual observations:*
4 12	*two entries per line (health and f)*
3 9	*The values 1-5 are used to represent the 5 distinct*
2 6	*health status responses.*
1 4	
run;	

title 'Frequency Distribution Table';

proc freq;	*Procedure call (Proc Freq to generate frequency distribution table)*
tables health;	*Specification of variable health*
weight f;	*weight f specifies variable containing frequencies (f)*
run;	

title 'Summary Statistics';

proc means;	*Procedure call (Proc Means to generate summary statistics)*
var health;	*Specification of variable health*
freq f;	*Specification of variable containing frequencies (f)*
run;	*End of procedure section.*

Summarizing Data

title 'Relative Frequency Histogram';
proc chart;

Procedure call (Proc Chart to generate relative frequency histogram)

vbar health/freq=f type=pct levels=5
space=0 width=10;

*Specification of variable **health**; vbar option requests a vertical histogram, freq=f specifies variable containing frequencies (f), type=pct requests a relative frequency histogram, levels=5 indicates the number of response options, space=0 requests no space between response options (appropriate for a histogram), and width=10 requests that the rectangles above each response option have width 10.*

run;

SAS EXAMPLE 3.6 Age Classes (ordinal variable)

Proc Freq (Frequency Distribution Table for Discrete Variable)
Proc Chart (Histogram for Ordinal Discrete Variable)

A sample of 51 individuals are selected for participation in a study of cardiovascular risk factors. The following data represent the ages of enrolled individuals, measured in years. Organize the subjects into age classes and summarize the age classes data using SAS.

60	62	63	64	64	65	65	65	65	65	65
66	66	66	66	66	67	67	67	68	68	68
70	70	70	71	71	72	72	73	73	73	73
73	75	75	75	75	76	76	77	77	77	77
77	79	82	83	85	85	87				

Program Code

options ps=64 ls=80;	*Formats output page*
data in;	*Beginning of Data Step*
input age;	*Inputs continuous variable **age***
if 60 le age le 64 then ageclass='60-64';	*Creates **ageclass** variable from continuous variable*
else if 65 le age le 69 then ageclass='65-69';	***age** (measured in years)*
else if 70 le age le 74 then ageclass='70-74';	*(The SAS function le=less than or equal to)*
else if 75 le age le 79 then ageclass='75-79';	
else if 80 le age le 84 then ageclass='80-84';	
else if 85 le age le 89 then ageclass='85-89';	
cards;	*Beginning of Raw Data section*
60	*actual observations (one observation per line,*
62	*51 lines total – See SAS Example 3.3)*
63	
85	
87	
run;	*End of Raw Data section*
title 'Frequency Distribution Table';	*Title for output of Proc Freq*
proc freq;	*Procedure call (Proc Freq)*
tables ageclass;	*Specification of variable **ageclass***
run;	
title 'Frequency Histogram';	
proc chart;	*Procedure call (Proc Chart to generate frequency histogram)*
vbar ageclass/levels=6	*Specification of variable **ageclass**; vbar option*
space=0 width=10;	*requests a vertical histogram, levels=6 indicates the number of response options, space=0 requests no space between response options (appropriate for a histogram)), and width=10 requests that the rectangles above each response option have width 10.*
run;	

Summarizing Data

Summary of SAS Procedures

The SAS procedures illustrated in this section are summarized in the table below. The SAS call statements for each procedure are shown with optional statements shown in italics. Users should refer to the examples in this section for complete descriptions of each procedure and associated options. A general description of each procedure is provided in the table below.

Procedure	Sample Procedure Call	Description
proc chart	proc chart; vbar x/*type=pct freq=f discrete*;	Generates (relative) frequency histogram for either continuous or discrete variables
proc freq	proc freq; tables x; *weight f;*	Generates frequency distribution table
proc means	proc means; var x1 x2; *freq f;*	Generates summary statistics (abbreviated version)
proc univariate	proc univariate *plot normal*; var x1; *freq f;*	Generates (extensive) summary statistics, Stem and Leaf plot, Box and Whisker plot

3.6   Problems

1.   Systolic blood pressures are recorded on a random sample of six males and are as follows:

156   85   103   92   108   128

a)   Compute the sample mean
b)   Compute the sample standard deviation
c)   Compute the sample median
d)   Compute the sample quartiles
e)   Generate a Box-Whisker plot

2.   Twenty five randomly selected,uncomplicated appendectomies lasted:

113, 118, 121, 123, 126, 128, 130, 135, 136, 137
138, 139, 140, 140, 142, 142, 142, 142, 143, 155,
157, 157, 158, 159, 164 minutes.

a)   Generate a frequency distribution table for the data using the following classes 110-119, 120-129, and so on.
b)   Generate a relative frequency histogram
c)   Generate a stem and leaf plot for the data

3.   The following data represent summary statistics on selling prices of diabetes home monitoring supplies from two different pharmaceutical companies:

	mean selling price	standard deviation
Company #1	$75	$ 4
Company #2	$56	$16

Use the empirical rule to describe the distributions of the selling prices of supplies from each company. How do the prices compare between companies (describe briefly) ?

4. An investigator is interested in examining the effects of alcohol on individual's abilities to perform physical activities. Twenty randomly selected individuals consumed a prescribed amount of alcohol, and their time to complete a specific physical activity was recorded. The data are given below:

Subject ID Number	Time In Minutes
1	12.5
2	14.0
3	16.1
4	27.2
5	14.5
6	22.6
7	15.9
8	16.2
9	10.1
10	29.1
11	12.5
12	12.6
13	9.9
14	12.8
15	11.1
16	23.2
17	14.5
18	32.6
19	16.9
20	9.0

a) Compute the sample mean
b) Compute the sample standard deviation
c) Compute the sample median
d) For this data, is the mean or the median a better measure of location ? Why?
e) Generate a frequency histogram for the data
f) The investigator suspects that there may be a difference in times to completion of these tasks under the influence of alcohol for males versus females. Subjects with identification numbers 1, 2, 5, 8, 9, 11, 13, 17, 19 and 20 are male while the remaining subjects are female. Generate and compare means, medians and standard deviations for the males and females, considered separately.

5. The following data reflect participants' scores on a standardized exam designed to measure medical literacy. Scores range from 0 to 100 with higher scores indicating better performance.

    91 94 83 84 46 90 77 85

    a) Compute the sample mean
    b) Compute the median
    c) Compute the sample variance
    d) Compute the sample standard deviation
    e) Compute the quartiles
    f) Display the Box-Whisker plot

6. Cholesterol levels for males aged 50 have a mean of 210 and a standard deviation of 22.8. Use the empirical rule to describe the distribution of cholesterol levels.

7. A study was recently conducted in the Boston area to investigate various aspects of AIDS complications and care. The study involved nearly 300 AIDS patients in the Boston area. The following data represent CD4 counts for 10 randomly selected AIDS patients involved in the study.

    75   135   210   240   86   100   125   59   63   100

    a) Compute the mean CD4 count for the sample
    b) Compute the median CD4 count
    c) Compute the range of CD4 counts
    d) Compute the standard deviation in the CD4 counts
    e) Which is the best measure of location (e.g., mean or median)? Which is the best measure of dispersion (e.g., standard deviation or interquartile range) ?

8.   The following is a reproduction of a figure presented in a medical journal describing the distribution of ages (in years) of participants involved in their research study of cardiovascular risk factors :

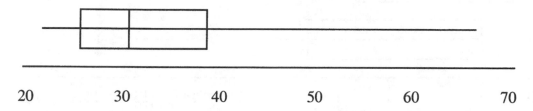

20   30   40   50   60   70

a)   What is the approximate value of the 75th percentile in the age distribution ?
b)   Is there any evidence of outliers in the distribution ?   (HINT: Use (3.11), the Q ± 1.5 IQR rule).

9.   The following is a frequency distribution table of systolic blood pressures (SBP) measured on a sample of female patients who are 50 years of age:

SBP	f
105	6
108	2
110	3
120	4
124	8
128	6
130	7
136	8
140	2
170	3
190	2
204	1

a)   Compute the sample median
b)   Compute the first and third quartiles
c)   Construct a Box-Whisker plot of the SBPs
d)   Are there outliers in the sample of SBPs ? Use (3.11), the Q ± 1.5 IQR rule).

Summarizing Data

10. A figure similar to the following was shown in a paper comparing visits to primary physicians over 12 months between Canadian and U.S. patients (* = Mean).

Fill in the following statements:

a) Half of the Canadian patients have fewer than _____ visits per year.
b) Twenty five percent of the U.S. patients had more than _____ visits per year.
c) Briefly compare the numbers of primary care visits per year between Canadian and U.S. patients.

11. The following reflect the times (in minutes) between prescription drop-off and pick-up at a local pharmacy:

    8   15   7   6   4   20   3   6   4   3

a) Calculate the mean time between drop-off and pick-up
b) Calculate the standard deviation in time between drop-off and pick-up
c) Calculate the median time between drop-off and pick-up

12. The time between successive routine primary care visits (in months) in a sample of patients from a local health maintenance organization (HMO) are shown below:

    6   18   12   9   10   14   6   5

a) Compute the sample mean
b) Compute the sample median
c) Compute the quartiles
d) Compute the sample standard deviation
e) Construct a Box-Whisker plot

Summarizing Data

13. The following data were taken from a paper published in a medical journal from a study investigating the effects of exercise in pregnant women.

Characteristics of Women at Baseline *

	Exercise Group (n=18)	Control Group (n=15)
Age (yrs)	31.1 ± 5.4	29.7 ± 4.9
Education (yrs)	17.1 ± 2.2	15.9 ± 2.1
Race or ethnic group (Number of Subjects)		
Non-Hispanic white	15	13
Hispanic	1	2
Asian	2	0

* Table entries are Numbers of Subjects or Means $\pm$ SD

a) Use the empirical rule to describe the distributions of ages of women in the exercise and control groups (separately).

b) Is there a difference in the ages of women in the two groups ? Briefly - base answer only on part (a).

14. A study is undertaken to investigate the self-reported health status among women with osteoporosis. Health status is measured on a scale from 0 to 100 with higher scores indicative of better health status. The following data are observed:

    35   48   70   42   80   57   74   39   40

a) Compute the sample mean
b) Compute the sample standard deviation
c) Compute the median
d) Compute the quartiles

15. The following figure represents the distribution of self-reported health status measured on a sample of women free of osteoporosis:

    a) Draw the Box Whisker plot reflecting the self-reported health status among women with osteoporosis (using the data from problem 14 below the box whisker plot describing the self-reported health status among women free of osteoporosis.

Free of Osteoporosis

With Osteoporosis

```
0      10     20     30     40     50     60     70     80     90    100
```

    b) How do the average self-reported health status scores compare between women with osteoporosis and those free of osteoporosis ?

    c) How do the self-reported health status scores compare between women with osteoporosis and those free of osteoporosis with respect to variation ?

Summarizing Data

16. A study is conducted to assess the extent to which patients who had coronary artery bypass surgery were maintaining their prescribed exercise programs. The following data reflect the numbers of times patients reported exercising over the previous month (4 weeks). For the purposes of this study, exercise was defined as moderate physical activity lasting at least 20 minutes in duration.

14	11	8	6	5	3
6	13	12	8	1	4

a) Compute the mean number of times patients exercised per month
b) Compute the median number of times patients exercised per month
c) Compute the standard deviation in the number of times patients exercised per month
d) If a patient reported exercising more than twice per week, that patient was considered adherent with respect to his/her exercise program. Classify each patient as adherent or not based on the number of times he/she reported exercising and generate a relative frequency histogram displaying the distribution of adherent and non-adherent patients.

17. The following data were collected from a random sample of 10 students who started an MPH program in the fall of 1998. The data reflect GRE scores, which range from 200-800 with higher scores indicative of better achievement.

520 680 470 560 510 610 670 560 525 475

a) Compute the sample mean
b) Compute the sample standard deviation
c) Compute the sample median
d) Compute the sample range

Summarizing Data

18.   Twenty five patients are randomly selected from the population of all patients in a particular health maintenance organization (HMO) with rheumatoid arthritis. Each patient is surveyed and asked to rate their HMO with respect to satisfaction with the care they receive for their arthritis. The satisfaction scores range from 0 to 100 with higher scores indicative of more satisfaction. The data are given below.

$$65 \quad 86 \quad 84 \quad 85 \quad 97 \quad 94 \quad 89 \quad 84 \quad 83 \quad 89$$
$$88 \quad 78 \quad 77 \quad 76 \quad 82 \quad 72 \quad 92 \quad 99 \quad 94 \quad 83$$
$$81 \quad 85 \quad 97 \quad 93 \quad 79$$

a)   Compute the sample mean
b)   Compute the sample standard deviation
c)   Compute the sample median
d)   Compute the quartiles
e)   Compute the sample range
f)   Generate a frequency distribution table to summarize the data. A frequency distribution table is most informative when there are between 6 and 12 classes or categories. For this example, suppose we wish to construct a frequency distribution table with 7 classes. In order to define the classes, the following formula is used to compute the width of each class (i.e., the range of values covered by each class): Width = range/(number of intervals).
g)   Construct a Stem and Leaf plot using stems defined by the classes given in the frequency distribution table in (f).

19. A study is conducted to assess the drinking behaviors of college seniors. A random
     sample of ten seniors is selected and each is asked the number of alcoholic drinks they
     consume in a typical week. The data are shown below.

| 20 | 40 | 25 | 12 | 16 | 8 | 8 | 5 | 10 | 0 |

a)   Compute the mean number of drinks per week.
b)   Compute the standard deviation in the number of drinks per week.
c)   Compute the range in the number of drinks per week.
d)   Compute the median number of drinks per week.

20. The following Box-Whisker plots display the numbers of alcoholic drinks consumed by
     male and female students in a typical week.

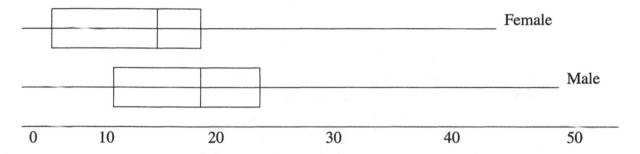

a)   Estimate the median number of drinks per week for males.
b)   Estimate the median number of drinks per week for females.
c)   Estimate the quartiles for males.
d)   Estimate the quartiles for females.
e)   Are there any outliers among the male scores? (HINT: Use the quartile rule)

21. The following displays the results of a standardized neuropsychological test administered to a sample of high school seniors. The test measures analytic skills and is scored on a scale of 0 to 100 with higher scores indicative of stronger analytic skills.

Males			Females
1	1		
8 8	2		
4 5 6	3		2 3 3 3 4 4 6 7 8
3 5 6 9	4		4 5 5 6 7 8 8 9 9
1 1 2 3	5		3 3 4 4 5 5 6 7
	6		1 1 1 2 2 3 3 4 4 4 5 5 5 7 7 8 9
3 4 5 5 5 7 8 9	7		2 2 3 3 3 4 5 6 6 7 8
4 5 6 6 6 8 9 9 9	8		2 2 3 6 7
1 2 2 2 3 3 3 3 4 4 4 4 4 4	9		1 3 4
0 0	10		

a) Compute the median scores for males and females.
b) Compute the quartiles for males and females
c) Compute the median for the total sample (males and females combined)
d) Are there any outliers among the male scores? (HINT: Use the quartile rule)
e) Based on your answer to (d), would the mean or median provide a better estimate of a typical result for males on the standardized exam?

22. A study is conducted to assess the number of cigarettes smoked in a typical day among college freshman who smoke. All freshman involved in the assessment report that they are regular smokers. The following reflect the reported numbers of cigarettes smoked per day:

20   40   25   12   16   8   8   5   10   12

a) Compute the mean number of cigarettes smoked.
b) Compute the standard deviation in the number of cigarettes smoked.
c) Compute the range in the number of cigarettes smoked.
d) Compute the median number of cigarettes smoked.

Summarizing Data

23.   A self-administered survey is fielded to measure the number of hours of exercise that
        patients with Type II diabetes get in a typical week. The following data are observed:

        5   3   2   0   2   7   4   3   4   6   0

        a)   Compute the sample mean
        b)   Compute the sample median
        c)   Compute the sample standard deviation
        d)   Compute the sample range

24.   The following plots describe the lengths of stay following total knee replacement surgery
        for patients in three different hospitals (denoted A, B and C below).

        a)   What is the median length of stay for patients in hospital B?
        b)   Which hospital has the shortest lengths of stay, on average, for patients
                undergoing knee replacement?
        c)   What is the range of lengths of stay in hospital C?
        d)   Are there any outliers in lengths of stay in hospital A? (HINT: Use the quartile
                rule to check for outliers)

25. In the study described in problem 24, suppose that patients in each hospital (A, B and C) were asked to rate their overall satisfaction with the total knee replacement surgery experience. The data are shown below:

Satisfaction	Hospital		
	A	B	C
Extremely Satisfied	62	50	20
Satisfied	20	30	16
Dissatisfied	14	12	17
Extremely Dissatisfied	9	8	42
Total Number of Patients	105	100	95

a) Generate frequency distribution tables for each hospital's satisfaction data, considered separately.

b) Generate a frequency distribution table for all hospitals satisfaction data combined.

SAS Problems: Use SAS to solve each of the following problems.

1.   Systolic blood pressures are recorded on a random sample of six males and are as follows:

156   85   103   92   108   128

Use SAS Proc Univariate to determine the following:

a)   the sample mean
b)   the sample standard deviation
c)   the sample median
d)   the sample quartiles

2.   Twenty five randomly selected,uncomplicated appendectomies lasted:

113, 118, 121, 123, 126, 128, 130, 135, 136, 137
138, 139, 140, 140, 142, 142, 142, 142, 143, 155,
157, 157, 158, 159, 164 minutes.

Use SAS Proc chart to generate a frequency distribution table for the data.

3.   The following data reflect participants' scores on a standardized exam designed to measure medical literacy.   Scores range from 0 to 100 with higher scores indicating better performance.

91   94   83   84   46   90   77   85

Use SAS Proc Means to determine the following:

a)   the sample mean
b)   the sample variance
c)   the sample standard deviation
d)   the minimum and maximum

Summarizing Data

4. A study was recently conducted in the Boston area to investigate various aspects of AIDS complications and care. The study involved nearly 300 AIDS patients in the Boston area. The following data represent CD4 counts for 10 randomly selected AIDS patients involved in the study.

75	135	210	240	86	100	125	59	63	100

    Use SAS Proc Univariate to determine the following:

    a)   the sample mean and median
    b)   the standard deviation and the interquartile range
    c)   Which is the best measure of location (e.g., mean or median)? Which is the best measure of dispersion (e.g., standard deviation or interquartile range) ?

5. The following is a frequency distribution table of systolic blood pressures (SBP) measured on a sample of female patients who are 50 years of age:

SBP	f
105	6
108	2
110	3
120	4
124	8
128	6
130	7
136	8
140	2
170	3
190	2
204	1

    Use SAS Proc Univariate to generate a Box-Whisker plot of the SBPs. Are there outliers in the sample of SBPs ?

Summarizing Data

6.   The following reflect the times (in minutes) between prescription drop-off and pick-up at a local pharmacy:

     8   15   7   6   4   20   3   6   4   3

     Use SAS Proc Univariate to determine the following:

     a)   the mean time between drop-off and pick-up
     b)   the standard deviation in time between drop-off and pick-up
     c)   the median time between drop-off and pick-up

7.   The time between successive routine primary care visits (in months) in a sample of patients from a local health maintenance organization (HMO) are shown below:

     6   18   12   9   10   14   6   5

     Use SAS Proc Means to determine the following:

     a)   the sample mean
     b)   the minimum and maximum
     c)   the sample standard deviation

8.   A study is conducted to assess the extent to which patients who had coronary artery bypass surgery were maintaining their prescribed exercise programs. The following data reflect the numbers of times patients reported exercising over the previous month (4 weeks). For the purposes of this study, exercise was defined as moderate physical activity lasting at least 20 minutes in duration.

     | 14 | 11 | 8 | 6 | 5 | 3 |
     | 6 | 13 | 12 | 8 | 1 | 4 |

     If a patient reported exercising more than twice per week, that patient was considered adherent with respect to his/her exercise program. Using SAS, Classify each patient as adherent or not based on the number of times he/she reported exercising and generate a relative frequency histogram (using SAS Proc Chart) displaying the distribution of adherent and non-adherent patients.

9. Twenty five patients are randomly selected from the population of all patients in a particular health maintenance organization (HMO) with rheumatoid arthritis. Each patient is surveyed and asked to rate their HMO with respect to satisfaction with the care they receive for their arthritis. The satisfaction scores range from 0 to 100 with higher scores indicative of more satisfaction. The data are given below.

65 86 84 85 97 94 89 84 83 89
88 78 77 76 82 72 92 99 94 83
81 85 97 93 79

Use SAS Proc Univariate to generate a Box-Whisker plot and a Stem and Leaf plot for the satisfaction data.

10. The following data reflect satisfaction scores measured in patients undergoing total knee replacement surgery in three different hospitals (A, B and C).

	Hospital		
Satisfaction	A	B	C
Extremely Satisfied	62	50	20
Satisfied	20	30	16
Dissatisfied	14	12	17
Extremely Dissatisfied	9	8	42
Total Number of Patients	105	100	95

Use SAS Proc Freq to generate a frequency distribution table for all hospitals satisfaction data combined. Use SAS Proc Chart to generate a frequency histogram for the data (all hospitals combined).

Introductory Applied Biostatistics

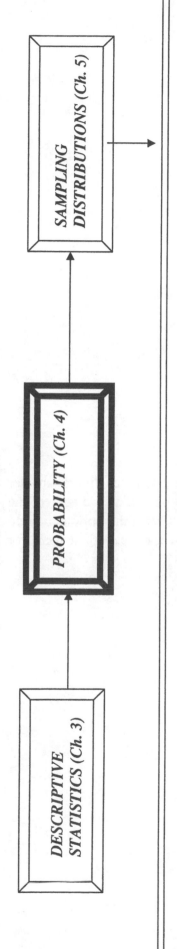

DESCRIPTIVE STATISTICS (Ch. 3) → PROBABILITY (Ch. 4) → SAMPLING DISTRIBUTIONS (Ch. 5)

STATISTICAL INFERENCE (Chapters 6-14)

Outcome Variable	Grouping Variable(s)/ Predictor(s)	Analysis	Chapter(s)
Continuous	-	Estimate μ, Compare μ to Known, Historical Value	6/13
Continuous	Dichotomous (2 Groups)	Compare Independent Means (Estimate/Test ($\mu_1-\mu_2$)) or the Mean Difference(μ_d)	7/13
Continuous	Discrete (> 2 Groups)	Test the Equality of K Means using Analysis of Variance ($\mu_1=\mu_2=\ldots\mu_k$)	10/13
Continuous	Continuous	Estimate Correlation or Determine Regression Equation	11/13
Continuous	Several Continuous or Dichotomous	Multiple Linear Regression Analysis	11
Dichotomous	-	Estimate p, Compare p to Known, Historical Value	8
Dichotomous	Dichotomous (2 Groups)	Compare Independent Proportions (Estimate/Test (p_1-p_2))	8/9
Dichotomous	Discrete (>2 Groups)	Test the Equality of k Proportions (Chi-Square Test)	8
Dichotomous	Several Continuous or Dichotomous	Multiple Logistic Regression Analysis	12
Discrete	Discrete	Compare Distributions Among k Populations (Ch-Square Test)	8
Time to Event	Several Continuous or Dichotomous	Survival Analysis	14

CHAPTER 4: Probability

4.1 Introduction

In Chapter 3 we discussed an array of statistics used to describe a sample. In this chapter we focus on the process of sampling from the population of interest. In particular, we address the experimental mechanisms and subsequent likelihood of sampling subjects from the population into a sample (See Figure 4.1). The sampling process is a critical part of the theoretical basis for statistical inference.

Figure 4.1   Sampling

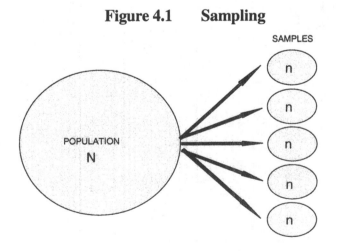

Probability

In Section 4.2 we provide background, and introduce vocabulary terms and definitions. In Section 4.3 we discuss the basic or first principles of probability. In Section 4.4 we address combinations and permutations. In Section 4.5 we discuss a popular discrete probability model, the binomial distribution model, and in Section 4.6 we discuss a popular continuous probability model, the normal distribution model. Throughout this chapter we use simple examples to illustrate the concepts of probability theory. It is probability theory which provides the basis for statistical inference (where we make generalizations about unknown population parameters based on a single sample) which is discussed in subsequent chapters. We summarize key formulas and concepts in Section 4.7. In Section 4.8 we use SAS to simulate probability distributions.

4.2 Background

4.2.1. Vocabulary

An *experiment* is defined as a process by which a measurement is taken or observations are made, or as a procedure that generates outcomes. An experiment may involve the selection of 3 individuals from a pool of 12 individuals to participate in a particular activity, or the administration of a psychological test to a human subject to record a response (e.g. completion of the test (Yes/No), or the time to complete the test in minutes).

There are two types of experiments, *deterministic and random*. In deterministic experiments the same outcome is observed each time the experiment is performed. In random experiments one of several possible outcomes is observed each time the experiment is performed. For example, consider an experiment in which a coin is dropped from the roof of a twenty story building. Each time the experiment is performed (i.e., a coin is dropped) the same outcome is observed, the coin falls to the earth due to gravity. This is an example of a deterministic experiment. Consider a second experiment in which a single individual is selected from a pool of 6 individuals. Each time the experiment is performed (i.e., an individual is selected), only one of the six candidate individuals is selected. This is an example of a random experiment. In this experiment we know that there are six possible outcomes, but do not know with certainty which individual will be selected on a given performance of the experiment. Probability theory is concerned with random experiments.

The enumeration of all possible outcomes of an experiment is called the *sample space*, denoted S. Consider the experiment involving a population of six individuals (N=6). Suppose each individual is assigned a unique identification number from 1 to 6. The sample space is the listing of all possible outcomes (or individuals) given by $S = \{1, 2, 3, 4, 5, 6\}$.

In probability we often look at certain characteristics or types of outcomes. For example, in the experiment which involves selecting individuals (or sampling) from the population, we might be interested in subjects with specific characteristics, such as female gender or ages over 65 years.

Collections of outcomes are called *events* and are usually denoted with capital letters (e.g., A, B, C). Individual outcomes are referred to as simple events.

Example 4.1. Consider a population of N=6 individuals in which each individual is assigned a unique identification number ranging from 1-6. Suppose we record gender and age on each member of the population. The population is described below.

Subject Number	Gender	Age
1	M	40
2	F	42
3	M	51
4	F	58
5	M	67
6	F	70

Suppose our experiment involves sampling or selecting one subject at random from the population. There are six possible outcomes of the experiment. Again, in most statistical applications we are generally not concerned with the probability that any particular individual is selected into a sample. Instead, we are concerned with selecting certain "types" of individuals. For example, if the population is comprised of 50% females, then a representative sample is one in which about half of the individuals are female.

Consider the following events, denoted A, B and C, and the outcomes or subjects which comprise each:

Event	Outcomes (i.e., Subject Numbers)
A = female subjects	{2,4,6}
B = male subjects	{1,3,5}
C = subjects over the age of 65	{5,6}

In probability we assign a numeric value, a *probability*, to each outcome (i.e., simple event) and to each event to denote the likelihood that the outcome or event occurs. There are a number of strategies for assigning probabilities. We use a strategy in which each outcome is considered to have an equal chance of occurring. This strategy is appropriate when outcomes are assumed to occur at random. Probabilities of outcomes and events (e.g., event A) are denoted as follows: P(outcome) and P(A), respectively. Probabilities are values between 0 and 1 (i.e., $0 \leq P(outcome) \leq 1$). A probability of 0 indicates that the outcome or event has no chance of occurring or is not possible, while a probability of 1 indicates that an outcome or event is certain to occur. The sum of the probabilities of the individual outcomes in any sample space is equal to 1 (i.e., Σ_S P(outcome)=1).

4.3 First Principles

Assuming that each outcome in an experiment is equally likely, the probability of each outcome is given by:

$$P(outcome) = 1/N \qquad (4.1)$$

where N denotes the total number of outcomes in the experiment.

Using the same logic, the probability of an event is given by:

$$P(event) = (\# \ outcomes \ in \ event)/N \qquad (4.2)$$

In Example 4.1 where we had a population of N=6 individuals and each individual was assigned a unique identification number ranging from 1-6 (shown below).

Subject Number	Gender	Age
1	M	40
2	F	42
3	M	51
4	F	58
5	M	67
6	F	70

Probability

The probability that any subject (e.g., subject assigned number 1 or 2) is selected is 1/6 by (4.1): $P(1) = P(2) = P(3) = P(4) = P(5) = P(6) = 1/6$.

Recall the events A, B and C, and the outcomes or subjects which comprise each:

Event	Outcomes (i.e., Subject Numbers)
A = female subjects	{2,4,6}
B = male subjects	{1,3,5}
C = subjects over the age of 65	{5,6}

Using formula 4.2: $P(A) = 3/6$, $P(B) = 3/6$, and $P(C) = 2/6$. The probability of selecting a female is 3/6=50%, the probability of selecting a male is 3/6=50% and the probability of selecting a subject over the age of 65 is 2/6=33%.

The *complement* of an event consists of all outcomes in the sample space that are not in the event. The complement of an event is denoted with a " ' " following the event label. For example, the complement of event A in Example 4.1 is denoted A' and consists of all outcomes in the sample space that are not in event A. Event A includes all female subjects. Therefore the complement of event A includes all subjects who are not female (i.e., subjects who are male): A'= {1, 3, 5}. Similarly, B'={2, 4, 6} and C' = {1, 2, 3, 4}. The probabilities of these events can be found by applying formula 4.2: $P(A')=3/6$, $P(B') = 3/6$, and $P(C') = 4/6$.

In probability theory there are a number of rules which can be applied in specific situations to generate probabilities. These rules can be extremely useful. However, we recommend using formula (4.2) wherever possible, as it is more intuitive in most cases. The complement rule is given below.

The *Complement Rule*: $P(A') = 1 - P(A)$ (4.3)

Using the Complement Rule, $P(A') = 1 - P(A) = 1 - 3/6 = 3/6$, $P(B') = 1 - P(B) = 1 - 3/6 = 3/6$, and $P(C') = 1 - P(C) = 1 - 2/6 = 4/6$.

The *union* of 2 events consists of outcomes in either event (or in both events). The union of 2 events is denoted by the symbol $\cup$ or with the word "or." For example, the union of events A and C consists of outcomes in event A, in event C or in both, and is denoted: $(A \cup C) = (A \text{ or } C)$. Recall event A consists of female subjects while event C consists of subjects over 65 years of age. The union of events A and C consists of individuals who are either female or over 65 years of age (or both): $(A \cup C) = (A \text{ or } C) = \{2, 4, 5, 6\}$. The union is displayed graphically in Figure 4.2 using a Venn Diagram.

Probability

Figure 4.2 Union of Events A and C: (A ∪ C) = (A or C)

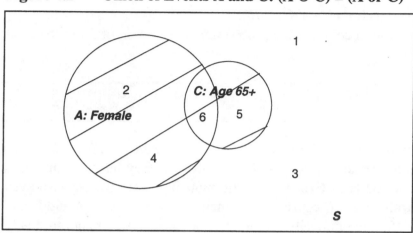

The probability of the union of 2 events can be computed using first principles, or formula (4.2): P(A ∪ C) = (A or C) = 4/6.

The *intersection* of 2 events consists of outcomes that are in both events. The intersection of 2 events is denoted by the symbol ∩ or with the word "and". For example, the intersection of events A and C consists of all individuals that are both in event A and in event C and is denoted: (A ∩ C) = (A and C). Event A consists of female subjects while event C consists of individuals over 65 years of age. The intersection of events A and C consists of the one female who is 70 years of age and assigned number 6: (A ∩ C) = (A and C) = {6}. The intersection is displayed graphically in Figure 4.3 using a Venn Diagram.

Figure 4.3 Intersection of Events A and C: (A ∩ C) = (A and C)

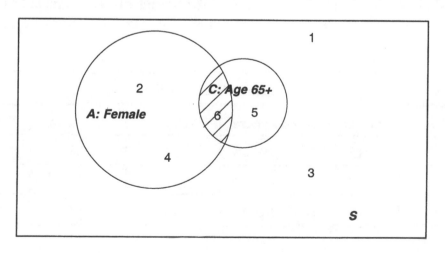

The probability of the intersection of 2 events can be computed using first principles, or formula (4.2): $P(A \cap C) = (A \text{ and } C) = 1/6$.

There is a rule to compute the probability of the union of 2 events called the addition rule, and it is given below.

$$\text{The } \textit{Addition Rule}: \ P(A \text{ or } B) = P(A) + P(B) - P(A \text{ and } B) \qquad (4.4)$$

Using the addition rule, $P(A \text{ or } C) = P(A) + P(C) - P(A \text{ and } C) = 3/6 + 2/6 - 1/6 = 4/6$.

Two events are said to be *mutually exclusive* if they have no outcomes in common. For example, events A and B in Example 4.1 are mutually exclusive because event A consists of all female subjects while event B consists of all male subjects. Events A and B have no individuals in common and therefore are mutually exclusive. Events A and C are not mutually exclusive because both events include the individual assigned number {6}. Events B and C are not mutually exclusive because both events include the individual assigned number {5}. When two events are mutually exclusive, the probability of their intersection is zero (e.g., For Example 4.1, $P(A \cap B) = P(A \text{ and } B) = 0$). We now consider a more realistic example.

Example 4.2: A survey is conducted to assess the background characteristics of patients seeking care at an allergy clinic. All patients who visit the clinic over a three month period are asked to complete a self-administered survey which contains questions to measure age, gender, ethnicity, educational levels and so on. Shown below are data collected on gender and educational level. The data are displayed in a cross-tabulation, or contingency table. The rows of the table indicate patients' gender while the columns of the table display patients' highest educational level. Educational level is presented in categories: 0-8 years, 9-12 years, 13-16 years, 17+ years and reflect patients' highest level of formal education. The 8 gender by educational level combinations (e.g., males with 0-8 years of education, males with 9-12 years of education) are called the cells of the table.

Gender	0-8 years	9-12 years	13-16 years	17+ years	TOTAL
		Educational Level			
Male	15	20	17	26	78
Female	30	42	31	27	130
TOTAL	45	62	48	53	208

Suppose a patient is selected at random, using formula (4.2), find the probability that:

a) the patient is male.

b) the patient has 9-12 years of education.

c) the patient is female and has 9-12 years of education.

d) the patient has 17+ years of education and is male.

e) the patient has at most 12 years of education.

$P(male) = P(M) = 78/208$

$P(\text{9-12 years}) = 62/208$

$P(\text{F and 9-12}) = 42/208$

$P(\text{17+ and M}) = 26/208$

$P(\leq \text{12 years})$
$= P(\text{0-8 or 9-12 years})$
$= 107/208$

From (a), the probability that a male is selected is $P(M) = 78/208 = 0.375$. The probability can be stated as a percent, for example there is a 37.5% chance that a male is selected at random from this population. Another interpretation is: 37.5% of the population is male.

Suppose we now want to compute the probability that a male is selected among those with 17+ years of education. That is, considering only those patients with the highest educational levels (17+ years), what is the probability of selecting a male ? This question can be addressed with *conditional probability*. Specifically, we are interested in the probability of an event (i.e., that a male is selected at random) among a specific subgroup of the population (i.e., among those with 17+ years of education). In probabilistic terms, we "condition on the given event" (i.e., the event that a patient has 17+ years of education) and then compute the probability of the event of interest (i.e., that a male is selected). We write the conditional event as follows: $P(\text{Male}|\text{17+ years})$, where the "|" is read as "given." The probability that a male is selected "given" that a patient with 17+ years of education is selected is computed using first principles, or by formula (4.2). $P(M|\text{17+ years}) = 26/53 = 0.491$. Notice that the conditional probability is not equal to the unconditional probability, $P(M|\text{17+ years}) = 0.491$, and $P(M) = 0.375$, respectively. The proportion of males among patients with 17+ years of education is higher than the proportion of males in the population as a whole.

There is a rule for computing conditional probabilities which may be useful in certain applications.

The *Conditional Probability Rule*: $P(A|B) = P(A \text{ and } B)/P(B)$ (4.5)

Using the data in Example 4.2, we illustrate the use of the conditional probability rule. We leave it to the reader to compute the following conditional probabilities using first principles, or formula (4.2):

Probability

P(M|9-12 years) = P(M and 9-12 years) / P(9-12 years) = (20/208)/(62/208) = 20/62
P(13-16 years|F) = P(13-16 years and F) / P(F) = (31/208)/(130/208) = 31/130

Two events are said to be *independent* if the probability of one is not "influenced" by the occurrence or nonoccurrence of the other. For example, suppose the data in Example 4.2 looked like the following:

Gender	0-8 years	9-12 years	13-16 years	17+ years	TOTAL
	Educational Level				
Male	25	30	25	25	105
Female	25	30	25	25	105
TOTAL	50	60	50	50	210

Using the data shown above, P(Male) = P(M) = 105/210 = 0.50. The probability that a male is selected among those patients with 17+ years of education is selected is : P(M|17+ years) = 25/50 = 0.50. In this data (N=210), the unconditional probability (i.e., P(M)) is equivalent to the conditional probability (i.e., P(M|17+ years)). Therefore, the two events (selecting a male and selecting a subject with 17+ years of education) are said to be independent since the occurrence of one event does not influence the probability of the other.

The definition of independence is given in (4.6):

Two events A and B are independent if: (4.6)

P(A) = P(A|B), or
P(B) = P(B|A), or equivalently

P(A ∩ B) = (A and B) = P(A)·P(B)

We illustrate the definition of independence by applying the last version of the definition. Using first principles (4.2) applied to the data shown above (N=210), P(M and 17+ years) = 25/210 = 0.119, P(M) = 105/210 and P(17+ years) = 50/210. To test if these two events (M and 17+ years of education) are independent, we must check whether the probability of their intersection, P(M and 17+ years), is equal to the product of their unconditional probabilities: i.e., Is P(M and 17+ years) = P(M)·P(17+ years) ?

Substituting the probabilities,

$$P(M) \cdot P(17+ \text{ years}) = (105/210) \cdot (50/210) = 0.119.$$

Thus, $P(M \text{ and } 17+ \text{ years}) = P(M) \cdot P(17+ \text{ years})$ and these events are independent by definition (4.6). The probability of selecting a male is the same, regardless of which educational subgroup we assess. This is only true for the data in which N=210, and is not the case for the data given in Example 4.2 where N=208. In Example 4.2, the probability a male is selected at random is 0.375. The probability that a male is selected among the subgroup with 17 or more years of education is 0.491. There is a higher proportion of males among the patients with the highest educational levels.

Example 4.3. Consider a screening tool comprised of a series of questions designed to assess whether patients exhibit symptoms of depression. Each participant completes the series of questions and his/her scores are summed producing a total score. Patients are then classified as at low, moderate or high risk for depression based on their total scores. In the present study each patient completes the screening tool and, in addition, undergoes an extensive psychiatric examination. Based on the psychiatric examination, patients are classified as clinically depressed or not. The following table summarizes the results of the study:

	Risk for Depression (Screening Tool)			
Clinical Assessment	Low	Moderate	High	TOTAL
Not Depressed	522	127	39	688
Depressed	28	18	11	57
TOTAL	550	145	50	745

a) What is the probability that a patient who is at low risk for depression (based on the screening tool) is clinically depressed? P(depressed|low risk) = 28/550 = 0.05 using (4.2).

b) What is the probability that a patient who is at moderate risk for depression (based on the screening tool) is clinically depressed? P(depressed|moderate risk) = 18/145 = 0.12 using (4.2).

c) What is the probability that a patient who is at high risk for depression (based on the screening tool) is clinically depressed? P(depressed|high risk) = 11/50 = 0.22 using (4.2).

d) Is the clinical assessment of depression independent of the patient's risk for depression (based on the screening tool)? From (a) – (c) we see that the likelihood that a patient is clinically depressed increases as a function of his/her risk. If we compare the unconditional and conditional probabilities for depression (4.6) we find the following: P(clinically depressed) = 57/ 745 = 0.077. The conditional probabilities of depression we computed in (a) – (c) are either below or above the unconditional probability of clinical depression for patients classified as low, moderate or high risk. Thus, the diagnosis of depression is not independent of the patient's risk for depression (based on the screening tool). The diagnosis of depression is related to the patient's risk (based on the screening tool); patients at higher risk are more likely to be clinically depressed.

Example 4.3 (continued). Suppose that 60% of the participants in the study described in Example 4.3 are male (P(male) = 0.60). In addition, suppose that 70% of the participants who are classified as clinically depressed are male (i.e., P(male|clinically depressed) = 0.70). Using the rules of probability, answer the following:

a) What proportion of participants are female?
P(female) = 1-P(male) = 1-0.60=0.40.

b) Are the events male and clinically depressed mutually exclusive?
Find P(male and clinically depressed). From the conditional probability rule we have the following: P(male|clinically depressed) = P(male and clinically depressed)/P(clinically depressed). From Example 4.3, P(clinically depressed) = 0.077. P(male and clinically depressed) = 0.70·0.077 = 0.0539. Because the probability is not zero, the events male and clinically depressed are not mutually exlcusive.

c) Are the events male and clinically depressed independent?
Because P(male|clinically depressed) = 0.70 and P(male)=0.60 (the conditional and unconditional probabilities are not equal), these two events are not independent.

d) Who is more likely to be clinically depressed – a male or a female in this study?
Find P(clinically depressed|male) and P(clinically depressed|female) and compare. P(clinically depressed|male)=P(male and clinically depressed)/P(male) = 0.0539/0.60 = 0.09. P(clinically depressed|female) = P(female and clinically depressed)/P(female) = 0.02/0.40 = 0.05. To see these computations more clearly, try to construct a two-by-two table (with gender and clinical depression as the rows and columns) and fill in the information. In this study, men are almost twice as likely to be classified as clinically depressed.

4.4 Combinations and Permutations

Consider the population of 6 individuals (N=6) in Example 4.1 where subjects are identified by unique numbers 1-6. There are several different strategies for selecting individuals from a population into a sample. We will describe several strategies here. It is important to know which sampling strategy is employed in any given application to ensure the validity of statistical inferences.

Issues which must be considered in determining a sampling strategy include: 1) whether individuals are sampled from a population with replacement or without replacement, and 2) whether the order in which individuals are sampled is important or not.

We illustrate each of these considerations and their impact on the sampling process using our population of 6 individuals from Example 4.1. Suppose for the population in Example 4.1 we generate all possible samples of size n=2. Suppose we sample with replacement (i.e., we select one individual at random from the population, record their unique subject number, and place them back into the population prior to the second selection) and consider the case where order is important (i.e., the samples (1, 2) and (2, 1) are considered two distinct samples). Thirty six samples of size n=2 are produced with this strategy and are given below.

Sampling with replacement, order important

(1,1)	(1,2)	(1,3)	(1,4)	(1,5)	(1,6)
(2,1)	(2,2)	(2,3)	(2,4)	(2,5)	(2,6)
(3,1)	(3,2)	(3,3)	(3,4)	(3,5)	(3,6)
(4,1)	(4,2)	(4,3)	(4,4)	(4,5)	(4,6)
(5,1)	(5,2)	(5,3)	(5,4)	(5,5)	(5,6)
(6,1)	(6,2)	(6,3)	(6,4)	(6,5)	(6,6)

Suppose we now sample with replacement and consider the case where order is not important (i.e., the samples (1, 2) and (2, 1) are considered the same sample). Twenty one samples of size n=2 are produced with this strategy and are given below.

Sampling with replacement, order not important

(1,1)	(1,2)	(1,3)	(1,4)	(1,5)	(1,6)
	(2,2)	(2,3)	(2,4)	(2,5)	(2,6)
		(3,3)	(3,4)	(3,5)	(3,6)
			(4,4)	(4,5)	(4,6)
				(5,5)	(5,6)
					(6,6)

Usually we are interested in samples where individuals do not appear more than once. Suppose we sample without replacement (i.e., we select one individual at random, record their unique subject number, keep them aside and make the second selection from remaining five subjects) and consider the case where order is important. Thirty samples of size n=2 are produced with this strategy and are given below.

Sampling without replacement, order important

	(1,2)	(1,3)	(1,4)	(1,5)	(1,6)
(2,1)		(2,3)	(2,4)	(2,5)	(2,6)
(3,1)	(3,2)		(3,4)	(3,5)	(3,6)
(4,1)	(4,2)	(4,3)		(4,5)	(4,6)
(5,1)	(5,2)	(5,3)	(5,4)		(5,6)
(6,1)	(6,2)	(6,3)	(6,4)	(6,5)	

Finally, suppose we sample without replacement and consider the case where order is not important. Fifteen samples of size n=2 are produced with this strategy and are given below.

Sampling without replacement, order not important

(1,2)	(1,3)	(1,4)	(1,5)	(1,6)
	(2,3)	(2,4)	(2,5)	(2,6)
		(3,4)	(3,5)	(3,6)
			(4,5)	(4,6)
				(5,6)

In most applications we sample without replacement, thereby ensuring that the sample is comprised of different individuals rather than the same individual selected more than once. There are formulas which are useful to determine the numbers of distinct samples which are produced under specific sampling strategies. When sampling without replacement, the number of distinct arrangements (i.e., order important) called *permutations* of n individuals from a population of size N is given by:

$$_N P_n = \frac{N!}{(N-n)!} \tag{4.7}$$

Where N! (N "factorial") = $N \cdot (N-1) \cdot (N-2) \cdot \ldots \cdot (2) \cdot (1)$.
(e.g., $5! = 5 \cdot 4 \cdot 3 \cdot 2 \cdot 1 = 120$, $3! = 3 \cdot 2 \cdot 1 = 6$, $1! = 1$ and $0!$ is defined as 1)

In Example 4.1 when sampling without replacement, the number of permutations (order important) of size 2 from the population of size 6 is given by:

$$_6 P_2 = \frac{6!}{(6-2)!} = \frac{6!}{4!} = \frac{6 \cdot 5 \cdot 4 \cdot 3 \cdot 2 \cdot 1}{4 \cdot 3 \cdot 2 \cdot 1} = 30$$

Probability

When sampling without replacement, the number of samples in which order is not important, or *combinations*, of n individuals from a population of size N is given by:

$$_NC_n = \begin{bmatrix} N \\ n \end{bmatrix} = \frac{N!}{n!(N-n)!} \qquad (4.8)$$

In Example 4.1 when sampling without replacement, the number of combinations of size 2 from the population of size 6 is given by:

$$_6C_2 = \begin{bmatrix} 6 \\ 2 \end{bmatrix} = \frac{6!}{2!(6-2)!} = \frac{6 \cdot 5 \cdot 4 \cdot 3 \cdot 2 \cdot 1}{2 \cdot 4 \cdot 3 \cdot 2 \cdot 1} = 15$$

Through the previous illustration we see that specific sampling strategies produce different numbers of samples. In general, we are less interested in the number of distinct samples produced by a particular sampling strategy and more interested in the attributes of subjects selected into specific samples. For example, suppose we are interested in the number of females selected into each sample under the last sampling strategy (sampling without replacement, order not important).

Suppose we let the variable X denote the number of females selected into each sample. X is called a *random variable* and takes on the values 0, 1 or 2 depending on the number of females (subjects assigned even identification numbers) selected into each sample. The samples are shown below along with the value of X = the number of females selected into each sample.

Sampling Strategy: Sampling without replacement, order not important

Sample	X = # females
(1,2)	1
(1,3)	0
(1,4)	1
(1,5)	0
(1,6)	1
(2,3)	1
(2,4)	2
(2,5)	1
(2,6)	2
(3,4)	1
(3,5)	0
(3,6)	1
(4,5)	1
(4,6)	2
(5,6)	1

Probability

In the population, 3 of 6 subjects, or 50% are female. In the samples shown above, females make up 0%, 50% or 100% of each sample. Three of 15 samples (20%) have no females, 9 of 15 samples (60%) have exactly one female and 3 of 15 samples (20%) have exactly 2 females. In statistical inference we will have only a <u>single</u> sample and attempt to draw inferences about a population based on that one sample. Based on the above, we are most likely (probability=0.60) to select a sample containing exactly one female. It is, however, possible that we select a sample containing no females or two females. It is important to understand the relationship between characteristics of samples and characteristics of populations to draw appropriate inferences. We will elaborate on this notion in the remainder of this and subsequent chapters.

Example 4.4. Consider a population consisting of 4 laboratory mice. Suppose that 2 are diseased and 2 are not. Let D1 and D2 denote the two diseased subjects and N1 and N2 denote the two non-diseased subjects. Suppose we take simple random samples of size n=2 from the population.

The following table displays the samples generated under three different sampling strategies: Sampling without replacement-order important, sampling without replacement-order not important and sampling with replacement-order important. Next to each sample we display the probability that each sample is selected, assuming that all samples are equally likely.

Sampling Without Replacement				**Sampling With Replacement**	
Ordered Samples	Prob.	Unordered Samples	Prob.	Ordered Samples	Prob.
N1,N2	1/12	N1,N2	1/6	N1,N1	1/16
N1,D1	1/12	N1,D1	1/6	N1,N2	1/16
N1,D2	1/12	N1,D2	1/6	N1,D1	1/16
N2,N1	1/12	N2,D1	1/6	N1,D2	1/16
N2,D1	1/12	N2,D2	1/6	N2,N1	1/16
N2,D2	1/12	D1,D2	1/6	N2,N2	1/16
D1,N1	1/12			N2,D1	1/16
D1,N2	1/12			N2,D2	1/16
D1,D2	1/12			D1,N1	1/16
D2,N1	1/12			D1,N2	1/16
D2,N2	1/12			D1,D1	1/16
D2,D1	1/12			D1,D2	1/16
				D2,N1	1/16
				D2,N2	1/16
				D2,D1	1/16
				D2,D2	1/16

Suppose we are interested in the likelihood that a diseased subject is selected into the sample. Let the *random variable* X denote the number of diseased subjects selected into each sample. In this example X takes on the values 0, 1 or 2 depending on the number of diseased subjects selected into each sample. The following tables display the samples along with the corresponding values for the random variable X under each sampling strategy.

Sampling Without Replacement					**Sampling With Replacement**	
Ordered Samples	X	Unordered Samples	X		Ordered Samples	X
N1,N2	0	N1,N2	0		N1,N1	0
N1,D1	1	N1,D1	1		N1,N2	0
N1,D2	1	N1,D2	1		N1,D1	1
N2,N1	0	N2,D1	1		N1,D2	1
N2,D1	1	N2,D2	1		N2,N1	0
N2,D2	1	D1,D2	2		N2,N2	0
D1,N1	1				N2,D1	1
D1,N2	1				N2,D2	1
D1,D2	2				D1,N1	1
D2,N1	1				D1,N2	1
D2,N2	1				D1,D1	2
D2,D1	2				D1,D2	2
					D2,N1	1
					D2,N2	1
					D2,D1	2
					D2,D2	2

The random variable X generates a *probability model or probability distribution*. The distribution has 3 possible values 0, 1 and 2 and corresponding probabilities under each different sampling strategy. The listing of the values of the random variable and corresponding probabilities constitutes a *probability distribution or model*. The probability distributions for the random variable X under each sampling strategy are given below:

Sampling Without Replacement				**Sampling With Replacement**	
Ordered Samples		Unordered Sample		Ordered Samples	
X	P(X)	X	P(X)	X	P(X)
0	2/12	0	1/6	0	4/16
1	8/12	1	4/6	1	8/16
2	2/12	2	1/6	2	4/16

Probability

Based on these probability distributions, we can now answer questions such as: What is the probability of selecting exactly two diseased subjects into a sample of size 2 from a population in which 50% are diseased? The answer depends on the exact sampling distribution. Answers are provided below, relative to the three sampling strategies shown above.

a) P(exactly 1 diseased subjects is selected) = P(X=1)
    P(X=1) = 8/12   P(X=1) = 4/6   P(X=1) = 8/16
b) P(no diseased subjects are selected) = P(X=0)
    P(X=0) = 2/12   P(X=0) = 1/6   P(X=0) = 4/16
c) P(at least 1 diseased subject is selected) = P(X=1 or X=2)
    P(X=1 or X=2) = 10/12   P(X=1 or X=2) = 5/6   P(X=1 or X=2) = 12/16

Notice that the probabilities that exactly 1 diseased subject is selected, that no diseased subjects are selected and that at least one diseased subject is selected differ depending on the sampling strategy employed. The order consideration does not influence probabilities (the first 2 sampling strategies produce equivalent probabilities), however, sampling with versus without replacement does influence results.

In order to construct the probability distributions, we enumerated every possible sample under a particular sampling strategy and computed the value of the random variable of interest (e.g., the number of diseased subjects selected into the sample). Example 4.4 involved a very small population (N=4) and small samples (n=2). When the population is large (as it is in most applications), it is practically impossible to enumerate every possible sample of a given size to develop a probability distribution.

To avoid having to enumerate every possible sample to develop a probability distribution, there are probability models which are based on specific attributes of the experiment under investigation. If a particular experiment satisfies the attributes of a specific probability model, that model can be used to compute probabilities. We present two of the most popular probability models in Sections 4.5 and 4.6. For each model we specify the attributes that distinguish it from other models and illustrate the computation of probabilities using the models by way of examples.

4.5 The Binomial Distribution

We now focus on a probability distribution model which is used in a variety of applications called the binomial probability distribution model. The *binomial distribution model* has three specific attributes:

     i)   each performance of the experiment, called a trial, results in only one of two possible outcomes, which we call either a success or a failure,

     ii)   the probability of a success on each trial is constant, denoted p, with $0 \le p \le 1$, and

     iii)   the trials are independent.

There are many applications that satisfy these attributes. For example, suppose we are concerned with the risk of mortality (i.e., death) associated with a particular surgical procedure. The experiment involves performing the surgical procedure on different patients. Each patient who undergoes the surgical procedure either survives or does not survive the procedure (attribute i). Suppose that for this procedure, the probability of surviving (or not surviving) is constant (attribute ii). (This may not be the case for some surgical procedures, as older patients or patients with comorbid conditions for example might be at higher risk for not surviving the procedure.) In this example, the trials (i.e., performances of the surgical procedure) are independent (attribute iii). The outcome observed in one patient (i.e., he/she survives or does not survive the surgical procedure) has no effect on the outcome observed in another patient.

In applications with these attributes, the following mathematical formula (4.11), the binomial distribution formula, can be used to generate probabilities. The binomial distribution is defined for the discrete random variable X which denotes the number of successes out of n trials.

$$P(X = x) = {}_nC_x \, p^x (1-p)^{n-x} \qquad (4.11)$$

where x = # successes of interest (x=0, 1, ..., n),
   p – P(success) on any trial,
   n = # trials, and
   ${}_nC_x$ = the number of combinations of x successes in n trials.

Recall n! ("n factorial") = n · (n-1) ·....2·1 (e.g. 5!=5·4·3·2·1=120, 3!=3·2·1=6, 1! =1, and 0! =1).

The binomial distribution formula (4.11) could be used, for example, to compute the probability that a specified number of patients, x, do not survive a particular surgical procedure, when it is applied to a collection of patients, n, with a known, constant risk of mortality (i.e., death), p.

Model (4.11) is equivalent to,

$$P(X = x) = \frac{n!}{x!(n-x)!} p^x (1-p)^{n-x}$$

The binomial distribution is an appropriate model for the number of successes in a simple random sample with replacement of size n drawn from a population where there are two possible outcomes (success and failure). We illustrate the use of the binomial distribution in Examples 4.5 and 4.6, and revisit Example 4.4 from the previous section.

Example 4.5. Suppose there is an antibiotic which has been shown to be 70% effective against a common bacteria.

a)   If the antibiotic is given to 5 unrelated individuals with the bacteria, what is the probability that it will be effective in exactly three?

In this experiment, we give each of five unrelated patients affected with a common bacteria the antibiotic and monitor whether or not it is effective. Assume that it is possible to classify the outcome of a course of therapy as effective (success) or not effective (failure). Assuming that the probability of success is constant (i.e., the probability that the antibiotic is effective is the same for each patient) and the trials are independent (i.e., the outcome observed in one patient does not influence the outcome observed in another), then this experiment satisfies the attributes of the binomial probability model and (4.11) can be used to compute the desired probability.

Using the binomial formula (4.11) with n=5, p=0.70 and x=3, the probability is :

$$P(X = 3) = \frac{5!}{3!(5-3)!} 0.70^3 (1-0.70)^{5-3} = 10(0.343)(0.09) = 0.3087.$$

Thus, there is a 30.87% chance that the antibiotic is effective in exactly 3 patients out of 5 (when the probability of effectiveness in a single patient is 70%).

b)   What is the probability that the antibiotic will be effective in all 5? Using (4.11) with n=5, p=0.70 and x=5 the probability is:

$$P(X = 5) = \frac{5!}{5!(5-5)!} 0.70^5 (1-0.70)^{5-5} = 1(0.1681)(1) = 0.1681.$$

c)   What is the probability that the antibiotic will be effective in none of the 5? Using (4.11) with n=5, p=0.70 and x=0 the probability is:

$$P(X = 0) = \frac{5!}{0!(5-0)!} 0.70^0 (1-0.70)^{5-0} = 1(1)(0.0024) = 0.0024.$$

In this binomial experiment, the sample space consists of the following outcomes S = {0, 1, 2, 3, 4, 5}. Exactly one of these outcomes will be observed each time the experiment is run. In this experiment, the probability of a success on any one trial (i.e., the probability that the antibiotic is effective) is 70%. Because the probability of success is high, we are much more likely to observe 3 or 5 successes out of 5 trials than we are to observe none (probabilities 0.3087, 0.1681 and 0.0024, respectively from above). In fact it is very unlikely that we will not observe any successes (less than a 1% chance) out of 5 trials.

Table 1 in the Appendix contains probabilities of the binomial distribution for specific values of n, p and x. The probabilities in Table 1 were computed using (4.11) for various

Probability

combinations of n, p and x. To use Table 1, locate n along the left hand margin, find x in the adjacent column and locate p across the top row. Find the probability in the appropriate row and column of the table. Table 1 provides probabilities of observing <u>exactly</u> x successes out of n trials for specific values of p.

If we locate n=5 and p=0.70 in Table 1 we find the following probabilities (a probability distribution):

X = Number of Successes	$P(X \mid n=5, p=0.70)$
0	0.0024
1	0.0284
2	0.1323
3	0.3087
4	0.3601
5	0.1681
	1.0

Table 1 can be used to determine probabilities for the binomial distribution. Table 1 contains probabilities for many binomial distributions (i.e., many combinations of n and p). However, there will be applications in which Table 1 will not contain the desired probabilities. For example, the values of p in Table 1 start at 0.10 and increase in increments of 0.10 to 0.90 (i.e., p = 0.10, 0.20, 0.30, ..., 0.90) and a specific application may involve p=0.27. Table 1 does not contain the desired probabilities and the binomial formula (4.11) is used to compute the exact probability.

Example 4.6. Consider an application similar to the one described in Example 4.5 involving a second antibiotic which has been shown to be only 50% effective against the same common bacteria. If 10 unrelated individuals with the bacteria take the antibiotic, what is the probability that it will be effective in more than eight ?

Again, we let X denote the number of successes (i.e., the number of patients in whom the antibiotic is effective), and we wish to determine $P(X > 8)$. The binomial formula and Table 1 produce the probability of observing <u>exactly</u> x success out of n. In this example we are interested in the probability of observing more than 8 successes. Before using the binomial formula (or Table 1), we must restate our problem in a format compatible with the binomial formula (and Table 1):

$$P(X > 8) = P(X=9 \text{ or } X=10) = P(X=9) + P(X=10).$$

We can now compute $P(X=9)$ and $P(X=10)$ using the binomial formula (4.11), applied twice.

$$P(X = 9) = \frac{10!}{9!(10-9)!} 0.50^9 (1 - 0.50)^{10-9} = 10(0.0020)(0.5) = 0.0098.$$

$$P(X = 10) = \frac{10!}{10!(10-10)!} 0.50^{10} (1 - 0.50)^{10-10} = 1(0.0001)(1) = 0.0010.$$

Thus, $P(X > 8) = 0.0098 + 0.0010 = 0.0108$ (See also Table 1).

Probability

In this application, there is a 1.1% chance that the antibiotic will be effective on more than eight patients when given to ten. The possible outcomes of this experiment are the values in the sample space S = {0, 1, 2, 3, 4, 5, 6, 7, 8, 9, 10}. Because the antibiotic is 50% effective for any given patient, we are more likely to observe 3, 4, 5 or 6 successes out 10 than we are to observe few (0, 1 or 2) or many (8, 9, 10) successes (See Table 1).

Recall Example 4.4 from Section 4.4 in which we had a population which consisted of four laboratory mice, two diseased and two non-diseased. In the example we selected two subjects at random. Each selection resulted in one of two possible outcomes: selection of a diseased subject or selection of a non-diseased subject. The probability that we selected a diseased subject was 0.50 under the sampling with replacement strategy. Note that sampling with replacement ensures constant probability of success (in this example, probability of selecting a diseased subject is 0.5 on each selection as long as we sample with replacement). Suppose we call the selection of a diseased subject a success. (In general, we call the outcome of interest a success and often, particularly in medical applications, the outcome of interest is not the healthy outcome (e.g., disease, mortality)). The random variable X under the sampling with replacement strategy illustrated in Example 4.4 is an example of a binomial random variable whose probability distribution is given below. Recall we developed this probability distribution by enumerating every possible sample, assigning the value of X (the number of diseased subjects selected into each sample) and summarizing:

X	P(X)
0	4/16 = 0.25
1	8/16 = 0.50
2	4/16 = 0.25

This probability distribution is an example of a binomial probability distribution with n=2 and p=0.5. The binomial distribution formula (and Table 1) give the same probabilities, for example with n=2 and p=0.5 the probability if selecting no diseased subjects (X=0) is:

$$P(X = 0) = \frac{2!}{0!(2-0)!} 0.50^0 (1 - 0.50)^{2-0} = 1(1)(0.25) = 0.25.$$

In the binomial distribution, it can be shown that the mean and variance of the random variable X (i.e., the mean and variance of the number of successes out of n binomial trials) are computed as follows:

$$\mu = np \qquad (4.12)$$

$$\sigma^2 = n\,p\,(1-p).$$

In Example 4.5, we involved 5 patients in the experiment and the probability of success was 70% for each patient. The possible outcomes of the experiment are given in the sample space S={0, 1, 2, 3, 4, 5}, each time the experiment is run, exactly one of these outcomes (number of successes) is observed. The mean (or expected) number of patients (out of 5) in whom the antibiotic is effective is $\mu = np = 5 \cdot 0.7 = 3.5$. We could never observe 3.5 successes in any performance of the experiment, instead we observe one of the values in the sample space (e.g., exactly 3 or 4 or 5 successes out of 5). The variance in the number of successes is $\sigma^2 = 5(0.7)(1-0.7) = 1.05$. The standard deviation is $\sigma = 1.02$. The mean number of successes represents the typical number of successes. In reviewing the probability distribution (shown in Example 4.5) we see that the most likely outcomes (those with the highest probabilities) of the experiment are 3 and 4 successes.

In Example 4.6 where the probability of success was 50% for each patient, the expected number of patients (out of 10) in whom the antibiotic is effective is $\mu = np = 10(0.5) = 5$. Again, any of the outcomes in the sample space can be observed on any performance of the experiment. We are most likely (or expect) to observe 5 successes out of 10 when p=0.50.

4.6 The Normal Distribution

The normal distribution is our second probability model and is the most widely used probability distribution for continuous random variables. A characteristic (continuous variable) is said to follow a normal distribution if the distribution of values is bell or mound shaped (Figure 4.4).

Figure 4.4   The Normal Distribution

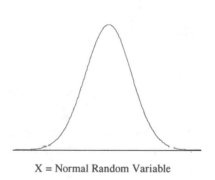

X = Normal Random Variable

Probability

The horizontal axis displays the values of the continuous normal random variable X. The vertical axis is scaled to accommodate the height of the curve at each value of the random variable which reflects the probability (or relative frequency) of observing that value. The total area under the normal curve is 1.0 as it is a probability distribution. The normal distribution is one where values in the center of the distribution are more likely (have higher probabilities) than values at the extremes.

The mathematical formula for the normal probability distribution is :

$$f(x) = \frac{1}{\sigma\sqrt{2\pi}} e^{-\frac{1}{2}\left[\frac{(x-\mu)}{\sigma}\right]^2} \tag{4.13}$$

where x is a continuous random variable ($-\infty < x < \infty$),

μ = mean of the random variable X, and

σ = standard deviation of the random variable X.

The normal probability distribution has the following attributes, each of which will be illustrated through examples.

Table 4.1   Attributes of Normal Distribution

i)   The normal distribution is symmetric about the mean (i.e., $P(X > \mu) = P(X < \mu) = 0.5$). See Figure 4.5 (a). A characteristic which follows a normal distribution is one in which there are as many values above the mean as below.

ii)   The mean = the median = the mode. This attribute follows directly from (i). If the distribution is symmetric at the mean, then half (50%) of the values are above the mean and half (50%) are below the mean. This is the definition of the median. Notice in Figure 4.5 (a) that the peak of the distribution is exactly in the center of the distribution (at the mean = median). The height of the curve indicates the probability (or relative frequency) of observations at each point. The peak indicates the most frequent value which is, by definition, the mode.

Only for normal distribution

iii)   The mean and variance, μ and σ^2, completely characterize the normal distribution. If we know that a particular characteristic follows a normal distribution, then we know everything about that distribution. The mean and variance are the only parameters required to compute probabilities about a normal distribution (See the normal distribution function (4.13)).

iv) $P(\mu - \sigma < X < \mu + \sigma) = 0.68$, $P(\mu - 2\sigma < X < \mu + 2\sigma) = 0.95$, $P(\mu - 3\sigma < X < \mu + 3\sigma) = 0.99$. See Figure 4.5(b). These three properties of a normal distribution are the basis for the Empirical Rule presented in Chapter 3. For any normal distribution, 68% of the values fall between the mean minus one standard deviation and the mean plus one standard deviation. Ninety five percent of the values fall between the mean minus two standard deviations and the mean plus two standard deviations, 99% of the values fall between the mean minus three standard deviations and the mean plus three standard deviations. If we plot the mean of a normal distribution (in the center of the X axis) and count three standard deviations in either direction, the third standard deviation away from the mean should be very close to the end of the distribution.

v)   P(a < X < b) = the area under the normal curve from a to b. See Figure 4.5(c). In a normal distribution, the area under the curve reflects probability. We will compute probabilities for normal random variables by determining desired areas under the normal curve.

If a random variable X follows a normal distribution with mean μ and variance σ^2, we write X ~ $N(\mu, \sigma^2)$, where "~" is read "is distributed as."

Figure 4.5 (a)-(c)   Properties of the Normal Distribution

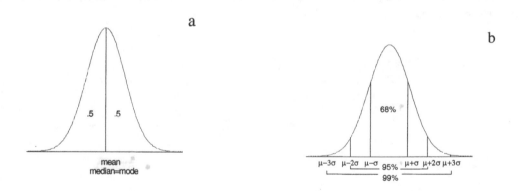

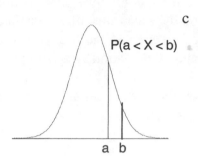

Example 4.7. Suppose males age 25 have an average height of 70" with a standard deviation of 2". Heights for a specific age and gender are approximately normally distributed. Let the random variable X denote height in inches for males age 25, then $X \sim N(70, 2^2)$. Using the attributes for normal random variables summarized in Table 1, the distribution is shown below. Notice that the mean (70) is in the middle of the distribution. We also count three standard deviation units (units of 2") in either direction, and at the third standard deviations above and below the mean we are very close to the ends of the distribution.

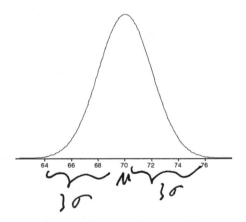

Suppose a male age 25 is selected at random. Use the attributes outlined in Table 4.1 to solve the following.

Probability

a) What is the probability that his height is more than 70" ? Because 70 is the mean of this normal distribution, we know from Table 4.1(i) that the distribution is symmetric about the mean:

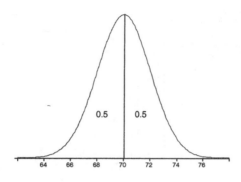

Therefore, $P(X > 70) = 0.5$.

b) What is the probability that his height is between 70" and 72" ? From Table 4.1(iv) we know that the probability that his height is between 68 and 72 is 0.68 (the area under the curve between the mean minus one standard deviation and the mean plus one standard deviation). In addition, the normal distribution is symmetric about the mean (70), so the area to the right of 70 is identical to the area to the left of 70:

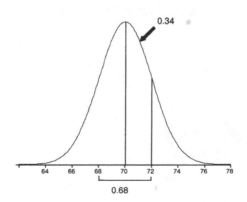

The desired probability is $P(70 < X < 72) = 0.34$.

Probability

c) What is the probability that his height exceeds 72" ? From Table 4.1(i) we know that the area under the curve above 70 (and the area under the curve below 70) is equal to 0.5. From part (b) we determined that the area under the curve between 70 and 72 was 0.34.

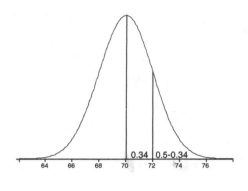

$$= P(X \gtrless 72)$$

The area under the curve above 72 can be determined by subtraction, $P(X > 72) = 0.5 - 0.34 = 0.16$

The attributes given in Table 4.1 are very useful so long as questions are focused on the mean and multiples of the standard deviation about the mean of a normal distribution. Of course, there will be other questions of interest. For example, suppose in Example 4.7 we wanted to know the probability that a male, age 25 has a height exceeding 71 inches? From parts (a) and (c) above we know that this probability is between 0.5 and 0.16 - but we cannot determine the exact probability using the attributes listed in Table 4.1.

To compute probabilities for a normal distribution, a table such as Table 2 in the Appendix is needed. Table 2 is a table of probabilities of the *standard normal distribution*. The standard normal distribution is the normal distribution with $\mu = 0$, and $\sigma = 1$ (Figure 4.6). The random variable Z is used to refer to the standard normal distribution (i.e., $Z \sim N(0,1)$).

Figure 4.6 The Standard Normal Distribution

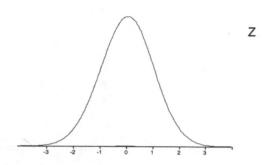

To calculate probabilities for any normal distribution X, we first standardize (i..e, subtract the mean and divide by the standard deviation) using (4.14) and then use Table 2: Probabilities of the Standard Normal Distribution.

reserved for standard Normal Distribution

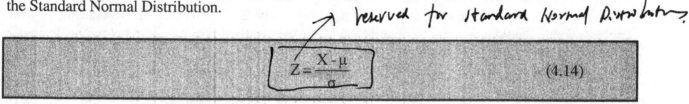

$$Z = \frac{X - \mu}{\sigma}$$

(4.14)

In Example 4.7, we posed the question, what is the probability that a male, age 25 has a height exceeding 71 inches. To determine this probability, we convert the problem to a problem about Z, the standard normal distribution using (4.14). Formula (4.14) converts a value from any normal distribution (X) to a value from the standard normal distribution (Z). The X value of interest is 71. It corresponds to a Z value of Z = (71-70)/2 = 0.5. The value 71 in a normal distribution with mean 70 and standard deviation 2 corresponds to Z = 0.5. The value 71 is one half (0.5) a standard deviation above the mean. The area of interest is shown below (indicated by "?") both for the original distribution (X) and the standard normal distribution (Z).

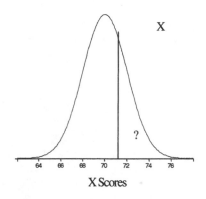

X Scores

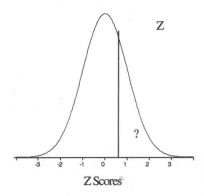

Z Scores

The areas under the two curves (above 71 in the normal distribution with mean 70, standard deviation 2 (X) and above 0.5 in the standard normal distribution (Z)) are identical; only the scales along the horizontal axes are different. The probability that Z exceeds 0.5 can be determined from Table 2. Table 2 contains areas (probabilities) under the standard normal curve below the desired Z value. If we locate Z=0.5 on Table 2 (Find 0.5 down the left hand margin (which contains the units and first decimal place of Z) and 0.00 across the top row (which contains the second decimal place of Z)) we read $P(Z < 0.5) = 0.6915$. We want $P(Z > 0.5) = 1 - 0.6915 = 0.3085$. Thus, 31% of the males have heights exceeding 71 inches. We illustrate the application of the normal probability distribution model, formula (4.14) and the use of Table 2 in the following example.

$$P(X > 71) =$$

Example 4.8. Systolic blood pressures are assumed to follow a normal distribution with a mean of 108 and a standard deviation of 14 (this is denoted X~N(108, 14^2)). From Table 4.1 we know that the distribution of systolic blood pressures looks like that presented in Figure 4.7.

Figure 4.7   Distribution of Systolic Blood Pressures $\mu = 108$, $\sigma = 14$

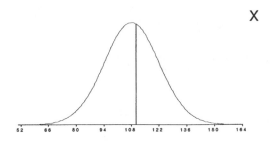

Probability

If an individual is selected at random from the population, find the following:

a)   The probability that his/her systolic blood pressure is below 112
      (i.e., P(X < 112)). This probability concerns the random variable X which is approximately
      normally distributed with μ=108 and σ=14. To find this probability we must first
      standardize using (4.14) and then use Table 2. We standardize the 112 as follows:

$$Z = (112 - 108)/14 = 0.29$$

The problem can be represented as follows:

$$P(X < 112) = P(Z < 0.29)$$

Table 2 contains probabilities for the standard normal distribution Z. The probabilities contained in Table 2 reflect the areas under the standard normal curve below each particular Z value. Here we wish to determine the area under the standard normal curve below 0.29. We locate the Z value 0.29 and read the probability. Using Table 2, the desired probability is 0.6141. The probability that an individual's systolic blood pressure is below 112 is 61.41%. Another interpretation is that 61.41% of the population have systolic blood pressures below 112.

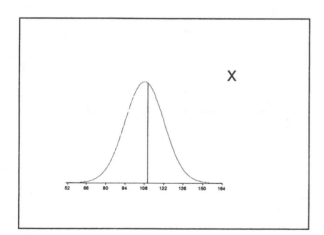

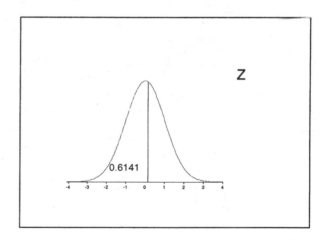

b)   The probability that his/her systolic blood pressure is below 120
(i.e., P(X < 120)). To find this probability we again standardize using (4.14) and then use
Table 2.

$$Z = (120 - 108)/14 = 0.86$$

The problem can be represented as follows:

$$P(X < 120) = P(Z < 0.86)$$

Now we wish to determine the area under the standard normal curve below 0.86. We locate the Z
value 0.86 and read the probability.   Using Table 2, the desired probability is 0.8051.   The
probability that an individual's systolic blood pressure is below 120 is 80.51% (or 80.51% of the
population have systolic blood pressures below 120).

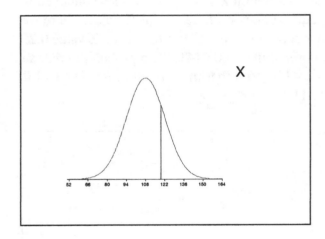

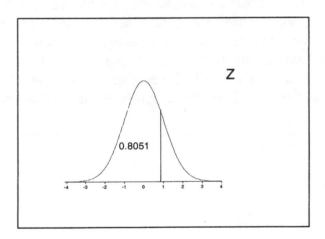

c)   The probability that his/her systolic blood pressure is above 120 (i.e., P(X > 120)).

In part (b), we standardized the value 120. The problem can be represented as follows:
$P(X > 120) = P(Z > 0.86)$. The desired probability is computed by subtraction (from part (b) we determined $P(Z < 0.86) = 0.8051$): $P(X > 120) = P(Z > 0.86) = 1 - 0.8051 = 0.1949$.

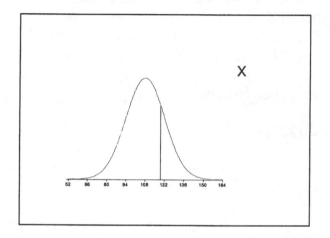
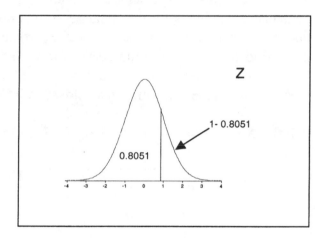

d)   The probability that his/her systolic blood pressure is above 122 (i.e., P(X > 122)).
We first standardize using (4.14):

$$Z = (122 - 108)/14 = 1$$

The problem can be represented as follows: $P(X > 122) = P(Z > 1)$. Using Table 2 we can determine $P(Z < 1) = 0.8413$. The desired probability is computed by subtraction: $P(X > 122) = P(Z > 1) = 1 - 0.8413 = 0.1587$.

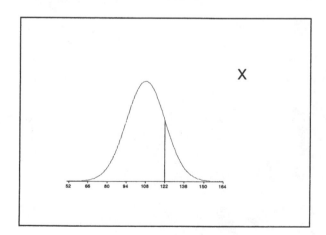
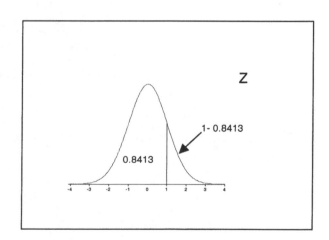

e)   The probability that his/her systolic blood pressure is between 100 and 108 (i.e., P(100 < X < 108)).

We first standardize the two X values of interest: Z=(100 - 108)/14 = -0.57, Z=(108 - 108)/14 = 0. The problem can be represented as follows: P(100 < X < 108) = P(-0.57 < Z < 0). Recall Table 2 contains probabilities for the standard normal distribution Z, and the probabilities reflect the areas under the standard normal curve below a particular Z value. If we locate the Z value -0.57 in Table 2 we find the area under the curve below –0.57 is 0.2843. We want the area under the curve between –0.57 and 0. If we locate the Z value 0 in Table 2 we find the area under the curve below 0 (because this is the mean, the area below 0 is 0.5). The desired probability is

P(100 < X < 108) = P(-0.57 < Z < 0) = 0.5 – 0.2843 = 0.2157.

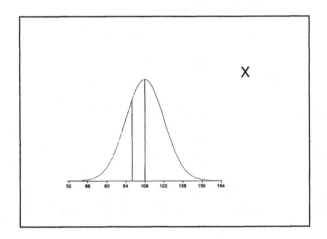
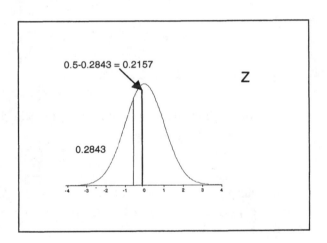

f)   The probability that his/her systolic blood pressure is between 104 and 118
      (i.e., P(104 < X < 118)).

We first standardize the two X values of interest: Z=(104 - 108)/14 = -0.29, Z=(118 - 108)/14 = 0.71. The problem can be represented as follows: P(104 < X < 118) = P(-0.29 < Z < 0.71). Again, Table 2 contains probabilities which reflect the areas under the standard normal curve below the selected Z value. To compute the desired probability, we enter Table 2 twice, once with the value 0.71 (to determine the area below 0.71) and once with the value -0.29 (to determine the area below –0.29). The problem can be solved as follows:

$$P(104 < X < 118) = P(-0.29 < Z < 0.71) = 0.7611 - 0.3839 = 0.3752.$$

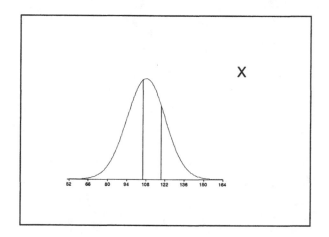

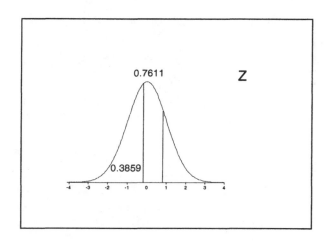

g)   The probability that his/her systolic blood pressure is between 120 and 125 (i.e., P(120 < X < 125)).

We first standardize the two X values of interest: Z=(120 - 108)/14 = 0.86, Z=(125 - 108)/14 = 1.21. The problem can be represented as follows: P(120 < X < 125) = P(0.86 < Z < 1.21). Again, Table 2 contains probabilities which reflect the areas under the standard normal curve below the selected Z value. To determine the desired probability, we need to enter Table 2 twice, once with the value 1.21 (to determine the area below 1.21) and once with the value 0.86 (to determine the area below 0.86). The desired probability is then determined by subtraction:

$$P(120 < X < 125) = P(0.86 < Z < 1.21) = 0.8869 - 0.8051 = 0.0818.$$

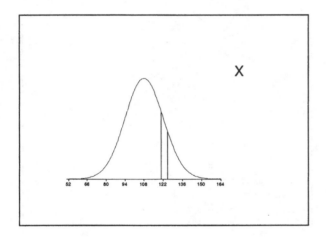 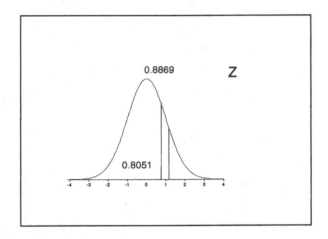

In determining probabilities for a normal distribution, it is always useful to draw a picture to display the desired probability (equivalent to the area under the normal curve). The picture is particularly useful as a check on the appropriateness of the solution. For example, in part (g) of Example 4.8 the desired probability is reflected by a narrow area under the normal curve. The solution was 0.0818, a small fraction of the total area under the normal curve. If an error was made in the computation and the solution was proposed as 0.818, it would not reflect this narrow area. It is not necessary, in general, to draw both the X and Z distributions as the areas are identical. We presented both for Example 4.8 to illustrate this notion.

4.6.1 Percentiles of the Normal Distribution

Recall from Chapter 3, that the kth *percentile* is defined as the <u>score</u> that holds k percent of the scores below it. For example the 90th percentile in a distribution is the score that holds 90 % of the scores below it. The first quartile, Q_1, is equivalent to the 25th percentile, the median is equivalent to the 50th percentile, and the third quartile, Q_3, is equivalent to the 75th percentile. For the normal distribution, the following formula (4.17) is used to compute percentiles.

$$X = \mu + Z\sigma \qquad (4.17)$$

where μ = mean of the random variable X,
σ = standard deviation of the random variable X, and
Z = value from the standard normal distribution for the desired percentile.

Table 4.2 provides the values from the standard normal distribution (Z) corresponding to commonly used percentiles.

Table 4.2   Percentiles of the Standard Normal Distribution

Percentile	Z
1	-2.326
2.5	-1.960
5	-1.645
10	-1.282
90	1.282
95	1.645
97.5	1.960
99	2.326

This table is even more precise than table 2, of course the values are not too many.

The percentiles of the standard normal distribution, Z, were computed using the following approach and Table 2. The same approach could be used to determine any other percentile of the standard normal distribution.

Suppose we wish to determine the 95th percentile of Z. By definition, the 95th percentile is the Z score which holds 95% of the Z scores below it. The following figure displays the approximate (to be determined) value of the 95th percentile of Z.

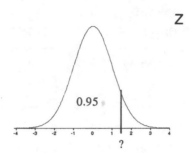

The probabilities in Table 2 reflect the areas under the standard normal curve below the selected Z value. We need to find the probability 0.95 (the area below the Z value we wish to determine, denoted "?" above) in Table 2 among the probabilities. The closest values are 0.9495 and 0.9505. The value we wish to locate (0.95) is in between. Using Table 2, the Z value corresponding to the probability 0.9495 is 1.64 (i.e., the area under the curve below 1.64 is 0.9495). Similarly, the Z value corresponding to 0.9505 is 1.65 (i.e., the area under the curve below 1.65 is 0.9505). The 95th percentile is the Z value which holds 95% of the area below it. The 95th percentile of the standard normal distribution is 1.645. Other percentiles of the standard normal distribution can be determined in this same manner.

Example 4.9. Consider the distribution of systolic blood pressures presented in Example 4.8. Systolic blood pressures were assumed to follow a normal distribution with $\mu = 108$ and $\sigma = 14$ (i.e., $X \sim N(108, 14^2)$).

a) Find the 5th percentile in the systolic blood pressures. Using (4.17) and the appropriate value from Table 4.2:
$$X = 108 + (-1.645)(14) = 108 - 23.0 = 85.0$$

b) Find the 90th percentile in the systolic blood pressures. Using (4.17) and the appropriate value from Table 4.2:
$$X = 108 + (1.282)(14) = 108 + 17.9 = 125.9$$

Thus, 5% of the systolic blood pressures arel at or below 85.0 while 90% of the systolic blood pressures are at or below 125.9 (i.e., 10% are above 125.9).

Probability

4.6.2 Normal Approximation to the Binomial   *did not go over this part*

In Section 4.5 we discussed the binomial probability distribution model and presented two methods for computing probabilities: i) the binomial distribution formula (4.11), and ii) Table 1: Probabilities of the Binomial Distribution. Both the binomial distribution formula and Table 1 produce the probability of observing exactly x successes out of n trials when the probability of success on a single trial is p. We illustrated applications in which we determined the probability of observing a range of values. For example, suppose that 10 binomial trials are performed and it is of interest to compute the probability of observing more than 5 successes (i.e., P(X > 5)). To compute the probability, we first enumerate the values of x (the number of successes out of n) of interest and then proceed to the binomial distribution formula or to Table 1. In this example, P(X > 5) = P(X=6) + P(X=7) + P(X=8) + P(X=9) + P(X=10). Each of these probabilities is computed with the binomial formula or using Table 1. The final answer is the sum of the five probabilities. When the number of trials (n) is large, probabilities of observing ranges of values are increasingly cumbersome to compute. There are alternatives, however, which utilize the techniques described in the previous section.

It can be shown that as the number of trials, n, in binomial applications becomes large, the distribution of the number of successes, x, is approximately normally distributed. Recall Example 4.5 which described an antibiotic which was 70% effective against a common bacteria. Figures 4.9(a) - 4.9(d) display the distributions of the number of successes, x, for n = 5, 10, 15 and 25 respectively. Notice that as n gets large (e.g., n=25) the distribution of the number of successes looks approximately normally distributed.

Figure 4.9(a) Binomial Distribution
n=5, p=0.7

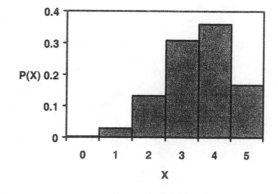

Figure 4.9(b) Binomial Distribution
n=10, p=0.7

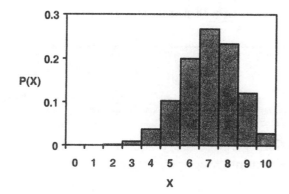

Probability

Figure 4.9(c) Binomial Distribution
n=15, p=0.7

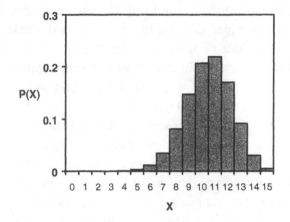

Figure 4.9(d) Binomial Distribution
n=25, p=0.7

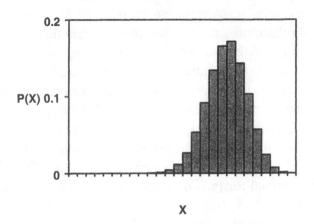

If the distribution of the number of successes is approximately normal, which occurs when n is large and p is not close to 0 or 1, then the normal probability distribution can be used to approximate binomial probabilities. In general, the normal approximation is appropriate if:

$$np \geq 5 \; and \; n(1-p) \geq 5.$$

The distribution displayed in Figure 4.9(d) is approximately normal and satisfies the above (n=25, p=0.7: np=25(0.7)=17.5 and n(1-p)=25(0.3)=7.5). For n=25, p=0.7 the normal distribution can be used to compute probabilities concerning the number of successes X. Recall, for the binomial distribution, the mean number of successes is $\mu = np$ and the standard deviation is $\sigma = \sqrt{np(1-p)}$. The following formula is used to convert values of the random variable X (reflecting the number of successes out of n binomial trials) into corresponding values of the standard normal distribution:

$$Z = \frac{X - np}{\sqrt{np(1-p)}} \qquad (4.18)$$

A continuity correction of 0.5 is generally applied prior to implementing (4.18). Formula (4.18) and the continuity correction are illustrated in Example 4.10.

Example 4.10. Consider Example 4.5 which described an antibiotic which was 70% effective against a common bacteria. Suppose that the antibiotic is given to 25 patients with the bacteria.

a)   What is the probability that the antibiotic is effective in more than 15 patients ?

We wish to compute P(X > 15). In order to solve the problem using the binomial distribution formula or Table 1, we first enumerate the values of X of interest: P(X > 15) = P(X=16) + P(X=17) + P(X=18) + P(X=19) + P(X=20) + P(X=21) + P(X=22) + P(X=23) + P(X=24) + P(X=25). Each of these probabilities could be computed by applying the binomial formula (4.14) or by using Table 1.   In either approach, particularly the former, the computations are cumbersome.   In this application, we satisfy the criteria which suggest that the distribution of the number of successes, X, is approximately normally distributed (i.e., np =25(0.7)=17.5 $\geq$ 5 and n(1-p)=25(0.3)=7.5 $\geq$ 5) with μ = 25(0.7) = 17.5 and $\sigma = \sqrt{np(1-p)} = \sqrt{25(0.7)(0.3)} = \sqrt{5.25} = 2.29$. The distribution of all possible outcomes for the binomial experiment with n=25 and p=0.7 was displayed in Figure 4.9(d). The desired probability is the sum of the areas in the shaded bars, shown below:

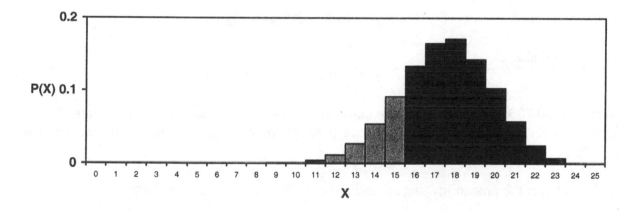

If we overlay a normal curve on the binomial histogram shown above, we can see the area under the normal curve which approximates the desired probability:

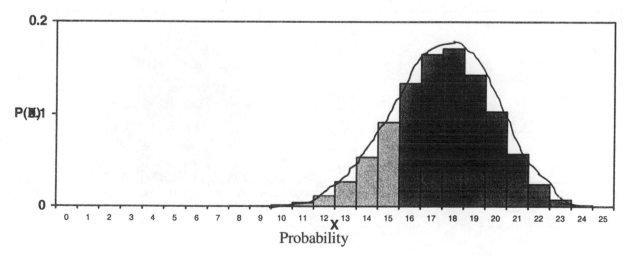

We wish to compute P(X > 15). In the figure with the normal curve overlaid, this is approximated by the area under the normal curve to the right of the vertical bar (above the right hand side of the bar over the value 15). Notice that each of the bars in the binomial histogram is centered above the respective X value. For example, the bar over the value X=15 starts at 14.5 and ends at 15.5, the bar over the value X=16 starts at 15.5 and ends at 16.5, and so on. The continuity correction essentially takes this into account. The binomial probability P(X > 15) is approximated by the area under the normal curve above the value 15.5: $P(X_{binomal} > 15) \approx P(X_{normal} > 15.5)$. To compute the probability that a normal variable with mean 17.5 and standard deviation 2.29 exceeds 15.5, we convert to the standard normal distribution (Z):

$$Z = \frac{15.5 - 17.5}{2.29} = -0.87$$

$$P(X > 15.5) = P(Z > -0.87).$$

Now, using Table 2:

$$P(Z > -0.87) = 1 - 0.1922 = 0.8078.$$

There is a 80.78% probability that the antibiotic is effective in more than 15 patients. This is an approximation to the exact solution, which can be determined using Table 1. The exact solution is 0.8105.

b)   What is the probability that the antibiotic is effective in at most 12 patients ?

We wish to compute P(X ≤ 12). Again, if we overlay a normal curve on the binomial histogram, we can see the area under the normal curve which approximates the desired probability:

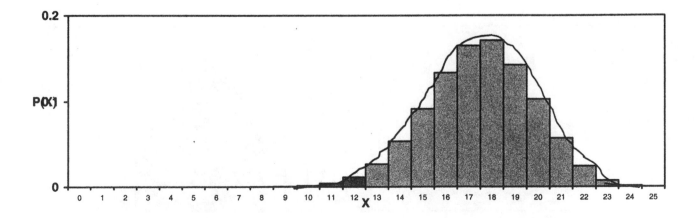

Probability

Adding the continuity correction:

$$P(X \leq 12) \approx P(X < 12.5)$$

Now, converting to the standard normal distribution:

$$Z = \frac{12.5 - 17.5}{2.29} = -2.18$$

$$P(X < 12.5) = P(Z < -2.18).$$

Using Table 2:

$$P(Z < -2.18) = 0.0146.$$

There is a 1.46% probability that the antibiotic is effective in at most 12 patients. (Using Table 1 the exact probability is 0.0175.)

4.7 Key Formulas

CONCEPT	FORMULA	DESCRIPTION
First Principles	$P(A) = \dfrac{\text{\# outcomes in A}}{\text{Total \# outcomes}}$	where A is an event defined in the sample space S
Complement Rule	$P(A') = 1 - P(A)$	where A' is the complement of A (i.e., contains all outcomes in S that are not in A)
Addition Rule	$P(A \cup B) = P(A)+P(B) \\ -P(A \cap B)$	where A and B are events defined in the sample space S
Mutually Exclusive Events	$P(A \cap B) = 0$	no outcomes in common
Conditional Probability Rule	$P(A \mid B) = \dfrac{P(A \cap B)}{P(B)}$	Probability of event A "given" event B
Independent Events	$P(A\mid B) = P(A) \\ P(A \cap B) = P(A)\cdot P(B)$	the probability of event A is not influenced by the occurrence or nonoccurrence of event B
Permutations	$_N P_n = \dfrac{N!}{(N-n)!}$	Number of distinct arrangements (order important) of size n from a population of size N
Combinations	$_N C_n = \dfrac{N!}{n!\,(N-n)!}$	Number of samples (order not important) of size n from a population of size N

$$\begin{cases} \mu = np \\ \sigma^2 = np(1-p) \end{cases}$$

135

4.7 Key Formulas (continued)

CONCEPT	FORMULA	DESCRIPTION
Binomial Probability Model	$P(X) = \dfrac{n!}{x!(n-x)!} p^x (1-p)^{n-x}$ *also table 1*	Probability of x successes out of n trials
Normal Probability Model	$f(x) = \dfrac{1}{\sigma \sqrt{2\pi}} e^{-\frac{1}{2}\frac{(x-\mu)^2}{\sigma}}$	Normal probability model
Standard Normal Distribution	$Z = \dfrac{x-\mu}{\sigma}$, *then use table* $Z \sim N(0, 1)$ $\mu = 0, \sigma = 1$	Normal distribution with $\mu = 0$, $\sigma^2 = 1$, Table 2
Percentiles of the Normal Distribution	$X = \mu + Z\sigma$	k^{th} percentile holds k% of scores below it (See Table 4.2)
Normal Approximation to the Binomial	$Z = \dfrac{(X \pm 0.5) - np}{\sqrt{np(1-p)}}$	Standardizes Binomial Random Variable X

Percentiles of the Standard Normal Distribution (or table 2)

percentile	Z
1	-2.326
2.5	-1.960
5	-1.645
10	-1.282
90	1.282
95	1.645
925	1.960
99	2.326

Probability

4.8 Application Using SAS

SAS has several functions which can generate random variables from specific probability distribution functions, such as the binomial and normal distributions. We illustrate the use of these functions as well as a function which can be used to determine specific values of the standard normal distribution, such as the 90th percentile.

A series of SAS programs and output are presented below which illustrate several different SAS functions. The probability distribution to which the function relates is noted in the header to each example. Also noted is the specific SAS function illustrated and a brief description of its use. Notes are provided to the right of the SAS programs (*in italics*) for orientation purposes and are not part of the programs. In addition, there are blank lines in the programs that follow which are solely to accommodate the notes. Blank lines and spaces can be used throughout SAS programs and are generally used to enhance readability.

The first example uses SAS to generate random variables from a binomial distribution with n=5 and p=0.7 (See Example 4.5). The second example uses SAS to generate random variables from a normal distribution with mean 108 and standard deviation 14 (See Example 4.8). The third example illustrates the computation of percentiles of the normal distribution using the data presented in Example 4.8.

For each example we present three components:

i) the SAS program code,
ii) the computer output, and
iii) a description of the relevant components of the computer output along with their interpretation.

EXAMPLE 4.1:   **Generating Random Variables from the Binomial Distribution**
Effectiveness of Antibiotic (Example 4.5)
Ranbin Function (Generates Random Variables from Binomial Distribution)

Suppose there is an antibiotic which has been shown to be 70% effective against a common bacteria. If the antibiotic is given to 5 individuals with the bacteria, what is the probability that it will be effective on exactly three ?

In this application, we will use SAS to generate random variables from the binomial distribution with n = 5 (number of trials), and p=0.70 (probability of success on any trial). We will generate many such random variables and in doing so estimate the theoretical probability distribution. The SAS program code is shown below.
i) Program Code

options ps=64 ls=80;	*Formats the output page to 64 lines in length, 80 columns in width*
data one;	*Beginning of Data Step (Data set name **one**).*
do i=1 to 5000;	*Beginning of a DO loop which will be repeated 5000 times. The index **i** counts the iterations.*
x=ranbin(21439,5,0.7);	*Generates a random variable, called **x**, from the binomial distribution with n=5, p=0.7. The first argument to the Ranbin function is the seed[1].*
output;	*Writes the generated variable, **x**, to data set **one**.*
end;	*End of DO loop.*
run;	*End of Data Step.*
proc means;	*Procedure call (Proc Means to generate summary statistics)*
var x;	*Specification of variable **x**.*
run;	*End of Procedure section*
proc freq;	*Procedure call (Proc Freq to generate frequency distribution=probability distribution)*
tables x;	*Specification of variable **x**.*
run;	*End of Procedure section*

Probability

(1) The seed is simply a random number which is used as a starting point for the random number generator (i.e., Ranbin). When the seed is changed, different values are generated.

ii) Computer Output

```
                        The MEANS Procedure
                        Analysis Variable : x

   N         Mean         Std Dev        Minimum          Maximum
 ------------------------------------------------------------------
 5000      3.4942000      1.0282888           0          5.0000000
 ------------------------------------------------------------------
```

```
                        The FREQ Procedure
                                     Cumulative      Cumulative
      x    Frequency    Percent      Frequency         Percent
    --------------------------------------------------------------
      0          15       0.30             15            0.30
      1         135       2.70            150            3.00
      2         686      13.72            836           16.72
      3        1529      30.58           2365           47.30
      4        1798      35.96           4163           83.26
      5         837      16.74           5000          100.00
```

iii) Interpretation

The first part of the output is from the Means Procedure. Shown are the number of observations in the sample (N=5000), the mean, standard deviation, minimum and maximum. In this application, the summary statistics are computed on the random variable X, which reflects the numbers of patients in whom the antibiotic is effective. Recall for the binomial distribution, the mean is given by: μ=np. In Example 4.5 n=5 and p=0.7. The theoretical mean is $\mu = 5 \cdot 0.7 = 3.5$. In this data set in which we generated 5000 random variables from a binomial distribution with n=5 and p=0.7, the observed mean is 3.494.

The next section of output displays a frequency distribution table for the random variable X. Notice that the values of X generated by SAS are between 0 and 5. The observed frequency distribution (based on 5000 observations) closely approximates the true binomial probability distribution (See Example 4.5). For example, from Table 1 in the Appendix: P(X=0)=0.0024, P(X=1)=0.0284, P(X=2)=0.1323, P(X=3)=0.3087, P(X=4)=0.3601, P(X=5)=0.1681. The observed distribution (shown in the SAS output) is 0.003, 0.0027, 0.137, 0.306, 0.360 and 0.167. Using the SAS output we can address the original question, If the antibiotic is given to 5 individuals, what is the probability that it will be effective in exactly 3 ? From the SAS output, P(X=3) = 0.306.

Probability

EXAMPLE 4.2: **Generating Random Variables from the Normal Distribution**
Systolic Blood Pressures (Example 4.8)
Rannor Function (Generates Random Variables from the Standard Normal Distribution)

Systolic blood pressures are assumed to follow a normal distribution with a mean of 108 and a standard deviation of 14.

In this application, we will use SAS to generate random variables from the standard normal distribution (Z) and then transform them into normal random variables with a mean of 108 and a standard deviation of 14.

i) Program Code

options ps=64 ls=80;	*Formats the output page to 64 lines in length, 80 columns in width*
data one;	*Beginning of Data Step (Data set name **one**).*
mu=108;	*Create a variable called **mu**, assign the constant 108*
sigma=14;	*Create a variable called **sigma**, assign the constant 14*
do i=1 to 10000;	*Beginning of a DO loop which will be repeated 10,000 times. The index **i** counts the iterations.*
z=rannor(13755);	*Generates a random variable, called z, from the standard normal distribution. The only argument to the Rannorm function is the seed[1].*
x=mu+(z*sigma);	*Creates a new variable, **x**, which is a linear function of z ($x = \mu + z \cdot \sigma$).*
output;	*Writes the generated variables, **x** and z, to data set **one**.*
end;	*End of DO loop.*
run;	*End of Data Step.*
proc chart;	*Procedure call (Proc Chart to generate relative frequency histogram)*
vbar z/type=pct;	*Specification of variable z, type=pct displays relative frequencies*
vbar x/type=pct;	*Specification of variable **x**, type=pct displays relative frequencies*

Probability

run;   *End of Procedure section*

proc means;   *Procedure call (Proc Means to generate summary*
                                        statistics)
  var z x;   *Specification of variables z and x.*
run;   *End of Procedure section*

(1) The seed is simply a random number which is used as a starting point for the random number generator (i.e., Rannor). When the seed is changed, different values are generated.

ii) Computer Output

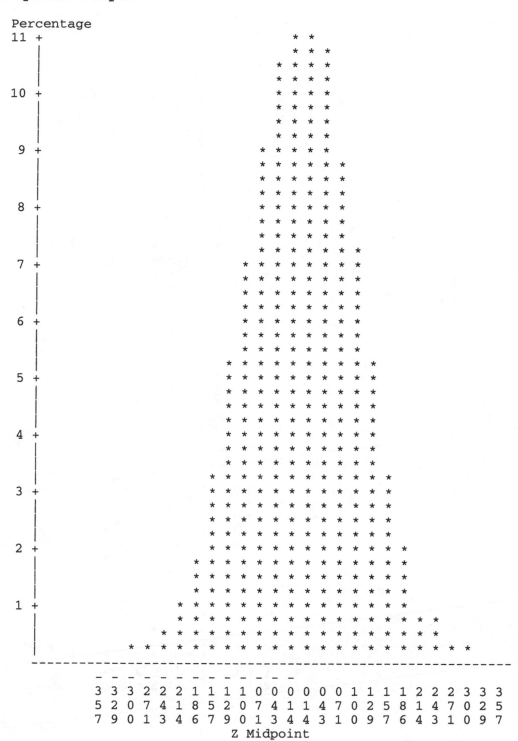

Probability

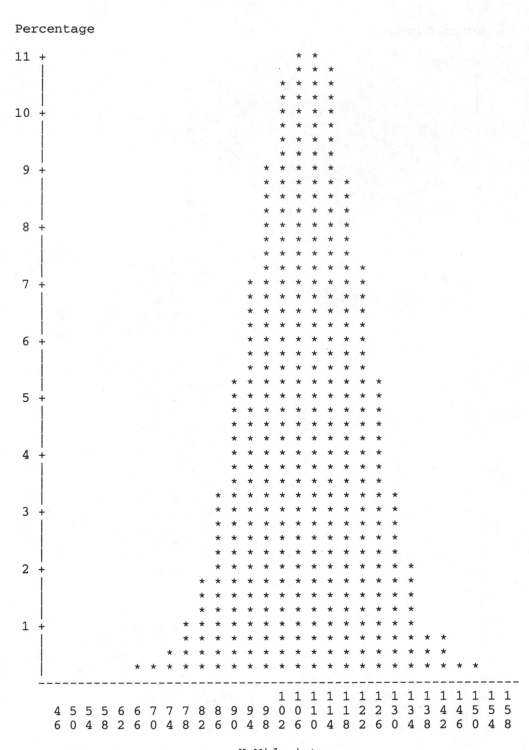

```
                           The MEANS Procedure

Variable        N          Mean        Std Dev        Minimum        Maximum
-----------------------------------------------------------------------------
z           10000   0.000430808      0.9914063     -4.4742924      3.6632627
x           10000   108.0060313     13.8796876     45.3599064    159.2856784
-----------------------------------------------------------------------------
```

iii) Interpretation

The first part of the output is a relative frequency histogram for the random variable z. The random variable z follows the standard normal distribution with a mean of 0 and a standard deviation of 1. The range of the observed values of z are listed along the horizontal axis, and range from -3.6 to 3.6. Although the figure produced by SAS Proc Chart is somewhat crude, it does resemble the normal distribution curve.

The second figure displays the relative frequency histogram for the random variable x. The random variable x follows a normal distribution with a mean of 108 and a standard deviation of 14. The range of the observed values of x are listed along the horizonal axis, and range from 45 to 159. Again, the figure produced by SAS Proc Chart is crude, but it does resemble the normal distribution curve. In fact it is identical to the relative frequency distribution of z, only the values along the horizontal axis are different.

The next section of output is from the Means Procedure. Shown are the number of observations in the sample (N=10,000), the mean, standard deviation, minimum and maximum. Summary statistics are computed on both random variables z and x. Notice that the observed mean of z is 0.00043, the theoretical mean of z is 0. The observed standard deviation of z is 0.99141, the theoretical standard deviation is 1. The observed mean of x is 108.006, the theoretical mean of x is 108. The observed standard deviation of z is 13.8797, the theoretical standard deviation is 14.

EXAMPLE 4.3:   **Computing Percentiles of the Normal Distribution**
Systolic Blood Pressuresc (Example 4.9)
Probit Function (Returns Percentiles of the Standard Normal Distribution)

In Example 4.9 we computed the 5th and 90th percentiles of the systolic blood pressures illustrated in the previous example.

In this application, we will use SAS to determine the 5th and 90th percentiles of the systolic blood pressures which are assumed to follow a normal distribution with a mean of 108 and a standard deviation of 14.

i) Program Code

options ps=64 ls=80;	*Formats the output page to 64 lines in length, 80 columns in width*
data one;	*Beginning of Data Step (Data set name **one**).*
mu=108;	*Create a variable called **mu**, assign the constant 108*
sigma=14;	*Create a variable called **sigma**, assign the constant 14*
z05=probit(0.05);	*Creates a variable called **z05**, which is assigned the 5th percentile of the standard normal distribution.*
z90=probit(0.90);	*Creates a variable called **z90**, which is assigned the 90th percentile of the standard normal distribution.*
x05=mu+(z05*sigma);	*Creates a new variable, **x05**, which is a linear function of **z05** ($x = \mu + z*\sigma$). **x05** is the 5th percentile of a normal distribution with a mean of 108 and a standard deviation of 14.*
x90=mu+(z90*sigma);	*Creates a new variable, **x90**, which is a linear function of **z90** ($x = \mu + z*\sigma$). **x90** is the 90th percentile of a normal distribution with a mean of 108 and a standard deviation of 14.*
run;	*End of Data Step.*
proc print;	*Procedure call (Proc Print to print raw data)*
run;	*End of Procedure section*

ii) Computer Output

OBS	MU	SIGMA	Z05	Z90	X05	X90
1	108	14	-1.64485	1.28155	84.9720	125.942

iii) Interpretation

In this application we computed several variables in the Data Step. Using Proc Print we printed the values of each variable. Notice that no variables were specified in the Proc Print call statement (i.e., no var statement). In such a case, SAS simply prints the values for all variables. Shown above are the values for each of the variables in the data set. SAS includes an observations number (labeled 'OBS') which appears in the first column. In this example there is only one observation (with multiple variables). The values of the remaining variables are shown with the variable names on top, printed in all caps. The 5th percentile of the standard normal distribution is -1.64485 (i.e., 5% of the distribution is at or below -1.64485). The 90th percentile of the standard normal distribution is 1.28155. The 5th percentile of the systolic blood pressures, assumed to follow a normal distribution with a mean of 108 and a standard deviation of 14, is 84.9720, and the 90th percentile of the systolic blood pressures is 125.942.

Summary of SAS Functions

The SAS functions illustrated are summarized in the table below. SAS functions are used in the Data Step. A general description of each function is provided in the table below.

Function	Sample Call Statement	Description
ranbin	$x = \text{ranbin}(seed,n,p)$;	Generates random variables from the Binomial Distribution with n trials and P(success)=p
rannorm	$z = \text{rannorm}(seed)$; $x = \mu + \text{rannorm}(seed) * \sigma$;	Generates random variables from the standard normal distribution (z) Transforms standard normal random variables to normal random variables with mean μ, standard deviation σ
probit	$z = \text{probit}(\alpha)$; $x = \mu + \text{probit}(\alpha) * \sigma$;	Determines percentiles of the standard normal distribution with lower tail area α Transforms percentile of the standard normal distribution to percentile of a normal distribution with mean μ and standard deviation σ

4.9 Problems

First Principles

1. You are assisting in the design of a study to evaluate the effects of social factors on individual's self-esteem. The study is to involve college-age individuals. The following table displays sociodemographic information on college-age individuals, taken from the 1990 census. The table contains proportions of individuals in each age by gender category:

	Age						
	17	18	19	20	21	22	Total
Male	0.05	0.07	0.10	0.06	0.11	0.06	0.45
Female	0.09	0.08	0.11	0.07	0.12	0.08	0.55
Total	0.14	0.15	0.21	0.13	0.23	0.14	

If an individual is sampled from the population of interest (described above), calculate the probability that the sampled individual is:

a) Female
b) At least 19 years of age
c) Female and age 18
d) Over 20 given he is male

2. The following table displays the numbers of defective and non-defective medical devices produced by three plants working for the same manufacturer.

	Plant A	Plant B	Plant C
Non-Defective	180	70	190
Defective	20	30	10

Suppose a medical device is selected at random, find the probability that:

a) The device is defective
b) The device is produced in Plant A
c) The device is defective and produced in Plant A or in Plant C
d) The device is defective given that it is produced in either Plant A or in Plant C

3. The following table classifies institutions of higher education in the United States by region and type :

	Public	Private
Northeast	266	555
Midwest	359	504
South	533	502
West	313	242

Define the following events :   P = Public Institution
N = Northeast, M = Midwest, S = South, W = West.

Suppose an institution is selected at random. Find the following:

a)   P(P)
b)   P(W')
c)   P(P' and N)
d)   P(P or S)
e)   P(P|M)

4. We wish to design a study to evaluate patient's satisfaction with the medical care they receive at a community health center. Before developing a sampling plan, we will attempt to understand the patients in the population according to their insurance status (which might influence their satisfaction with medical care) and gender (investigators have shown that men tend to report more satisfaction with medical care than women). Some descriptive information are available on each patient who has been seen in the community health center within the last 3 years. Assume that patients who have been seen in the last 3 years constitute our population of interest

Gender	Insurance Status		No
	Private Health Insurance	Medicaid	Insurance
Female	250	452	208
Male	128	680	157

Probability

a) What proportion of the population have no insurance ?
b) What proportion of the population are female and have no insurance ?
c) Among the patients with no insurance, what proportion are female ?
d) Are the events "No Insurance" and "Female" independent ? Justify.

5. The probability that patients are treated (T) <u>given</u> they have a particular disease (D) is 0.8. Thirty percent of all patients have the disease. The proportion of all patients not treated (T') is 0.76.

a) What proportion of all patients have the disease and are treated ?
b) What proportion of all patients are treated ?
c) Are the events T' and D mutually exclusive ?
d) Are the events T and D independent ?

6. The following data were collected from the 1997 administrative records of a local community center.

Age	Primary Diagnosis			
	Diabetes	Asthma	Arthritis	Cardiac
30-39	27	56	20	30
40-49	32	32	25	24
50-59	30	14	43	43
60 +	29	7	65	41

a) What proportion of patients had a primary diagnosis of asthma in the 1997 records?
b) What proportion of patients age 40-49 had primary diagnosis of asthma in the 1997 records?
c) What proportion of patients with a primary diagnosis of asthma in 1997 were 40-49 years of age?
d) What proportion of patients with a primary diagnosis of diabetes in 1997 were at least 50 years of age?

7. The following table summarizes ages (organized into groups) and disease stages (I - IV with higher stages indicative of more advanced disease) of patients seen in the oncology clinic in the summer of 1998:

Probability

Age	Disease Stage			
	I	II	III	IV
18-39	70	55	40	32
40-59	10	31	45	51
60 years or older	5	12	48	84

a)   What proportion of patients have stage III disease?
b)   What proportion of patients have stage II disease and are 60 years of age or older?
c)   What proportion of patient 60 years of age or older have stage III or IV disease?
d)   Are the events A=stage IV disease and B=60 years of age or older independent?

8.   A ventilation perfusion scan is used to generate the probability of pulmonary embolism (PE) in patients. Based on the results of the scan, patients are classified as high probability for PE, moderate probability for PE or low probability for PE. Consider the following data, classified by patient's age:

		Age			total
		45-54	55-64	65-74	
Scan	High Prob. for PE	13	15	20	48
	Moderate Prob. for PE	14	18	17	49
	Low Prob. for PE	25	20	31	76
total		52	53	68	173

If a patient is selected at random, find the probability that:

a)   s/he has a high probability for PE
b)   s/he is at least 55 years of age
c)   s/he has a high probability for PE given s/he is at least 55 years of age
d)   Are the events, high probability for PE and at least 55 years of age independent ? Justify.

Probability

9. The following table cross classifies primary care physicians by their gender and the size of their patient panel.

Gender	Size of Patient Panel				Total
	< 100	100-299	300-499	500 +	Total
Male	65	225	310	256	856
Female	78	120	650	308	1156
Total	143	345	960	564	2012

If a physician is selected at random, find the probability that:

a) A female who has a panel of 500 or more patients is selected.
b) A male from among those with 500 or more patients is selected.
c) What proportion of the physicians have panel sizes of at least 300 patients?
d) What proportion of females have panel sizes of at least 300 patients?
e) What proportion of physicians with the largest panels are female?

10. Consider the following probabilities:

$P(A) = 0.5$, $P(B) = 0.3$, $P(C)=0.1$, $P(A|C) = 1$, $P(A \text{ and } B) = 0.2$.

a) Find P(A or B)
b) Suppose that B and C are mutually exclusive, find P(B or C)
c) Find P(A and C)
d) Are A and C independent? Justify.
e) Are B and C independent? Justify. (Hint: See part (b))

11. The following table classifies patients in a research study of cardiovascular risk factors according to their gender and whether they have a family history of cardiovascular disease.

Gender		History of Cardiovascular Disease	
		No	Yes
	Female	120	28
	Male	165	34

Compute the following probabilities.

Probability

a)   P(family history of cardiovascular disease)
b)   P(family history of cardiovascular disease|male)
c)   P(family history of cardiovascular disease|female)
d)   Are family history of cardiovascular disease and gender independent? Justify.

12.   The following table displays medication adherence (defined as the percent of doses taken over the past month measured to the nearest whole number), classified by the time since initiation of therapy.

	Percent of Doses Taken		
Time Since Initiation of Therapy	< 70%	70%-89%	90% - 100%
Less than 3 months	18	58	24
Between 3 and 12 months	33	40	27
More than 12 months	21	27	52

a)   What proportion of patients take 70% or more of their prescribed doses?
b)   What proportion of patients on medication therapy for more than 12 months take 70% or more of their prescribed doses?
c)   What proportion of patients on medication therapy for less than 3 months take 70% or more of their prescribed doses?
d)   What is the nature of the relationship between medication adherence and time since initiation of medication therapy? (In words, be brief please).

13.   A recent report described a clinical trial which was conducted comparing an existing medication for asthma to a newly developed medication for asthma with respect to efficacy. In the trial, all patients who were seen in an asthma clinic at a particular hospital were approached and asked to participate in the trial. The following table classifies patients according to gender, eligibility and enrollment status.

	Eligibility-Enrollment Status			
Gender	Ineligible	Eligible – Enrolled	Eligible – Refused to Enroll	TOTAL
Male	35	80	20	135
Female	60	70	35	165
TOTAL	95	150	55	300

a)   What proportion of patients approached were eligible for the trial?
b)   What proportion of patients approached were ineligible?

c)   What proportion of male patients enrolled?

d)   What proportion of ineligible patients were female?

e)   Are gender and eligibility status independent?

Binomial Distribution

14.   Researchers claim that a new treatment for chronic bronchitis is 80% effective in reducing symptoms. If the new treatment is given to 12 patients suffering from bronchitis, what is the probability that:

      a)   The treatment is effective in exactly 10 patients ?

      b)   The treatment is effective in exactly 10 patients if the true effectiveness rate is 82%?

      c)   If the true effectiveness rate is 80%, how many patients would you expect to have reduced symptoms ?

15.   A political pollster knows from a very elaborate poll that 60% of voters favor stricter regulations on Medicaid. If 9 voters are randomly selected, find the probability that:

      a)   Exactly 3 voters favor stricter regulations on Medicaid.

      b)   At least 3 voters favor stricter regulations on Medicaid.

      c)   The mean number of voters who favor stricter regulations on Medicaid.

      d)   If the probability that voters favor stricter regulations on Medicaid is really 55%, find the probability that exactly 3 favor stricter regulations on Medicaid.

16.   A longitudinal study is conducted requiring patients to follow-up with research associates every month for assessments. The probability that a patient fails to follow-up in a given month is 10%. A pilot study is conducted to assess feasibility involving 20 patients. What is the probability that at most 3 patients fail to follow-up in the first month?

17.   Data from a national survey conducted in 1992 reported that 10% of all women over age 50 changed their primary care physicians at least once during a three year period.

      a)   Assuming that this rate applies to all women over 50, what is the probability that less than 3 of 10 women over 50 years of age change their primary care physicians within 3 years ?

      b)   If the true rate is higher than 10% - would the probability in (a) increase or decrease ? Justify. (Be brief but complete).

Probability

c)   If a primary care physician's practice included 650 women over the age of 50 - how many would be expected to leave the practice (i.e., change primary care physicians) over 3 years ?

18.   An HMO is considering a marketing plan to increase the numbers of patients who get flu shots, particularly patients with chronic diseases. Before implementing such a plan, some preliminary analyses are conducted. Available data indicate that 40% of patients with chronic diseases get flu shots.

a)   If 10 patients with chronic diseases are sampled, what is the probability that at least 4 get flu shots regularly?

b)   What is the probability that at most 8 get flu shots regularly?

c)   If a particular clinic (which is part of the HMO) serves 5250 patients with chronic disease, how many would be expected to get flu shots ?

19.   A article was published recently suggesting that persons who exercise regularly can reduce their risk of major clinical events (e.g., diabetes, cardiovascular disease) by up to 50% in some instances. It is believed that only 30% of adult Americans exercise on a regular basis.

a)   If 8 adult Americans are analyzed, what is the probability that more than half of them exercise regularly?

b)   If 8 adult Americans are analyzed, what is the probability that less than half of them exercise regularly?

c)   How many of 8 would you expect to exercise regularly?

20.   Suppose we plan to conduct a clinical trial and will recruit patients from a hospital clinic. In this clinic, approximately 10 new patients are seen in each session. We expect 70% of patients approached to be eligible for the new trial.

a)   What is the probability that at least 7 patients will be eligible per clinic session?

b)   How many patients would you expect to be eligible each clinic session?

Probability

21. Data show that 20% of all patients who make appointments at a primary care clinic never show up. If in a given clinic session, 15 patients make appointments,

    a) What is the probability that at most 4 patents do not show up?
    b) What is the probability that all patients show up?
    c) How many patients would be expected to not show up?
    d) How many patients would be expected to show up?
    e) What is the probability that all patients show up if the true "no show" percentage is 18%?

22. A study is proposed to evaluate whether a new medication is effective in controlling the symptoms of asthma. Prior to mounting the full study, a pilot study is conducted to test specific aspects of the study protocol. A sample of 10 patients with asthma are enrolled in the pilot study. Each patient agrees to take the study medication and to report back to the study evaluation team in 30 days for a clinical evaluation. Similar studies report that 90% of all patients complete follow-up visits when they are scheduled for one month from the initial contact.

    a) What is the probability that at least 8 patients show up for the clinical assessment?
    b) What is the probability that all patients show up for the clinical assessment?
    c) What is the probability that no more than 4 patients DO NOT show up for the clinical assessment?
    d) What is the probability that all show up if the true show rate is 87% ?

23. Using the data in problem 22, determine the expected (or mean) number of patients that would show up for clinical assessments if 10 were enrolled. What if 100 were enrolled?

Normal Distribution

24.   The number of telephone calls to emergency phone numbers per hour is assumed to follow a normal distribution with a mean of 7.5 calls per hour with a standard deviation of 2.1 calls per hour.

a)   What is the probability that more than 12 calls are received per hour?
b)   What is the probability that between 5 and 10 calls are received per hour?
c)   Complete the following statement: 90% of the time, the number of calls received per hour is less than

25.   Durations of clinical research studies awarded by a particular Federal agency are assumed to follow a normal distribution with a mean of 24 months and a standard deviation of 2.5 months.

a)   Find the probability that a study is awarded with a duration of more than 22 months.
b)   Find the probability that a study is awarded with a duration between 15 and 25 months.
c)   Fill in the following statement. 90% of studies awarded have durations between _____ and _____ months. (consider the middle 90% of the distribution).

26.   Graduate Record Exam (GRE) scores are approximately normally distributed with a mean of 500 and a standard deviation of 100.

a)   What proportion of individuals score between 580 and 620?
b)   What proportion of individuals score between 400 and 800?
c)   What is the 90th percentile score?

27.   Weights, in pounds, for specific gender and age subgroups are assumed to follow a normal distribution. The mean weight for males age 25 years is 160 pounds with a standard deviation of 19.2 pounds. What proportion of males age 25 have weights:

a)   exceeding 140 pounds ?
b)   between 140 and 160 pounds ?
c)   Compute the weight which separates the top 10% of males age 25 from the remaining males.

Probability

28.   Total cholesterol levels are assumed to follow a normal distribution for males and
      females in specific age groups (e.g., 20-39, 40-59). The following parameters are
      available on total cholesterol levels measured on patients enrolled in a particular HMO.

Gender, Age Group	Mean	Standard Deviation
Female, 30-49	185	28
Male 30-49	192	24

   a)   What proportion of females, 30-49 years of age have total cholesterol levels
        exceeding 200?
   b)   What proportion of males, 30-49 years of age have total cholesterol levels
        exceeding 200?
   c)   Compute the 90Th percentile of total cholesterol levels among females, 30-49
        years of age.
   d)   Compute the 90Th percentile of total cholesterol levels among males, 30-49 years
        of age.

29.   Prior to medical visits, patients are asked to complete a self-administered medical history
      form. The time it takes each patient to complete the form is approximately normally
      distributed with a mean of 12.4 minutes and a standard deviation of 2.1 minutes.

   a) What proportion of patients complete the medical history form in 10 minutes or less?
   b) What proportion of patients take 20 minutes or more to complete the medical history
      form?
   c) What is the 90th percentile of the time to complete the medical history form?

30.   Total serum cholesterol levels for individuals 65 years of age and older are assumed to
      follow a normal distribution with a mean of 182 and a standard deviation of 14.7.

   a)   What proportion of individuals 65 years of age and older have cholesterol levels
        of 175 or more ?
   b)   What proportion of individuals 65 years of age and older have cholesterol levels
        between 150 and 175 ?
   c)   If the top 10% of the cholesterol levels are assumed to be abnormally high - what
        is the upper limit of the normal range ?

Probability

31.   Data are collected in a national survey and weighted to reflect the population of all U.S. adults. Suppose that the mean Quality of Life (QOL) score is 70 with a standard deviation of 10. The range of QOL scores is 0-100, with higher scores indicative of better QOL.

    a)   What proportion of U.S. adults have QOL scores of 90 or higher ?
    b)   What is the 90th percentile of the QOL scores ?

32.   Heights of children are approximately normally distributed at each age and gender. If the mean height for 2 year old males is 28 inches with a standard deviation of 2.4 inches, find the following:

    a)   The probability that a 2 year old male is more than 26 inches.
    b)   The probability that a 2 year old male is between 30 and 35 inches.
    c)   Suppose the smallest 10% and largest 10% of heights are considered "abnormal".
    d)   What range of heights are "normal" ?

33.   A census of persons recovering from lower extremity fractures finds they work a mean of 8 hours per week with a standard deviation of 2 hours per week while in the first 6 months of recovery. One year post-injury, these persons work a mean of 12 hours per week with a standard deviation of 3.5 hours per week.

    a)   What proportion of persons work at least 10 hours per week during the first 6 months of recovery?
    b)   What proportion of persons work at least 10 hours per week one year post-injury?
    c)   What is the median number of hours worked during the first 6 months of recovery?
    d)   What is the 90th percentile in the number of hours worked per week for persons one year post-injury?

34. Body mass index (BMI) is computed as the ratio of weight in kilograms to height in meters squared. The distribution of BMI is approximately normal for specific gender and age groups. For females aged 30-39 the mean BMI is 24.5 with a standard deviation of 3.3.

   a)   What proportion of females aged 30-39 have BMI over 25?
   b)   Persons with a BMI of 30 or greater are considered obese. What proportion of females aged 30-39 are obese?
   c)   Suppose we classify females aged 30-39 in the top 10% of the BMI distribution as high risk. What is the threshold for classifying a female as high risk?

35. Based on a set of risk factors such as systolic blood pressure, total cholesterol level and smoking status, risks for developing cardiovascular disease (CVD) are calculated. The risk scores range from 0 to 100 with higher scores indicative of increased risk. Suppose for men over age 50, the CVD risk scores are approximately normally distributed with a mean of 30 and standard deviation of 8.5.

   a)   What proportion of men over age 50 have CVD risk scores exceeding 25?
   b)   What is the 90th percentile of the CVD risk scores?
   c)   If a man age 65 has a CV risk score of 50, what percentile is he in?

SAS Problems: Use SAS to solve each of the following problems.

1.   Use the SAS Ranbin function to generate a probability distribution for a binomial random variable with n=10 and p=0.25. Run 5000 replications (See SAS EXAMPLE 4.1), and use the probability distribution to find:

   a) $P(X = 2)$
   b) $P(X < 2)$
   c) $P(X > 2)$

2.   Use the SAS Probit function to find the following percentiles for a normal random variable with mean 500 and standard deviation 100 (See SAS EXAMPLE 4.3).

   a)   5th Percentile
   b)   50th Percentile
   c)   80th Percentile
   d)   95th Percentile

3.   Use the SAS Rannor function to generate a probability distribution for a normal random variable with mean 65 and standard deviation 8.6. Use Proc Chart to generate a relative frequency histogram for this variable. (See SAS EXAMPLE 4.3).

4.   Use the SAS Ranbin function to generate a probability distribution for a binomial random variable with n=50 and p=0.25. Run 5000 replications (See SAS EXAMPLE 4.1), and use SAS Proc Chart to generate a relative frequency histogram for this variable. Does the variable look approximately normally distributed?

5.   Use the SAS Rannor function to generate a probability distribution for a normal random variable with mean 50 and standard deviation 2. Use the SAS Probit function to find the 5th, 10th, 50th, 90th, and 95th percentiles. Run also a Proc Univariate and compare results.

Introductory Applied Biostatistics

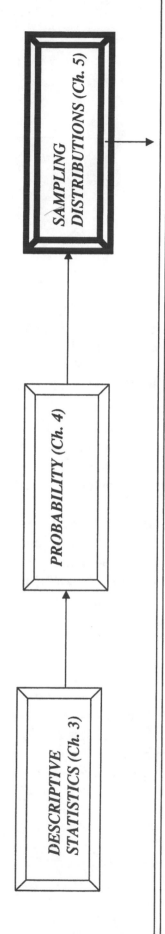

DESCRIPTIVE STATISTICS (Ch. 3) → **PROBABILITY (Ch. 4)** → **SAMPLING DISTRIBUTIONS (Ch. 5)**

STATISTICAL INFERENCE (Chapters 6-14)

Outcome Variable	Grouping Variable(s)/ Predictor(s)	Analysis	Chapter(s)
Continuous	–	Estimate μ, Compare μ to Known, Historical Value	6/13
Continuous	Dichotomous (2 Groups)	Compare Independent Means (Estimate/Test $(\mu_1-\mu_2)$) or the Mean Difference(μ_d)	7/13
Continuous	Discrete (> 2 Groups)	Test the Equality of K Means using Analysis of Variance ($\mu_1=\mu_2=\ldots\mu_k$)	10/13
Continuous	Continuous	Estimate Correlation or Determine Regression Equation	11/13
Continuous	Several Continuous or Dichotomous	Multiple Linear Regression Analysis	11
Dichotomous	–	Estimate p, Compare p to Known, Historical Value	8
Dichotomous	Dichotomous (2 Groups)	Compare Independent Proportions (Estimate/Test (p_1-p_2))	8/9
Dichotomous	Discrete (>2 Groups)	Test the Equality of k Proportions (Chi-Square Test)	8
Dichotomous	Several Continuous or Dichotomous	Multiple Logistic Regression Analysis	12
Discrete	Discrete	Compare Distributions Among k Populations (Ch-Square Test)	8
Time to Event	Several Continuous or Dichotomous	Survival Analysis	14

CHAPTER 5: Sampling Distributions

5.1   Introduction
5.2   Background
5.3   The Central Limit Theorem
5.4   Key Formulas
5.5   Applications Using SAS
5.6   Problems

5.1 Introduction

In the previous chapter we introduced the concept of probability, we presented rules for computing probabilities of outcomes and events and discussed two probability models.   The purpose of that discussion was to provide the mathematical background to understand the process of sampling subjects at random from a population. In statistical inference, which is addressed in the remaining chapters of this book, we will discuss the techniques of drawing inferences about unknown population parameters (e.g., the population mean μ) based on sample statistics (e.g., the sample mean $\overline{X}$).   In this chapter we discuss the relationships between sample statistics and population parameters. In particular, we focus on the relationship between the sample mean and the population mean. The statistical inference techniques described in subsequent chapters are based on this relationship.

In Section 5.2 we provide background and in Section 5.3 we present the Central Limit Theorem which is the mathematical theorem that formalizes the relationship between the sampling distribution of the sample mean and the population mean.   In Section 5.4 we summarize key formulas and in Section 5.5 we illustrate the relationship between the sampling distribution of the sample mean and the population mean using SAS.

5.2   Background

Suppose we have a population with mean μ and standard deviation σ. For now we do not specify any distributional form for the population, such as the normal distribution. Suppose we take simple random samples (s.r.s.) of size n from this population (we illustrated this process with our example in Chapter 2), and we compute the sample mean for each simple random sample. The collection of all possible sample means is called the *sampling distribution of the sample means*. If we compute the mean and standard deviation over all possible sample means, the following results hold.

For simple random samples *with* replacement:

$$\mu_{\overline{X}} = \mu \, , \qquad \sigma_{\overline{X}} = \frac{\sigma}{\sqrt{n}} \qquad\qquad (5.1)$$

where $\mu_{\overline{X}}$ and $\sigma_{\overline{X}}$ denote the mean and standard deviation of the sample means, respectively. The standard deviation of the sample means, $\sigma_{\overline{X}}$, is called the *standard error*.

For simple random samples *without* replacement:

$$\mu_{\overline{X}} = \mu \, , \qquad \sigma_{\overline{X}} = \frac{\sigma}{\sqrt{n}} \sqrt{\frac{N-n}{N-1}} \qquad\qquad (5.2)$$

If either the population size (N) is large or the ratio of the sample size to the population size (n/N) is small (e.g., less than 0.10), the quantity $\sqrt{\frac{N-n}{N-1}}$ will be close to one and the standard error in (5.2) is identical to that given in (5.1). In most applications, this is the case as the population is almost always very large. We illustrate these formulas through Examples 5.1 and 5.2.

Example 5.1. Suppose we have a small population of 5 patients with Type II (adult onset) diabetes and measure each patient's glycosolated hemoglobin (blood sugar) level. The population data are shown below.

$$8.9 \qquad 9.1 \qquad 10.4 \qquad 11.0 \qquad 10.1$$

The population mean is $\mu = \frac{\sum X}{N} = 9.9$ and the population standard deviation is $\sigma = \sqrt{\frac{\sum (X-\mu)^2}{N}} = 0.79$. (These parameters were computed using formulas (2.1) and (2.3) presented in Chapter 2.)

Suppose we take simple random samples of size 3 from the population. For illustration purposes, consider all possible samples of size n=3 from the population without replacement (i.e., once an individual is selected into the sample, they cannot be selected again). There are ten samples of size n=3 ($_5C_3 = 5 \cdot 4/2 = 10$) in which no individual appears more than once. The ten samples and the sample means computed on each sample are shown in Table 5.1.

Table 5.1.   Simple Random Samples of Size n=3 and Sample Means

Sample (X_1, X_2, X_3)	Sample Mean $\overline{X}$
8.9, 9.1, 10.4	9.47
8.9, 9.1, 11.0	9.67
8.9, 9.1, 10.1	9.37
8.9, 10.4, 11.0	10.10
8.9, 10.4, 10.1	9.80
8.9, 11.0, 10.1	10.00
9.1, 10.4, 11.0	10.17
9.1, 10.4, 10.1	9.87
9.1, 11.0, 10.1	10.07
9.1, 11.0, 10.1	10.50

The right hand column is the sampling distribution of the sample means. Notice that the sample means range from 9.37 to 10.50. The mean and standard deviation of this population (of all possible sample means based on simple random samples of size 3 without replacement) can be computed directly using (2.1) and (2.3), respectively: $\mu_{\overline{X}} = 9.9$ and $\sigma_{\overline{X}} = 0.32$. By (5.2), the mean of the ten sample means is $\mu_{\overline{X}} = \mu = 9.9$. The standard error, or standard deviation of the sample means, by (5.2) is:

$$\sigma_{\overline{X}} = \frac{\sigma}{\sqrt{n}} \sqrt{\frac{N-n}{N-1}} = \frac{0.79}{\sqrt{3}} \sqrt{\frac{5-3}{5-1}} = 0.32 .$$

Results (5.1) and (5.2) are very useful for statistical inference in which we make inferences about a population parameter (e.g., μ) based on a sample statistic (e.g., $\overline{X}$). Results (5.1) and (5.2) indicate that the sample mean is an *unbiased* estimator of the population mean (i.e., on the average, the sample mean is equal to μ) regardless of whether samples are taken with or without replacement. The sampling strategy, however, does affect the variation in the sample means (i.e., the standard error $\sigma_{\bar{x}}$). As noted, when the population size (N) is large, this effect is negligible.

Example 5.2. Suppose we have a population with mean 100, and standard deviation 10. (Note that the form of the distribution of the population is not specified.) If we take simple random samples of size n = 4 with replacement, the following are true by (5.1):

$$\mu_{\bar{x}} = \mu = 100$$

$$\sigma_{\bar{x}} = \frac{\sigma}{\sqrt{n}} = \frac{10}{\sqrt{4}} = \frac{10}{2} = 5$$

Suppose we increase our sample size to n = 25 :

$$\mu_{\bar{x}} = \mu = 100$$

$$\sigma_{\bar{x}} = \frac{\sigma}{\sqrt{n}} = \frac{10}{\sqrt{25}} = \frac{10}{5} = 2$$

Notice that under each sampling strategy the mean of the sample means is equal to the population mean. However, when the sample size is increased from 4 to 25 the standard error (or variation in the sample means) is reduced from 5 to 2. When the samples are larger, there is less variation in the sample means; the sample means are more tightly clustered about the population mean.

5.3 The Central Limit Theorem

The following mathematical theorem formalizes the statements and concepts presented in the previous section. This theorem is perhaps the most important theorem in statistics.

Central Limit Theorem (5.3)

Suppose we have a population with mean μ and standard deviation σ. If we take simple random samples of size n with replacement from the population, for large n,

the sampling distribution of the sample means is approximately normally distributed with:

$$\mu_{\bar{x}} = \mu, \quad \sigma_{\bar{x}} = \frac{\sigma}{\sqrt{n}}$$

where, in general, $n \geq 30$ is sufficiently large.

Sampling Distributions

The Central Limit Theorem (CLT) is important because in statistical inference we will make inferences about the population mean (μ) based on the value of a <u>single</u> sample mean. The CLT states that for large samples (n ≥ 30), the distribution of the sample means is approximately normal. In Chapter 4, we computed probabilities about normal random variables by standardizing (transforming to Z) and using Table 2. The CLT tells us that we can use that same process to compute probabilities about the sample mean. This will be useful in statistical inference because when we make inferences about population parameters based on sample statistics we will attach probability statements that quantify the precision in our inferences.

To reinforce the results of the Central Limit Theorem we now present three illustrations. In each illustration we display the population distribution along with the sampling distributions of the sample means based on samples of size 5, 15, 30 and 50. We display the distributions graphically and present parameters associated with each in tabular form. The illustrations are presented in Figures 5.2 - 5.4. Figures 5.2 and 5.3 depict non-normal populations, while Figure 5.4 illustrates the normal population distribution. Figures 5.2 (a) - 5.2 (d) display the sampling distributions of the sample mean based on s.r.s. of size n=5, n=15, n=30 and n=50, respectively, from the population displayed in Figure 5.2. Similar graphs are shown for sampling distributions from the populations displayed in Figures 5.3 and 5.4. Notice for the normal population distribution (Figure 5.4), the sampling distributions of the sample mean are normal for each sample size considered. In the non-normal cases (Figures 5.2 and 5.3) the sampling distributions of the sample means approach normality as the sample size increases (e.g., n ≥ 30).

Figure 5.2 Uniform Population

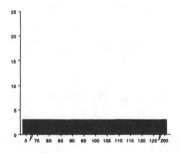

Sampling Distributions

Table 5.2 Population Parameters

Population	Sample Size	Mean	Std.Dev
Population Distribution	-	$\mu = 100$	$\sigma = 57.7$
Sampling Distribution	5	$\mu_{\overline{X}} = 100$	$\sigma_{\overline{X}} = 25.9$
Sampling Distribution	15	$\mu_{\overline{X}} = 100$	$\sigma_{\overline{X}} = 14.9$
Sampling Distribution	30	$\mu_{\overline{X}} = 100$	$\sigma_{\overline{X}} = 10.5$
Sampling Distribution	50	$\mu_{\overline{X}} = 100$	$\sigma_{\overline{X}} = 8.2$

The mean of the uniform population shown in Figure 5.2 is 100 with a standard deviation of 57.7. When simple random samples are drawn the means of the sampling distributions for samples of each size are equal to 100. The standard deviations of the sample means, or the standard errors, decrease as the sample size increases (See Figures 5.2 (a)-(d)). Notice how the shapes of the sampling distributions of the sample means start to look approximately normal for sample sizes of 30 or larger.

Figure 5.2(a) Sampling Distribution of $\overline{X}$, n=5

Figure 5.2(b) Sampling Distribution of $\overline{X}$, n=15

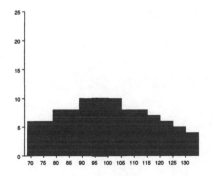

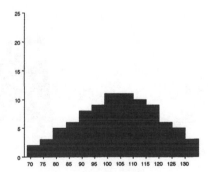

Sampling Distributions

**Figure 5.2(c) Sampling Distribution
of X̄, n=30**

**Figure 5.2(d) Sampling Distribution
of X̄, n=50**

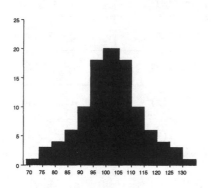

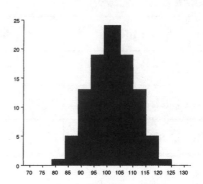

We now consider a second non-normal population, one which is skewed to the right, most observations are clustered at the low end of the distribution.

Figure 5.3 Skewed Population

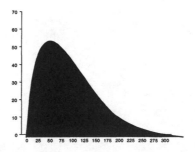

Table 5.3 Population Parameters

Population	Sample Size	Mean	Std.Dev
Population Distribution	-	$\mu = 100$	$\sigma = 100$
Sampling Distribution	5	$\mu_{\overline{X}} = 100$	$\sigma_{\overline{X}} = 44.8$
Sampling Distribution	15	$\mu_{\overline{X}} = 100$	$\sigma_{\overline{X}} = 25.8$
Sampling Distribution	30	$\mu_{\overline{X}} = 100$	$\sigma_{\overline{X}} = 18.2$
Sampling Distribution	50	$\mu_{\overline{X}} = 100$	$\sigma_{\overline{X}} = 14.2$

The mean of the skewed population shown in Figure 5.3 is 100 with a standard deviation of 100. When simple random samples are drawn the means of the sampling distributions for samples of each size are equal to 100. The standard deviations of the sample means, or the standard errors, decrease as the sample size increases (See Figures 5.3 (a)-(d)). Again, notice how the shapes of the sampling distributions of the sample means start to look approximately normal for sample sizes of 30 or larger.

Figure 5.3(a) Sampling Distribution of $\overline{X}$, n=5

Figure 5.3(b) Sampling Distribution of $\overline{X}$, n=15

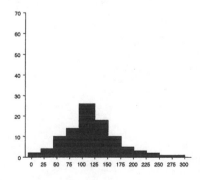

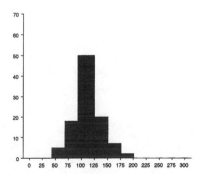

Sampling Distributions

Figure 5.3(c)   Sampling Distribution
of $\overline{X}$, n=30

Figure 5.3(d)   Sampling Distribution
of $\overline{X}$, n=50

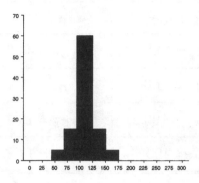

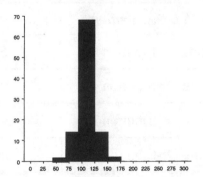

The last example involves a normal population (Figure 5.4). Notice that the sampling distributions of the sample means are approximately normally distributed for even small sample sizes (e.g., n=5). In examples 5.2 and 5.3 the sampling distributions of the sample mean looked normal only when the sample size was 30 or greater.

Figure 5.4 Normal Population

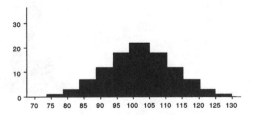

Sampling Distributions

Table 5.4 Population Parameters

Population	Sample Size	Mean	Std.Dev
Population Distribution	-	$\mu = 100$	$\sigma = 10$
Sampling Distribution	5	$\mu_{\overline{X}} = 100$	$\sigma_{\overline{X}} = 4.5$
Sampling Distribution	15	$\mu_{\overline{X}} = 100$	$\sigma_{\overline{X}} = 2.6$
Sampling Distribution	30	$\mu_{\overline{X}} = 100$	$\sigma_{\overline{X}} = 1.8$
Sampling Distribution	50	$\mu_{\overline{X}} = 100$	$\sigma_{\overline{X}} = 1.4$

Figure 5.4(a) Sampling Distribution of $\overline{X}$, n=5

Figure 5.4(b) Sampling Distribution of $\overline{X}$, n=15

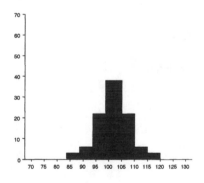

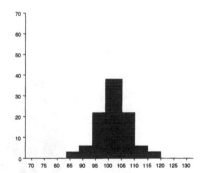

Sampling Distributions

Figure 5.4(c) Sampling Distribution
of $\overline{X}$, n=30

Figure 5.4(d) Sampling Distribution
of $\overline{X}$, n=50

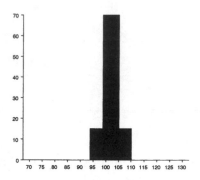

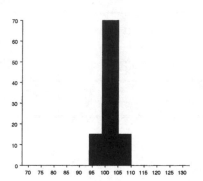

Example 5.3: Telephone calls placed to a drug hotline during the hours of 9-5 weekdays have a mean length of 5 minutes with a standard deviation of 5 minutes. The distribution of all calls (i.e., the population) is of the form:

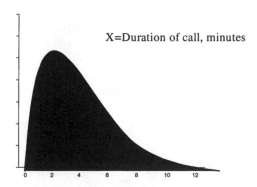

Suppose the population is very large, and we take simple random samples of 3 calls (n=3). The following are true by (5.1):

$$\mu_{\overline{X}} = 5, \quad \sigma_{\overline{X}} = \frac{5}{\sqrt{3}} = 2.89$$

Similarly, if we take s.r.s. of size n = 100, then by (5.1):

$$\mu_{\overline{X}} = 5, \quad \sigma_{\overline{X}} = \frac{5}{\sqrt{100}} = 0.5.$$

By the Central Limit Theorem (5.3) the distribution of the sample means for samples of size n=100 is approximately normal as shown below:

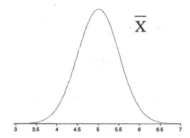

The distribution of the lengths of phone calls (X) to the drug hotline for all callers (the population) was not normally distributed. The population distribution was skewed to the right, with the majority of calls lasting a shorter time and fewer calls lasting a longer time. When we look at collections of 100 calls at a time (samples of n=100), the distribution of the mean length of calls ($\overline{X}$) is normally distributed (shown above).

     Most of the topics addressed in subsequent chapters deal with statistical inference based on a single sample from the population of interest. We summarize the sample and then make inferences about unknown population parameters (e.g., μ) based on sample statistics (e.g., $\overline{X}$). Without knowing the population distribution, as long as the sample is sufficiently large (usually size 30 is sufficient), we can appeal to the Central Limit Theorem to make probabalistic statements about the relationship between the sample mean and the unknown population mean. For example, the Central Limit Theorem states that the sample mean is approximately normally distributed with mean and standard deviation $\mu_{\overline{X}}$ and $\sigma_{\overline{X}}$, respectively. We can make statements about the distribution of sample means of size n by transforming to the standard normal distribution using (5.4) and using Table 2 (See Chapter 4, Section 6).

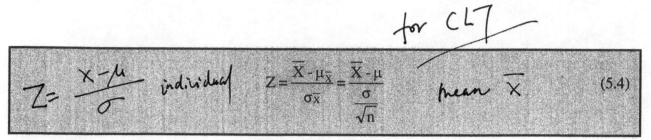

The following example illustrates the use of the CLT and formula (5.4).

Example 5.4: Suppose we have normal population with $\mu = 100$, and $\sigma = 10$. If we take a simple random sample of size 225, find the probability that the sample mean falls between 99 and 102. The problem of interest is stated as follows:

$$P(99 < \overline{X} < 102).$$

By the Central Limit Theorem, we know that the distribution of the sample mean is approximately normal with $\mu_{\overline{X}} = \mu$ and $\sigma_{\overline{X}} = \dfrac{\sigma}{\sqrt{n}}$. Because n is large (n=225) we can appeal to the Central Limit Theorem and transform the problem concerning $\overline{X}$ into a problem concerning the standard normal distribution Z.

For $\overline{X} = 99$:

$$Z = \frac{99 - 100}{10/\sqrt{225}} = -1.5$$

For $\overline{X} = 102$:

$$Z = \frac{102 - 100}{10/\sqrt{225}} = 3$$

Thus, $P(99 < \overline{X} < 102) = P(-1.5 < Z < 3)$.

Using Table 2,

$$P(-1.5 < Z < 3) = 0.9987 - 0.0668 = 0.9319.$$

Therefore, the probability that the sample mean based on 225 observations falls between 99 and 102 is 93.2%. If we were to use the sample mean (based on a sample size of 225) to estimate the unknown population mean, this result suggests that there is a high probability that the sample mean is close to the population mean.

   Suppose instead that an <u>individual</u> is selected at random from the population. Find the probability that his/her score falls between 99 and 102 (i.e., find P(99 < X < 102)). Because X is distributed as a normal random variable with mean 100 and standard deviation 10, we can transform the problem into a problem concerning Z using (4.14) (i.e., $Z = \dfrac{X - \mu}{\sigma}$).

For X = 99:

$$Z = \frac{99 - 100}{10} = -0.1$$

For X = 102:

$$Z = \frac{102 - 100}{10} = 0.2$$

Thus, P(99 < X < 102) = P(-0.1 < Z < 0.2).

Using Table 2:

$$P(-0.1 < Z < 0.2) = 0.5398 - 0.4207 = 0.1191.$$

Therefore, the probability that a single observation falls between 99 and 102 is only 11.91%.

Example 5.5: Suppose we have a population with $\mu = 50$, $\sigma = 10$. (Notice that no distributional form is specified for the population.) Suppose we take a simple random sample of size 100, find the probability that the sample mean is between 48 and 51. Using (5.1), (5.3) and Table 2.

We want to compute P(48 < $\overline{X}$ < 51). We first standardize each value :

$$Z = \frac{48 - 50}{10/\sqrt{100}} = -2$$

$$Z = \frac{51 - 50}{10/\sqrt{100}} = 1$$

Thus, P(48 < $\overline{X}$ < 51) = P(-2 < Z < 1) = 0.9772 - 0.1587 = 0.8185

The following example illustrates the application of the Central Limit Theorem for statistical inference.

Example 5.6. Suppose we wish to estimate the mean of a population (μ) whose standard deviation is known and equal to 12. Suppose a simple random sample of 100 individuals is selected from the population. Find the probability that the sample mean is no more than 2 units from the population mean.

In this example we do not know the population mean and wish to estimate it based on a single random sample. Here we ask the question, what is the likelihood that the sample mean is close to the true population mean (which is unknown). We define close as within 2 units in either direction.

Because the sample size is large (n=100), we can appeal to the Central Limit Theorem which indicates that the distribution of the sample means is approximately normal with $\mu_{\overline{X}} = \mu$ (See below).

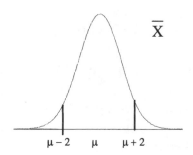

We wish to find the following:

$$P(\mu - 2 < \overline{X} < \mu + 2)$$

Standardizing using (5.4):

$$Z = \frac{(\mu - 2) - \mu}{12/\sqrt{100}} = \frac{-2}{1.2} = -1.67$$

$$Z = \frac{(\mu + 2) - \mu}{12/\sqrt{100}} = \frac{2}{1.2} = 1.67$$

The above is equal to: $P(-1.67 < Z < 1.67) = 0.9525 - 0.0475 = 0.9050$.

Sampling Distributions

There is a 90.5% probability that the sample mean, based on 100 observations, will be within 2 units of the true mean, μ. There is a very high likelihood (exceeding 90%) that the sample mean will be close to the population mean (within 2 units in either direction). If the goal of the application was to generate an estimate of the unknown population mean, the above suggests that the value of the sample mean (based on a sample of size n=100) would be quite close to the unknown population mean.

Would you feel comfortable using the value of the sample mean as an estimate of the unknown population mean ? We will explore these types of applications in further detail in the next several chapters.

5.4 Key Formulas

CONCEPT	FORMULA	DESCRIPTION
Simple Random Sampling With Replacement	$\mu_{\overline{X}} = \mu$, $\sigma_{\overline{X}} = \dfrac{\sigma}{\sqrt{n}}$	Mean and Standard Error of Sampling Distribution
Simple Random Sampling Without Replacement	$\mu_{\overline{X}} = \mu$, $\sigma_{\overline{X}} = \dfrac{\sigma}{\sqrt{n}}\sqrt{\dfrac{N-n}{N-1}}$	Mean and Standard Error of Sampling Distribution
Standardize a Sample Mean (If CLT applies, e.g., n$\geq$ 30)	mean: $Z = \dfrac{\overline{X} - \mu_{\overline{X}}}{\sigma_{\overline{X}}} = \dfrac{\overline{X} - \mu}{\dfrac{\sigma}{\sqrt{n}}}$	Standardizes Sample Mean

comparing with individual: $z = \dfrac{x - \mu}{\sigma}$

5.5   Applications Using SAS

In this section we review another SAS function, one which generates random variables from the uniform probability distribution. We take repeated samples from the uniform distribution, compute sample means and generate the sampling distribution of the sample means. Specifically, using SAS we illustrate the results of the Central Limit Theorem presented in Section 5.3 for the uniform distribution.

A SAS program is presented below which illustrates the use of the SAS Ranuni function which is used to generate random variables from the uniform probability distribution. Notes are provided to the right of the SAS program (*in italics*) for orientation purposes and are not part of the program. In addition, there are blank lines in the programs that follow which are solely to accommodate the notes. Blank lines and spaces can be used throughout SAS programs and are generally used to enhance readability.

We present a single example which is related to the example given in Chapter 5, presented in Figure 5.2. For the example we present three components:

i)   the SAS program code,
ii)   the computer output, and
iii)   a description of the relevant components of the computer output along with their interpretation.

SAS EXAMPLE 5.1   Uniform Population, Simple Random Samples of Size n=30
Ranuni Function (Generates Random Variables from the Uniform Distribution)

Consider a population which follows a uniform distribution, like the one illustrated in Figure 5.2. The uniform distribution is a continuous distribution. Suppose for this example that the population covers the interval 0 to 1. Using SAS we illustrate the results of the Central Limit Theorem presented in Section 5.3.

In this application, we will generate 30 random variates from the uniform distribution, which is equivalent to drawing 30 observations at random from a uniform population. We will compute the sample mean for each set of 30 observations, and will then summarize the distribution of the sample means. We will generate 10,000 samples of size n=30. Due to the large number of samples our observed probability distribution of the sample means will closely approximate the theoretical sampling distribution. The SAS program code is shown below.

i) Program Code

```
options ps=66 ls=80;
```
Formats the output page to 66 lines in length, 80 columns in width

```
data pop;
```
*Beginning of Data Step (Data set name **pop**).*

```
do i=1 to 10000;
```
*Beginning of a DO loop which will be repeated 10,000 times. The index **i** counts the iterations.*

```
   x=ranuni(13755);
```
*Generates a random variable, called **x**, from the uniform distribution. The only argument to the Ranuni function is the seed[1].*

```
   output;
```
*Writes the generated variable, **x**, to data set **pop**.*

```
end;
```
End of DO loop.

```
run;
```
End of Data Step.

```
title 'population distribution';
```
Title for output

```
proc chart;
```
Procedure call (Proc Chart to generate relative frequency histogram)

```
   vbar x/type=pct;
```
*Specification of variable **x**, type=pct displays relative frequencies*

```
run;
```
End of Procedure section

```
proc means;
```
Procedure call (Proc Means to generate summary statistics)

```
   var x;
```
*Specification of variable **x**.*

```
run;
```
End of Procedure section

Sampling Distributions

data samples;	*Beginning of Data Step (Data set name **samples**).*
do i=1 to 10000;	*Beginning of a DO loop which will be repeated 10,000 times. The index **i** counts the iterations.*
sumx=0;	*Create a variable called **sumx**, initialized to 0.*
do n=1 to 30;	*Beginning of a DO loop which will be repeated 30 times (for each value of **i**). The index **n** counts the iterations.*
x=ranuni(13755);	*Generates a random variable, called **x**, from the uniform distribution.*
sumx+x;	*Increase the value of **sumx** by the value of **x**.*
end;	*End of DO loop indexed by **n**.*
xbar=sumx/30;	*Create a variable called **xbar**, computed by dividing **sumx** by 30.*
output;	*Writes the generated variable, **xbar**, to data set **samples**.*
end;	*End of DO loop indexed by **i**.*
run;	*End of Data Step.*
title 'sampling distribution of the sample means';	*Title for output.*
proc chart;	*Procedure call (Proc Chart to generate relative frequency histogram)*
vbar xbar/type=pct;	*Specification of variable **xbar**, type=pct displays relative frequencies*
run;	*End of Procedure section*
proc means;	*Procedure call (Proc Means to generate summary statistics)*
var x;	*Specification of variable **xbar**.*
run;	*End of Procedure section*

(1) The seed is simply a random number which is used as a starting point for the random number generator (i.e., Ranuni). When the seed is changed, different values are generated.

ii)   *Computer Output*

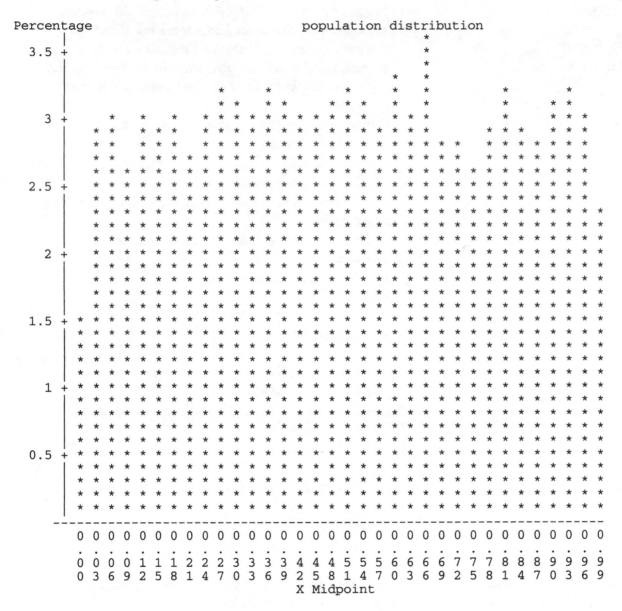

population distribution

Analysis Variable : X

N	Mean	Std Dev	Minimum	Maximum
10000	0.5014338	0.2860532	0.000025837	0.9999899

sampling distribution of the sample means

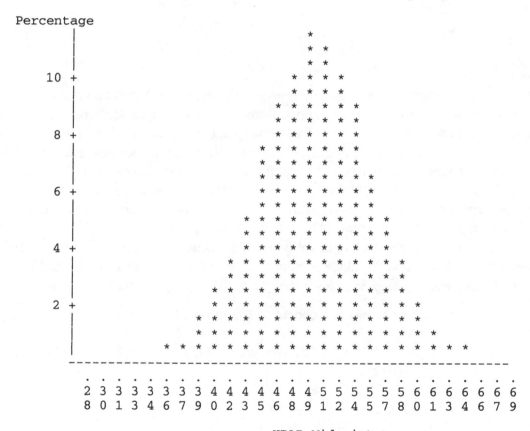

XBAR Midpoint

Sampling Distributions

sampling distribution of the sample means

Analysis Variable : XBAR

N	Mean	Std Dev	Minimum	Maximum
10000	0.5011598	0.0527851	0.2880476	0.7361218

iii) Interpretation

The first part of the output is a relative frequency histogram for the random variable X, which follows a uniform distribution over the interval 0 to 1. In theory, the distribution should be perfectly rectangular. Our simulated distribution has some spikes in various places. The summary statistics for the random variable X, the population variable, are given in the next section. The mean of X is 0.501, the standard deviation is 0.2861, and the range is from 0.000026 to 0.999989.

The next two sections of the output relate the the variable xbar - which is the sample mean based on 30 randomly selected observations from the uniform population. Notice that the distribution of xbar is approximately normal. Although the figure produced by SAS Proc Chart is somewhat crude, it does resemble the normal distribution. The mean of the variable xbar is 0.501, the standard deviation is 0.0528, and the range is from 0.2880 to 0.7361. Recall from the Central Limit Theorem that for sufficiently large samples (usually $n \geq 30$), that the following are true:

$\mu_{\bar{x}} = \mu$, and $\sigma_{\bar{x}} = \dfrac{\sigma}{\sqrt{n}}$. The mean of xbar is identical to the mean of x (0.501). The theoretical

standard deviation of xbar $= 0.2861/\sqrt{30} = 0.0522$ (assuming that $\sigma = 0.2861$). We observed a standard deviation of xbar equal to 0.0528.

Summary of SAS Function

The SAS functions illustrated are summarized in the table below. SAS functions are used in the Data Step. A general description of each function is provided in the table below.

Function	Sample Call Statement	Description
ranuni	x = ranuni(*seed*);	Generates random variables from the Uniform Distribution with range 0 -1

5.6   Problems

1.   Scores on a standardized exam are assumed to follow a normal distribution with a mean of 100 and standard deviation of 32.

    a)   If a simple random sample of 5 exams are selected, what is the mean exam score? What is the standard error of the scores ?

    b)   If a simple random sample of 50 exams are selected, what is the mean exam score? What is the standard error of the scores ?

    c)   If a simple random sample of 500 exams are selected, what is the mean exam score? What is the standard error of the scores ?

    d)   Compare the results of (a)-(c).

2.   The mean number of calories ingested per day for females aged 25 is 2750 with a standard deviation of 276. If a simple random sample of 100 females is selected, what is the probability that the mean number of calories ingested is more than 2800 ?

3.   A dispensing machine is set to produce 1 pound lots of a particular compound. The machine is fairly accurate, producing mean weights of lots equal to 1.0 pound with a standard deviation of 0.12 pounds. Thirty five lots are randomly selected:

    a)   What is the expected weight of the sample (i.e., find $\mu_{\overline{X}}$) ?

    b)   What is the standard error in the weights (i.e., find $\sigma_{\overline{X}}$) ?

    c)   Find the probability that the mean weight is greater than 1 pound.

    d)   Find the probability that the mean weight is less than 0.95 pounds.

4.   Following cardiac surgery patients are encouraged to exercise regularly (assume that regular exercise is defined as exercising on 3 or more days per week). A physician suspects that patients exercise regularly immediately following cardiac surgery, but tend to reduce, even stop exercising completely over time. An investigation is planned to estimate the mean number of weeks that patients exercise regularly following cardiac surgery. Assume that the standard deviation in the number of weeks cardiac patients exercise regularly following surgery is 6.3 weeks.

a) If a sample of 40 cardiac patients are followed, and the number of weeks in which each patient exercises regularly is recorded, what is the probability that the sample mean will be no more than 1 week higher than the true mean ?

b) If the sample is increased to 100 cardiac patients - what is the probability that the sample mean will be no more than 1 week higher than the true mean ?

5. We wish to estimate the mean cholesterol level for Type II diabetic patients free of cardiovascular disease. Suppose the standard deviation is known and equal to 20 ($\sigma = 20$). A random sample of 50 Type II diabetic patients free of cardiovascular disease are selected. What is the probability that their mean cholesterol level will be no more than 4 units above the true mean ?

6. We wish to estimate the mean gestational age (in days) in high risk pregnancies. If we sample 50 high risk pregnancies, and if the standard deviation is 4 days, what is the probability that the point estimate is within 1 day of the true mean gestational age.

7. Data are collected in a national survey and weighted to reflect the population of all U.S. adults. Suppose that the mean Quality of Life (QOL) score is 70 with a standard deviation of 10. The range of QOL scores is 0-100, with higher scores indicative of better QOL. What is the probability that the mean QOL score in a random sample of 40 adults is 72 or higher ?

8. Total serum cholesterol levels for individuals 65 years of age and older are assumed to follow a normal distribution with a mean of 182 and a standard deviation of 14.7.

a) If a random sample of 20 individuals 65 years or older is selected, what is the probability that their mean total serum cholesterol level is between 180 and 185 ?

b) Suppose that the mean total serum cholesterol level for individuals less than 65 years of age is 170 with a standard deviation of 26.8. If a random sample of 40 individuals less than 65 years of age is selected, what is the probability that their mean total serum cholesterol level is between 180 and 185 ?

Age Group	n	mean
65 Years of Age or Older	20	182
Less than 65 Years of Age	40	170

9. An article reported that patients in care for HIV have CD4 tests every 3 months, on average, with a standard deviation of 1.5 months. What is the probability that in a sample of 40 patients, the mean time between tests exceeds 3.5 months?

10. We want to estimate the mean of a population. A random sample of subjects is selected and the sample mean is computed. What is the probability that the sample mean is within 3 units of the true mean is the standard error is 1.8?

11. Suppose we measure length of hospital stay (in days) for patients undergoing a particular surgical procedure. The mean is 3 days with a standard deviation of 3.2 days.

    a. Does it appear that the length of stay data follow a normal distribution? Justify briefly.
    b. If a sample of 35 patients are selected, what is the probability that their mean length of stay is less than 2 days?
    c. If a sample of 35 patients are selected, what is the probability that their mean length of stay is more than 5 days?

SAS Problem: Use SAS to solve the following problem.

1. Use the SAS Ranuni function to generate random variables from a uniform distribution over the range of 0–1. Take samples of size n=10, n=20 and n=30 and plot the distributions of the sample means based on these different sample sizes. In addition, compute the mean and standard error of each distribution. Run 5000 replications (See SAS EXAMPLE 5.1) for each sampling distribution.

2. Use the SAS Rannor function to generate random variables from a normal distribution with mean 100 and standard deviation 8. Take samples of size n=5, n=10, n=20 and n=30 and plot the distributions of the sample means based on these different sample sizes. In addition, compute the mean and standard error of each distribution. Run 5000 replications for each sampling distribution. Do the sampling distributions look normally distributed?

3. Use the SAS Probit function to find the following percentiles for the mean of the sampling distribution generated in problem 1 for samples of size 30 (See SAS EXAMPLE 4.3).

    a) 5th Percentile
    b) 50th Percentile
    c) 80th Percentile
    d) 95th Percentile

4. Use the SAS Probit function to find the following percentiles for the mean of the sampling distribution generated in problem 2 for samples of size 10

    a) 5th Percentile
    b) 50th Percentile
    c) 80th Percentile
    d) 95th Percentile

5. Use the SAS Ranexp function to generate random variables from an exponential distribution (The SAS ranexp function is similar to the SAS ranuni function in that the only input required is a seed, e.g., ranexp(15243).) Take samples of size n=10, n=20 and n=30 and plot the distributions of the sample means based on these different sample sizes. Also plot the population distribution. What does the exponential distribution look like (i.e., the population) ? What characteristics might follow an exponential distribution? Run 5000 replications (See SAS EXAMPLE 5.1) for each sampling distribution.

Introductory Applied Biostatistics

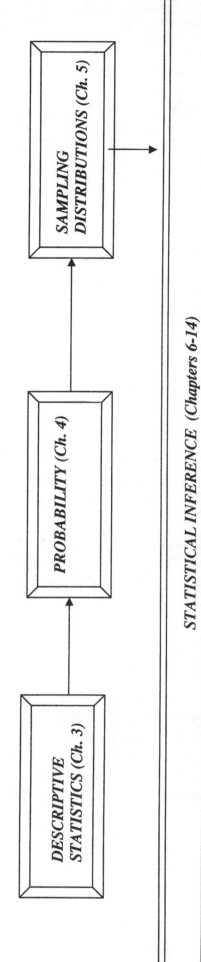

DESCRIPTIVE STATISTICS (Ch. 3) → **PROBABILITY (Ch. 4)** → **SAMPLING DISTRIBUTIONS (Ch. 5)**

STATISTICAL INFERENCE (Chapters 6-14)

Outcome Variable	Grouping Variable(s)/ Predictor(s)	Analysis	Chapter(s)
Continuous	-	**Estimate μ, Compare μ to Known, Historical Value**	**6/13**
Continuous	Dichotomous (2 Groups)	Compare Independent Means (Estimate/Test $(\mu_1-\mu_2)$) or the Mean Difference(μ_d)	7/13
Continuous	Discrete (> 2 Groups)	Test the Equality of K Means using Analysis of Variance ($\mu_1=\mu_2=\ldots\mu_k$)	10/13
Continuous	Continuous	Estimate Correlation or Determine Regression Equation	11/13
Continuous	Several Continuous or Dichotomous	Multiple Linear Regression Analysis	11
Dichotomous	-	Estimate p, Compare p to Known, Historical Value	8
Dichotomous	Dichotomous (2 Groups)	Compare Independent Proportions (Estimate/Test (p_1-p_2))	8/9
Dichotomous	Discrete (>2 Groups)	Test the Equality of k Proportions (Chi-Square Test)	8
Dichotomous	Several Continuous or Dichotomous	Multiple Logistic Regression Analysis	12
Discrete	Discrete	Compare Distributions Among k Populations (Ch-Square Test)	8
Time to Event	Several Continuous or Dichotomous	Survival Analysis	14

CHAPTER 6: Statistical Inference: Procedures for μ

6.1   Introduction

In this chapter we introduce the techniques of statistical inference. The techniques described here are concerned with the mean (μ) of a population. Similar techniques will be discussed in subsequent chapters applied to other parameters (e.g., the population proportion p, the difference between two population means $\mu_1 - \mu_2$, and so on).

There are two broad areas of statistical inference, *estimation* and *hypothesis testing*. In estimation, the population parameter (in this case μ) is unknown. A random sample is drawn from the population of interest and sample statistics are used to generate estimates of the unknown parameter. In hypothesis testing an explicit statement or hypothesis is generated about the population parameter. Again, a random sample is drawn from the population of interest, sample statistics are analyzed and determined to either support or reject the hypothesis about the parameter.

In Section 6.2 we introduce the techniques of estimation applied to the population mean μ. We present vocabulary and notation (Section 6.2.1), followed by a series of examples (Section 6.2.2). In Section 6.2.3 we discuss issues regarding precision in estimation and sample size requirements to achieve certain levels of precision. In Section 6.3 we introduce the techniques of hypothesis testing concerning the population mean μ. We present vocabulary and notation (Section 6.3.1), followed by a series of examples (Section 6.3.2). In Section 6.3.3 we introduce the concept of statistical power and sample size requirements to achieve certain levels of power. In Section 6.4 we summarize key formulas and in Section 6.5 we provide SAS program code used to perform statistical applications throughout this chapter.

6.2   Estimating μ

The goal in estimation is to make valid inferences about the population parameter based on a single random sample from the population. In the previous chapters we discussed the theoretical basis for statistical inference; in particular we described properties of sampling distributions of sample statistics. The relationship between sample statistics and population parameters is based on the sampling distribution of the sample statistics. With regard to estimating the population mean μ, we are particularly concerned with the relationship between the distribution of the sample mean $\overline{X}$ and μ.

6.2.1   Vocabulary and Notation

There are two types of estimates for population parameters: *point estimates* and *confidence interval estimates*. A point estimate for a population parameter is the "best" single number estimate of that parameter. A confidence interval estimate is a range of values for the population parameter with a level of confidence attached (e.g., 95% confidence that the range or interval contains the unknown parameter).

The point estimate for the population mean μ is the "best" single-valued estimate of the population mean derived from the sample. The "best" single number estimate for the population mean is the value of the sample mean. Estimates of a parameter are denoted by placing a "^" (read "hat") above the parameter notation. For example, the point estimate for the population mean is denoted as follows:

[handwritten: $\tilde{\mu}$: estimate]

$$\hat{\mu} = \overline{X} \tag{6.1}$$

In Chapters 2 and 5 we learned that the sample mean is an unbiased estimator of the population mean (i.e., on average, the sample mean is equal to the population mean $\mu_{\overline{X}} = \mu$). Unbiasedness is a desirable property in a point estimate. Consider the following example.

Example 6.1: Suppose the administration of a particular suburban hospital wants to estimate the mean waiting time (in minutes) for patients in their Emergency Room (ER) during weekends. In order to estimate the unknown mean waiting time, we take a random sample of patients visiting this ER on weekends and record the times that each patient waits. Suppose that there are resources available for this study to sample 100 patients. For each patient selected into the sample, the waiting time is recorded, measured as the number of minutes between the time he/she enters the ER and the time that he/she is seen by a clinician in the ER. Once the data are collected on each of the

100 patients, summary statistics for the waiting times can be produced. The mean waiting time in our sample is 37.85 minutes. A point estimate of the mean waiting time in the ER during weekends is given by (6.1).

$$\hat{\mu} = \overline{X} = 37.85$$

where μ = mean waiting time in the ER during weekends.

Suppose that the point estimate is reported to the administration who are disappointed that the waiting time is so long. The administration then decides to take a second look. Suppose that a second investigation is conducted in a similar fashion, involving a second (and distinct) sample of patients visiting the ER on weekends. The mean waiting time in the second sample is 33.75 minutes. A point estimate of the mean waiting time in the ER during weekends based on the second sample is 33.75 minutes.

Which point estimate is more appropriate ? From both the administration and a patient's point of view, the second estimate is more appealing - which one is better statistically ? Both are point estimates (assuming that both samples are random samples from the population of interest), but they are not the same. The point estimate alone is usually not sufficient in an estimation problem. In particular, it is generally of interest to assess how close the point estimate is to the true (unknown) mean μ.

Suppose the second sample involves 35 patients. Intuitively, one would think that a sample involving more patients would produce a more precise estimate of the unknown parameter. In Chapter 5 we presented the Central Limit Theorem which states that if we take simple random samples (s.r.s.) with replacement from a population, then the distribution of the sample mean is approximately normal with $\mu_{\overline{X}} = \mu$ and $\sigma_{\overline{X}} = \frac{\sigma}{\sqrt{n}}$. Suppose that the standard deviation in waiting times on weekends is known to be 9.5 minutes (i.e., $\sigma=9.5$). (In practice the value of the population standard deviation is often unknown. We generally know nothing about the parameters of the population (e.g., μ, σ). For now assume that the standard deviation is known. Later we will present examples in which the population standard deviation is unknown.) From the Central Limit Theorem we know that the distributions of sample means of size 100 (sample 1, $\overline{X} = 37.85$) and sample means of size 35 (sample 2, $\overline{X} = 33.75$) are approximately normal with

$$\mu_{\overline{X}} = \mu,\ \sigma_{\overline{X}} = \frac{\sigma}{\sqrt{n}} = \frac{9.5}{\sqrt{100}} = \frac{9.5}{10} = 0.95, \text{ and}$$

$$\mu_{\overline{X}} = \mu,\ \sigma_{\overline{X}} = \frac{\sigma}{\sqrt{n}} = \frac{9.5}{\sqrt{35}} = \frac{9.5}{5.9} = 1.61,$$

respectively. The distributions of the sample means based on samples of size 100 and size 35 are displayed in Figures 6.1a and 6.1b, respectively.

Figure 6.1a Distribution of $\overline{X}$ **for n=100,** $\sigma_{\overline{X}} = 0.95$

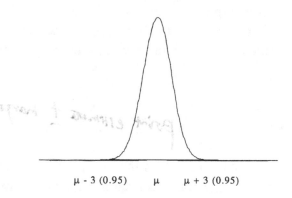

μ - 3 (0.95)   μ   μ + 3 (0.95)

Figure 6.1b Distribution of $\overline{X}$ **for n=35,** $\sigma_{\overline{X}} = 1.61$

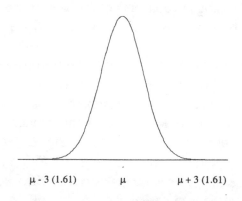

μ - 3 (1.61)   μ   μ + 3 (1.61)

Notice that there is less variation in the distribution of sample means when the sample size is 100 as compared to the distribution when the sample size is 35. On average, the sample means based on n=100 observations are 0.95 units from the population mean μ, while the sample means based on n=35 observations are 1.61 units from the population mean. The variation in the sample means, denoted $\sigma_{\overline{X}}$, is called the *standard error.* The standard error is a measure of the variation in the sample means, specifically the extent to which the $\overline{X}$'s vary around $\mu_{\overline{X}} = \mu$. As the sample size increases the standard error decreases, resulting in a more precise estimate of the population mean.

In Example 6.1 we described two samples and two point estimates for the unknown mean waiting time. In practice, a single sample would be selected and analyzed. We described two replications to illustrate the concept that the point estimate provides very valuable information, but alone is generally not sufficient in estimation problems. Once a point estimate is determined, the next issue of concern is how close that estimate is to the true, unknown value. The standard error addresses this issue, and the confidence interval incorporates both the point estimate and the standard error.

6.2.2 Confidence Intervals for μ *point estimate ± margin of error*

A confidence interval (CI) is a range of values that are likely to cover the true parameter (e.g., μ). CIs start with the point estimate, our best single number estimate, and build in a component which addresses how close (or how far) the point estimate is from the true, unknown parameter. This component is called the *margin of error.*

In all statistical applications we must accept the fact that we will not be 100% accurate in estimating a population parameter (e.g., μ) based on a sample statistic (e.g., $\overline{X}$). In statistical inference we make statements about a population parameter based on analysis of only a subset (a random sample) of the population. Using the concepts of probability we discussed in Chapter 4 and the properties of sampling distributions we discussed in Chapter 5 we can quantify the margin of error in each application.

The general form of a confidence interval is: point estimate $\pm$ margin of error. In developing confidence intervals we select a level of confidence which reflects the likelihood that the confidence interval contains the true, unknown parameter. Usually likelihoods of 90%, 95% and 99% are chosen for confidence intervals, although theoretically any likelihood could be selected. With a 95% confidence interval, for example, we are 95% confident that the range of values contains or covers the true, unknown parameter (e.g., μ). With a 99% confidence interval, we are 99% confident that the range of values contains or covers the true population parameter. In the following we derive the 95% confidence interval.

Recall the standard normal distribution Z presented in Chapter 4. The following statement is true for the standard normal distribution:

$$P(-1.96 < Z < 1.96) = 0.95 \qquad (6.2)$$

From the Central Limit Theorem (Chapter 5) we know that for large samples (usually $n \geq 30$ is sufficient), the sample means are approximately normally distributed with $\mu_{\bar{x}} = \mu$ and $\sigma_{\bar{x}} = \dfrac{\sigma}{\sqrt{n}}$ (See Figures 6.1a and 6.1b). Thus, the following statement holds for samples that are sufficiently large:

$$Z = \frac{\bar{X} - \mu_{\bar{x}}}{\sigma_{\bar{x}}}. \tag{6.3}$$

Substituting (6.3) into (6.2) gives the following:

$$P\left(-1.96 < \frac{\bar{X} - \mu_{\bar{x}}}{\sigma_{\bar{x}}} < 1.96\right) = 0.95 \tag{6.4}$$

Using algebra, (6.4) is equivalent to:

$$P(\bar{X} - 1.96\,\sigma_{\bar{x}} < \mu_{\bar{x}} < \bar{X} + 1.96\,\sigma_{\bar{x}}) = 0.95 \tag{6.5}$$

Because $\mu_{\bar{x}} = \mu$, (6.5) indicates that the probability that the interval between ($\bar{X} - 1.96\,\sigma_{\bar{x}}$) and ($\bar{X} + 1.96\,\sigma_{\bar{x}}$) contains the mean μ is 0.95, or 95%. Confidence intervals are based on this result.

The *95% confidence interval* for μ is given by:

$$\bar{X} \quad \pm \quad 1.96\,\sigma_{\bar{x}}$$

or equivalently:

$$\bar{X} \quad \pm \quad 1.96\,\frac{\sigma}{\sqrt{n}},$$

where $\bar{X}$ is the point estimate and $1.96\dfrac{\sigma}{\sqrt{n}}$ is the margin of error (1.96 reflects the fact that we chose a 95% confidence level and $\dfrac{\sigma}{\sqrt{n}}$ is the standard error).

Using the data in Example 6.1 (sample 1), the 95% confidence interval for the mean waiting time in the ER during weekends is:

$$37.85 \quad \pm \quad 1.96\frac{9.5}{\sqrt{100}}$$

$$37.85 \quad \pm \quad 1.96\,(0.95\,)$$

$$37.85 \quad \pm \quad 1.86$$

$$(35.99 \text{ to } 39.71)$$

Thus, we are 95% confident that the mean waiting time in the ER during weekends is between 35.99 and 39.71 minutes. The point estimate is 37.85 minutes and the margin of error is 1.86 minutes.

Following the same strategy, the 95% confidence interval based on the second sample of size n=35 is: 33.75 ± 3.15, or 30.60 to 36.90. Notice that this second interval is wider (30.60 to 36.90 as compared to 35.99 to 39.71 above) due to the smaller sample size (35 as compared to 100).

Suppose we construct 95% confidence intervals for the mean waiting time in the ER during weekends (μ) based on 40 different random samples of the same size (e.g., n=100). The intervals change from sample to sample depending on the value of the observed sample mean. The intervals cover the population mean approximately 95% of the time (Figure 6.2).

Figure 6.2   Interpreting Confidence Intervals (95% CI for μ)

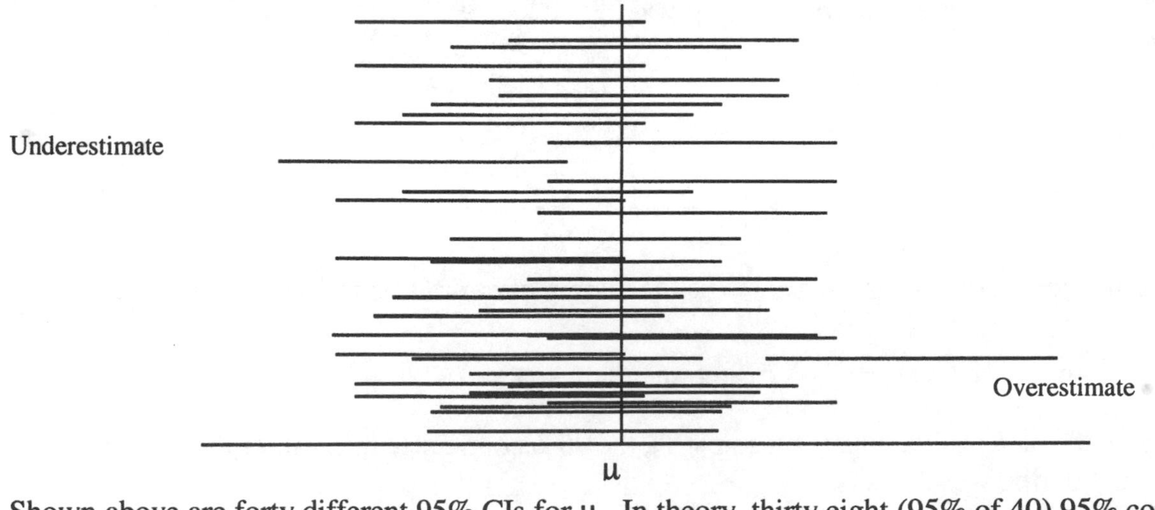

Shown above are forty different 95% CIs for μ. In theory, thirty eight (95% of 40) 95% confidence intervals will cover the true mean μ. In practice we take <u>one</u> random sample and develop a single

Statistical Inference for μ

confidence interval. We never know, with certainty, if our interval covers the true mean or not. It may be that our interval overestimates μ (i.e., the whole interval is above μ), or that the interval underestimates μ (i.e., the whole interval is below μ). Theoretically, using a 95% confidence interval, there is a 0.95 probability that any interval will cover the true mean.

Typically, confidence intervals are based on 80%, 90%, 95% or 99% confidence levels, although it is possible to construct confidence intervals for any level (from 0% to 100%). The general form of the confidence interval for μ is:

$$\overline{X} \pm Z_{1-\frac{\alpha}{2}} \frac{\sigma}{\sqrt{n}} \tag{6.6}$$

where $Z_{1-\alpha/2}$ is the value from the standard normal distribution with area in the lower tail (or area below it) equal to $1-\alpha/2$ corresponding to the $100(1-\alpha)\%$ confidence interval. Here α refers to the total area in the tails of the standard normal distribution. Figure 6.3 illustrates the value(s) from the standard normal distribution corresponding to the 95% confidence interval. In the 95% confidence interval (i.e., total tail area = 0.05, $100(1-0.05)\%=95\%$ confidence interval), $Z_{1-\alpha/2} = Z_{1-0.05/2} = Z_{0.975} = 1.96$.

Figure 6.3: $Z_{1-\alpha/2}=Z_{0.975}$ for 95% Confidence Interval

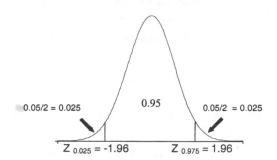

Table 6.1 contains values from the standard normal distribution for commonly used confidence levels. The entries in the table are taken from the standard normal distribution table (Table 2). If a confidence level and corresponding value of $Z_{1-\alpha/2}$ do not appear in Table 6.1, the

Statistical Inference for μ

standard normal distribution table can be used to extract the appropriate value of $Z_{1-\alpha/2}$ (See Figure 6.3).

Table 6.1: Values of $Z_{1-\alpha/2}$ for Confidence Intervals

Confidence Level	$Z_{1-\alpha/2}$	α*
99.99%	3.819	0.0001
99.9%	3.291	0.001
99%	2.576	0.01
95%	1.960	0.05
90%	1.645	0.10
80%	1.282	0.20

* α refers to the total area in the tails of the standard normal distribution.

Notice that as the confidence level increases, the value of $Z_{1-\alpha/2}$ increases and therefore so does the margin of error. In selecting the appropriate confidence level, investigators must evaluate the tradeoff between higher levels of confidence and larger margins of error (i.e., wider confidence intervals).

Example 6.2: Suppose we wish to estimate the mean age at which patients with hypertension are diagnosed using a 95% confidence interval. We randomly select 12 subjects with diagnosed hypertension and record the age at which they were diagnosed. The following data are observed:

32.8   40.0   41.0   42.0   45.5   47.0   48.5   50.0   51.0   52.0   54.0   59.2

(Ages are computed based on dates of birth and are computed to the nearest tenth of a year.)

We assume that age at diagnosis is approximately normally distributed (i.e., there is some mean value (unknown) and most patients are diagnosed around that mean value, while fewer are diagnosed at older or younger ages). The mean age at diagnosis in the sample is 47. A point estimate for the age at diagnosis among all hypertensives is 47. Suppose the standard deviation in the age at diagnosis of hypertension among all hypertensives is known to be 7.2 (i.e., $\sigma = 7.2$). A 95% confidence interval (CI) for the true age at diagnosis is calculated as follows.

$$\overline{X} \pm Z_{1-\frac{\alpha}{2}} \frac{\sigma}{\sqrt{n}}$$

$$47 \pm 1.96 \frac{7.2}{\sqrt{12}}$$

$$47 \pm 4.1 \quad \longleftarrow \quad \text{you can leave like this.}$$

$$(42.9, 51.1)$$

We are 95% confident that the mean age at diagnosis of hypertension is between 42.9 and 51.1. The point estimate is 47 years with a margin of error of 4.1 years.

SAS Example 6.2. The following output was generated using SAS Proc Means with an option to generate a 95% confidence interval. A brief interpretation appears after the output.

SAS Output for Example 6.2

```
                    The MEANS Procedure

    Analysis Variable : age_dx Age at diagnosis of hypertension

   N          Mean        Std Dev        Minimum        Maximum
  -----------------------------------------------------------------
  12      46.9166667      7.1598417     32.8000000     59.2000000
  -----------------------------------------------------------------

    Analysis Variable : age_dx Age at diagnosis of hypertension

                  Lower 95%        Upper 95%
                  CL for Mean      CL for Mean
                 ----------------------------------
                  42.3675203       51.4658131
                 ----------------------------------
```

Interpretation of SAS Output for Example 6.2

The SAS Means Procedure is used to generate abbreviated descriptive statistics on a continuous variable. The default statistics are the sample size n = 12, the sample mean $\overline{X}$ = 46.9, the sample standard deviation s = 7.2, the minimum 32.8 and the maximum 59.2. Here we requested that SAS generate a 95% confidence interval for the mean. The result is 42.4 to 51.5.

Statistical Inference for μ

The discrepancy between the interval we computed by hand in Example 6.2 and the interval computed by SAS in SAS Example 6.2 is largely due to rounding (SAS generally carries 8 decimal places through computations).

Example 6.3: Suppose we wish to estimate the mean systolic blood pressure (SBP) of members of a health maintenance organization (HMO) who are 25 years of age using a 90% confidence interval. We select a random sample of size n = 50 members of the HMO who are 25 years of age. This sample has a mean SBP of 120 with a standard deviation of 14.2.

In Example 6.3, the population standard deviation (σ) is unknown. This is the case in most estimation problems (i.e., neither population parameter is known). If the sample size is large (n $\geq$ 30 is usually sufficient) then the sample standard deviation (s) can be used to estimate σ (i.e., $\hat{\sigma}$ = s) in the confidence interval. In this case, the confidence interval is given by:

$$\bar{X} \pm Z_{1-\frac{\alpha}{2}} \frac{s}{\sqrt{n}} \tag{6.7}$$

Substituting our sample statistics, and using Table 6.1 to locate the appropriate value from the standard normal distribution, we have the following 90% confidence interval:

$$120 \pm 1.645 \frac{14.2}{\sqrt{50}}$$

$$120 \pm 3.30$$

$$(116.7, 123.3)$$

We are 90% confident that the mean SBP among 25 year old members of the HMO is between 116.7 and 123.3.

When the population standard deviation (σ) is unknown and the sample size is small (n <30), the Central Limit Theorem no longer applies and we cannot assume that the distribution of the sample mean is approximately normal. In this case (6.3) no longer holds and we cannot use $Z_{1-\alpha/2}$ in the formula for our confidence interval (6.6). If we can assume that the characteristic under investigation is approximately normally distributed in the population, then we can use a value from the t distribution in place of Z in (6.7). The t distribution is similar to the standard normal distribution (also bell shaped and symmetric), except in the tail areas. The shape of the t distribution depends on the exact sample size (See Figure 6.4).

t values for confidence intervals can be found in Table 3 in the Appendix. Tabled entries are indexed by their degrees of freedom (df). In the case of the confidence interval for μ, the degrees of freedom (df) are computed by df = n-1. As the degrees of freedom increase (with increasing sample size, n), the t values approach the standard normal values, Z (See Figure 6.4).

Figure 6.4. t Distributions for n=5, n=10, and n=20

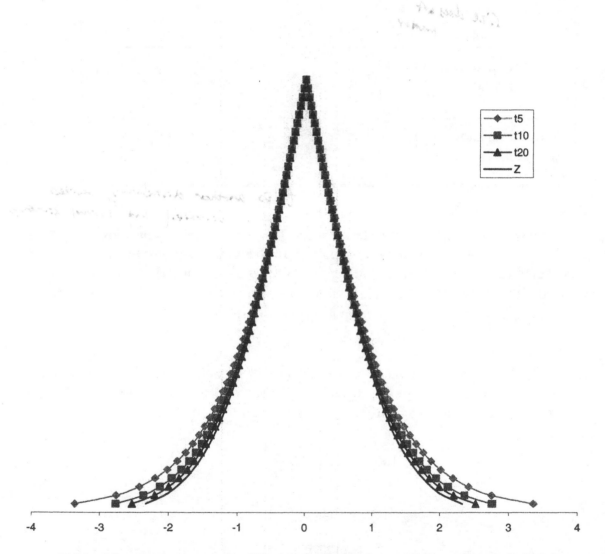

Table 6.2 summarizes three different formulas for confidence intervals for the population mean and the conditions under which each should be applied.

Statistical Inference for μ

Table 6.2 Confidence Intervals for μ

Assumptions: Normal distribution or large samples ($n \geq 30$) Simple random samples	
Case 1. σ known *size does not matter*	$\overline{X} \pm Z_{1-\frac{\alpha}{2}}\dfrac{\sigma}{\sqrt{n}}$
Case 2. σ unknown and $n \geq 30$	$\overline{X} \pm Z_{1-\frac{\alpha}{2}}\dfrac{s}{\sqrt{n}}$
Case 3. σ unknown and $n < 30$	$\overline{X} \pm t_{1-\frac{\alpha}{2}}\dfrac{s}{\sqrt{n}}, \quad df = n-1$

t is another distribution, besides binomial and normal distribution

Example 6.4: Suppose we wish to estimate the mean I.Q. for all twelve year olds using a 95% confidence interval. We select a random sample of 16 twelve year olds whose mean I.Q. score is 106 with a standard deviation of 12.4. Assuming that I.Q.s are approximately normally distributed for a specific age group, because σ is unknown and the sample size is small (i.e., $n < 30$), the appropriate confidence interval for μ is the formula given in Case 3:

$$\overline{X} \pm t_{1-\frac{\alpha}{2}}\frac{s}{\sqrt{n}}, \quad df = n-1$$

We first need to find the appropriate value from Table 3 for 95% confidence. In order to locate the appropriate value, we need to calculate the degrees of freedom: $df = n - 1 = 16 - 1 = 15$. The value from Table 3 for 95% confidence is $t = 2.131$. Now substituting the sample data:

$$106 \pm 2.131\frac{12.4}{\sqrt{16}}$$

Thus,

$$106 \pm 6.606$$

$$(99.4, 112.6)$$

Statistical Inference for μ

We are 95% confident that the mean I.Q. score for all twelve year olds is between 99.4 and 112.6.

Example 6.5: We wish to estimate the mean number of visits to primary care physicians over 3 years for all patients with Type II (adult onset) diabetes using a 99% confidence interval. A sample of 65 patients with diabetes are followed over a 3-year period and the number of visits each makes to their primary care physician is meaured. The mean number of visits to primary care physicians is 16 with a standard deviation of 1.4. In this example, σ is unknown and the sample size is large (i.e., $n \geq 30$), therefore this is an example of Case 2.

$$\overline{X} \pm Z_{1-\frac{\alpha}{2}} \frac{s}{\sqrt{n}}$$

Substituting our sample statistics and the appropriate value from Table 6.1:

$$16 \pm 2.576 \frac{1.4}{\sqrt{65}}$$

$$16 \pm 0.45$$

$$(15.5, 16.5)$$

We are 99% confident that the mean number of visits to primary care physicians over a three year period for diabetic patients is between 15.5 and 16.5.

In Example 6.5 the level of confidence is high, yet the width of the confidence interval is very small. This is primarily due to the magnitude of the standard error, or the variation in the sample means (i.e., $\sigma_{\overline{X}} = 1.4/\sqrt{65} = 0.174$). There is very little variation in the sample means suggesting that the number of visits to primary care physicians are very homogeneous among diabetic patients.

6.2.3   Precision and Sample Size Determination

The examples in the previous section illustrate estimation techniques for the population mean μ when presented with a random sample from the population of interest. In each example, sample statistics and sometimes population parameters (i.e., σ) were given. In order to construct confidence intervals for μ we first reviewed the available information and based on that information we chose the appropriate formula from Table 6.2. Once the appropriate formula was selected, the computations were relatively straightforward. Looking back over Examples 6.1 through 6.5 we see that the margins of error in the confidence intervals vary widely from example to example, depending both on the variation in the characteristic under investigation (σ or s) and the sample size. The margin of error is small when the sample size is large and/or when the standard error is small.

In *experimental design* statisticians are concerned with determining the numbers of subjects and the sampling strategy to be employed in a particular application so as to satisfy specific precision criteria in the statistical inference phase of the analysis. In some instances the number of subjects selected depends solely on practical considerations (e.g., time and/or financial constraints). In other applications, there is more flexibility such that the sample size is determined based on statistical considerations. Consider the following scenario.

Suppose we wish to estimate the mean of a population μ, and it is important to produce an estimate that is within 5 units of the true mean with 95% confidence. The statistician concerned with experimental design takes these criteria (i.e., margin of error = 5 units with 95% confidence) and determines the number of subjects required to produce such an estimate. Example 6.6 illustrates the experimental design aspect of an estimation problem.

Example 6.6. In Example 6.2 we generated a 95% confidence interval estimate for the mean age at which patients with hypertension were diagnosed. Our interval was based on a sample of n=12 subjects and had a margin of error of 4.1 years. Suppose that we wanted a more precise estimate, for example an interval with a margin of error not exceeding 2 years. In applications in which one desires a confidence interval estimate for the mean of a population, there are formulas that can be used to determine the necessary sample size. In order to design the experiment, specifically to determine the sample size, the following questions must be answered:

(i)   How much error can be tolerated in the estimate (i.e., how close must the estimate be to the true mean μ) ?

(ii)   What level of confidence is desired ?

An investigator with substantive knowledge of the application at hand is in the best position to address question (i). With respect to question (ii), the investigator must decide whether 90%, 95%, 99% or some other level will satisfy his/her needs.

Suppose that the investigator decides that the estimate must be within 2 years of the true mean to be useful. The investigator decides that a 95% confidence level is sufficient. (This aspect of the design is relatively simple, as 95% confidence is the standard confidence level used in most applications.)

The general form of the confidence interval for estimating μ is given by (6.6):

$$\overline{X} \pm Z_{1-\frac{\alpha}{2}} \frac{\sigma}{\sqrt{n}}$$

(In some applications we substitute s for σ, in other applications we use a value from the t distribution in place of Z in the above. Here we use the general formula and make adjustments as necessary in the examples that follow.)

The general form of the confidence interval for estimating μ is equivalent to:

$$\overline{X} \pm E$$

where E is the margin of error:

$$E = Z_{1-\alpha/2} \frac{\sigma}{\sqrt{n}} \tag{6.8}$$

The primary issue in this design application is to determine the number of subjects, n, required to satisfy pre-specified precision criteria. In particular, we want to determine the number of subjects required in order to generate an estimate for μ (the mean age at diagnosis) with a maximum margin of error of 2 years with 95% confidence.

Using algebra, we can solve for the sample size, n, in equation (6.8). We first multiply both sides by the square root of n:

$$\sqrt{n}\, E = Z_{1-\frac{\alpha}{2}}\, \sigma.$$

Now divide both sides by the margin of error:

$$\sqrt{n} = Z_{1-\frac{\alpha}{2}} \frac{\sigma}{E}.$$

Squaring both sides produces:

$$n = \left(\frac{Z_{1-\frac{\alpha}{2}}\, \sigma}{E} \right)^2 \qquad \text{no t here to calculate size} \tag{6.9}$$

Equation (6.9) produces the <u>minimum</u> number of subjects required to ensure a margin of error equal to E in the confidence interval for μ with the specified level of confidence (reflected in $Z_{1-\alpha/2}$). Notice that equation (6.9) calls for the standard deviation (σ) of the population characteristic under investigation. We can uses the value observed in Example 6.2, as it was based on a similar study. The number of subjects which must be sampled in order to produce an estimate for μ the mean age at diagnosis, with E = 2 years and 95% confidence is given by (6.8):

$$n = \left(\frac{Z_{1-\frac{\alpha}{2}} \sigma}{E} \right)^2 = \left(\frac{1.96\,(7.2)}{2} \right)^2 = (7.056)^2 = 49.79$$

Because (6.9) produces the minimum number of subjects to satisfy the specified criteria we always round up to the next integer. Therefore, 50 individuals with hypertension must be sampled and their age at diagnosis measured in order to generate a 95% confidence interval estimate for the mean age at diagnosis with a margin of error of no more than 2 years. With a sample size of 50, the resultant confidence interval for the mean age will be of the form: $\overline{X} \pm 2$ (where 2 is the margin of error). The exact value of the sample mean will be computed once the appropriate data are collected.

In design applications where the goal is to determine the sample size required to produce a confidence interval estimate for an unknown population mean, formula (6.9) can be applied. In order to apply (6.9), however, three inputs are required: the desired confidence level (reflected in $Z_{1-\alpha/2}$), the standard deviation of the characteristic under study (σ) and the margin of error (E). The confidence level is usually the easiest to determine as in most applications, the standard 95% confidence level is used ($Z_{1-\alpha/2}$ =1.960). The standard deviation of the characteristic under study can be difficult to determine. Recall, in these applications we are determining the number of subjects required so as to produce a confidence interval estimate with a specified level of precision for an unknown population mean μ. As we discussed in the previous section, it is often the case that the population standard deviation (σ) is unknown. If the value of σ is unknown, there are several options for generating a reasonable estimate: we can substitute a value derived from a similar application reported in the literature; we can generate an estimate based on a previous application (i.e., using historical data); we can conduct a pilot study (e.g., select a non-random, convenience sample and compute the sample standard deviation s); or we can use an educated "guess." Because estimation of the sample size is not a problem of statistical inference, it is acceptable to generate a best guess. In doing so, however, one should always err on the conservative side. One method of generating a best guess is based on the Empirical Rule presented in Chapter 3. The Empirical Rule states that for unimodal distributions, approximately 68% of the observations fall between $\mu - \sigma$ and $\mu + \sigma$, approximately 95% of the observations fall between $\mu - 2\sigma$ and $\mu + 2\sigma$, and almost all of the observations fall between $\mu - 3\sigma$ and $\mu + 3\sigma$. Thus the range is equal to about 6 standard deviations. Based on the Empirical Rule, a <u>conservative</u> estimate of the standard deviation is given by the second part of the Empirical Rule:

$$\hat{\sigma} = \frac{\text{range}}{4} \tag{6.10}$$

To use (6.10) to estimate the standard deviation, we must specify the range. The range is the difference between the minimum and maximum values which can often be approximated easily.

Consider the following example in which the population standard deviation is unknown and an estimate is used in (6.9) to determine the sample size requirements.

Example 6.7. Suppose the administration of a hospital wants to estimate the mean time it takes for patients to get from one department to another on the hospital grounds (e.g., radiology to surgery, radiology to orthopedics) for appointment scheduling purposes. How large a sample would be required to ensure that the margin of error in the estimate of the mean time it takes for patients to get from one department to another is no more than 5 minutes with 95% confidence ?

Suppose in this application, we know nothing about the time it takes for patients to get from one department to another on the hospital grounds. The goal of the present analysis is to determine the number of subjects required to generate an estimate of the mean time it takes, and here we have no idea of what the variation in times might be (i.e., σ is unknown).

Suppose that we cannot find any other, comparable studies and there are no historical data available. In order to generate an estimate of the variation in time it takes to get from one department to another we conduct a pilot study. In the pilot study we survey a "convenience" (or non-random) sample of n = 10 patients traveling from department to department on hospital grounds. For each patient we record the time it takes to get from one department to another in minutes. Using the formulas provided in Chapter 3, we compute the sample standard deviation s using the n=10 observations. Suppose we find s=17 minutes.

We can now implement formula (6.9) to determine the number of subjects required to estimate the mean time it takes for patients to get from one department to another with a margin of error of 5 minutes with 95% confidence:

$$n = \left(\frac{Z_{1-\frac{\alpha}{2}}\sigma}{E}\right)^2 = \left(\frac{1.96\,(17)}{5}\right)^2 = (6.664)^2 = 44.4$$

A random sample of size 45 patients would ensure that the estimate of the mean time to travel between departments is within 5 minutes of the true mean with 95% confidence.

What would happen to the sample size estimate if we chose a smaller margin of error? Suppose in the previous application, we wanted to estimate the mean time it takes for patients to get from one department to another with a margin of error of 2 minutes with 95% confidence:

$$n = \left(\frac{Z_{1-\frac{\alpha}{2}}\sigma}{E}\right)^2 = \left(\frac{1.96\,(17)}{2}\right)^2 = (16.667)^2 = 277.6$$

A random sample of size 278 patients would ensure that the estimate of the mean time to travel between departments is within 2 minutes of the true mean with 95% confidence. To ensure a smaller margin of error, more subjects are needed in the sample.

In design applications such as those described in Examples 6.6 and 6.7, investigators will often consider several scenarios. The different scenarios might reflect different margins of error. As we stated previously, there are often both practical and statistical considerations that must be evaluated in determining the sample size for a particular application. In the previous application, it may be feasible to sample as many as 125 subjects. Statistical considerations suggest that a sample of size 45 would ensure a margin of error of no more than 5 minutes, but 278 subjects are required to ensure a margin of error of no more than 2 units? What level of precision could we achieve with 125 subjects?

SAS Example 6.7. The following output was generated using SAS to determine the sample size required to ensure different margins of error with 95% confidence. A brief interpretation appears after the output.

```
SAS Output for Example 6.7
```

Obs	c_level	z	sigma	e	n
1	0.95	1.95996	17	1	1111
2	0.95	1.95996	17	2	278
3	0.95	1.95996	17	3	124
4	0.95	1.95996	17	4	70
5	0.95	1.95996	17	5	45
6	0.95	1.95996	17	10	12

Interpretation of SAS Output for Example 6.7

There is no SAS procedure specifically designed to determine the number of subjects required to produce a confidence interval estimate with a specific margin of error and confidence level. However, SAS can be used to program the appropriate formula (6.9). Once the formula is programmed, different scenarios can be evaluated easily. In the example shown above, six scenarios are considered (denoted Obs 1-6, respectively), related to SAS EXAMPLE 6.7. In each scenario the confidence level (c_level) is set at 0.95 (95% which is reflected in $Z_{1-\alpha/2} = 1.96$) and the standard deviation (sigma) is set at 17. The scenarios differ in terms of the margins of error (E) specified. Here we considered margins of error 1, 2, 3, 4, 5 and 10 in scenarios 1-6, respectively. In scenario 1 (Obs=1), the confidence level (c_level) is set at 0.95 (95%), the standard deviation (sigma) is set at 17 and the margin of error (E) is specified at 1. SAS produces a sample size of 1,111 which is the required number of subjects to generate a 95% confidence interval for the mean time to get from one department to another with a margin of error equal to 1 minute. As the margin of error increases, fewer subjects are required. For example, when the margin of error is 2 (Obs=2), 278 subjects are required to generate a 95% confidence interval for the mean time to get from one department to another and when the margin of error is 10 (Obs=6), only 12 subjects are required to generate a 95% confidence interval for the mean time to get from one department to another. This output would be weighed against practical constraints to determine the most appropriate sample size for the application. With 125 subjects, a margin of error of 3 minutes can be achieved.

6.3 Hypothesis Tests About μ

We now describe the second area of statistical inference called hypothesis testing. In hypothesis testing an explicit statement or hypothesis is generated about the population parameter. A random sample is drawn from the population of interest, sample statistics are analyzed and determined to either support or reject this hypothesis about the parameter. Based on the sample statistics, we determine the likelihood that the null hypothesis is true. As in the case of estimation, the theoretical basis for hypothesis testing is the relationship between the distribution of sample statistics and population parameters.

6.3.1 Vocabulary and Notation

There are a number of new vocabulary terms in hypothesis testing. These terms will be used in discussing hypothesis tests about μ (presented in this chapter), and about other parameters (e.g., p, μ_1-μ_2) presented in subsequent chapters. We use an example in this section to introduce these terms, along with their definitions. Section 6.3.2 contains a number of examples which are presented more concisely and utilize these terms and concepts.

Example 6.8: A large, national study was conducted in 1998 and reported that the mean systolic blood pressure for males aged 50 was 130 with a standard deviation of 15. In 1999, an investigator hypothesizes that due to increased stress in the workplace, faster paced lifestyles and poorer nutritional habits, systolic blood pressures have increased. In order to test this hypothesis (i.e., that the mean systolic blood pressure has increased in 1999), we set up two competing hypotheses, called the *null and alternative hypotheses*, denoted H_0 and H_1, respectively.

$$H(ov) \quad H_0: \mu = 130 \qquad\qquad (6.11)$$
$$H_1: \mu > 130$$

The null hypothesis reflects the "no change" or "no effect" situation (i.e., systolic blood pressures are no different in 1999 and have a mean of 130), while the alternative or research hypothesis reflects the investigator's claim (i.e., that the mean systolic blood pressure has increased in 1999).

In order to test the hypotheses (6.11), a random sample is selected from the population of interest. Suppose we have resources to sample n=108 males age 50 in 1999. We record the systolic blood pressure on each male and generate summary statistics. Of primary interest is the mean systolic blood pressure in the sample ($\overline{X}$).

Consider the following scenarios and their interpretations with respect to the hypotheses (6.11):

(a) Suppose the mean systolic blood pressure for the sample of n=108 males selected in 1999 is $\overline{X}$ = 130. The mean systolic blood pressure in the 1999 sample is identical to the population mean from 1998, which would lead us to believe that the null hypothesis, H_0, is most likely true (i.e., the population mean in 1999 is not changed and equal to 130). Notice that we cannot say with certainty that the null hypothesis is true because we have only a sample of males age 50 in 1999. However, because the sample was selected at random we would expect the mean systolic blood pressure among all males age 50 in 1999 to be close to the observed 130.

(b) Suppose $\overline{X}$ = 150. The mean systolic blood pressure in the 1999 sample is substantially higher than the population mean from 1998, which would lead us to believe that the alternative hypothesis, H_1, is most likely true (i.e., the population mean in 1999 is greater than 130).

(c) Suppose $\overline{X}$ = 135. The mean systolic blood pressure in the 1999 sample is numerically higher (135 > 130) than the population mean from 1998, but is it solely due to chance fluctuation? Remember, we are evaluating statistics based on a random sample of 108 subjects from the population of interest, and sampling always involves some error. In order to evaluate whether the observed sample mean ($\overline{X}$ = 135) provides evidence of a true increase in the underlying population mean, we must address the question: How likely are we to observe a sample mean of 135 from a population with μ = 130 ? (This probability can be computed using the Central Limit Theorem and techniques we described in Chapter 5.)

It is scenario (c) which illustrates the crux of the hypothesis testing problem. To conduct the test of hypothesis (i.e., to decide whether the null or alternative hypothesis is more likely true), we must determine a *critical value* such that if our sample mean is less than the critical value we will conclude that H_0 is true (i.e., μ =130), and if our sample mean is greater than the critical value we will conclude that H_1 is true (i.e., μ > 130).

Instead of determining critical values for $\overline{X}$ which would be specific to each application (because $\overline{X}$ depends on the unit of measurement), we instead appeal to the Central Limit Theorem. In particular, for samples of sufficient size (n $\geq$ 30) we know that the sample mean is approximately normally distributed with mean $\mu_{\overline{X}}$ and standard deviation $\sigma_{\overline{X}}$. Assuming that H_0 is true (or "under the null hypothesis") we can standardize $\overline{X}$, producing a Z score:

$$Z = \frac{\overline{X} - \mu_{\overline{X}}}{\sigma_{\overline{X}}} = \frac{\overline{X} - \mu_0}{\frac{\sigma}{\sqrt{n}}} \tag{6.12}$$

where μ_0 is the mean specified in H_0 (i.e., μ_0 = 130).

From Chapter 4 we know the properties of Z. If Z (6.12) is close to zero, which occurs when $\overline{X}$ is close to 130, we suspect that H_0 is most likely true. When Z is large, which occurs when $\overline{X}$ is larger than 130, we suspect that H_1 is most likely true. In hypothesis testing, we need to determine the point at which Z is "too large." That point is called the *critical value of Z*.

In Example 6.8 we know that σ=15 and n=108. Therefore, under the null hypothesis (i.e., when μ=130), the distribution of the sample means is approximately normal with mean $\mu_{\overline{X}}$ = μ = 130 and standard deviation $\sigma_{\overline{X}}$ = $\sigma/\sqrt{n}$ = 15/$\sqrt{108}$ = 1.44 (Figure 6.5).

Figure 6.5 Distribution of $\overline{X}$ Under H$_0$: $\mu = 130$ ($\sigma_{\overline{X}}$=1.44)

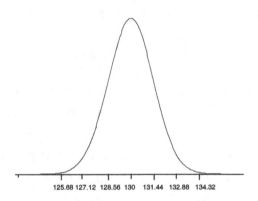

125.68 127.12 128.56 130 131.44 132.88 134.32

Under H$_0$: $\mu = 130$, it is possible to observe any value of $\overline{X}$ displayed in Figure 6.5. However, we know that it is unlikely (i.e., the probability is small) that $\overline{X}$ will take on a value in the tails of the distribution. For example, it is very unlikely to observe values of $\overline{X}$ exceeding 132.88 (which is two standard deviations above the mean: $\mu_{\overline{X}} + 2\sigma_{\overline{X}}$). Recall from Chapter 5, $P(\overline{X} > 132.88) = P(Z > 2) = 1 - 0.9772 = 0.0228$. Therefore if we observe a sample mean which exceeds 132.88 and we reject the null hypothesis in favor of the alternative, the probability that we are making a mistake in rejecting is only 2.28%. However, if we reject the null hypothesis for values $\overline{X}$ which exceed 131.44, the probability that we are making a mistake in rejecting H$_0$ is $P(\overline{X} > 131.44) = P(Z > 1) = 1 - 0.8413 = 0.1587$. We must decide what level of error (specifically error of this type where we incorrectly reject the null hypothesis) that we can tolerate in the analysis. A 15.87% probability of making this type of mistake may be too high.

In hypothesis testing, investigators must select a *level of significance*, denoted α, which is defined as the probability of rejecting H$_0$ when H$_0$ is true. The level of significance is generally in the range of 0.01 to 0.10, though any value from 0-1.0 can be selected. Because the level of significance reflects the likelihood of drawing an erroneous conclusion (i.e., $\alpha = P(\text{reject } H_0|H_0 \text{ true})$), small levels of significance are purposely selected.

Once a level of significance is selected, a *decision rule* is formulated. The decision rule is a formal statement of the criteria used to draw a conclusion in the hypothesis test. For example, suppose we select a level of significance of 5% (i.e., $\alpha = 0.05$, we allow a 5% chance of rejecting H_0 when H_0 is true). The critical value and decision rule are displayed graphically in Figure 6.6. The region in the tail end of the curve (shaded) is called the *rejection region*.

Figure 6.6 Decision Rule for H_0: $\mu=130$ vs. H_1: $\mu>130$, $\alpha=0.05$

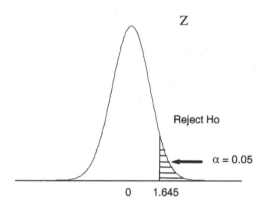

The decision rule is given by:

remember 90% CL is 1.645, here is one tail.

$$\text{Reject } H_0 \text{ if } Z \geq 1.645 \qquad\qquad (6.13)$$
$$\text{Do Not Reject } H_0 \text{ if } Z < 1.645$$

Once the decision rule is in place, we compute the value of the *test statistic*. The test statistic is given in (6.12) and is computed using the observed sample data. Suppose in Example 6.8, $\overline{X} = 135$. The test statistic is given below:

$$Z = \frac{\overline{X} - \mu_0}{\frac{\sigma}{\sqrt{n}}} = \frac{135 - 130}{\frac{15}{\sqrt{108}}} = 3.46$$

The final step in the test of hypothesis is to compare the test statistic to the decision rule to draw a conclusion. In Example 6.8 the test statistic falls in the rejection region and therefore we reject H_0 because $3.46 \geq 1.645$. A final statement is made concerning our findings relative to the research or alternative hypothesis. Such a statement is as follows. We have significant evidence, $\alpha = 0.05$, that the mean systolic blood pressure for males aged 50 in 1999 has increased from 130.

Table 6.3 summarizes the steps involved in the test of hypothesis procedure. The steps are displayed with reference to a test about the population mean μ. The same steps will be used to test hypotheses concerning other parameters.

Table 6.3 Tests of Hypothesis Concerning μ

Step	Example
1. Set up hypotheses. Select level of significance.	$H_0: \mu = \mu_0$ $H_1: \mu > \mu_0$ [1] $\alpha = 0.05$
2. Select appropriate test statistic.	$$Z = \dfrac{\overline{X} - \mu_0}{\dfrac{\sigma}{\sqrt{n}}}$$
3. Generate decision rule.	Reject H_0 if $Z \geq Z_{1-\alpha}$ [1] Do Not Reject H_0 if $Z < Z_{1-\alpha}$
4. Compute the value of the test statistic.	
5. Draw a conclusion about H_0 by comparing the test statistic (4) to the decision rule (3). Report findings relative to H_1. If we reject H_0, we compute the exact level of significance called the p-value.	

[1] NOTE: In hypothesis testing, there are three types of alternative or research hypotheses. The test of hypothesis presented in Example 6.8 is called an *upper tailed test* due to the direction (i.e., >) specified in the alternative hypothesis and the location of the rejection region (See Figure 6.6). The other two types of tests are called *lower tailed tests* and *two tailed tests*. In lower tailed tests the alternative hypothesis is of the form: $H_1: \mu < \mu_0$, while in two tailed tests the alternative hypothesis is of the form: $H_1: \mu \neq \mu_0$. In lower tailed tests the research hypothesis indicates a decrease in the mean value, while in two tailed tests the research hypothesis indicates a difference (either an increase or a decrease) in the mean value.

Table 6.4 contains critical values of Z for lower, upper, and two-tailed tests. The general form of the hypotheses are presented along with the form of the decision rule for each type of test.

Table 6.4   Important Values For Statistical Inference

Lower Tailed Tests			
Hypotheses	α	Z_α	**Decision Rule**
$H_0: \mu = \mu_0$ $H_1: \mu < \mu_0$	.0001 .001 .005 .010 .025 .050 .100	-3.719 -3.090 -2.576 -2.326 -1.960 -1.645 -1.282	Reject H_0 if $Z \leq Z_\alpha$
Upper Tailed Tests			
Hypotheses	α	$Z_{1-\alpha}$	**Decision Rule**
$H_0: \mu = \mu_0$ $H_1: \mu > \mu_0$	.0001 .001 .005 .010 .025 .050 .100	3.719 3.090 2.576 2.326 1.960 1.645 1.282	Reject H_0 if $Z \geq Z_{1-\alpha}$
Two Tailed Tests			
Hypotheses	α	$Z_{1-\alpha/2}$	**Decision Rule**
$H_0: \mu = \mu_0$ $H_1: \mu \neq \mu_0$	.0001 .001 .010 .050 .100 .200	3.819 3.291 2.576 1.960 1.645 1.282	Reject H_0 if $Z \leq -Z_{1-\alpha/2}$ or if $Z \geq Z_{1-\alpha/2}$

Table 6.5 summarizes three different formulas for test statistics in tests concerning the population mean μ and the conditions under which each should be applied.

Table 6.5 Test Statistics For Tests Concerning μ

Assumptions: Normal distribution or large sample ($n \geq 30$)	
Simple random sample	
Case 1. σ known	$$Z = \frac{\overline{X} - \mu_0}{\frac{\sigma}{\sqrt{n}}}$$
Case 2. σ unknown and $n \geq 30$	$$Z = \frac{\overline{X} - \mu_0}{\frac{s}{\sqrt{n}}}$$
Case 3. σ unknown and $n < 30$	$$t = \frac{\overline{X} - \mu_0}{\frac{s}{\sqrt{n}}}, \ df = n - 1$$

Notice in Case 3, where σ is unknown and the sample size is small ($n < 30$), the test statistic no longer follows a normal distribution (Z). Instead, the test statistic follows a t distribution. In this case the critical value(s) must be selected from the t table (Table 3). Use of this test statistic and selection of appropriate critical value(s) are illustrated in Example 6.9 in the next section.

In every test of hypothesis we draw a conclusion to either reject H_0 (in favor of H_1) or not to reject H_0. When the test indicates a rejection of H_0 a very strong statement is made. However, when the test does not indicate a rejection of H_0 a very weak statement is made, for the following reasons.

Suppose the true value of the population mean (or any parameter under investigation) is known (See Table 6.6). If in reality H_0 is true (first row of Table 6.6) and we do not reject H_0 based on our test of hypothesis, a correct decision is made. However, if in reality H_0 is true and we reject H_0 based on our test of hypothesis, an error is committed. This particular type of error is called a Type I error and its probability is given by $\alpha = P(\text{Type I Error}) = P(\text{Reject } H_0 | H_0 \text{ True})$. If in reality

H_0 is false (second row of Table 6.6) and we do not reject H_0 based on our test of hypothesis, an error is committed. This particular type of error is called a Type II error and its probability is given by $\beta = P(\text{Type II Error}) = P(\text{Do Not Reject } H_0 | H_0 \text{ False})$. If in reality H_0 is false and we reject H_0 based on our test of hypothesis, a correct decision is made.

Table 6.6 Errors In Tests of Hypothesis

	Statistical Test	
True State of Nature	*Do Not Reject H_0*	*Reject H_0*
H_0 True	Correct Decision	Type I Error
H_0 False	Type II Error	Correct Decision

Because the true state of the parameter is unknown in practice, we never know which of the four cells of Table 6.6 we fall into in a given application. However, because we select a level of significance, α, *a priori* we know that the probability of committing a Type I error is small. When our test of hypothesis indicates a rejection of H_0 (second column of Table 6.6) we can be confident that a correct decision has been made, since the chance of committing a Type I error is small. When our test of hypothesis does not indicate a rejection of H_0 (first column of Table 6.6) we cannot be so confident that a correct decision has been made, because the chance of committing a Type II error is unknown. In such a case a weaker statement is made at the conclusion of the test (See Example 6.9).

6.3.2   Hypothesis Tests About μ

In the following we present examples to illustrate the test of hypothesis technique. We cover lower tailed and two-tailed tests.

Example 6.9. Suppose that the mean cholesterol level for males age 50 is 241. An investigator wishes to examine whether cholesterol levels are significantly reduced by modifying diet only slightly. A random sample of 12 patients agree to participate in the study and follow the modified diet for 3 months. After 3 months, their cholesterol levels are measured and summary statistics are produced on the n=12 subjects. The mean cholesterol level in the sample is 235 with a standard deviation of 12.5. Based on the data, is there statistical evidence that the modified diet reduces cholesterol ? Run the appropriate test using a 5% level of significance. We will now run the test of hypothesis using the 5-step procedure outlined in Table 6.3.

1. Set up hypotheses.

In this example, we wish to test whether the diet <u>reduces</u> cholesterol levels. The mean cholesterol level (without modifying diet) is taken to be 241 for males age 50. The null hypothesis reflects the "no change" or "no effect" situation (i.e., cholesterol levels are unchanged and their mean is 241 as reported), and the alternative or research hypothesis reflects the situation where the diet is effective in reducing or lowering cholesterol levels. This is an example of a lower-tailed test.

$H_0: \mu = 241$
$H_1: \mu < 241$ *so this is one tail, note the difference in t table.* $\alpha = 0.05$

2. Select appropriate test statistic.

The appropriate test statistic is selected from Table 6.5 based on the information available. In this example, we have a sample of size n = 12 and we do not know the population standard deviation σ (i.e., the standard deviation in cholesterol levels among all males age 50). Here we have the sample standard deviation s = 12.5 (computed on the data collected from the 12 participants).

Because σ is unknown and the sample is small (n<30), this is an example of Case 3, and the test statistic is:

$$t = \frac{\overline{x} - \mu_0}{\dfrac{s}{\sqrt{n}}}$$

3. Decision Rule.

The decision rule for any hypothesis testing application depends on 3 factors: i) whether the test is an upper, lower or two-tailed test, ii) the level of significance α, and iii) the form of the test statistic (e.g., Z or t as determined in step 2). Here we have a lower tailed test, $\alpha = 0.05$, and we are using a t statistic. The critical value is found in Table 3 (as opposed to Table 6.4 which contains critical values for Z). In order to determine the appropriate value, we first compute the degrees of freedom: df = 12 - 1 = 11. In Table 3, we locate the 5% level of significance (across the top row labeled one-sided α) and read down to 11 degrees of freedom. The appropriate critical value is t = 1.796. We can now formulate the decision rule:

since the Reject H_0 if $t \leq -1.796$
Do Not Reject H_0 if $t > -1.796$

is lowertail,

we use -1.796

4. Test Statistic.

In this step we substitute our sample data into the appropriate formula for the test statistic, selected in step 2. Recall the value for μ_0 is 241 (as specified in H_0):

$$t = \frac{\overline{x} - \mu_0}{\dfrac{s}{\sqrt{n}}} = \frac{235 - 241}{\dfrac{12.5}{\sqrt{12}}} = -1.66$$

5. Conclusion.

In the final step, we draw a conclusion by comparing the test statistic (computed in step 4) to the decision rule (displayed in step 3). We do not reject H_0 because $-1.66 > -1.796$. We do not have significant evidence, $\alpha = 0.05$, to show that the diet reduces cholesterol levels. Notice in our concluding statement that we do not state an "acceptance" of H_0 since it is possible that a Type II error has been committed. It is possible that the diet truly does lower cholesterol levels and it may be the case that our sample size is too small to ensure a high likelihood of detecting the reduction. This issue will be discussed in the next section where we consider experimental design issues related to tests of hypothesis.

Example 6.10. In a managed care organization, the mean starting salary for males in entry level, non-clinical positions is $29,500. We want to determine whether the mean starting salary for females in similar entry-level, non clinical positions is different from $29,500. In order to make this assessment, we randomly select 10 females whose job titles and responsibilities fit our criteria and we record their starting salaries (in $000s). The data are given below.

32	27	31	27	26
26	30	22	25	36

Is the mean starting salary for females significantly different from $29,500 ? Use $\alpha = 0.05$.

Statistical Inference for μ

1.   Set up hypotheses.

In this example we wish to test whether the mean starting salary for females is <u>different</u> from the reported starting salary for males in similar positions. This is an example of a two-sided test.

H_0: $\mu = 29.5$
H_1: $\mu \neq 29.5$ $\alpha = 0.05$

2.   Select appropriate test statistic.

The appropriate test statistic is selected from Table 6.5 based on the information available. In this example, we have a sample of size n = 10 and we do not know the population standard deviation σ (i.e., the standard deviation in starting salaries for all females in entry-level, non clinical positions).

Because σ is unknown and the sample is small (n<30), this is an example of Case 3, and the test statistic is:

$$t = \frac{\bar{x} - \mu_0}{\dfrac{s}{\sqrt{n}}}$$

3.   Decision Rule.

Here we have a two-tailed test, $\alpha = 0.05$, and we are using a t statistic. The critical value is found in Table 3 (as opposed to Table 6.4 which contains critical values for Z). In order to determine the appropriate value, we first compute the degrees of freedom: df = 10 - 1 = 9. In Table 3, we locate the 5% level of significance (across the top row labeled two-sided α) and read down to 9 degrees of freedom. The appropriate critical value is t = 2.262. We can now formulate the decision rule:

Reject H_0 if t $\leq$ -2.262 or if t $\geq$ 2.262
Do Not Reject H_0 if -2.262 < t < 2.262

4.   Test Statistic.

In this step we substitute our sample data into the appropriate formula for the test statistic, selected in step 2. In this example we are given raw scores as opposed to summary statistics so we must first compute summary statistics using the formulas presented in Chapter 3.

X	X^2
32	1024
27	729
31	961
27	729
26	676
26	676
30	900
22	484
25	625
36	1296
$\Sigma X = 282$	$\Sigma X^2 = 8100$

$$\overline{X} = \frac{\Sigma X}{n} = \frac{282}{10} = 28.2$$

$$s^2 = \frac{\Sigma x^2 - \frac{(\Sigma x)^2}{n}}{n-1} = \frac{8100 - \frac{(282)^2}{10}}{9}$$

$$s^2 = 16.4, \quad s = \sqrt{16.4} = 4.0$$

The summary statistics are: $n = 10$ $\overline{X} = 28.2$ $s = 4.0$.

We can now compute the test statistic:

$$t = \frac{\overline{x} - \mu_0}{\frac{s}{\sqrt{n}}} = \frac{28.2 - 29.5}{\frac{4.0}{\sqrt{10}}} = -1.02$$

5. Conclusion.

In the final step, we draw a conclusion by comparing the test statistic (computed in step 4) to the decision rule (displayed in step 3). Do not reject H_0 because $-2.262 < -1.02 < 2.262$. We do not have significant evidence, $\alpha = 0.05$, to show that the mean starting salary for females is significantly different from \$29,500. Are the starting salaries the same?

Statistical Inference for μ

SAS Example 6.10. The following output was generated using SAS Proc Means with an option to conduct a test of hypothesis. A brief interpretation appears after the output.

SAS will conduct a one-sample test of hypothesis in its Means procedure. However, in the one-sample test SAS assumes that the test of interest is $H_0: \mu = 0$ vs. $H_1: \mu \neq 0$. In this application (and in most others), we are not interested in testing if the mean of the analytic variable is zero. In this example we want to test if the mean starting salary among females is 29.5 (in $000s, which is equal to $29,500). In order to use SAS to test the desired hypotheses we create a new variable, which we call TESTSTAT and it is simply our original analytic variable (i.e., starting salary) minus 29.5. Using the variable TESTSTAT we can use SAS to test the desired hypotheses. In particular, if the mean of TESTSTAT is significantly different from zero, then we can conclude that the mean salary is significantly different from 29.5 (since TESTSTAT is simply equivalent to salary - 29.5). Conversely, if the mean of TESTSTAT is not significantly different from zero, then we can conclude that the mean salary is not significantly different from 29.5.

SAS Output for Example 6.10

```
                        The MEANS Procedure

Variable  Label                   N       Mean       Std Dev     Std Error
---------------------------------------------------------------------------
salary    Annual Salary in $000s  10   28.2000000   4.0496913   1.2806248
teststat                          10   -1.3000000   4.0496913   1.2806248
---------------------------------------------------------------------------

              Variable  Label                   t Value    Pr > |t|
              -----------------------------------------------------
              salary    Annual Salary in $000s    22.02     <.0001
              teststat                            -1.02      0.3366
              -----------------------------------------------------
```

Interpretation of SAS Output for Example 6.10

To illustrate the default procedure and the necessary modifications, we asked SAS to analyze both the original data (variable named salary) and our created variable (teststat, which is equal to salary-29.5). SAS first provides the number of observations (n=10) and then summary statistics on each analysis variable. The mean salary is 28.2 while the mean of teststat is −1.3 (equal to 28.2-29.5). The standard deviations of salary and teststat are identical as subtracting a constant (e.g., 29.5) from a variable does not influence its standard deviation or variance. The standard error (i.e., $s/\sqrt{n}$) of both salary and teststat is 1.2806. The test statistic for salary (computed by taking the ratio of the mean (minus zero as SAS assumes μ=0) to the standard error) is 22.02. However, this is the test statistic for testing if the mean salary is zero. The test of interest is whether the mean salary is 29.5 or not. The test statistic for this test is t = -1.0151. If we were conducting this test of hypothesis by hand (as we did in Example 6.10), then we would compare the value of the test statistic to an appropriate critical value from the t distribution table (Table 3) with 9 degrees of freedom to draw a conclusion. SAS instead produces a value, called a *p -value*, which allows us to draw a conclusion. The p - value for this test is 0.3366. The p- value is defined as the exact level of significance of the data. In other words, the p-value is the smallest level of significance which would still lead to rejection of the null hypothesis. In this example, we could reject H_0 only if the level of significance was 0.3366 (or larger). Any level of significance smaller than 0.3366 would not lead to rejection. We generally use a 0.05 level of significance. Therefore, we would not reject H_0.

Interpreting P-Values From SAS

The following rule can be used to draw conclusions in tests of hypotheses based on p - values:

$$\text{Reject } H_0 \text{ if p - value} \leq \alpha \qquad (6.14)$$

where α is the level of significance selected for the test (e.g., 0.05, 0.01).

NOTE: It is important that the level of significance, α, is selected prior to the implementation of the test. It is inappropriate to select a level of significance so as to ensure a particular conclusion in a test (i.e., based on the observed p - value).

Consider the following example which illustrates the computation of p-values in tests of hypotheses.

Suppose we wish to perform the following two-sided test at $\alpha = 0.05$:

$$H_0: \mu = 30$$
$$H_1: \mu \neq 30.$$

Suppose that n = 25, $\overline{X}$ = 35, and s = 10. We perform the test using SAS and SAS produces a test statistic t = 2.50 and p-value = 0.0194. The two-sided p-value is displayed graphically below.

The p-value is the probability of observing a value exceeding the value of the test statistic (i.e., $t > 2.50$ or $t < -2.50$).

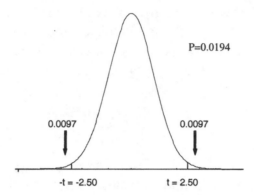

If this test were conducted by hand, critical values would be selected corresponding to the pre-selected level of significance α. The critical values for a two-sided test with $\alpha = 0.05$ are shown below. Because the test statistic $t = 2.50$ exceeds the critical value $t = 2.064$ (See figure below) we would reject H_0 in favor of H_1 and conclude that the mean is significantly different from 30.

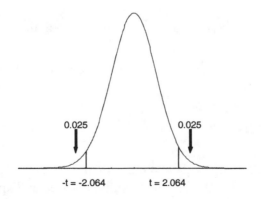

Statistical Inference for μ

The same conclusion is reached by comparing the p-value to the level of significance using the rule (6.14). Because $p = 0.0194 \leq \alpha = 0.05$ we reject H_0. (Note: for the same test if $\alpha = 0.01$, we would not reject H_0.)

Notice that when comparing the test statistic to a critical value, actual t statistics are compared, whereas when comparing the p-value to the level of significance, areas in the tails of the distribution are compared.

For the one sample tests of means (H_0: $\mu = \mu_0$ versus H_1: $\mu > \mu_0$, H_1: $\mu < \mu_0$, or H_1: $\mu \neq \mu_0$) SAS provides two-sided p-values. If a two-sided test is desired then the rule given in (6.14) is applied directly to draw a conclusion. However, if a one-sided test is desired the rule must be modified as follows:

$$\text{Reject } H_0 \text{ if } (p\text{-value}/2) \leq \alpha \qquad (6.15)$$

where α is the one-sided level of significance selected by the investigator.

For example, suppose we wish to perform the following upper-tailed test at $\alpha = 0.05$:

$$H_0: \mu = 30$$
$$H_1: \mu > 30.$$

SAS produces $p = 0.0197$ (shown above). Using rule (6.15) (i.e., Reject Ho if (p-value/2) $\leq \alpha$), we would reject Ho since $0.0097 \leq 0.05$.

Computing P-Values by Hand

Example 6.8 Revisited: In Example 6.8 we ran the following test at a 5% level of significance.

$$H_0: \mu = 130$$
$$H_1: \mu > 130$$

The decision rule we used was given by:

Reject H_0 if $Z \geq 1.645$
Do Not Reject H_0 if $Z < 1.645$

We computed a test statistic of $Z = \dfrac{\overline{X} - \mu}{\dfrac{\sigma}{\sqrt{n}}} = \dfrac{135 - 130}{\dfrac{15}{\sqrt{108}}} = 3.46$ and rejected H_0. In this example we could have selected a smaller level of significance and still reached the same conclusion. The following table displays various levels of significance, α, and their associated critical values for upper tailed Z tests. The boldface entry is the level of significance, 0.05, used in the test.

α	Z
0.0001	3.791
0.001	3.090
0.005	2.576
0.01	2.326
0.025	1.960
0.05	**1.645**

upper tail

We ran the test at $\alpha = 0.05$ and rejected H_0 because $3.46 \geq 1.645$. The p-value is the smallest level of significance α where we still reject H_0. Using the table above, we examine each level of significance smaller than 0.05 to determine if we could still reject H_0 at that level. For example, at $\alpha = 0.025$ we still reject H_0 because $3.46 \geq 1.960$. At $\alpha = 0.01$ we also reject H_0 because $3.46 \geq 2.326$, at $\alpha = 0.005$ we reject H_0 because $3.46 \geq 2.576$, at $\alpha = 0.001$ we reject H_0 because $3.46 \geq 3.090$, but at $\alpha = 0.0001$ we cannot reject H_0 because $3.46 < 3.791$. Therefore the smallest level of significance where we still reject H_0 is 0.001. The exact significance of this data, or the p-value is 0.001. If we run this analysis using SAS, SAS produces an exact p-value. Our hand computations produce only an approximate value (in fact the exact p-value is between 0.0001 and 0.001). To reflect the idea that this is an approximate p-value, sometimes the value is reported as p<0.001.

6.3.3   Power and Sample Size Determination

In hypothesis testing there are 2 types of errors that can be committed, a Type I error (i.e., reject H_0 when H_0 is true), or a Type II error (i.e., do not reject H_0 when H_0 is false). In Section 6.3.1 we introduced α = P(Type I Error) = P(Reject $H_0|H_0$ true) and β = P(Type II Error) = P(Do not reject $H_0|H_0$ false). In each test of hypothesis we specify α, purposely choosing small values (e.g., 0.01, 0.02, 0.05 or 0.10) so that the P(Type I error) is controlled. The probability of a Type II error, β, is difficult to control as it depends on several factors. In fact, one of the factors on which β

depends is α: β decreases as α increases. Therefore one must weigh the choice of a lower P(Type II error) which is desirable, against a higher level of significance which is undesirable.

In hypothesis testing, we are concerned with the *power* of a test, defined as 1-β. The power of a test is defined, in words, as its ability to "detect" or reject a false null hypothesis. Specifically, power is defined as:

$$\text{Power} = 1 - \beta = P(\text{Reject } H_0 | H_0 \text{ False}) \tag{6.16}$$

As power increases, β decreases, resulting in a better test. Power (and therefore β) is a complicated function of 3 components:

      i)   n = sample size
      ii) α = level of significance = P(Type I error)
      iii)   ES = the Effect Size = the standardized difference in means specified
                  under H_0 and H_1

The power of a particular test is higher (or better) with a larger sample size, a larger level of significance and relative to a larger effect size. In this section we introduce the concept of statistical power as it applies to the one sample test of hypothesis about μ and present a simple application.

Suppose we are interested in the following test.

      H_0: $\mu=100$
      H_1: $\mu > 100$ $\alpha=0.05$.

To conduct the test we select a random sample of subjects from the population of interest and analyze summary statistics, in particular $\overline{X}$. Under the null hypothesis (i.e., if $\mu=100$), the distribution of sample means is as follows.

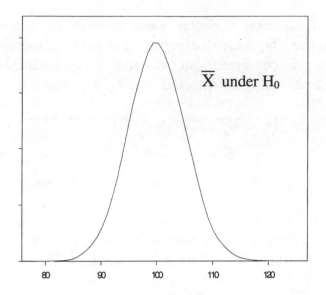

Suppose we want to determine the power of the test if the true mean is 110. The following displays the distributions of the sample mean under the null and alternative hypotheses.

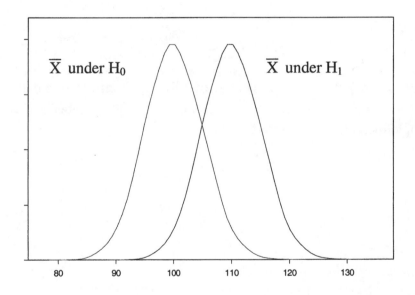

Statistical Inference for μ

We now add α=P(Type I error) = P(Reject H_0|H_0 true) and β=P(Type II error) = P(Do not reject H_0|H_0 false), and power = 1-β=P(Reject H_0|H_0 false) to the figure showing the distributions of the sample mean under H_0 (μ=100) and H_1 (μ=110).

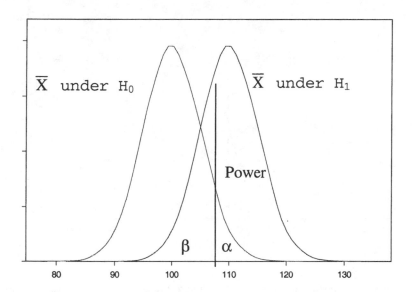

In the above, α is the are under the leftmost curve (H_0 true) where we reject H_0, β is the area under the rightmost curve (H_1 true) where we do not reject H_0, and power is 1-β or the remaining area under the rightmost curve (H_1 true).

Below we provide the formulas to compute the power of the test. Before moving to the computations, we investigate the impact of each component on power. What happens to the power if we increase our level of significance? Suppose in the above that α=0.05. What happens to the power if we increase α to 0.10?

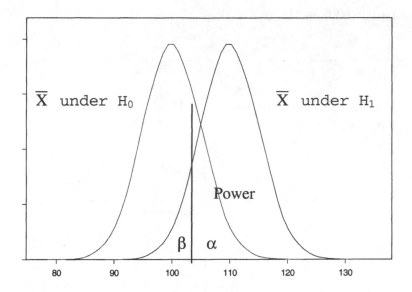

Notice that power increases when α increases. A better test is a more powerful test, however, it is unlikely that one would want to increase α so as to ensure a higher power in a test. Again, power is related to three components: sample size, α and the effect size. The most appropriate component to modify is the sample size, if possible. A larger sample size ensures a more powerful test. We will illustrate this in the next example. What is the impact of effect size on power? Suppose the mean under the alternative hypothesis is not 110 but instead 120. What happens to power? This situation is illustrated below. In the following we reset α to 0.05.

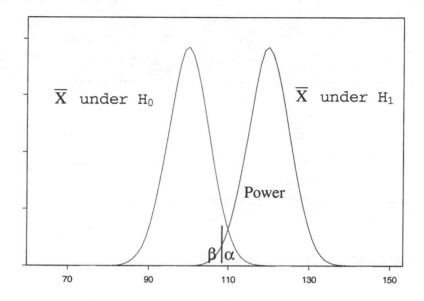

Notice that the power increases as the difference in means under H_0 and H_1 increases. This difference is called the detectable difference and power increases as the alternative diverges from the null.

We now illustrate the calculation of power (or the determination of the area under the curve when H_1 is true). The calculations require the specification of the sample size, the level of significance and the detectable difference and the use of the standard normal distribution table (Table 2). The following formula is used to determine power of a two-sided test for μ.

$$\text{Power} = P\left(Z > Z_{1-\frac{\alpha}{2}} - \frac{|\mu_0 - \mu_1|}{\sigma/\sqrt{n}}\right) \tag{6.17}$$

where μ_0 is the mean under H_0,
       μ_1 is the mean under H_1,
       σ is the standard deviation of the characteristic of interest,
       n is the sample size, and
       $Z_{1-\alpha/2}$ is the Z value with lower tail area $1-\alpha/2$.

Statistical Inference for μ

The following formula is used to determine power of a one-sided test for μ.

$$\text{Power} = P(Z > Z_{1-\alpha} - \frac{|\mu_0 - \mu_1|}{\sigma/\sqrt{n}}) \tag{6.18}$$

where μ_0 is the mean under H_0,

μ_1 is the mean under H_1,

σ is the standard deviation of the characteristic of interest,

n is the sample size, and

$Z_{1-\alpha}$ is the Z value with lower tail area $1-\alpha$.

Example 6.13. Suppose we wish to conduct the following test.

$H_0: \mu = 80$

$H_1: \mu \neq 80$ $\qquad \alpha = 0.05.$

Find the power of the test if $\mu = 85$. Suppose that n=20 and $\sigma = 9.5$. Using (6.17),

$$\text{Power} = P(Z > Z_{1-\frac{\alpha}{2}} - \frac{|\mu_0 - \mu_1|}{\sigma/\sqrt{n}}) = \text{Power} = P(Z > 1.96 - \frac{|80 - 85|}{9.5/\sqrt{20}}) = P(Z > 1.96 - 2.36)$$

$$= P(Z > -0.40) = 1 - 0.3446 = 0.6554.$$

There is a 65% probability that this test will detect a difference of 5 units in means with n=20 at $\alpha = 0.05$. The power is the same if $\mu = 75$.

What if the sample size is increased to n=50?

$$\text{Power} = P(Z > Z_{1-\frac{\alpha}{2}} - \frac{|\mu_0 - \mu_1|}{\sigma/\sqrt{n}}) = \text{Power} = P(Z > 1.96 - \frac{|80 - 85|}{9.5/\sqrt{50}}) = P(Z > 1.96 - 3.72)$$

$$=P(Z > -1.76) = 1 - 0.0392 = 0.9608.$$

There is a 96% probability that this test will detect a difference of 5 units in means (in either direction) with n=50 at α=0.05.

Suppose we go back to n=20 and consider a 3 point difference in means (i.e., μ_1=83):

$$\text{Power} = P(Z > Z_{1-\frac{\alpha}{2}} - \frac{|\mu_0 - \mu_1|}{\sigma/\sqrt{n}}) = \text{Power} = P(Z > 1.96 - \frac{|80-83|}{9.5/\sqrt{20}}) = P(Z > 1.96 - 1.41)$$

$$=P(Z > 0.55) = 1 - 0.7088 = 0.2912.$$

There is only a 29% probability that this test will detect a difference of 3 units in means (in either direction) with n=20 at α=0.05. This is a small difference in means and with a small sample size we are not very likely to detect it. A larger sample would be required to ensure a higher probability of detecting such a difference. In the following we describe techniques for determining sample size required to ensure a specified power.

In many applications the number of subjects that can be sampled depends on financial and/or time constraints. In other cases the investigators can choose a sample large enough to ensure a certain level of power. As described in Section 6.2.3 techniques from experimental design can be employed to determine the number of subjects required to achieve a certain level of power prior to mounting the study.

The sample size required to ensure a specific level of power in a <u>two-sided test</u> is given below:

$$n = \left(\frac{Z_{1-\frac{\alpha}{2}} + Z_{1-\beta}}{ES}\right)^2 \qquad (6.19)$$

where $Z_{1-\alpha/2}$ is the value from the standard normal distribution with lower tail area equal to $1-\alpha/2$,
$Z_{1-\beta}$ is the value from the standard normal distribution with lower tail area equal to $1-\beta$, and
ES is the effect size defined as the standardized difference in means under the null and alternative hypotheses (6.20):

$$ES = \frac{|(\mu_1 - \mu_0)|}{\sigma} \qquad (6.20)$$

where μ_0 is the mean specified in H_0,

μ_1 is the mean specified in H_1, and

σ is the population standard deviation of the characteristic under investigation.

The sample size required to ensure a specific level of power in a <u>one-sided test</u> is given below:

$$n = \left(\frac{Z_{1-\alpha} + Z_{1-\beta}}{ES} \right)^2 \qquad (6.21)$$

where $Z_{1-\alpha}$ is the value from the standard normal distribution with lower tail area equal to $1-\alpha$,

$Z_{1-\beta}$ is the value from the standard normal distribution with lower tail area equal to $1-\beta$, and

ES is the effect size (6.20).

In order to implement formulas (6.19) and (6.21) to compute the sample size required to ensure a certain level of power to detect a specified difference in means, several inputs are required. First, we must specify the level of significance α. This is usually straightforward as $\alpha=0.05$ is considered standard. Second, we must specify β = P(Type II error). In many experimental design applications, β is set to 0.20 which reflects 80% power. With $\beta = 0.20$ there is an 80% chance of rejecting a false null hypothesis relative to a specific effect size. (In some instances, β is set to 0.10 which reflects 90% power.) The third input, the effect size, is the most difficult to specify. The effects size reflects the magnitude of a clinically meaningful difference in means. The effect size is best determined by an expert in the substantive area under investigation. In order to compute the effect size we also need to quantify the variation in the characteristic under investigation (σ). If no such value exists, the same options outlined in section 6.2.3 can be used to determine a reasonable approximation. The following example illustrates the computations.

Example 6.14. Suppose we wish to conduct the following two-sided test at a 5% level of significance.

$$H_0: \mu = 100$$
$$H_1: \mu \neq 100$$

Suppose that a difference of 5 units in the mean score is considered a clinically meaningful difference. If the true mean is less than 95 or greater than 105 we do not want to erroneously accept the null hypothesis. How many subjects would be required to ensure that the probability of detecting a 5 unit difference is 80% (i.e., power = 0.80) ? For this example suppose we know that $\sigma = 9.5$.

Because we wish to conduct a two-sided test, formula (6.19) is appropriate. First, we compute the effect size (6.20) by substituting the mean specified in H_0 ($\mu_0=100$), the mean we wish to detect when H_0 is false (here we could use either $\mu_1=95$ or $\mu_1=105$, both produce the same result), and the standard deviation $\sigma = 9.5$:

$$ES = \frac{|(105 - 100)|}{9.5} = 0.526$$

We now substitute the effect size into formula (6.19):

$$n = \left(\frac{Z_{0.975} + Z_{0.80}}{0.526} \right)^2$$

We now use the standard normal distribution table (Table 2) to determine $Z_{0.975}$ and $Z_{0.80}$. By definition, $Z_{0.975}$ is the Z value which holds 0.975 below it (or 0.025 above it in the upper tail) and $Z_{0.80}$ is the Z value which holds 0.80 below it in the standard normal distribution . These values are shown in the figure below.

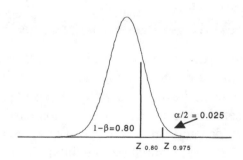

Statistical Inference for μ

Using Table 2 and the techniques we described in Chapter 4 we determine $Z_{0.975} = 1.96$ and $Z_{0.80} = 0.84$. We now substitute these values:

$$n = \left(\frac{1.96 + 0.84}{0.526} \right)^2 = (5.323)^2 = 28.33$$

A sample of size 29 (again we always round up to the next integer) will ensure that power = 0.80 (or 80%, $\beta = 0.20$) to detect a 5 point difference in means. If the true mean is at least 5 units different from 100, there is an 80% chance that the test will lead to rejection of the null hypothesis H_0: $\mu = 100$.

How many subjects would be required to ensure a power of 80% to detect a difference of 3 units ? This and other scenarios can be investigated by substituting into the formulas shown above or by using SAS.

SAS Example 6.14. The following output was generated using SAS to determine the sample size required to ensure a specified power (we considered scenarios with 80% and 90% power) and differences in means of 5 and 3 units. A brief interpretation appears after the output.

SAS Output for Example 6.14

Obs	alpha	beta	mu0	mu1	sigma	power	es	n_2	n_1
1	0.05	0.2	100	105	9.5	0.8	0.52632	29	23
2	0.05	0.1	100	105	9.5	0.9	0.52632	38	31
3	0.05	0.2	100	103	9.5	0.8	0.31579	79	62
4	0.05	0.1	100	103	9.5	0.9	0.31579	106	86

Interpretation of SAS Output for Example 6.14

There is no SAS procedure specifically designed to determine the number of subjects required to detect a specific effect in the mean of a population. Similar to the approach taken in SAS EXAMPLE 2, SAS can, however, be used to program appropriate formulas. Once the formulas are implemented, users can evaluate different scenarios easily. In the example shown above, four scenarios are considered (denoted Obs 1-4, respectively). Scenario 1 corresponds to the situation presented in Example 6.14. Four variables are input into the program; the level of significance (alpha), the power (power), the mean under the null hypothesis (mu0), the mean under the alternative hypothesis (mu1) and the standard deviation (sigma). Several variables are created in the program and the values of all variables are printed in the output. A description of the variables and an interpretation of results follows.

In scenario 1 (Obs=1), the level of significance (alpha) is set at 0.05 (5%), the probability of Type II error, β, is computed to be 0.20 (20%), the mean under the null hypothesis (mu0) was specified as 100, the mean under the alternative hypothesis was specified as 105 (mu1), the standard deviation (sigma) was specified at 9.5, and the power (power) was specified at 0.80 (80%). The effect size (es) was computed by dividing the absolute value of the difference in means under the null and alternative hypothesis by the standard deviation. In scenario 1 (OBS=1), the effect size is 0.52632. Twenty nine subjects (n_2) are required to ensure that the probability of detecting a 5 unit difference in means is 80% (i.e., power = 0.80), with a <u>two sided</u> level of significance of 5%. Twenty three subjects (n_1) are required to ensure that the probability of detecting a 5 unit difference in means is 80%, with a <u>one sided</u> level of significance of 5%.

In scenario 2 (Obs=2) we increase the power to 0.90 (90%). Thirty eight subjects (n_2) are required to ensure that the probability of detecting a 5 unit difference in means is 90%, with a <u>two sided</u> level of significance of 5%. Thirty one subjects (n_1) are required to ensure that the probability of detecting a 5 unit difference in means is 90%, with a <u>one sided</u> level of significance of 5%.

In scenario 3 (Obs=3) we input a power of 0.80 (80%) but decrease the mean under the alternative hypothesis (mu1) to 103, which decreases the effect size (ES) to 0.31579. Seventy nine subjects (n_2) are required to ensure that the probability of detecting a 3 unit difference in means is 80%, with a <u>two sided</u> test and 5% level of significance. Sixty two subjects (n_1) are required to ensure that the probability of detecting a 3 unit difference in means is 80%, with a <u>one sided</u> test and 5% level of significance.

In scenario 4 (Obs=4) we input a power of 0.90 (80%) and consider the mean under the alternative hypothesis (mu1) as 83. One hundred and six subjects (n_2) are required to ensure that the probability of detecting a 3 unit difference in means is 90%, with a <u>two sided</u> test and 5% level of significance. Eighty six subjects (n_1) are required to ensure that the probability of detecting a 3 unit difference in means is 90%, with a <u>one sided</u> test and 5% level of significance.

Notice that more subjects are required to ensure a higher power. More subjects are required to detect a smaller effect size. These results would be weighed against practical constraints to determine the most appropriate sample size for the application.

(handwritten, top right)

$$1). \quad \bar{X} \pm Z_{1-\frac{\alpha}{2}} \frac{\sigma}{\sqrt{n}} \qquad \sigma \text{ known}$$

$$3). \quad \bar{x} \pm t_{1-\frac{\alpha}{2}} \frac{s}{\sqrt{n}}, \quad df = n-1$$
$$\sigma \text{ unknown } \& \quad n < 30$$

6.4 Key Formulas

(handwritten) point estimate $\pm$ margin of error
$(\bar{X})$ (E)

APPLICATION	NOTATION/FORMULA	DESCRIPTION
Confidence Interval Estimate for μ	$\tilde{\mu} = \bar{x}$, $\quad$ 2). $\bar{X} \pm Z_{1-\frac{\alpha}{2}} \frac{s}{\sqrt{n}}$ $\quad$ *(hw)* σ unknown & $n \gtrsim 30$	See Table 6.2 for alternate formulas (Find Z in Table 6.1)
Find n to Estimate μ	$n = \left(\dfrac{Z_{1-\frac{\alpha}{2}}\sigma}{E}\right)^2$, $\quad$ *(hw)* $E = Z_{1-\frac{\alpha}{2}} \frac{\sigma}{\sqrt{n}}$	Sample size to ensure margin of error E with confidence level reflected in Z
Test Statistic for H_0: $\mu = \mu_0$	*(hw)* 2). $\quad Z = \dfrac{\bar{X} - \mu_0}{s/\sqrt{n}}$ $\quad$ *(hw)* σ unknown & $n \gtrsim 30$	See Table 6.3 for hypothesis testing procedure, Table 6.4 for critical values of Z and Table 6.5 for alternate formulas
Find power for test of H_0: $\mu = \mu_0$	$\text{Power} = P\left(Z > Z_{1-\frac{\alpha}{2}} - \dfrac{\lvert \mu_0 - \mu_1 \rvert}{\sigma/\sqrt{n}}\right)$	Power of two-sided test for mean
Find n to Test H_0: $\mu = \mu_0$	$n = \left(\dfrac{Z_{1-\frac{\alpha}{2}} Z_{1-\beta}}{ES}\right)^2$, where $ES = \dfrac{\lvert \mu_1 - \mu_0 \rvert}{\sigma}$	Sample size to detect ES with power $= 1 - \beta$
p-value	Reject H_0 if $p \leq \alpha$	Exact significance of the data

(handwritten, bottom)

$$1) \quad Z = \dfrac{\bar{x} - \mu_0}{\frac{\sigma}{\sqrt{n}}} \qquad \sigma \text{ known}$$

$$3). \quad t = \dfrac{\bar{x} - \mu_0}{\frac{s}{\sqrt{n}}}, \quad df = n-1 \qquad \sigma \text{ unknown } \& \quad n < 30$$

6.5   Statistical Computing

Following are the SAS programs which were used to generate the confidence interval, perform the test of hypothesis and compute required sample sizes in the examples in Sections 6.2 and 6.3. The SAS procedures used and brief descriptions of their use are noted in the header to each example. Notes are provided to the right of the SAS programs (*in italics*) for orientation purposes and are not part of the programs. In addition, there are blank lines in the programs that follow which are solely to accommodate the notes. Blank lines and spaces can be used throughout SAS programs to enhance readability. A summary of the SAS procedures used in the examples is provided at the end of this section.

Confidence Intervals for μ

SAS EXAMPLE 6.2   Age at Diagnosis of Hypertension

We randomly select 12 subjects with diagnosed hypertension and record the age at which they were diagnosed. Ages, measured in years, are recorded on each subject and are listed below. Generate a 95% confidence interval for the mean at which patients with hypertension are diagnosed using SAS.

| 32.8 | 40.0 | 41.0 | 42.0 | 45.5 | 47.0 | 48.5 | 50.0 | 51.0 | 52.0 | 54.0 | 59.2 |

Program Code

Code	Description
options ps=62 ls=80;	*Formats the output page to 62 lines in length and 80 columns in width*
data in;	*Beginning of Data Step*
input age_dx;	*Inputs variable **age_dx***
label age_dx='Age at diagnosis of hypertension';	*Attaches descriptive label to variable **age_dx***
cards;	*Beginning of Raw Data section.*
32.8	*actual observations*
40.0	
41.0	
42.0	
45.5	
47.0	
48.5	
50.0	
51.0	
52.0	
54.0	
59.2	
run;	
proc means n mean std min max alpha=0.05 clm;	*Procedure call. Proc Means generates summary statistics for continuous variables. Certain statistics are requested along with a 95% confidence interval by the 'clm' option (See (iii) Interpretation)*
var age_dx;	*Specification of variable **age_dx***
run;	*End of procedure section.*

Determine the Number of Subjects Required to Generate a Confidence Interval for μ
SAS EXAMPLE 6.7 Time to Travel Between Hospital Departments

Compute the number of subjects required to generate a 95% confidence interval estimate of the mean time it takes patients to travel from one department to another on hospital grounds. We wish to estimate the mean within 5 minutes of the true value with 95% confidence. Assume σ=17 (based on a pilot study).

Program Code

```
options ps=62 ls=80;
```
Formats the output page to 55 lines in length and 80 columns in width

```
data in;
```
Beginning of Data Step
```
  input c_level e sigma;
```
Inputs 3 variables c_level,   e   and sigma.

```
z=-probit((1-c_level)/2);
```
Determines the (positive) value from the standard normal distribution (z) with tail area (1-c_level)/2

```
tempn=(z*sigma/e)**2;
```
Creates a temporary variable, called tempn, determined by formula (6.8).

```
n=ceil(tempn);
```
Computes a variable n using the ceil function which computes the smallest integer greater than tempn.

```
/*  Input the following information (required)
   c_level: Confidence Level: range 0.0 to 1.0 (e.g., 0.95),
   e: Margin of Error, and
```
Beginning of comment section

```
   sigma: Standard Deviation    */
```
End of comment
```
cards;
```
Beginning of Raw Data section
```
0.95  1 17
```
actual observations
```
0.95  2 17
```
values of c_level, e and sigma
```
0.95  3 17
```
on each line
```
0.95  4 17
0.95  5 17
0.95 10 17
run;
proc print;
```
Procedure call. Print to display input and computed variables.
```
  var c_level z sigma e n;
run;
```
End of procedure section.

Tests of Hypothesis About μ

SAS EXAMPLE 6.10 Starting Salaries

In a managed care organization, the mean starting salary for males in entry level, non-clinical positions is $29,500. Starting salaries for a random sample of 10 females in similar positions are given below (in $000s).

| 32 | 27 | 31 | 27 | 26 | 26 | 30 | 22 | 25 | 36 |

Is the mean starting salary for females significantly different from $29,500 ? Use a 5% level of significance.

Program Code

options ps=62 ls=80;	*Formats the output page to 62 lines in length and 80 columns in width*
data in;	*Beginning of Data Step.*
input salary;	*Inputs variable salary.*
teststat=salary-29.5;	*Creates a new variable, called teststat, by subtracting 29.5 (m_0) from each salary.*
label salary='Annual Salary in $000s';	*Attaches a descriptive label to salary.*
cards;	*Beginning of Raw Data section.*
32	*actual observations*
27	
31	
27	
26	
26	
30	
22	
25	
36	
run;	
proc means n mean std stderr t prt;	*Procedure call. Proc Means generates summary statistics for continuous variables. Certain statistics are requested- t statistics and p-values for conducting the tests of hypothesis (See (iii) Interpretation)*
var salary teststat;	*Specification of variables.*
run;	*End of procedure section.*

Determine the Number of Subjects Required to Detect a Specific Effect Size in a Test of Hypothesis About μ

SAS EXAMPLE 6.14 Sample Size Requirements

We wish to conduct the following test at the 5% level of significance:

$$H_0: \mu = 100 \quad \text{vs.} \quad H_1: \mu \neq 100$$

How many subjects would be required to ensure that the probability of detecting a 5 (or a 3) unit difference is 80% (i.e., power = 0.80) ? Also consider scenarios with 90% power and assume that $\sigma = 9.5$.

Program Code

options ps=62 ls=80;	*Formats the output page to 62 lines in length and 80 columns in width*
data in;	*Beginning of Data Step.*
input alpha power mu0 mu1 sigma;	*Inputs 5 variables **alpha, power, mu0, mu1** and **sigma**.*
z_alpha2=probit(1-alpha/2);	*Determines the value from the standard normal distribution with tail area 1-**alpha**/2 (See $Z_{a/2}$ above)*
z_alpha1=probit(1-alpha);	*Determines the value from the standard normal distribution with tail area 1-**alpha** (See Z_a above)*
beta=1-power;	*Computes **beta**.*
z_beta=probit(beta);	*Determines the value from the standard normal distribution with tail area **beta** (See Z_b above)*
es=abs(mu1-mu0)/sigma;	*Computes the effect size (**es**)*
tempn_2=((z_alpha2+z_beta)/es)**2;	*Creates a temporary variable, called **tempn_2**, determined by formula (6.14).*
tempn_1=((z_alpha1+z_beta)/es)**2;	*Creates a temporary variable, called **tempn_1**, determined by formula (6.16).*

n_2=ceil(tempn_2);	*Computes a variable **n_2** using the ceil function which computes the smallest integer greater than **tempn_2**.*
n_1=ceil(tempn_1);	*Computes a variable **n_1** using the ceil function which computes the smallest integer greater than **tempn_1**.*

```
/*
    Input the following information (required)
    alpha: Level of Significance: range 0.0 to 1.0 (e.g., 0.05),
    power: Power: range 0.0 to 1.0 (e.g., 0.80), and
    mu0: Mean Under the Null Hypothesis,
    mu1: Mean Under the Alternative Hypothesis, and
    sigma: Standard Deviation
*
/
cards;
0.05 0.80 100 105 9.5
0.05 0.90 100 105 9.5
0.05 0.80 100 103 9.5
0.05 0.90 100 103 9.5
run;

proc print;
   var alpha beta mu0 mu1 sigma power es n_2 n_1;
run;
```

Beginning of comment section

End of comment section
Beginning of Raw Data section
actual observations

Procedure call. Print to display computed variables.
End of procedure section.

Summary of SAS Procedures and Functions

The SAS Means Procedure is used to generate confidence intervals for μ and to run test of hypothesis about μ. The specific options to generate a confidence interval and to perform a test of hypothesis are shown in italics below. Users should refer to the examples in this section for complete descriptions of the procedure and specific options. A general description of the procedure and options is provided in the table below.

Procedure	Sample Procedure Call	Description
proc means	proc means *n mean std min max alpha=0.05 clm*; var x;	Generates a 100(1-*alpha*)% confidence interval for μ
proc means	proc means *n mean std stderr t prt*; var teststat;	Conducts a one sample test of hypothesis (H_0: $\mu=0$ vs. H_1: $\mu \neq 0$)

The following SAS function was also used in this section in the programs to determine sample size requirements.

Function	Sample Call Statement	Description
Probit	Z_alpha = probit(*alpha*);	Determines the value from the standard normal distribution with lower tail area equal to *alpha*.

6.6 Problems

Estimation

1. A hospital researcher wishes to estimate the mean weight of full-term newborns. A random sample of 13 full-term newborns had a mean birthweight of 7.3 pounds with a standard deviation of 2.2 pounds. Compute a 95% confidence interval for the mean weight of all full-term newborns.

2. A newspaper article estimated that the mean cost for preventive dental care was $165 per year. They indicated that the estimate was based on a sample of 100 people and that the margin of error was 2.8. Assuming a 95% confidence level was used, calculate the value of the standard error.

3. A manufacturer of tobacco products plans to market a new brand of cigarette. The regulatory commission needs to know the mean tar content for the new brand. A random sample of 25 of the new brand of cigarettes is analyzed, which have a mean tar content of 10.98 milligrams with a standard deviation of 0.604 milligrams. Construct a 95% confidence interval for the mean tar content for the new brand.

4. Suppose we wish to design a study to investigate the effects of loud music on teenagers abilities to concentrate. We know from previous studies that the standard deviation of time to complete this task is 3.4 minutes. How many subjects would be required to ensure with 90% confidence that the generated estimate is within 1 minute of the true mean time required ?

5. The administrators of an Emergency Room (ER) at a local hospital wish to estimate the mean number of overtime hours worked by employees during December, typically a very busy month. A random sample of 15 employees worked an average of 27 overtime hours with a standard deviation of 4.2 hours. Estimate the mean number of overtime hours among all employees of the ER during December using a 90% confidence interval.

6.   A marketing research firm has put together a health survey which it would like to administer to individuals in malls across the state. They need to estimate the mean time it takes to complete the survey so that recruiters can give potential participants an idea of their time commitments. A random sample of 40 individuals are administered the survey. The mean time to complete the survey is 14.6 minutes with a standard deviation of 2.8 minutes. Estimate the mean time to complete the survey using a 95% confidence interval.

7.   Suppose we need to estimate the mean blood glucose levels of diabetic patients following a specific treatment regimen. To be useful to clinicians, the estimate must be within 5 units with 95% confidence. If the standard deviation in blood glucose levels is known to be $\sigma=15.6$, how many subjects be required to produce such an estimate ?

8.   A local health maintenance organization (HMO) wishes to estimate the mean age of its members for marketing purposes.

  a)   How large a sample would be required to generate an estimate that is within 5 years of the true mean with 90% confidence? Assume $\sigma = 8.6$.

  b)   Suppose this HMO has a budget which allows them to randomly sample 50 individuals in their HMO for their marketing survey. Suppose the mean age in the sample is 58.2 with a standard deviation of 7.4. Estimate the mean age of all members using a 90% confidence interval.

9.   Adherence to antiretroviral therapy is critical in patients with HIV disease. Suppose we wish to estimate the mean number of doses missed per month among HIV patients currently on antiretroviral therapy. A random sample of 25 patients on antiretroviral therapy is selected and each patient reports the number of doses missed over the previous month. The mean number of doses missed is 4.7 with a standard deviation of 1.2. Estimate the mean number of doses missed per month among all HIV patients using a 95% confidence interval.

10.   The following table appeared in a recent article which summarized a study investigating the relationship between body mass index (defined as the ratio of weight in kilograms to height in meters squared) and dietary habits:

Characteristics of Study Sample (n=100)	Mean $\pm$ SE
Body Mass Index	25 $\pm$ 0.67
Total Calories Per Day	1875 $\pm$ 36.4
Total Fat Grams Per Day	21 $\pm$ 0.31

The study summarized above involved subjects 18 years of age or older. Suppose we wish to design a study to investigate the relationship between body mass index and dietary habits among individuals aged 15-17 years.

a) How many subjects would be required to ensure that the estimate of the mean body mass index among 15-17 year olds is within 2 units of the true mean with 95% confidence ?

b) How many subjects would be required to ensure that the estimate of the mean number of total calories per day among 15-17 year olds is within 50 units of the true mean with 95% confidence ?

c) What is the smallest sample size that will ensure that both criteria (a) & (b) are satisfied?

11. A study is conducted to assess the extent to which patients who had coronary artery bypass surgery were maintaining their prescribed exercise programs. The following data reflect the numbers of times patients reported exercising over the previous month (4 weeks). For the purposes of this study, exercise was defined as moderate physical activity lasting at least 20 minutes in duration.

14	11	8	6	5	3
6	13	12	8	1	4

a) Construct a 95% confidence interval for the mean number of times patients exercise following surgery.

b) How many subjects would have been required in (a) to ensure that the margin of error was no more than 4 with 95% confidence?

Statistical Inference for μ

12. A community health center wants to assess the alcohol consumption of its patients. A random sample of 100 patients are selected to participate in the assessment. Each patient is surveyed with respect to his/her alcohol consumption. One survey item measures the number of alcoholic drinks consumed per drinking occasion. The mean number of alcoholic drinks consumed per occasion is 3.2 with a standard deviation of 2.6. Construct a 95% confidence interval for the mean number of alcoholic drinks consumed per occasion among all patients of the community health center.

13. The following data were presented summarizing the background characteristics of 50 participants in a research study evaluating HIV medications.

Patient Characteristics	Mean (SD)/%
Age, years	37.6 (6.8)
% Male	75.1%
Education, years	13.6 (2.4)
Annual Income, $000s	31.4 (5.2)
CD4 Cell Count, cells/μl	376 (94)

a) Construct a 95% confidence interval estimate of the mean CD4 cell count.
b) How many subjects would be required to estimate the mean CD4 cell count within 15 cells of the true value with 95% confidence?

14. A clinical study is conducted to estimate the proportion of patients who relapse following treatment for drug addiction. Twenty of 85 patients relapsed in the study. The mean number of days until relapse was 31 with a standard deviation of 3.8. Estimate the mean number of days to relapse among patients who do relapse using a 95% confidence interval.

15. We wish to design a study to estimate the mean weight loss (in pounds) among participants in a study of a newly developed weight loss drug. How many subjects would be required to ensure that the estimate is within 2 pounds of the true mean with 95% confidence ?

A published study of a similar drug reported the following:

	Number of Subjects	Mean	Variance	Median	Range
Weight Loss (pounds)	125	53	14.4	6.1	0-15

16. A recent article appeared in *Medical Care* describing the health status of HIV-infected patients. Based on the Centers for Disease Control classification algorithm, patients were classified into one of three HIV disease stages: Asymptomatic, Symptomatic or AIDS. Demographic characteristics of the study sample were provided in the article and are summarized below:

Demographic Characteristics by HIV Disease Stage

	Asymptomatic	Symptomatic	AIDS
n	40	49	71
Mean age ($\pm$ SE)	37.7 ± 1.3	36.1 ± 1.2	36.6 ± 1.0
Male (%)	70	61	68

a) Is there a relationship between disease stage and age ? If yes, what is the nature of the relationship ? Explain (briefly).

b) Compute a 90% confidence interval for the mean age of Asymptomatic patients.

17. The following data were collected from a random sample of 10 asthmatic children enrolled in a research study and reflect the number of days each child missed school during the past 3 months:

6   12   14   3   2   4   7   8   10   6

Statistical Inference for μ

a) Compute the sample mean
b) Compute the sample standard deviation
c) Construct a 95% confidence interval for the mean number of days asthmatic children miss from school during a 3 month period.
d) How many children would be required in (c) to ensure that the margin of error in the estimate is 1 day with 95% confidence ? (Use (b) to estimate σ.)

18. A dentist has recently started a new private practice and wants to estimate how long patients are waiting, on average, in the waiting room before their appointments. He randomly selects 50 patients for observation and records how many minutes each waits in minutes. The mean waiting time is 18 minutes with a standard deviation of 3.6 minutes.

a) Estimate the mean waiting time among all patients using a 95% confidence interval.
b) Based on the interval, could you conclude that the mean waiting time is significantly different from 20 minutes? Justify briefly.

19. Consider the following table which appeared in a recent publication:

Table 1. Description of study sample (n=85)

Characteristic	Mean (SD) or %
Age (years)	38 (5.7)
Gender: % Female	46%
Race: % White	86%
Education (years)	11.5 (4.3)

a) What is the standard error in the ages ?
b) Construct a 90% confidence interval estimate for the mean age in the study population.
c) How many subjects would be required to ensure that the margin of error in the confidence interval of (b) is no more than 2 units with 90% confidence?
d) Suppose we wish to estimate the mean educational level using a 95% confidence interval. How many subjects would be required to ensure that the margin of error is no more than 1 year?
e) How many subjects should be enrolled in a study to satisfy the requirements of BOTH (c) and (d)?

Statistical Inference for μ

20. Currently, there are highly effective medications available for patients infected with HIV. However, in order to achieve the maximum benefits of therapy, patients must be extremely adherent with respect to pill taking. Suppose we wish to estimate medication adherence among HIV infected persons new to medication therapy (initiated within the past 3 months). Adherence is measured as the percent of prescribed doses taken over the past month (e.g., 100% = perfect adherence - all doses taken, 50%=half of prescribed doses taken). A random sample of 75 HIV infected patients new to medication therapy agree to participate in the study. Each reports their medication regimen and the doses they took over the past month. The mean percent adherence is 78% with a standard deviation of 7.2%. Construct a 95% confidence interval estimate of the mean percent adherence for all HIV infected patients new to medication therapy.

Hypothesis Testing

21. The mean lung capacity for non-smoking males aged 50 is 2 liters. An investigator wants to examine if the mean lung capacities are significantly lower among former smokers of similar backgrounds (i.e., males aged 50 who smoked in the past and are not currently smokers). A random sample of 60 former smokers is selected. Their lung capacities have a mean of 1.8 liters with a standard deviation of 0.27 liters. Run the appropriate test at a 5% level of significance.

22. Among private Universities in the United States, the mean ratio of students to professors is 35.2 (i.e., 35.2 students for each professor), with a standard deviation of 8.8.

   a) What is the probability that in a random sample of 50 private Universities that the mean student to professor ratio exceeds 38?

   b) Suppose a random sample of 50 Universities is selected and the observed mean student to professor ratio is 38. Is there evidence that the reported mean ratio actually exceeds 35.2 ? Use $\alpha = 0.05$.

23. The recommended daily allowance (RDA) of iron for adult females under the age of 51 is 18 mg. We wish to test if females under age 51 are, on average, getting less than 18 mg. A random sample of 48 females between the ages of 18 and 50 are selected. The average iron intake is 16.4 mg with a standard deviation of 4.1 mg. Run the appropriate test at the 5% level of significance.

24. If a statistical test is performed and H_0 is rejected at $\alpha = 0.01$, will it also be rejected at $\alpha = 0.05$?

25. A journal article reported that the mean hospital stay following a particular surgical procedure in 1991 was 7.1 days. A researcher feels that the mean hospital stay in 1992 should be less due to initiatives aimed at reducing health care costs. A random sample of 40 patients undergoing the same surgical procedure in 1992 had a mean length of stay of 6.85 days with a standard deviation of 7.01 days. Run the appropriate statistical test at $\alpha = 0.05$.

26. A consumer group is investigating a producer of diet meals to examine if their pre-packaged meals actually contain the advertised 6 ounces of protein in each package. Based on the following data, is there any evidence that the meals do not contain the advertised amount of protein. Run the appropriate test at a 5% level of significance.

5.1	4.9	6.0	5.1	5.7	5.5	4.9	6.1	6.0	5.8
5.2	4.8	4.7	4.2	4.9	5.5	5.6	5.8	6.0	6.1

27. An article reported that patients in care for HIV have CD4 tests every 3 months, on average. There is concern that there is a longer lag between tests at Boston Medical Center. To test the concern, a random sample of 15 patients currently in care for HIV are selected and the time between their two most recent CD4 tests is recorded. The mean time between tests is 3.9 months with a standard deviation of 0.4 months. Run the appropriate test at a 5% level of significance.

28.   A study reports that the mean systolic blood pressure for patients with a history of cardiovascular disease is 125 with a standard deviation of 15. We wish to design a study to evaluate an experimental medication for reducing blood pressure. How many subjects would be required to detect a 10 unit reduction in systolic blood pressure with 80% power? Assume that a two sided test will be run at a 5% significance level.

29.   We wish to test the hypothesis that the mean weight for females, 5'8" is 140 pounds. Assuming $\sigma = 15$, using a 5% level of significance and with $n = 36$, find the power of the test if $\mu = 150$. (Use a two-sided test of hypothesis.)

30.   We wish to run the following test: Ho: $\mu=100$ vs. H_1: $\mu \neq 100$ at $\alpha = 0.05$. If $\sigma = 10$, how large a sample would be required so that $\beta = 0.04$ if $\mu = 110$?

31.   In a normal population with $\sigma = 5$, we wish to test Ho: $\mu = 12$ vs. H_1: $\mu \neq 12$ at $\alpha = 0.05$. With a sample of 64 subjects what is the probability of rejecting Ho if $\mu = 14$? If $\mu = 9$?

32.   Results of an industry survey in the computer software field finds that the mean number of sick days taken by employees is 9.4 per year with a standard deviation of 2.7 per year. A local computer software company feels its employees take significantly fewer sick days per year. A random sample of 15 employees is selected from the local company and attendance records are reviewed. The following data represent the numbers of sick days taken by these employees over the past year:

8	10	5	0	6	9	5	15
5	4	3	2	0	4	15	

Run the appropriate test at a 5% level of significance.

Statistical Inference for μ

33.   An analysis is conducted to compare the mean GRE scores among seniors in a local university to the national average of 500. Use the following SAS output shown below to address the questions below:

```
Variable    N     Mean         Std Dev      Std Error    T        Prob>|T|
-----------------------------------------------------------------------------
GRE         250   512.0463595  86.2844894   5.4571103    93.835   0.0001
TESTSTAT    250   12.0463595   86.2844894   5.4571103    2.207    0.0282
-----------------------------------------------------------------------------
```

a)   Is the mean GRE score among the local seniors significantly different from the national average? Justify briefly (show all parts of the test).

b)   Can we say that the mean GRE score among the local seniors is significantly higher than the national average? Support your conclusion with data (show all parts of the test).

34.   An academic medical center surveyed all of its patients in 1998 to assess their satisfaction with medical care. Satisfaction was measured on a scale of 0 to 100 with higher scores indicative of more satisfaction. The mean satisfaction score in 1998 was 84.5. Several quality improvement initiatives were implemented in 1999 and the medical center is wondering whether the initiatives increased patient satisfaction. A random sample of 125 patients seeking medical care in 1999 were surveyed using the same satisfaction measure. Their mean satisfaction score was 89.2 with a standard deviation of 17.4. Is there evidence of a significant improvement in satisfaction? Run the appropriate test at the 5% level of significance.

SAS Problems: Use SAS to solve the following problems.

1. A study is conducted to assess the extent to which patients who had coronary artery bypass surgery were maintaining their prescribed exercise programs. The following data reflect the numbers of times patients reported exercising over the previous month (4 weeks). For the purposes of this study, exercise was defined as moderate physical activity lasting at least 20 minutes in duration.

14	11	8	6	5	3
6	13	12	8	1	4

    Use SAS Proc Means to generate summary statistics on the numbers of times patients exercised over the previous month and a 95% confidence interval estimate for the mean number of times patients exercised following surgery.

2. We wish to design a study to estimate the mean of a population. We wish to consider several scenarios and to estimate the required sample size for each. Use SAS to determine the sample sizes required for each scenario. Consider margins of error of 5, 10, and 20; Confidence levels of 90% and 95%; Standard deviations of 55 and 65 (for a total of 3x2x2 = 12 scenarios).

3. The following data were collected from a random sample of 10 asthmatic children enrolled in a research study and reflect the number of days each child missed school during the past 3 months:

6	12	14	3	2	4	7	8	10	6

    Use SAS Proc Means to generate summary statistics and request a 95% confidence interval for the mean.

4.   A consumer group is investigating a producer of diet meals to examine if their pre-packaged meals actually contain the advertised 6 ounces of protein in each package.

5.1	4.9	6.0	5.1	5.7	5.5	4.9	6.1	6.0	5.8
5.2	4.8	4.7	4.2	4.9	5.5	5.6	5.8	6.0	6.1

Use SAS Proc Means to generate summary statistics on the ounces of protein contained in the packaged meals. In addition, run a test to determine if there any evidence that the meals do not contain the advertised amount of protein. Run the appropriate test at a 5% level of significance.

5.   We wish to design a study to test the following hypotheses: H_0: $\mu=100$ Vs. H_1: $\mu \neq 100$. We wish to consider several scenarios and to estimate the required sample size for each. Use SAS to determine the sample sizes required for each scenario. Consider means under the alternative hypothesis of 90, 95 and 120; Levels of significance of 0.05 and 0.01; Standard deviations of 7 and 10 (for a total of 3x2x2 = 12 scenarios).

6.   Results of an industry survey in the computer software field finds that the mean number of sick days taken by employees is 9.4 per year with a standard deviation of 2.7 per year. A local computer software company feels its employees take significantly fewer sick days per year. A random sample of 15 employees is selected from the local company and attendance records are reviewed. The following data represent the numbers of sick days taken by these employees over the past year:

8	10	5	0	6	9	5	15
5	4	3	2	0	4	15	

Use SAS Proc Means to generate summary statistics on the numbers of sick days taken by employees. In addition, run a test to determine if there any evidence that the employees take fewer than 9.4 sick days per year. Run the appropriate test at a 5% level of significance. (NOTE: SAS performs a two-sided test, make the adjustment in the p-value to draw your conclusion.)

Statistical Inference for μ

Introductory Applied Biostatistics

DESCRIPTIVE STATISTICS (Ch. 3) → PROBABILITY (Ch. 4) → SAMPLING DISTRIBUTIONS (Ch. 5)

STATISTICAL INFERENCE (Chapters 6-14)

Outcome Variable	Grouping Variable(s)/ Predictor(s)	Analysis	Chapter(s)
Continuous	-	Estimate μ, Compare μ to Known, Historical Value	6/13
Continuous	**Dichotomous (2 Groups)**	**Compare Independent Means (Estimate/Test ($\mu_1-\mu_2$)) or the Mean Difference(μ_d)**	**7/13**
Continuous	Discrete (> 2 Groups)	Test the Equality of K Means using Analysis of Variance ($\mu_1=\mu_2=...\mu_k$)	10/13
Continuous	Continuous	Estimate Correlation or Determine Regression Equation	11/13
Continuous	Several Continuous or Dichotomous	Multiple Linear Regression Analysis	11
Dichotomous	-	Estimate p, Compare p to Known, Historical Value	8
Dichotomous	Dichotomous (2 Groups)	Compare Independent Proportions (Estimate/Test (p_1-p_2))	8/9
Dichotomous	Discrete (>2 Groups)	Test the Equality of k Proportions (Chi-Square Test)	8
Dichotomous	Several Continuous or Dichotomous	Multiple Logistic Regression Analysis	12
Discrete	Discrete	Compare Distributions Among k Populations (Ch-Square Test)	8
Time to Event	Several Continuous or Dichotomous	Survival Analysis	14

CHAPTER 7: Statistical Inference: Procedures for $(\mu_1 - \mu_2)$

7.1   Introduction

We now describe statistical inference procedures when there are two comparison groups and the outcome of interest is a continuous variable. In such a case we compare means between groups. In Chapter 6 we described statistical inference procedures for a single sample (estimation of an unknown mean and tests of hypothesis about the mean of the population). Two sample problems are extremely common. For example, recall Example 6.9 from the previous chapter in which a diet was evaluated for it ability to reduce cholesterol levels. In the example a single sample of 12 individuals were placed on the diet and followed for 3 months. At the end of the 3 month observation period, cholesterol levels were measured and compared against a known (or historical) value. In that example, we did not find statistically significant evidence of a reduction in cholesterol attributable to the diet. In that test we made the assumption that the mean cholesterol level for males age 50 not following the diet was 241. We observed a mean cholesterol level in our sample of 235. The assumption that the mean cholesterol level would be 241 in persons not following the diet may or may not be a valid assumption. There are different study designs that we could have used that might have given a better assessment of the impact of the diet on cholesterol. For example we could have used a concurrent comparison group (instead of a historical comparison). This type of study involves selecting a group of individuals appropriate for the study (e.g., males age 50) and randomly assigning them to one of two groups. (Later we will describe in detail the procedures for assigning individuals at random to comparison groups.) One group of individuals follows the diet while the comparison group does not. At the end of 3 months we compare the mean cholesterol levels between comparison groups. If the groups are similar (and this is related to random assignment) except that one group followed the diet and the other did not, then differences in cholesterol can be attributed to the diet. This is an example of what we call a two independent sample procedure in which the two comparison groups are physically distinct (i.e., they are comprised of different individuals). Another study design for the assessment of the effect of the diet on cholesterol involves our 12 subjects but before starting the diet we measure their initial (sometimes called baseline) cholesterol levels. Each individual then follows the diet for 3 months and we measure a final cholesterol level. The focus in this design is how much each individual changes over time. If individuals' cholesterol levels drop from where they started we conclude that the diet is effective in reducing cholesterol. This is an example of what we call a two dependent samples procedure in which each individual serves as his own control. In this chapter we will describe two independent and two dependent samples procedures in detail. The most appropriate design for a specific application depends on a variety of factors including the treatment under investigation and characteristics of the study subjects. These details will be discussed in subsequent chapters.

The techniques described in this chapter are concerned with the difference between two means. The techniques for estimating the difference between two means as well as the techniques for testing if two means are significantly different (or if one is larger than the other) are identical in principle to the techniques described in Chapter 6 which were concerned with the mean of a single population μ.

The assumptions necessary for valid applications of the techniques and formulas that follow are:

(i)   Random samples from the populations under consideration

(ii)   Large samples ($n_i \geq 30$, where i=1, 2) or normal populations

In two independent sample procedures, the parameter of interest is the difference in population means: $(\mu_1 - \mu_2)$. Confidence intervals in two independent sample applications are concerned with estimating $(\mu_1 - \mu_2)$, the <u>difference</u> in means, as opposed to the <u>value</u> of either mean as was the case in the one sample estimation problems. The same is true in tests of hypotheses in the two sample case. Both the null and research or alternative hypothesis are concerned with the difference in means (e.g., H_0: $\mu_1 - \mu_2 = 0$ (no difference in means) vs. H_1: $\mu_1 - \mu_2 \neq 0$ (means are different)).

In two dependent sample procedures, the parameter of interest is the mean difference score: μ_d. Confidence intervals in two dependent sample applications are concerned with estimating μ_d, the mean <u>difference</u> score.

Techniques and formulas for estimation and test of hypotheses concerning the difference between two independent means and the mean difference depend on the specific attributes of the application. We classify these attributes into 4 cases (Table 7.1).

Statistical Inference for $(\mu_1 - \mu_2)$

Table 7.1 Procedures Concerning $(\mu_1 - \mu_2)$

Case 1.	Two Independent Populations - Population Variances Known (i.e., σ_1^2 and σ_2^2 known)
Case 2.	Two Independent Populations - Population Variances Unknown But Assumed to be Equal
Case 3.	Two Independent Populations - Population Variances Unknown and Possibly Unequal
Case 4.	Two Dependent Populations - The Data are Matched or Paired

Cases 1, 2 and 3 apply when the two populations are independent. Going back to the example to evaluate the diet on cholesterol levels, two independent samples procedures would be appropriate if individuals were assigned to either the diet group or to a comparison group. The outcome variable measured on each subject is their cholesterol level after 3 months on the assigned treatment (diet or not), which is a continuous variable. Interest lies in comparing mean cholesterol levels for persons on the diet with those who are not. The treatment group defines two distinct (non-overlapping) populations.

A second example might involve a comparison of the mean cholesterol levels for persons following a special diet as compared to persons taking medication to reduce cholesterol. Again, the outcome variable measured on each subject is their cholesterol level. The treatment or grouping variable in this example is the assigned treatment (diet or medication). Assuming that patients are assigned to follow the diet OR to receive medication, the treatment defines two distinct (non-overlapping) populations.

Case 4 applies when the two populations are dependent (sometimes referred to as matched or paired). For example, suppose we again wish to assess the effect of the diet on cholesterol levels. In order to assess whether the diet is effective in reducing cholesterol, we recruit a sample of subjects into the study. At the outset, we record their cholesterol levels (a continuous variable). Each subject then participates in the diet program and after 3 months we again measure cholesterol (a continuous variable). We now have two samples of data: a sample of values reflecting baseline or initial cholesterol levels, and a sample of values reflecting cholesterol levels after 3 months on the diet. In this application, different from the two described above, two measurements were taken on each individual. We again have two samples of data, however, the samples are matched by individual. This is an example of a Case 4 application. The conduct of two dependent samples procedures are discussed in detail in Section 7.2.4.

It is extremely important to recognize the difference between a two independent samples application and a two dependent samples application in order to apply the appropriate statistical analysis. Incorrectly classifying an application could result in incorrect inferences. Using a series of examples we will illustrate the difference between these types of applications in the sections that follow.

Examples of confidence intervals and test of hypothesis for each case are provided in Section 7.2. Section 7.2.1 contains examples of Case 1. Section 7.2.2 contains examples of Case 2. Sections 7.2.3 and 7.2.4 contain examples of Cases 3 and 4, respectively. Power and sample size determination are discussed in Section 7.3. Key formulas are summarized in Section 7.4 and in Section 7.5 we provide SAS program code used to perform statistical computing applications presented in this chapter.

7.2   Statistical Inference Concerning $(\mu_1 - \mu_2)$

Similar to the applications described in Chapter 6 involving the mean of a single population, statistical inference procedures for the difference in two means can be classified as either estimation or hypothesis testing applications. (Although we present these as distinct applications, there is a clear relationship between them which we will discuss through examples in this section.)

Estimation in two sample applications is concerned with estimating $(\mu_1 - \mu_2)$, the <u>difference</u> in means between groups. Tests of hypotheses in the two sample case are also concerned with the difference in means (e.g., H_0: $\mu_1 - \mu_2 = 0$ (no difference in means) vs. either H_1: $\mu_1 - \mu_2 \neq 0$ (means are different), or H_1: $\mu_1 > \mu_2$ (the mean of population 1 is larger than the mean of population 2), or H_1: $\mu_1 < \mu_2$ (the mean of population 1 is smaller than the mean of population 2)). We now present a series of examples to illustrate estimation and hypothesis testing techniques for applications classified as Cases 1, 2, 3 or 4.

7.2.1 Case 1: Two Independent Populations - Population Variances Known

Case 1 applies when there are two independent populations being compared with respect to their means and the population variances (σ_1^2 and σ_2^2) are known. The two independent populations can be defined on the basis of a characteristic inherent to the subjects under study (e.g., male gender versus female gender , age $<$ 30 years versus age $\geq$ 30 years) or by design (e.g., medication A versus medication B, treatment plan 1 versus treatment plan 2). When an application satisfies the attributes of Case 1 the test statistic for testing H_0: $\mu_1 - \mu_2 = 0$ (equivalent to H_0: $\mu_1 = \mu_2$) and the confidence interval for estimating $(\mu_1 - \mu_2)$ are given in Table 7.2.

two ~~~s ——————— *hypothesis testing*

Table 7.2 Case 1: Two Independent Populations - Population Variances Known

Attributes	Test Statistic	Confidence Interval
σ_1^2 and σ_2^2 known	$Z = \dfrac{\overline{X}_1 - \overline{X}_2}{\sqrt{\dfrac{\sigma_1^2}{n_1} + \dfrac{\sigma_2^2}{n_2}}}$	$(\overline{X}_1 - \overline{X}_2) \pm Z_{1-\frac{\alpha}{2}} \sqrt{\dfrac{\sigma_1^2}{n_1} + \dfrac{\sigma_2^2}{n_2}}$

We now illustrate the use of these formulas through examples.

Example 7.1. Following total knee replacement (TKR) surgery, physical therapy is initiated almost immediately. Depending on a variety of factors (e.g., patients' physical abilities, age, health insurance coverage), patients receive varying numbers of physical therapy sessions following TKR surgery. As a means of assessing the effectiveness of physical therapy, patients are given periodic walking tests. One particular test involves measuring the distance (in feet) that a patient can walk independently (i.e., unassisted). We wish to compare two physical therapy programs with respect to how far patients can walk independently. The first program involves four physical therapy sessions which take place one hour per day over four days following surgery. The second program involves two physical therapy sessions which take place two hours per day on the first and second days following surgery. The outcome of interest, the number of feet patients can walk independently, is measured on the fifth day following surgery for all patients.

To compare the programs, patients are randomly assigned to either the four-day or the two-day program. After completing the assigned program, the number of feet each can walk independently is measured on day 5. A random sample of 18 patients undergoing physical therapy for four days following TKR walked a mean of 47.2 feet independently. A second random sample of 16 patients undergoing physical therapy for only two days following TKR walked a mean of 24.6 feet independently. We will compare the programs by constructing a 95% confidence interval for the difference in mean walking distances between patients undergoing physical therapy for four versus two days following TKR. Suppose in this application that the variations in walking distances are known: $\sigma_1^2 = 6.4$ feet and $\sigma_2^2 = 4.2$ feet. (It is more common in practice to encounter applications in which the population variances are unknown. We will illustrate the procedures when the populations variances are unknown in the next two sections.) Because we have two independent populations and the population variances are known, this application is an example of

Case 1. The appropriate formula for the confidence interval for the difference in population means is:

$$(\overline{X}_1 - \overline{X}_2) \pm Z_{1-\frac{\alpha}{2}} \sqrt{\frac{\sigma_1^2}{n_1} + \frac{\sigma_2^2}{n_2}}$$

The data layout can be represented as follows:

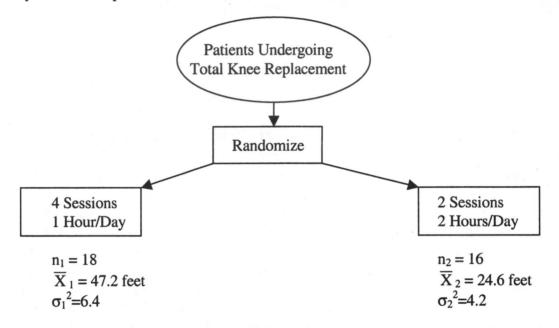

We now substitute the available data and the appropriate Z value for 95% confidence from Table 2A in the Appendix (Z Values for Confidence Intervals, this table was also called Table 6.1 in Chapter 6):

$$(47.2 - 24.6) \pm 1.96 \sqrt{\frac{6.4}{18} + \frac{4.2}{16}}$$

$$22.6 \pm 1.96 \,(0.786)$$

$$22.6 \pm 1.54$$

$$(21.06, 24.14)$$

Statistical Inference for $(\mu_1 - \mu_2)$

We are 95% confident that the difference in mean walking distances between patients undergoing physical therapy for four versus two days is between 21.06 and 24.14 feet.

Again, in two sample procedures, the confidence interval estimates the <u>difference</u> in means, as opposed to the value of either mean (as was the case in the one sample applications). In this example, a point estimate for the difference in mean walking distances between patients undergoing physical therapy for four versus two days is 22.6 feet, and the 95% confidence interval for the difference in means is 21.06 to 24.14 feet. If you could choose either the four-day or the two-day program following TKR, which physical therapy program would you select? Why?

There is a relationship between confidence interval estimates and tests of hypothesis. In Example 7.1 we generated a 95% confidence interval estimate for $(\mu_1 - \mu_2)$. If we were to run a test of hypothesis, the hypotheses would be of the form: $H_0 : \mu_1 = \mu_2$ Vs. $H_1 : \mu_1 \neq \mu_2$ There is an alternate representation: $H_0 : \mu_1 - \mu_2 = 0$ Vs. $H_1: \mu_1 - \mu_2 \neq 0$. If the confidence interval estimate contains the value specified in H_0, then we do not reject H_0 in a test of hypothesis. In Example 7.1 we can conclude that there is a significant difference in mean walking distances between patients undergoing physical therapy for four versus two days (because the confidence interval estimate does not include 0).

Example 7.2. Management in a hospital hypothesizes that employees in non-supervisory positions take significantly more sick days per calendar year than employees in supervisory positions. To assess this claim, a test of hypothesis is planned. The two groups under investigation are defined by job classification (non-supervisory versus supervisory). We take random samples of employees from each job classification and record the number of sick days that each took in the last calendar year. A random sample of 100 employees in non-supervisory positions took a mean of 6.7 sick days in the last calendar year while a random sample of 100 employees in supervisory positions took a mean of 4.8 sick days in the last calendar year. The variance in the number of sick days among non-supervisory employees is known to be 2.2 days while the variance in the number of sick days among supervisors is known to be 3.1 days. Notice that the mean number of sick days reported by employees in non-supervisory positions is larger than the mean number reported by employees in supervisory positions (6.7 versus 4.8). The question is "Is this evidence of a <u>significantly higher</u> mean in the population of all non-supervisory employees as compared to supervisors?" Run the appropriate test at a 5% level of significance.

NOTE: The same 5 steps used in testing hypotheses concerning μ (outlined in Table 6.3) are used in the tests of hypotheses concerning $(\mu_1 - \mu_2)$.

1. Set up hypotheses.

    The null hypothesis in the two sample problem again reflects the "no change" or "no effect" situation. The alternative hypothesis here reflects the claim that the mean number of sick days taken by non-supervisory employees is <u>higher</u> than the mean number of sick days taken by supervisors.

$$H_0 : \mu_1 = \mu_2 \qquad\qquad \alpha = 0.05$$
$$H_1 : \mu_1 > \mu_2$$

where μ_1 = the mean number of sick days per calendar year among employees in non-supervisory positions and μ_2 = the mean number of sick days per calendar year among employees in supervisory positions.

2. Select appropriate test statistic.

    Because we have two independent populations and the population variances are known, this is an example of Case 1 (See Table 7.2), so the test statistic is:

$$Z = \frac{\overline{X}_1 - \overline{X}_2}{\sqrt{\dfrac{\sigma_1^2}{n_1} + \dfrac{\sigma_2^2}{n_2}}}$$

3. Decision Rule.

    The decision rule in two sample tests of hypothesis depends on the same three factors: i) whether the test is upper, lower or two-tailed, ii) the level of significance, and iii) the form of the test statistic. Here we have an upper tailed test, $\alpha=0.05$ and we are using a Z statistic. The critical value of Z is found in Table 2B in the Appendix (Z Values for Tests of Hypothesis, this table was also called Table 6.4 in Chapter 6):

    Reject H_0 if $Z \geq 1.645$
    Do Not Reject H_0 if $Z < 1.645$

3. Test Statistic.

    We now substitute the available data:

Statistical Inference for $(\mu_1 - \mu_2)$

$$Z = \frac{\overline{X}_1 - \overline{X}_2}{\sqrt{\dfrac{\sigma_1^2}{n_1} + \dfrac{\sigma_2^2}{n_2}}} = \frac{6.7 - 4.8}{\sqrt{\dfrac{2.2}{100} + \dfrac{3.1}{100}}} = \frac{1.9}{0.23} = 8.26$$

5. Conclusion. Reject H_0 because $8.26 \geq 1.645$. We have significant evidence, $\alpha = 0.05$, to show that employees in non-supervisory positions take significantly more sick days per calendar year than employees in supervisory positions (i.e., $\mu_1 > \mu_2$).

What is the p-value for this application? Use Table 2B to determine the smallest level of significance α we could have selected and still rejected H_0. Because the test statistic is so large, we could have selected $\alpha = 0.0001$ and still rejected H_0. Thus $p < 0.0001$.

7.2.2 Case 2: Two Independent Populations - Population Variances Unknown but Assumed Equal

Case 2 applies when there are two independent populations and the population variances (σ_1^2 and σ_2^2) are unknown but assumed to be equal. The sample variances (s_1^2 and s_2^2) are used as estimates and the formulas for test statistics and confidence intervals concerning ($\mu_1 - \mu_2$) are given in Table 7.3.

Table 7.3 Case 2: Population Variances Unknown but Assumed to be Equal

Attributes	Test Statistic	Confidence Interval
σ_1^2 and σ_2^2 unknown but assumed to be equal, $n_1 \geq 30$ and $n_2 \geq 30$ *and*	$Z = \dfrac{\overline{X}_1 - \overline{X}_2}{S_p \sqrt{\dfrac{1}{n_1} + \dfrac{1}{n_2}}}$	$(\overline{X}_1 - \overline{X}_2) \pm Z_{1-\frac{\alpha}{2}} S_p \sqrt{\dfrac{1}{n_1} + \dfrac{1}{n_2}}$
σ_1^2 and σ_2^2 unknown but assumed to be equal, $n_1 < 30$ or $n_2 < 30$ *or*	$t = \dfrac{\overline{X}_1 - \overline{X}_2}{S_p \sqrt{\dfrac{1}{n_1} + \dfrac{1}{n_2}}}$ $df = n_1 + n_2 - 2$	$(\overline{X}_1 - \overline{X}_2) \pm t_{1-\frac{\alpha}{2}} S_p \sqrt{\dfrac{1}{n_1} + \dfrac{1}{n_2}}$ $df = n_1 + n_2 - 2$

Statistical Inference for ($\mu_1 - \mu_2$)

where S_p is the pooled estimate of the common standard deviation:

$$S_p = \sqrt{\frac{(n_1-1)\,s_1^2 + (n_2-1)\,s_2^2}{n_1 + n_2 - 2}}$$

$$S_1 \leq S_p \leq S_2$$

or

$$S_2 \leq S_p \leq S_1$$

NOTE : $\quad$ S_p^2 is the pooled estimate of the common variance, defined as a weighted average of the sample variances (s_1^2 and s_2^2). The weights are determined by the sample sizes (n_1 and n_2). If the sample sizes are equal (i.e., $n_1=n_2$, then Sp reduces to

$$S_p = \sqrt{\frac{s_1^2 + s_2^2}{2}}.$$

In Case 2 we assume that the population variances are equal in the two comparison groups. This is a reasonable assumption in many applications. In this case we an estimate the unknown population variance by pooling data from both samples. In order to count more heavily the sample variance derived from the larger group, a weighted mean is used.

Example 7.3. A new curriculum has been implemented across medical schools which is designed to improve medical school students' analytic skills. An evaluation committee is concerned that the new curriculum may be differentially effective among male and female students. To evaluate the new curriculum, random samples of male and female students who completed the new curriculum are selected and given a test to assess their analytic skills. The following data are observed which denote the numbers of analytic problems correctly solved by the students:

Statistic	Males	Females
Sample Size	15	12
Mean	15.8	12.4
Standard Deviation	4.2	3.6

Use the data to test if there is a significant <u>difference</u> in the mean numbers of analytic problems correctly solved by male and female students using a 5% level of significance.

1. Set up hypotheses.

Here we wish to test whether there is a significant <u>difference</u> in means, therefore a two-sided alternative is used.

$$H_0 : \mu_1 = \mu_2 \qquad\qquad \alpha = 0.05$$
$$H_1 : \mu_1 \neq \mu_2$$

where μ_1 = mean number of analytic problems correctly solved by male students and
μ_2 = mean number of analytic problems correctly solved by female students.

2. Select appropriate test statistic. Here we assume that the variation in the numbers of analytic problems correctly solved is the same for male and female students (i.e., assume $\sigma_1^2 = \sigma_2^2$). Notice that the sample standard deviations ($s_1 = 4.2$ and $s_2 = 3.6$) are similar in magnitude, evidence that the assumption regarding the equality of population variances is appropriate.

Because of this assumption regarding equality of population variances and due to the small sample sizes ($n_1 < 30$ and $n_2 < 30$), the test statistic is:

$$t = \frac{\overline{X}_1 - \overline{X}_2}{S_p \sqrt{\dfrac{1}{n_1} + \dfrac{1}{n_2}}}$$

where S_p is the pooled estimate of the common standard deviation (i.e., $\sigma_1 = \sigma_2 = \sigma$).

3. Decision Rule:

Here we have a two-sided test, $\alpha=0.05$ and we are using a t statistic. The critical value of t is found in Table 3. In order to determine the appropriate critical value, we first compute degrees of freedom. $df = n_1 + n_2 - 2 = 15 + 12 - 2 = 25$. The critical value is $t = 2.060$.

Reject H_0 if $t \geq 2.060$ or if $t \leq -2.060$
Do Not Reject H_0 if $-2.060 < t < 2.060$

4. Test Statistic.
First, we compute S_p:

$$S_p = \sqrt{\frac{(n_1 - 1)\, s_1^2 + (n_2 - 1)\, s_2^2}{n_1 + n_2 - 2}}$$

Statistical Inference for ($\mu_1 - \mu_2$)

$$S_p = \sqrt{\frac{14\,(4.2)^2 + 11\,(3.6)^2}{15 + 12 - 2}} = \sqrt{15.581} = 3.95$$

(Notice that S_p, the pooled estimate of the common standard deviation, falls in between the values of the two sample standard deviations.)

Now, the test statistic:

$$t = \frac{15.8 - 12.4}{3.95\sqrt{\dfrac{1}{15} + \dfrac{1}{12}}} = \frac{3.4}{1.56} = 2.18$$

5. Conclusion: Reject Ho because $2.18 \geq 2.060$. We have significant evidence, $\alpha = 0.05$, to show that there is a significant difference in the mean number of analytic problems correctly solved by male and female students. The p-value for this test is determined using Table 3. With 25 degrees of freedom, the critical value for $\alpha=0.05$ is 2.060 (See Step 3). The next smallest level of significance in Table 3 (for df=25) is 0.02 which has a corresponding critical value of 2.485. Because we would not reject H_0 at $\alpha=0.02$, p=0.05. If this test were run on SAS, an exact p-vlaue would be computed. The exact p-value is between 0.02 and 0.05 (thus we often report p<0.05).

NOTE: The above test is appropriate for testing the equality of 2 independent population means when variances are assumed to be equal. The assumption regarding the equality of population variances must be evaluated carefully. If it is not reasonable to assume that the population variances are equal between comparison groups then the techniques described in the next section must be employed. However, if the sample sizes are equal (i.e., $n_1 = n_2$), the formulas presented in this section are *robust*. Formulas and techniques are robust if they maintain their statistical properties under violations of assumptions (e.g., if the population variances are not equal). As a rule of thumb, if the sample sizes are unequal but the sample variances are close in value, defined as $0.5 \leq \dfrac{s_1^2}{s_2^2} \leq 2$, then the formulas presented in this section (Case 2) can be used.

Statistical Inference for $(\mu_1 - \mu_2)$

Example 7.4. An investigation is undertaken to examine the average times to relief from headache pain under 2 entirely different treatments: Medication vs. Relaxation Treatment. Patients suffering from chronic headaches are enrolled in a study and randomly assigned to one of the two treatments under investigation. Patients are instructed to either take the assigned medication or to perform the relaxation exercises at the onset of their next headache. They are also instructed to record the time, in minutes, until the headache pain is resolved. Fifteen subjects are assigned to the medication treatment and report a mean time to relief of 33.8 minutes with a variance of 2.85 minutes. A second random sample of 15 subjects are assigned to the relaxation treatment, and report a mean time to relief of 22.4 minutes with a variance of 3.07 minutes.

The data layout is as follows:

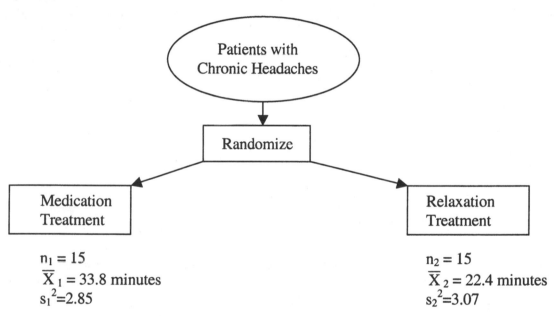

$n_1 = 15$
$\overline{X}_1 = 33.8$ minutes
$s_1^2 = 2.85$

$n_2 = 15$
$\overline{X}_2 = 22.4$ minutes
$s_2^2 = 3.07$

Are these sample means evidence of a statistically significant difference in means in the populations? Run the appropriate test to assess whether there is a significant <u>difference</u> in the mean time to relief under the two different treatments using a 5% level of significance.

1. Set up hypotheses.

   $H_0 : \mu_1 = \mu_2$
   $H_1 : \mu_1 \neq \mu_2$ $\qquad$ $\alpha = 0.05$

Statistical Inference for $(\mu_1 - \mu_2)$

where μ_1 = mean time to relief with medication treatment and
μ_2 = mean time to relief with relaxation treatment .

2. Select appropriate test statistic. Assuming that population variances are equal (i.e., $\sigma_1^2 = \sigma_2^2$, since $n_1 = n_2$), the test statistic is :

$$t = \frac{\overline{X}_1 - \overline{X}_2}{S_p\sqrt{\dfrac{1}{n_1} + \dfrac{1}{n_2}}}$$

3. Decision Rule.

Here we have a two-sided test, $\alpha=0.05$ and are using a t statistic. The appropriate critical value is found in Table 3. We first compute degrees of freedom: df = $n_1 + n_2 - 2 = 15 + 15 - 2 = 28$.

Reject H_0 if $t \geq 2.048$ or if $t \leq -2.048$
Do Not Reject H_0 if $-2.048 < t < 2.048$

4. Test Statistic.
First, we compute S_p:

$$S_p = \sqrt{\frac{(n_1 - 1)\, s_1^2 + (n_2 - 1)\, s_2^2}{n_1 + n_2 - 2}}$$

$$S_p = \sqrt{\frac{14\,(2.85) + 14\,(3.07)}{15 + 15 - 2}} = \sqrt{2.96} = 1.72$$

(Notice that S_p^2 falls in between the values of the two sample variances. Because the sample sizes are equal here, the following can also be used $S_p = \sqrt{\dfrac{s_1^2 + s_2^2}{2}} = \sqrt{\dfrac{2.85 + 3.07}{2}} = 1.72$)

Now, the test statistic:

$$t = \frac{33.8 - 22.4}{1.72\sqrt{\dfrac{1}{15} + \dfrac{1}{15}}} = \frac{11.4}{0.63} = 18.10$$

Statistical Inference for $(\mu_1 - \mu_2)$

5. Conclusion: Reject Ho since $18.10 \geq 2.048$. We have significant evidence, $\alpha = 0.05$, to show that the mean time to relief from headache pain is different under medication as compared to relaxation treatment. This test statistic is so large that $p < 0.0001$.

7.2.3   Case 3: Two Independent Populations - Population Variances Possibly Unequal

In Case 3, we are again concerned with estimating the difference between two independent population means or conducting a test concerning the equality of two independent population means. Population variances, however, are not known and cannot be assumed to be equal (as in Case 2). In order to determine if the population variances are equal, we conduct a *preliminary test* of H_0: $\sigma_1^2 = \sigma_2^2$ against H_1: $\sigma_1^2 \neq \sigma_2^2$. It is necessary to conduct this preliminary test concerning the population variances in order to determine the appropriate formula for estimating the difference in means or for the test statistic in the two independent samples test of means. If, based on the preliminary test, we determine that the population variances are equal then the formulas given in the previous section (Case 2) can be used. If, however, we find that the population variances are not equal, then the formulas given in this section (Case 3) are used.

The preliminary test is conducted in the same fashion as the tests for means (i.e., following the same 5 steps used in tests of hypotheses concerning μ and tests of hypotheses concerning (μ_1 - μ_1)). The preliminary test for equality (or homogeneity) of variances is outlined below and summarized in Table 7.4. The hypotheses to be tested in the preliminary test are:

$$H_0: \sigma_1^2 = \sigma_2^2$$
$$H_1: \sigma_1^2 \neq \sigma_2^2$$

The preliminary test is always a two-sided test, as we are solely interested in determining whether the population variances are equal or not for the purposes of selecting the appropriate test statistic (or confidence interval formula) for evaluating the difference in means (μ_1 - μ_2).

NOTE: In some applications, a larger level of significance is chosen for the preliminary test (e.g., $\alpha = 0.10$ or 0.15).

The test statistic is determined by the ratio of the sample variances which follows an F distribution:

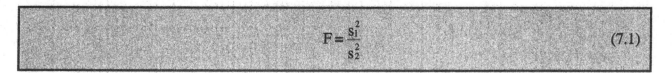

$$F = \frac{s_1^2}{s_2^2} \tag{7.1}$$

Statistical Inference for (μ_1 - μ_2)

The F distribution has two degrees of freedom, denoted df_1 and df_2, called the numerator and denominator degrees of freedom, respectively:

$$df_1 = n_1 - 1 \quad \text{(numerator degrees of freedom), and}$$
$$df_2 = n_2 - 1 \quad \text{(denominator degrees of freedom).}$$

If the test statistic, F, is close to one (which occurs when s_1^2 and s_2^2 are approximately equal in value), then H_0 is most likely true (i.e., $\sigma_1^2 = \sigma_2^2$). However, if F is significantly smaller or larger than unity, then H_1 is most likely true. In order to make a decision about H_0 or H_1, critical values from the F distribution must be determined to formulate the decision rule.

The F distribution is an asymmetric distribution (Figure 7.1) which takes on values greater than or equal to zero.

Figure 7.1 F Distribution

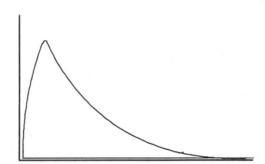

The decision rule is given by:

Reject H_0 if $F \leq 1/F_{1-\alpha/2}(df_2, df_1)$ or if $F \geq F_{1-\alpha/2}(df_1, df_2)$
Do Not Reject H_0 if $1/F_{1-\alpha/2}(df_2, df_1) < F < F_{1-\alpha/2}(df_1, df_2)$

Figure 7.2 displays the critical values in the preliminary test for equality of variances (always a two sided test):

Statistical Inference for $(\mu_1 - \mu_2)$

Figure 7.2 Critical Values in the Preliminary Test

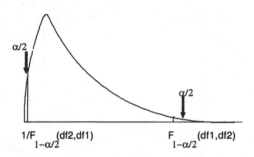

For example, suppose we conduct a preliminary test of the equality of population variances (i.e., H_0: $\sigma_1^2 = \sigma_2^2$ vs. H_1: $\sigma_1^2 \neq \sigma_2^2$) using a 5% level of significance based on samples of size $n_1 = 20$ and $n_2 = 20$. The critical values from the F distribution are found in Table 4 and are given below for $df_1 = n_1 - 1 = 20 - 1 = 19$ and $df_2 = n_2 - 1 = 20 - 1 = 19$:

$$F_{0.975}(19, 19) = 2.51 \qquad\qquad 1/F_{0.975}(19, 19) = 1/2.51 = 0.40$$

The critical values from the F distribution table (Table 4) with numerator degrees of freedom equal to 19 and denominator degrees of freedom equal to 19 are 0.40 and 2.51.

The steps involved in the preliminary test of variances are summarized in Table 7.4.

Table 7.4 Preliminary Test For Homogeneity of Variances

Step	Example
1. Set up hypotheses.	$H_0: \sigma_1^2 = \sigma_2^2$ $H_1: \sigma_1^2 \neq \sigma_2^2$ [1]
Select level of significance. [2]	$\alpha = 0.05$
2. Select appropriate test statistic.	$F = \dfrac{s_1^2}{s_2^2}$
3. Generate decision rule.	Reject H_0 if $F \leq 1/F_{1-\alpha/2}(df_2, df_1)$ or if $F \geq F_{1-\alpha/2}(df_1, df_2)$ Do Not Reject H_0 if $1/F_{1-\alpha/2}(df_2, df_1) < F < F_{1-\alpha/2}(df_1, df_2)$
4. Compute the value of the test statistic.	
5. Draw a conclusion by comparing the test statistic (4) to the decision rule (3). Report findings relative to H_1.	

NOTES: (1) In the preliminary test, we are not concerned with upper or lower tailed tests. The preliminary test is conducted solely to determine if the population variances are equal or not for the purposes of selecting the correct formula for statistical inference concerning the difference in population means.

(2) Larger levels of significance (e.g., $\alpha = 0.10, 0.15$) are sometimes used in the preliminary test.

If the null hypothesis in the preliminary test is not rejected (i.e., $\sigma_1^2 = \sigma_2^2$) then the formulas given in Table 7.3 (Case 2) are used for the *main test* of $H_0: \mu_1 = \mu_2$ or for the confidence interval for $(\mu_1 - \mu_2)$. If the null hypothesis in the preliminary test is rejected (i.e., $\sigma_1^2 \neq \sigma_2^2$) then the formulas given in Table 7.5 (Case 3) are used for the *main test* of $H_0: \mu_1 = \mu_2$ or for the confidence interval for $(\mu_1 - \mu_2)$.

Table 7.5 Case 3: Two Independent Populations - Population Variances Unequal

Attributes	Test Statistic	Confidence Interval
σ_1^2 and σ_2^2 unknown but assumed to be unequal, $n_1 \geq 30$ and $n_2 \geq 30$	$Z = \dfrac{\overline{X}_1 - \overline{X}_2}{\sqrt{\dfrac{s_1^2}{n_1} + \dfrac{s_2^2}{n_2}}}$	$(\overline{X}_1 - \overline{X}_2) \pm Z_{1-\frac{\alpha}{2}} \sqrt{\dfrac{s_1^2}{n_1} + \dfrac{s_2^2}{n_2}}$
σ_1^2 and σ_2^2 unknown but assumed to be unequal, $n_1 < 30$ or $n_2 < 30$	$t = \dfrac{\overline{X}_1 - \overline{X}_2}{\sqrt{\dfrac{s_1^2}{n_1} + \dfrac{s_2^2}{n_2}}}$ $df = \dfrac{\left(\dfrac{s_1^2}{n_1} + \dfrac{s_2^2}{n_2}\right)^2}{\dfrac{\left(\dfrac{s_1^2}{n_1}\right)^2}{n_1 - 1} + \dfrac{\left(\dfrac{s_2^2}{n_2}\right)^2}{n_2 - 1}}$	$(\overline{X}_1 - \overline{X}_2) \pm t_{1-\frac{\alpha}{2}} \sqrt{\dfrac{s_1^2}{n_1} + \dfrac{s_2^2}{n_2}}$ $df = \dfrac{\left(\dfrac{s_1^2}{n_1} + \dfrac{s_2^2}{n_2}\right)^2}{\dfrac{\left(\dfrac{s_1^2}{n_1}\right)^2}{n_1 - 1} + \dfrac{\left(\dfrac{s_2^2}{n_2}\right)^2}{n_2 - 1}}$

The formula for the degrees of freedom (df) for both the critical value(s) in the test of hypothesis and to determine the appropriate value for the confidence interval is based on the Welch-Satterthwaite solution (Welch BL (1938). The significance of the difference between two means when the population variances are unequal. *Biometrika*, 29, 350-362., Satterthwaite FE (1946). An approximate distribution of estimates of variance components. *Biometrics Bulletin*, 2, 110-114). Using this formula for the degrees of freedom allows for the use of the t distribution table (Table 3) in applications involving these formulas (Case 3).

To simplify the computations of the degrees of freedom, consider the following:

$$c_i = \frac{s_i^2}{n_i}, \quad \text{where } i = 1,2 \tag{7.2}$$

Statistical Inference for $(\mu_1 - \mu_2)$

Then,

$$df = \frac{(c_1 + c_2)^2}{\dfrac{c_1^2}{n_1 - 1} + \dfrac{c_2^2}{n_2 - 1}}$$
(7.3)

We illustrate the use of this formula in the following examples. In Example 7.5 we illustrate the use of the Case 3 formulas and then illustrate the complete procedure (i.e., the preliminary test followed by the main test or estimation) in the subsequent examples.

Example 7.5. Consider an experiment involving the comparison of the mean heart rates following 30 minutes of aerobic exercise among females aged 20 to 24 years as compared to females aged 25-30 years. For this experiment ten second heart rates are recorded on each participant following 30 minutes of intense aerobic exercise and converted to beats per minute (i.e., heart rate per 60 seconds). The sample data are given below:

Statistic	Age: 20-24	Age: 25-30
Sample Size	15	10
Mean Heart Rate	146.22	141.10
Variance in Heart Rates	40.0	10.0

Use the data to test if there is a significant <u>difference</u> in the mean heart rates following 30 minutes of aerobic exercise among females aged 20-24 years as compared to females aged 25-30 years using a 5% level of significance. Assume that the variations in heart rates are not equal between comparison groups. (Notice that the sample variances are quite different (40.0 vs. 10.0, $s_1^2/s_2^2 = 4$ which is outside of the range 0.5-2), suggesting a true difference in population variances.)

1. Set up hypotheses.

$H_0: \mu_1 = \mu_2$

$H_1: \mu_1 \neq \mu_2$ $\qquad \alpha = 0.05$

where μ_1 = mean heart rate for females 20-24 years of age, and

μ_2 = mean heart rate for females 25-30 years of age.

2. Select appropriate test statistic.

(Because population variances are assumed to be unequal, this is an example of Case 3.)

$$t = \frac{\overline{X}_1 - \overline{X}_2}{\sqrt{\dfrac{s_1^2}{n_1} + \dfrac{s_2^2}{n_2}}}$$

3. Decision Rule.

To determine the appropriate critical value, we first compute the degrees of freedom:

$$df = \frac{\left(\dfrac{s_1^2}{n_1} + \dfrac{s_2^2}{n_2}\right)^2}{\dfrac{\left(\dfrac{s_1^2}{n_1}\right)^2}{n_1 - 1} + \dfrac{\left(\dfrac{s_2^2}{n_2}\right)^2}{n_2 - 1}}$$

To simplify the computations, using (7.2) we first compute:

$$c_i = \frac{s_i^2}{n_i}, \quad c_1 = 40/15 = 2.67, \quad c_2 = 10/10 = 1$$

Then:

$$df = \frac{(c_1 + c_2)^2}{\dfrac{c_1^2}{n_1 - 1} + \dfrac{c_2^2}{n_2 - 1}} = \frac{(2.67 + 1)^2}{\dfrac{2.67^2}{14} + \dfrac{1^2}{9}} = \frac{13.47}{0.62} = 21.73 = 22$$

Using the t distribution table (Table 3), the two-sided critical value with 22 degrees of freedom is t=2.074. The decision rule is:

Reject H_0 if $t \geq 2.074$ or if $t \leq -2.074$
Do Not Reject H_0 if $-2.074 < t < 2.074$

Statistical Inference for $(\mu_1 - \mu_2)$

4. Test Statistic.

$$t = \frac{\overline{X}_1 - \overline{X}_2}{\sqrt{\frac{s_1^2}{n_1} + \frac{s_2^2}{n_2}}} = \frac{146.22 - 141.10}{\sqrt{\frac{40}{15} + \frac{10}{10}}} = \frac{5.12}{1.92} = 2.67$$

5. Conclusion. Reject H_0 since $2.67 > 2.074$. We have significant evidence, $\alpha = 0.05$, to show that there is a difference in the mean heart rates following 30 minutes of aerobic exercise for females aged 20-24 years as compared to females aged 25-30 years. For this example, p=0.02.

We now illustrate the preliminary test for homogeneity of variances followed by a test of hypothesis concerning means of two independent populations.

Example 7.6. Random samples of eleven male high school students and twelve female high school students are selected within a particular school district for an investigation. Students' scores on a public health awareness test are recorded and descriptive statistics are given below. The test is scored on a scale of 0-1000 with higher scores indicative of more awareness.

Statistic	Males	Females
Sample Size	11	12
Mean SAT Score	560.0	554.2
Standard Deviation in SAT Scores	133.1	129.4

Test if the male students score significantly <u>higher</u> than the female students on the public health awareness test within this school district using a 5% level of significance.

1. Set up hypotheses.
   H_0: $\mu_1 = \mu_2$
   H_1 : $\mu_1 > \mu_2$ $\qquad \alpha = 0.05$

where μ_1 = mean test score for males and μ_2 = mean test score for females.

2. Select appropriate test statistic.
In order to determine if this application falls into Case 2 or Case 3, a preliminary test of the equality of population variances must be conducted.

1. Set up hypotheses.

H_0: $\sigma_1^2 = \sigma_2^2$
H_1: $\sigma_1^2 \neq \sigma_2^2$ $\alpha = 0.05$

2. Select appropriate test statistic.

$$F = \frac{s_1^2}{s_2^2}$$

3. Decision Rule.

$df_1 = n_1 - 1 = 11 - 1 = 10$ (numerator degrees of freedom)
$df_2 = n_2 - 1 = 12 - 1 = 11$ (denominator degrees of freedom)
$F_{0.975}(10,11) = 3.53$ and $F_{0.975}(11,10) = 3.72$

Reject Ho if $F \leq 1/3.72 = 0.269$ or if $F \geq 3.53$
Do Not Reject Ho if $0.269 < F < 3.53$

4. Test Statistic.

$$F = \frac{s_1^2}{s_2^2} = \frac{(133.1)^2}{(129.4)^2} = 1.06$$

5. Conclusion. Do not reject Ho since $0.269 < 1.06 < 3.53$. We do not have significant evidence to show that $\sigma_1^2 \neq \sigma_2^2$. Therefore, for the purposes of this test of means, we assume that the population variances are equal (i.e., $\sigma_1^2 = \sigma_2^2$) and apply the test statistic given under Case 2:

$$t = \frac{\overline{X}_1 - \overline{X}_2}{s_p \sqrt{\dfrac{1}{n_1} + \dfrac{1}{n_2}}}$$

3. Decision Rule. $df = n_1 + n_2 - 2 = 11 + 10 - 2 = 21$

Reject H_0 if $t \geq 1.721$
Do Not Reject H_0 if $t < 1.721$

Statistical Inference for $(\mu_1 - \mu_2)$

4. Test Statistic.
We first compute Sp:

$$S_p = \sqrt{\frac{(n_1-1)\,s_1^2 + (n_2-1)\,s_2^2}{n_1+n_2-2}}$$

$$S_p = \sqrt{\frac{10\,(133.1)^2 + 11\,(129.4)^2}{11+12-2}} = \sqrt{17253.11} = 131.4$$

Now the test statistic:

$$t = \frac{560.0 - 554.2}{131.4\sqrt{\dfrac{1}{11}+\dfrac{1}{12}}} = \frac{5.8}{54.8} = 0.11$$

5. Conclusion: Do not reject Ho since $0.11 < 1.721$. We do not have significant evidence, $\alpha = 0.05$, to show that the male students score significantly higher than the female students on the public health awareness test within this school district.

NOTE: The two sample t tests (Case 2 and Case 3) concerning ($\mu_1 - \mu_2$) are generally robust (i.e., insensitive to violations in assumptions such as normality and/or equality of variances) when the sample size are equal (i.e., $n_1 = n_2$). However, the F test for homogeneity of variances is sensitive to violations in the normality assumption. For example, if the analytic variable is not normally distributed and the F test is applied, the actual level of significance, α, may exceed the specified level (e.g. 0.05).

SAS Example 7.6. The following output was generated using SAS Proc Ttest which conducts a two independent samples test of hypothesis. The same procedure automatically produces a preliminary test of the homogeneity of variances. A brief interpretation appears after the output.

The TTEST Procedure
Statistics

Variable	Class	N	Lower CL Mean	Mean	Upper CL Mean	Lower CL Std Dev	Std Dev
test	female	12	471.93	554.17	636.41	91.692	129.44
test	male	11	470.57	560	649.43	93.011	133.12
test	Diff (1-2)		-119.7	-5.833	108.06	100.94	131.2

Statistics

Variable	Class	Upper CL Std Dev	Std Err	Minimum	Maximum
test	female	219.77	37.365	260	750
test	male	233.61	40.136	370	770
test	Diff (1-2)	187.5	54.767		

T-Tests

Variable	Method	Variances	DF	t Value	Pr > |t|
test	Pooled	Equal	21	-0.11	0.9162
test	Satterthwaite	Unequal	20.7	-0.11	0.9163

Equality of Variances

Variable	Method	Num DF	Den DF	F Value	Pr > F
test	Folded F	10	11	1.06	0.9215

Interpretation of SAS Output for Example 7.6

In the top section of the output, SAS provides summary statistics on the analytic variable (test score) for each comparison group (females and males) and then for the differences in means (females-males). The summary statistics include the sample sizes, the sample mean (Mean) and 95% confidence intervals (CI) for the population means of each group and for the difference in means (the limits of the CI for the mean are labeled lower CL mean and upper CL mean), standard deviations and 95% confidence intervals for the population standard deviations of each group and for the differences (the limits of the CI for the standard deviation are labeled lower CL Std Dev and Upper CL Std Dev), standard errors ($s/\sqrt{n}$), minimums and maximums.

In the next section of the output SAS performs the test of hypothesis for equality of means. SAS actually carries out two different tests, one in which the population variances are assumed to be equal (we called this Case 2) and one in which the population variances are assumed to be unequal (we called this Case 3). SAS uses the formulas we summarized in Tables 7.5 and 7.3 for unequal and equal variances, respectively. The values of the test statistics appear under the column headed "t Value", and just before these SAS displays the degrees of freedom. Again, these are computed using the formulas from Tables 7.5 and 7.3. Finally, SAS provides two-sided p-values (assuming that the alternative hypothesis is H_1: $\mu_1 \neq \mu_2$).

The user must decide which analysis (unequal or equal variances) is most appropriate. To aid in this decision, SAS provides a preliminary test of the homogeneity of variances (i.e., H_0: $\sigma_1^2 = \sigma_2^2$ vs H_1: $\sigma_1^2 \neq \sigma_2^2$). The results of the preliminary test appear at the bottom of the SAS Ttest output in the section titled "Equality of Variances." SAS provides an F statistic (computed by taking the ratio of the sample variances). SAS computes F by dividing the larger sample variance by the smaller, regardless of the group (1 or 2) designation. Therefore, the F statistic produced by SAS is always greater than or equal to 1. In this case, since the sample variance among males is larger: $F = (133.1)^2/(129.4)^2 = 1.06$. The degrees of freedom associated with F are given by: $df_1 = n_1 - 1 = 11 - 1 = 10$ and $df_2 = n_2 - 1 = 12 - 1 = 11$. For the preliminary test, SAS produces the probability of observing a value of F greater than the value of the test statistic (denoted Prob > F). Since we are generally interested in a two sided test for the preliminary test (i.e., H_1: $\sigma_1^2 \neq \sigma_2^2$), the following rule should be applied to draw a conclusion: Reject H_0 if p-value $\leq \alpha_F/2$, where p - value is the one-sided p - value produced by SAS, and α_F is the (two-sided) level of significance selected for the preliminary test. In this case we do not Reject H_0 since the p - value, 0.9215, is larger than $0.05/2 = 0.025$. Therefore the population variances are not significantly different and this example is considered an example of Case 2.

Because the preliminary test suggested that this example is an example of Case 2 (i.e., equal variances), we look at the output for the equal variances case, and the test statistic is t = -0.11 with 21 degrees of freedom ($n_1 + n_2 - 2$). For the main test, SAS produces a two-sided p - value. In this example, the p - value is 0.9162. Because we are interested in a one-sided test, the following rule should be applied: Reject H_0 if (p - value/2) $\leq \alpha$. In this example we do not reject H_0 since (p - value/2) = (0.9162/2) = 0.4581 > 0.05. We do not have significant evidence to show that male students score significantly higher than female students on the public health awareness test within this school district. (NOTE: The test statistic produced by SAS differs somewhat from that computed in Example 7.6 due to the fact that SAS is using more decimal places in computations.)

Example 7.7. Suppose we wish to estimate the difference in the mean numbers of Emergency Room (ER) visits in 12 months among children with asthma under age 5 as compared to children aged 6-10. For the purposes of this investigation our analyses are restricted to children who are free of any other chronic conditions (i.e., they suffer from asthma alone). The following data are collected on random samples of 65 children under age 5 and 50 children aged 6-10:

Age	Number of patients	Mean number of ER visits	Variance in number of ER visits
≤ 5 Years	65	6.4	4.1
6-10 Years	50	3.2	2.6

Use the data to construct a 95% confidence interval for the difference in the mean numbers of ER visits in 12 months among children with asthma under age 5 as compared to children aged 6-10.

In order to select the appropriate formula for the confidence interval for ($\mu_1 - \mu_2$) we need to determine if the population variances are equal (Case 2) or not (Case 3) using the preliminary test for equality of variances.

1.　　Set up hypotheses.

$$H_0: \sigma_1^2 = \sigma_2^2$$
$$H_1: \sigma_1^2 \neq \sigma_2^2 \qquad \alpha = 0.05$$

2. Select appropriate test statistic.

$$F = \frac{s_1^2}{s_2^2}$$

3. Decision Rule.

$df_1 = n_1 - 1 = 65 - 1 = 64$ (numerator degrees of freedom)
$df_2 = n_2 - 1 = 50 - 1 = 49$ (denominator degrees of freedom)
$F_{0.975}(64,49) = 1.76$ and $F_{0.975}(49,64) = 1.65$

Reject H_0 if $F \leq 1/1.65 = 0.61$ or if $F \geq 1.76$
Do Not Reject H_0 if $0.61 < F < 1.76$

4. Test Statistic.

$$F = \frac{s_1^2}{s_2^2} = \frac{4.1}{2.6} = 1.58$$

Statistical Inference for ($\mu_1 - \mu_2$)

5.   Conclusion. Do not reject Ho since $0.61 < 1.58 < 1.76$. We do not have significant evidence to show that $\sigma_1^2 \neq \sigma_2^2$. Therefore, the appropriate formula for the confidence interval for $(\mu_1 - \mu_2)$ is given under Case 2:

$$(\overline{X}_1 - \overline{X}_2) \pm Z_{1-\frac{\alpha}{2}} S_p \sqrt{\frac{1}{n_1} + \frac{1}{n_2}}$$

We first compute S_p:

$$S_p = \sqrt{\frac{(n_1 - 1) s_1^2 + (n_2 - 1) s_2^2}{n_1 + n_2 - 2}}$$

$$S_p = \sqrt{\frac{64\,(4.1) + 49\,(2.6)}{65 + 50 - 2}} = \sqrt{3.45} = 1.86$$

The confidence interval is:

$$(6.4 - 3.2) \pm 1.96\,(1.86) \sqrt{\frac{1}{65} + \frac{1}{50}}$$

$$3.2 \pm 1.96\,(0.348)$$

$$3.2 \pm 0.682$$

$$(2.52,\ 3.88)$$

We are 95% confident that the difference in the mean numbers of Emergency Room visits in 12 months is between 2.52 and 3.88 for children with asthma under age 5 as compared to children aged 6-10. If we conducted a test of H_0: $\mu_1 = \mu_2$ against H_1: $\mu_1 \neq \mu_2$, would we reject H_0 (Use the confidence interval) ?

Example 7.8. The health department at a major University is interested in whether there is a difference in the mean number of visits to the student health center between college freshman and sophomores. The following data are collected on random samples of freshman and sophomores, respectively over the course of one academic year:

Year in School	Number of Students	Mean number of visits to the student health center	Variance in number of visits to the student health center
Freshman	50	2.8	4.40
Sophomores	60	4.1	1.96

Test if there is a significant __difference__ in the mean number of visits to the student health center between university freshman and sophomores using a 5% level of significance. In this example we consider the number of visits a continuous variable.

1. Set up hypotheses.

$H_0: \mu_1 = \mu_2$
$H_1: \mu_1 \neq \mu_2$ $\alpha = 0.05$

where μ_1 = mean number of visits to the student health center among university freshman, and
μ_2 = mean number of visits to the student health among university sophomores.

2. Select appropriate test statistic.
In order to determine if this example falls into Case 2 or Case 3, a preliminary test of the equality of population variances must be conducted.

1. Set up hypotheses.

$H_0: \sigma_1^2 = \sigma_2^2$
$H_1: \sigma_1^2 \neq \sigma_2^2$ $\alpha = 0.05$

2. Select appropriate test statistic.

$$F = \frac{s_1^2}{s_2^2}$$

Statistical Inference for $(\mu_1 - \mu_2)$

3. Decision Rule.

$df_1 = n_1 - 1 = 50 - 1 = 49$ (numerator degrees of freedom)
$df_2 = n_2 - 1 = 60 - 1 = 59$ (denominator degrees of freedom)
$F_{0.975}(49,59) = 1.70$ and $F_{0.975}(59,49) = 1.76$

Reject H_0 if $F \leq 1/1.76 = 0.568$ or if $F \geq 1.70$
Do Not Reject H_0 if $0.571 < F < 1.70$

4. Test Statistic.

$$F = \frac{s_1^2}{s_2^2} = \frac{4.40}{1.96} = 2.24$$

5. Conclusion. Reject H_0 since $2.24 > 1.70$. We have significant evidence to show that $\sigma_1^2 \neq \sigma_2^2$. Therefore, for the purposes of this test of means, we apply the test statistic given under Case 3:

$$Z = \frac{\overline{X}_1 - \overline{X}_2}{\sqrt{\dfrac{s_1^2}{n_1} + \dfrac{s_2^2}{n_2}}}$$

3. Decision Rule.

Using Table 2B, the decision rule is:

Reject H_0 if $Z \geq 1.960$ or if $Z \leq -1.960$
Do Not Reject H_0 if $-1.960 < Z < 1.960$

4. Test Statistic.

$$Z = \frac{\overline{X}_1 - \overline{X}_2}{\sqrt{\dfrac{s_1^2}{n_1} + \dfrac{s_2^2}{n_2}}} = \frac{2.8 - 4.1}{\sqrt{\dfrac{4.40}{50} + \dfrac{1.96}{60}}} = \frac{-1.30}{0.370} = -3.51$$

5. Conclusion. Reject Ho since $-3.51 \leq -1.960$. We have significant evidence, $\alpha = 0.05$, to show that there is a significant difference in the mean number of visits to the student health center between university freshman and sophomores. For this test, p=0.001 (See Table 2B).

Statistical Inference for $(\mu_1 - \mu_2)$

SAS Example 7.8. The following output was generated using SAS Proc Ttest. A brief interpretation appears after the output.

SAS Output for Example 7.8

```
                        The TTEST Procedure

                             Statistics
                        Lower CL            Upper CL  Lower CL
Variable  Class       N    Mean    Mean       Mean    Std Dev  Std Dev
visits    freshman    50  2.8046  3.4071     4.0096   1.7709     2.12
visits    sophmore    60  4.1457  4.5059     4.8661   1.1819    1.3944
visits    Diff (1-2)      -1.767  -1.099     -0.43    1.5543    1.761

                             Statistics
                        Upper CL
         Variable  Class      Std Dev   Std Err   Minimum    Maximum
         visits    freshman    2.6418   0.2998    -1.548     7.7675
         visits    sophmore    1.7007    0.18      1.4388    8.0719
         visits    Diff (1-2)  2.0318   0.3372

                               T-Tests
         Variable  Method        Variances     DF    t Value    Pr > |t|
         visits    Pooled        Equal        108     -3.26      0.0015
         visits    Satterthwaite Unequal      81.9    -3.14      0.0023

                        Equality of Variances
            Variable   Method     Num DF   Den DF   F Value    Pr > F
            visits     Folded F     49       59      2.31      0.0022
```

Interpretation of SAS Output for Example 7.8

In the top section of the output, SAS provides summary statistics on the analytic variable (Numer of Visits) for each comparison group (freshman and sophmores) and then for the differences in means (freshman-sophmores). The summary statistics include the sample sizes, the sample mean (Mean) and 95% confidence intervals (CI) for the population means of each group and for the difference in means (the limits of the CI for the mean are labeled lower CL mean and upper CL mean), standard deviations and 95% confidence intervals for the population standard deviations of each group and for the differences (the limits of the CI for the standard deviation are

Statistical Inference for $(\mu_1 - \mu_2)$

labeled lower CL Std Dev and Upper CL Std Dev), standard errors $(s/\sqrt{n})$, minimums and maximums.

In the next section of the output SAS performs the test of hypothesis for equality of means. SAS actually carries out two different tests, one in which the population variances are assumed to be equal (we called this Case 2) and one in which the population variances are assumed to be unequal (we called this Case 3). SAS uses the formulas we summarized in Tables 7.5 and 7.3 for unequal and equal variances, respectively. The values of the test statistics appear under the column headed "t Value", and just before these SAS displays the degrees of freedom. Again, these are computed using the formulas from Tables 7.5 and 7.3. Finally, SAS provides two-sided p-values (assuming that the alternative hypothesis is H_1: $\mu_1 \neq \mu_2$).

The user must decide which situation (unequal or equal variances) is most appropriate. To aid in this decision, SAS provides a preliminary test of the homogeneity of variances (i.e., H_0: $\sigma_1^2 = \sigma_2^2$ vs H_1: $\sigma_1^2 \neq \sigma_2^2$). The results of the preliminary test appear at the bottom of the SAS Ttest output in the section titled "Equality of Variances." SAS provides an F statistic (computed by taking the ratio of the sample variances). SAS computes F by dividing the larger sample variance by the smaller, regardless of the group (1 or 2) designation. Therefore, the F statistic produced by SAS is always greater than or equal to 1. In this case, since the sample variance among freshmen is larger: $F = (2.12)^2/(1.39)^2 = 2.31$. The degrees of freedom associated with F are given by: $df_1 = n_1 - 1 = 50\text{-}1 = 49$ and $df_2 = n_2 - 1 = 60\text{-}1 = 59$. For the preliminary test, SAS produces the probability of observing a value of F greater than the value of the test statistic (denoted Prob $> F$). Since we are generally interested in a two sided test for the preliminary test (i.e., H_1: $\sigma_1^2 \neq \sigma_2^2$), the following rule should be applied to draw a conclusion: Reject H_0 if p-value $\leq \alpha_F/2$, where p - value is the one-sided p - value produced by SAS, and α_F is the (two-sided) level of significance selected for the preliminary test. In this case we Reject H_0 since the p - value, 0.0022, is smaller than 0.05/2 = 0.025. Therefore the population variances are significantly different and this example is considered an example of Case 3.

Because the preliminary test suggested that this example is an example of Case 3 (i.e., unequal variances), we use the output for the unequal variances case, and the test statistic is t = -3.14 with 81.9 degrees of freedom (we computed df=83 and the difference is due to rounding). For the main test, SAS produces a two-sided p - value. In this example, the p - value is 0.0023. We are interested in a two-sided test, the following rule should be applied: Reject H_0 if p - value $\leq \alpha$. In this example we reject H_0 since 0.0022 < 0.05. We have significant evidence to show that there is a significant difference in the mean number of visits to the student health center between university freshmen and sophomores. (NOTE: Some of the calculations produced by SAS differ somewhat from those we computed in Example 7.8 due to the fact that SAS is using more decimal places in computations.)

7.2.4 Case 4: Two Dependent Populations - The Data Are Matched or Paired

The techniques presented in the previous sections were for estimating or testing for a difference between two independent population means. In each application, random samples were selected from each population and sample statistics were compared to draw inferences about the difference in population means ($\mu_1 - \mu_2$). Case 4 is concerned with the comparison of two dependent (matched or paired) population means. The following example (Example 7.9) illustrates the distinction between applications involving two independent as compared to two dependent populations.

Example 7.9. An investigator is concerned that fraudulent claims have been made about new diet pills. The manufacturers of the pills claim that the pills are 100% effective in inducing weight loss within four weeks. To evaluate the claim, the investigator randomly selects a sample of individuals who are overweight and interested in weight loss for an evaluation. Subjects are randomly assigned to either the active treatment group or to a control group. Subjects in the active treatment group take the new diet pills as directed. Subjects in the control group take a *placebo*. A placebo is an inert substance designed to look exactly like the active treatment (in this case, exactly like the new diet pills). The subjects in the control group are instructed to take the placebo according to the same protocol as subjects in the treatment group. After four weeks, subject's weights are measured and compared. The hypotheses of interest are: H_0: $\mu_{treatment} = \mu_{control}$ vs. H_1: $\mu_{treatment} < \mu_{control}$. The alternative hypothesis reflects the situation in which the diet pills (treatment) are effective for weight loss. It is possible that the mean weight among subjects in the control group ($\mu_{control}$) will be less than the mean weight among subjects in the treatment group ($\mu_{treatment}$) after four weeks for a variety of reasons which may be unrelated to the effectiveness or ineffectiveness of the treatment.

A more efficient design for an application of this type is the two dependent samples test in which a single sample of subjects is drawn and evaluated. Each subject is weighed at the outset, instructed to follow the treatment (i.e., take the new diet pills as directed) and then weighed again after four weeks of treatment. Similar to the applications presented in Sections 7.2.1 through 7.2.3, two samples of data are analyzed. In this example, two weights are measured on each subject, one pre-treatment and the second at four weeks post-treatment. However, the samples are dependent or matched by subject. The dependent samples test is a technique used to remove extraneous variation. Examples 7.10 and 7.11 illustrate the technique. The formulas for the test statistics and confidence intervals when the samples are matched or paired are given in Table 7.6

Table 7.6 Case 4: Two Dependent Populations - The Data Are Matched or Paired

Attributes	Test Statistic	Confidence Interval
Samples are matched or paired, n (# pairs) ≥ 30	$Z = \dfrac{\overline{X}_d - \mu_d}{\dfrac{S_d}{\sqrt{n}}}$	$\overline{X}_d \pm Z_{1-\frac{\alpha}{2}} \dfrac{S_d}{\sqrt{n}}$
Samples are matched or paired, n (# pairs) < 30	$t = \dfrac{\overline{X}_d - \mu_d}{\dfrac{S_d}{\sqrt{n}}}$ df = n - 1	$\overline{X}_d \pm t_{1-\frac{\alpha}{2}} \dfrac{S_d}{\sqrt{n}}$ df=n - 1
where $\overline{X}_d$, S_d are the mean and standard deviation of the <u>difference scores.</u>		

Example 7.10. A nutrition expert is examining a weight loss program to evaluate its effectiveness (i.e., if participants lose weight on the program). Ten subjects are randomly selected for the investigation. Each subject's initial weight is recorded, they follow the program for six weeks, and they are again weighed. The data are given below:

Subject	Initial Weight	Final Weight
1	180	165
2	142	138
3	126	128
4	138	136
5	175	170
6	205	197
7	116	115
8	142	128
9	157	144
10	136	130

Statistical Inference for $(\mu_1 - \mu_2)$

The test of interest is:

H_0: $\mu_{\text{initial}} = \mu_{\text{final}}$

H_1: $\mu_{\text{initial}} > \mu_{\text{final}}$

where μ_{initial} = mean initial weight, and

μ_{final} = mean final weight.

The samples of initial and final weights are matched by participants. That is, for each participant an initial and a final weight is recorded and these two measurements are specific to each individual.

The two samples (initial weights and final weights) cannot be analyzed separately as in the two independent samples case (Sections 7.2.1 - 7.2.3). When the samples are matched or paired, we instead analyze difference scores. Difference scores can be computed by subtracting the first measurement from the second or vice versa. In this example, it is most intuitive to compute difference scores by subtracting the final weight from the initial weight. Computing differences in this fashion allows for a more intuitive interpretation of the differences, in this case differences reflect the weight lost by each participant over six weeks.

The following table displays the sample data, along with difference scores (d) and differences squared (d^2) which are required to compute the variance in difference scores:

Subject	Initial Weight	Final Weight	Difference (d)	Difference^2 (d^2)
1	180	165	15	225
2	142	138	4	16
3	126	128	-2	4
4	138	136	2	4
5	175	170	5	25
6	205	197	8	64
7	116	115	1	1
8	142	128	14	196
9	157	144	13	169
10	136	130	6	36
			66	740

The summary statistics on the difference scores (d) are given below:

Statistical Inference for $(\mu_1 - \mu_2)$

$$\overline{X}_d = \frac{\Sigma d}{n} = \frac{66}{10} = 6.6$$

$$s_d^2 = \frac{\Sigma d^2 - (\Sigma d)^2/n}{n-1} = \frac{740 - (66)^2/10}{9}$$

$$s_d^2 = 33.82 \qquad s_d = \sqrt{33.82} = 5.82$$

The test of hypotheses concerning the mean difference of two dependent populations follow the same 5 steps illustrated in previous applications.

1. Set up hypotheses.

$H_0 : \mu_d = 0$ $\qquad\qquad \alpha = 0.05$

$H_1 : \mu_d > 0$

where μ_d = mean difference in weights (or mean weight loss).

2. Select appropriate test statistic.
Because n=10, the test statistic is:

$$t = \frac{\overline{X}_d - \mu_d}{\dfrac{s_d}{\sqrt{n}}}$$

3. Decision rule. $\qquad$ df = n - 1 = 10 - 1 = 9

Reject H_0 if t $\geq$ 1.833
Do Not Reject H_0 if t < 1.833

4. Compute test statistic.

$$t = \frac{\overline{X}_d - \mu_d}{\dfrac{s_d}{\sqrt{n}}} = \frac{6.6 - 0}{\dfrac{5.82}{\sqrt{10}}} = 3.59$$

5. Conclusion: Reject Ho since 3.59 $\geq$ 1.833. We have significant evidence, $\alpha = 0.05$, to show that the mean weight loss following 6 weeks of the program is significantly greater that zero. For this test p=0.005 (See Table 3). A point estimate for the mean weight loss in the program is 6.6 pounds.

Statistical Inference for $(\mu_1 - \mu_2)$

SAS Example 7.10. SAS will conduct a two dependent samples test of hypothesis in its Means procedure using a similar approach to that used in the one sample test of hypotheses illustrated in Chapter 6. Recall, in the one-sample test SAS assumes that the test of interest is H_0: $\mu = 0$ vs. H_1: $\mu \neq 0$. In this application, we are interested in testing if the mean **difference** in weights is significantly greater than zero. In order to use the SAS Means procedure, we first create a new variable for each subject, which is simply the difference in weights computed by subtracting the final weight from the initial weight (i.e., difference (we call this variable DIFF in the SAS program) = initial – final weight). The difference score can be interpreted as the number of pounds lost over the course of the investigation, in this example 6 weeks. The output from the SAS means procedure is shown below. A brief interpretation appears after the output.

SAS Output for Example 7.10

```
                         The MEANS Procedure
                       Analysis Variable : diff
     N          Mean        Std Dev        Std Error     t Value    Pr > |t|
    ------------------------------------------------------------------------
    10       6.6000000      5.8156876      1.8390819       3.59      0.0059
    ------------------------------------------------------------------------
```

Interpretation of SAS Output for Example 7.10

SAS provides the number of observations (N=10). Again, SAS uses upper case, but this does not denote a population size. The mean of DIFF (i.e., the mean difference in initial and final weights, or the mean number of pounds lost) is 6.6. The standard deviation in DIFF is 5.8157 and the standard error (i.e., $s_d/\sqrt{n}$) is 1.8391. The test statistic (computing by taking the ratio of the mean DIFF to its standard error) is 3.590. The two-sided p-value is 0.0059. Because we are interested in a one-sided test, the following rule should be applied: Reject H_0 if p - value/2 < α. We reject H_0 since 0.0059/2 = 0.003 < α = 0.05. Therefore, we have significant evidence, $\alpha = 0.05$, to show that the mean weight loss following 6 weeks of the program is significantly greater that zero.

Example 7.11. There are continuing concerns about patient's appropriate use of over-the-counter medications. A particular concern involved the effects of over-the-counter antihistamines taken (inappropriately) in combination with alcohol on functional abilities. A study was undertaken to investigate these effects, according to a *cross-over design*. In a cross-over design each participant is given each treatment under investigation (as opposed to only a single treatment as was the case in previous examples).

In this investigation, each subject has functional ability measured (defined as time in seconds to complete a physical task) in the presence of alcohol and the antihistamine, and also has functional ability measured in the presence of alcohol and a placebo. The order of treatments (i.e., alcohol and antihistamine, alcohol and placebo) is randomly assigned to eliminate *carryover effects*. Carryover effects occur when subjects learn from one treatment and therefore improve under the second treatment solely due to the carryover from the first treatment and not due to the second treatment itself.

One hundred subjects are involved in the investigation. Differences in times to complete the task were taken as follows: Time under the influence of alcohol and antihistamine - Time under the influence of alcohol and placebo. Summary statistics on the differences in times to complete the task are as follows:

Number of subjects (n)	100
Mean difference in times (i.e., increase in time due to antihistamine)	25 seconds
Standard deviation in difference in times	20 seconds

The test of interest is as follows.

1. Set up hypotheses.

$H_0: \mu_d = 0$ $\qquad\qquad \alpha = 0.05$
$H_1: \mu_d \neq 0$

2. Test statistic (since the sample size is large, n = 100).

$$Z = \frac{\overline{X}_d - \mu_d}{\frac{s_d}{\sqrt{n}}}$$

3. Decision Rule.

Statistical Inference for $(\mu_1 - \mu_2)$

Reject H_0 if $Z \geq 1.96$ or if $Z \leq -1.96$
Do not Reject H_0 if $-1.96 < Z < 1.96$

4. Test Statistic

$$Z = \frac{\overline{X}_d - \mu_d}{\frac{s_d}{\sqrt{n}}} = \frac{25}{\frac{20}{\sqrt{100}}} = 12.5$$

5. Conclusion: Reject H_0 since $12.5 \geq 1.96$. We have significant evidence, $\alpha = 0.05$, to show that there is a difference in the time to complete a physical task under the influence of alcohol and antihistamine as compared to alcohol and a placebo. For this example, p=0.0001.

7.4   Power and Sample Size Determination

In Chapter 6 we introduced the concepts of precision (in estimation) and statistical power (in hypothesis testing), respectively. In the two sample applications concerning $(\mu_1 - \mu_2)$ these same concepts are of interest.

As in the one sample applications, power (i.e., Power = $1 - \beta$ = P(Reject $H_0|H_0$ False)) depends on 3 components:

        i) n_i = sample sizes (i=1,2)
        ii) α = level of significance = P(Type I error)
        iii)   ES = the Effect Size = the standardized difference in means specified under H_0 and H_1

The relationship between each component (i)-(iii) and statistical power is the same in the two sample case as it was in the one sample case. In the two sample applications the hypotheses are (assuming a two sided test):

        H_0: $\mu_1 = \mu_2$
        H_1: $\mu_1 \neq \mu_2$

or equivalently,

        H_0: $\mu_1 - \mu_2 = 0$
        H_1: $\mu_1 - \mu_2 \neq 0$.

where μ_1 = mean of population 1 and μ_2 = mean of population 2.

The Effect Size (ES) is defined as the standardized difference in the values of the parameter of interest specified under H_0 and H_1. In the two sample applications, the parameter of interest is ($\mu_1 - \mu_2$). The ES is defined as follows:

$$ES = \frac{\left|(\mu_1 - \mu_2)_{H_1} - (\mu_1 - \mu_2)_{H_0}\right|}{\sigma} \tag{7.4}$$

Under H_0: $\mu_1 - \mu_2 = 0$, therefore the ES reduces to:

$$ES = \frac{\left|(\mu_1 - \mu_2)_{H_1}\right|}{\sigma} = \frac{\left|(\mu_1 - \mu_2)\right|}{\sigma} \tag{7.5}$$

where σ = the common standard deviation (i.e., $\sigma_1 = \sigma_2 = \sigma$).

The power of a test is higher (or better) with larger sample sizes (n_1 and n_2), a larger level of significance and relative to a larger effect size. The following formula is used to determine the power of a two independent samples test with a two-sided alternative (i.e., H_0: $\mu_1 = \mu_2$ Vs. H_1: $\mu_1 \neq \mu_2$):

$$Power = P(Z > Z_{1-\frac{\alpha}{2}} - \frac{\left|\mu_1 - \mu_2\right|}{\sqrt{2\sigma^2/n}}) \tag{7.6}$$

where $\mu_1 - \mu_2$ is the difference in means under H_1,

σ is the common standard deviation of the characteristic of interest,

n is the common sample size (i.e., $n_1 = n_2 = n$), and

$Z_{1-\alpha/2}$ is the Z value with lower tail area $1-\alpha/2$.

We restrict our attention to situations where the sample sizes are equal ($n_1 = n_2$), there are formulas which can compute power for samples of unequal size.

Example 7.12. Suppose we wish to conduct the following test.

$$H_0: \mu_1 = \mu_2$$
$$H_1: \mu_1 \neq \mu_2 \qquad \alpha=0.05.$$

Find the power of the test if the difference in means is 2 units. Suppose that the standard deviation of the characteristic of interest is 3, and equal samples of size 20 are available.

$$\text{Power} = P(Z > Z_{1-\frac{\alpha}{2}} - \frac{|\mu_1 - \mu_2|}{\sqrt{2\sigma^2/n}}) \; = \; P(Z > 1.96 - \frac{2}{\sqrt{2(3)^2/20}}) \; = P(Z > 1.96\text{-}2.11)$$

$$= P(Z > \text{-}0.15) = 1 - 0.4404 = 0.5596.$$

There is a 56% probability that this test will detect a difference of 2 units in means with samples of size 20 at a 5% level of significance.

Suppose for the same problem we are interested in the probability of detecting a difference of 3 units in means. Find the power of the test with all other conditions the same as above.

$$\text{Power} = P(Z > Z_{1-\frac{\alpha}{2}} - \frac{|\mu_1 - \mu_2|}{\sqrt{2\sigma^2/n}}) \; = \; P(Z > 1.96 - \frac{3}{\sqrt{2(3)^2/20}}) \; = P(Z > 1.96\text{-}3.16)$$

$$= P(Z > \text{-}1.20) = 1 - 0.1151 = 0.8849.$$

There is an 88% probability that this test will detect a difference of 3 units in means with samples of size 20 at a 5% level of significance.

Example 7.13. Suppose we conducted a two sided test of the equality of mean cholesterol levels under two competing treatments and failed to reject the null hypothesis at $\alpha=0.05$. The sample data and an outline of the test are shown below. What is the probability that we committed a Type II error (i.e., find β)?

Summary Statistics	Treatment 1	Treatment 2
Sample Size	20	20
Mean Cholesterol Level	205	190
Standard Deviation	42	36

Statistical Inference for $(\mu_1 - \mu_2)$

$H_0: \mu_1 = \mu_2$

$H_1: \mu_1 \neq \mu_2$ $\qquad$ $\alpha=0.05$.

Assuming equal variances (Case 2),

Reject H_0 if $t \geq 2.042$ or if $t \leq -2.042$.

$$S_p = \sqrt{\frac{42^2 + 36^2}{2}} = 39.1,$$

$$t = \frac{25 - 190}{39.1\sqrt{\frac{1}{20} + \frac{1}{20}}} = 1.2$$

Do not reject H_0 because $-2.042 \leq 1.2 \leq 2.042$.

To find, β we fist compute power:

$$\text{Power} = P(Z > Z_{1-\frac{\alpha}{2}} - \frac{|\mu_1 - \mu_2|}{\sqrt{2\sigma^2/n}}) \ = \ P(Z > 1.96 - \frac{15}{\sqrt{2(39.1)^2/20}}) \ = P(Z > 1.96 - 1.21)$$

$=P(Z > 0.75) = 1 - 0.7734 = 0.2266$.

So, β = 1-0.2266 = 0.7734. There is a very high probability that we committed a Type II error (relative to a difference of 15 units in mean cholesterol levels). Are the two treatments the same with respect to their effect on cholesterol levels?

In many applications the number of subjects (i.e., n_1 and n_2) that can be involved depends on financial, logistic and/or time constraints. In other cases, the sample size required to ensure a certain level of power can be determined relative to alternative hypotheses of importance. The sample size required to ensure a specific level of power in a <u>two-sided two independent samples test</u> is given below:

$$n_i = 2\left(\frac{Z_{1-\frac{\alpha}{2}} + Z_{1-\beta}}{ES}\right)^2 \qquad (7.7)$$

where n_i is the minimum number of subjects required in sample i (i=1,2), $Z_{1-\alpha/2}$ is the value from the standard normal distribution with two-sided tail area equal to $1-\alpha/2$, $Z_{1-\beta}$ is the value from the standard normal distribution with a tail area equal to $1-\beta$, and ES is the effect size (7.5).

The sample size required to ensure a specific level of power in a <u>one-sided two independent samples test</u> is given below:

$$n_i = 2\left(\frac{Z_{1-\alpha} + Z_{1-\beta}}{ES}\right)^2 \tag{7.8}$$

where n_i is the minimum number of subjects required in sample i (i=1,2), $Z_{1-\alpha}$ is the value from the standard normal distribution with one-sided tail area equal to $1-\alpha$, $Z_{1-\beta}$ is the value from the standard normal distribution with a tail area equal to $1-\beta$, and ES is the effect size (7.5).

The following example illustrates the use of formula (7.7).

Example 7.14. A certain lung capacity measurement varies from day to day with a standard deviation of 0.3 liters. As individuals age from 30 to 50 years, their lung capacities decrease. Mean lung capacities decrease 0.02 liters/year for non-smokers and 0.04 liters per year for smokers. Over 20 years, mean lung capacities decrease 0.4 (0.02 X 20) liters for non-smokers and 0.8 (0.04 X 20) liters for smokers.

How many 30 year-old smokers and non-smokers should be followed for 20 years for there to be a 90% chance of recognizing the difference at $\alpha = 0.05$? Assume a two-sided test will be conducted.

The formula to determine sample sizes is given by (7.7):

$$n_i = 2\left(\frac{Z_{1-\frac{\alpha}{2}} + Z_{1-\beta}}{ES}\right)^2$$

The ES is (7.5):

$$ES = \frac{|\mu_1 - \mu_2|}{\sigma} = \frac{0.4}{0.3} = 1.33$$

At $\alpha = 0.05$, $Z_{1-\alpha/2} = Z_{0.975} = -1.96$. Similarly, for power=0.90, $\beta = 0.10$, therefore $Z_{1-\beta} = -1.282$.

Substituting:

$$n_i = 2\left[\frac{(1.96 + 1.282)}{1.33}\right]^2 = 11.88$$

Thus $n_1 = n_2 = 12$ subjects (24 total) are needed.

Example 7.15. Consider the application described in Example 7.13 where two treatments were compared for their effect on cholesterol levels. How many subjects would have been required in the study to ensure an 80% chance of detecting a difference of 20 units in cholesterol levels between treatments at $\alpha=0.05$? Assume that a two-sided test will be conducted. We use Sp to estimate the variation in cholesterol levels.

The ES is (7.5):

$$ES = \frac{|\mu_1 - \mu_2|}{\sigma} = \frac{20}{39.1} = 0.5$$

Since $\alpha = 0.05$, $Z_{1-\alpha/2} = Z_{0.975} = -1.96$. Similarly, for power=0.80, $\beta = 0.20$, therefore $Z_{1-\beta} = -0.84$.

Substituting:

$$n_i = 2\left[\frac{(1.96 + 1.282)}{10.5}\right]^2 = 62.72$$

Thus $n_1 = n_2 = 63$ subjects (126 total) are needed.

SAS Example 7.14. The following output was generated using SAS to determine the sample sizes (per group) required to ensure a specified power (we considered scenarios with 80% and 90% power) and differences in means of 0.4 and 0.3 units. A brief interpretation appears after the output.

OBS	ALPHA	BETA	Z_ALPHA2	Z_BETA	MU1	MU2	SIGMA	POWER	ES	N_2
1	0.05	0.1	1.95996	1.28155	0.8	0.4	0.3	0.9	1.33333	12
2	0.05	0.1	1.95996	1.28155	0.8	0.5	0.3	0.9	1.00000	22
3	0.05	0.2	1.95996	0.84162	0.8	0.4	0.3	0.8	1.33333	9
4	0.05	0.2	1.95996	0.84162	0.8	0.5	0.3	0.8	1.00000	16

Interpretation of SAS Output for Example 7.14

There is no SAS procedure specifically designed to determine the number of subjects required to detect a specific difference in the means of two independent populations. Similar to the program we developed in Chapter 6 to determine the sample size required to detect a specific effect size in the one sample test of hypothesis, we use SAS to program appropriate formulas. Once the formulas are implemented, users can evaluate different scenarios easily. In the example shown above, four scenarios are considered (denoted OBS 1-4, respectively). Scenario 1 corresponds to the situation presented in Example 7.14. Four variables are input into the program; the level of significance (alpha), the power (power), the mean for groups 1 (mu1), the mean for group 2 (mu2) and the standard deviation (sigma). Several variables are created in the program and the values of all variables are printed in the output. A description of the variables and an interpretation of results follows.

In scenario 1 (OBS=1), the level of significance (alpha) is set at 0.05 (5%), the probability of Type II error, β, is computed to be 0.10 (10%), the mean in group 1 (mu1) as specified as 0.8 the mean in group 2 was specified as 0.4 (mu2), the standard deviation (sigma) was specified at 0.3, and the power (power) was specified at 0.90 (90%). The effect size (es) was computed by dividing the absolute value of the difference in means between groups by the standard deviation. In scenario 1 (OBS=1), the effect size is 1.33. Twelve subjects (n_2) are required per group to ensure that the probability of detecting a 0.4 unit difference in means is 90% (i.e., power = 0.90), with a two sided level of significance of 5%. In scenario 2 (OBS=2) we change the mean in group 2 to 0.5. The effect size is reduced to 1.00 and a total of 22 subjects are required per group to ensure that the probability of detecting a 0.3 unit difference in means is 90% (i.e., power = 0.90), with a two sided level of significance of 5%.

In scenarios 3 and 4 (OBS=3 and 4) we consider the same scenarios and reduce the power to 0.80 (80%). The result is that fewer subjects are required. Nine and sixteen subjects are required per group, respectively, to ensure that the probability of detecting a 0.4 and 0.3 unit difference in means is 80%, with a <u>two sided</u> level of significance of 5%.

7.4 Key Formulas

Note: CI must be two tail but t/Z might be one tail if > or <.

APPLICATION	NOTATION/FORMULA	DESCRIPTION
Confidence Interval Estimate for $(\mu_1 - \mu_2)$	$n_1 \geq 30$ & $n_2 \geq 30$: $(\overline{X}_1 - \overline{X}_2) \pm Z_{1-\frac{\alpha}{2}} Sp\sqrt{\dfrac{1}{n_1} + \dfrac{1}{n_2}}$ *Case 2* $n_1 < 30$ or $n_2 < 30$: $t_{1-\frac{\alpha}{2}}$ $df = n_1 + n_2 - 2$	CI for two independent samples - See Tables 7.2, 7.3 and 7.5 for alternate formulas (Find Z in Table 2A)
Test Statistic for $H_0: \mu_1 = \mu_2$	$n_1 \geq 30$ & $n_2 \geq 30$: $Z = \dfrac{\overline{X}_1 - \overline{X}_2}{Sp\sqrt{\dfrac{1}{n_1} + \dfrac{1}{n_2}}}$ *Case 2* $n_1 < 30$ or $n_2 < 30$: t $df = n_1 + n_2 - 2$	Test statistic for two independent samples - See Tables 7.2, 7.3 and 7.5 for alternate formulas (Find critical Z in Table 2B)
Confidence Interval Estimate for μ_d	$n < 30$: $t_{1-\frac{\alpha}{2}}$ $df = n-1$ $n \geq 30$: $\overline{X}_d \pm Z_{1-\frac{\alpha}{2}} \dfrac{s_d}{\sqrt{n}}$ *Case 4*	CI for two dependent samples - See Table 7.6.
Test Statistic for $H_0: \mu_d = 0$	$n \geq 30$: $Z = \dfrac{\overline{X}_d - \mu_d}{s_d/\sqrt{n}}$ *Case 4* $n < 30$: t $df = n-1$	Test statistic for two dependent samples - See Table 7.6.
Find Power for test of $H_0: \mu_1 = \mu_2$	$Power = P\left(Z > Z_{1-\frac{\alpha}{2}} - \dfrac{|\mu_1 - \mu_2|}{\sqrt{2\sigma^2/n}}\right)$	Power of two sided test of the equality of means
Find n_i to Test $H_0: \mu_1 = \mu_2$	$n_i = 2\left[\dfrac{Z_{1-\frac{\alpha}{2}} + Z_{1-\beta}}{ES}\right]^2$, where $ES = \dfrac{|\mu_1 - \mu_2|}{\sigma}$	Sample sizes to detect ES with power $=1-\beta$

$$S_p = \sqrt{\frac{(n_1 - 1)S_1^2 + (n_2 - 1)S_2^2}{n_1 + n_2 - 2}} \qquad \text{if } n_1 = n_2, \; S_p = \sqrt{\frac{S_1^2 + S_2^2}{2}}$$

$$(\text{so } S_1 \leq S_p \leq S_2 \text{ or } S_2 \leq S_p \leq S_1)$$

7.5   Statistical Computing

Following are the SAS programs which were used to run the two independent samples tests (Cases 2 and 3), the two dependent samples test (Case 4), and to estimate the sample sizes to detect a specified effect size. The SAS procedures used and brief descriptions of their use are noted in the header to each example. Notes are provided to the right of the SAS programs (*in italics*) for orientation purposes and are not part of the programs. In addition, there are blank lines in the programs that follow which are solely to accommodate the notes. Blank lines and spaces can be used throughout SAS programs to enhance readability. A summary of the SAS procedures used in the examples is provided at the end of this section.

Two Independent Samples Test
SAS EXAMPLE 7.6   Compare Mean SAT Scores Between Men and Women

Random samples of eleven male high school students and twelve female high school students are selected within a particular school district. Students' scores on a public health awareness test are recorded Use the following data to test if the male students score significantly <u>higher</u> than the female students on the public health awareness test within this school district. Use SAS to run the appropriate test at a 5% level of significance.

Men	540	520	510	640	720	440	370	600	670	770	380	
Women	420	630	750	260	470	520	63	540	30	620	610	670

Program Code

options ps=62 ls=80;	*Formats the output page to 62 lines in length and 80 columns in width*
data in;	*Beginning of Data Step*
input gender $ test;	*Inputs variables **gender** (a character variable) and **test** score (numeric).*
cards;	*Beginning of Raw Data section.*
male 540	*actual observations*
male 520	
male 510	

Statistical Inference for ($\mu_1 - \mu_2$)

```
male 640
male 720
male 440
male 370
male 600
male 670
male 770
male 380
female 420
female 630
female 750
female 260
female 470
female 520
female 630
female 540
female 530
female 620
female 610
female 670
run;

proc ttest;

   class gender;

   var test;

run;
```

Procedure call Proc Ttest to run two independent samples test of means.
*Specification of grouping variable **gender**.*
*Specification of analytic variable **test**.*
End of procedure section.

Two Independent Samples Test
SAS EXAMPLE 7.8 Compare Mean Numbers of Visits to Health Center Between Freshman and Sophmores

The health department at a major University is interested in whether there is a difference in the mean number of visits to the student health center between college freshman and sophomores. Data are collected from each of 50 freshman and 60 sophomores reflecting the number of visits each made to the health center during the academic year. Use the data test if there is a significant <u>difference</u> in the mean number of visits to the student health center between university freshman and sophomores. Use SAS to run the appropriate test at a 5% level of significance.

NOTE: The raw data are not shown here and are abbreviated in the SAS program below.

Program Code

options ps=62 ls=80;	*Formats the output page to 62 lines in length and 80 columns in width*
data in;	*Beginning of Data Step*
input year $ visits;	*Inputs variables **year** (a character variable) and number of **visits** (numeric).*
cards;	*Beginning of Raw Data section.*
freshman 4	*actual observations*
freshman 0	
freshman 6	
.	
.	
.	
sophmore 2	
sophmore 1	
sophmore 0	
.	
.	
.	
run;	

Statistical Inference for $(\mu_1 - \mu_2)$

```
proc ttest;
```
Procedure call Proc Ttest to run two independent samples test of means.

```
  class year;
```
Specification of grouping variable **year.**

```
  var visits;
```
Specification of analytic variable **visits.**

```
run;
```
End of procedure section.

Two Dependent Samples Test

SAS EXAMPLE 7.10   Test the Effectiveness of a Weight Loss Program

A nutrition expert is examining a weight loss program to evaluate its effectiveness (i.e., if participants lose weight on the program). Ten subjects are randomly selected for the investigation. Each subject's initial weight is recorded, they follow the program for six weeks, and they are again weighed. The data are given below:

Subject	Initial Weight	Final Weight
1	180	165
2	142	138
3	126	128
4	138	136
5	175	170
6	205	197
7	116	115
8	142	128
9	157	144
10	136	130

Use SAS to test if there is evidence of a significant weight loss. Run the appropriate test at a 5% level of significance.

Program Code

options ps=62 ls=80;   *Formats the output page to 62 lines in length and 80 columns in width*

data in;   *Beginning of Data Step*
  input initial final;   *Inputs variables **initial** and **final** weight for each subject. Because both weights were recorded on the same subject (matched/paired data), they are input from the same record. Distinct records reflect different (independent) observations.*

Statistical Inference for $(\mu_1 - \mu_2)$

diff=final-initial;	*Creates new variable **diff** by taking the difference between the **final** and **initial** weights.*
cards;	*Beginning of Raw Data section.*
180 165	*actual observations*
142 138	
126 128	
138 136	
175 170	
205 197	
116 115	
142 128	
157 144	
136 130	
run;	
proc means n mean std stderr t prt;	*Procedure call Proc Means to run one sample test. Here we request specific options for testing (t and prt generate the test statistic and p-value, respectively).*
var diff;	*Specification of analytic variable **diff.***
run;	*End of procedure section.*

Determine the Number of Subjects Required to Detect a Specific Effect Size in a Test of Hypothesis About ($\mu_1 - \mu_2$)

SAS EXAMPLE 7.14 Sample Size Requirements

We wish to conduct the following test at the 5% level of significance:

$$H_0: \mu_1=\mu_2 \quad \text{vs.} \quad H_1: : \mu_1 \neq \mu_2$$

If the mean of group 1 is 0.8, how many subjects would be required to ensure that the probability of detecting a 0.4 (or a 0.3) unit difference is 80% (i.e., power = 0.80) ? Also consider scenarios with 90% power and assume that $\sigma = 0.3$.

Program Code

Code	Description
options ps=62 ls=80;	*Formats the output page to 62 lines in length and 80 columns in width*
data in;	*Beginning of Data Step.*
input alpha power mu1 mu2 sigma;	*Inputs 5 variables* **alpha, power, mu1, mu2** *and* **sigma**.
z_alpha2=probit(1-(alpha/2));	*Determines the value from the standard normal distribution with tail area 1-**alpha**/2 (See $Z_{\alpha/2}$ above)*
beta=1-power;	*Computes* **beta**.
z_beta=probit(beta);	*Determines the value from the standard normal distribution with tail area **beta** (See Z_b above)*
es=abs(mu1-mu2)/sigma;	*Computes the effect size (**es**)*
tempn_2=2*((z_alpha2+z_beta)/es)**2;	*Creates a temporary variable, called* **tempn_2**, *determined by formula (7.7).*
n_2=ceil(tempn_2);	*Computes a variable **n_2** using the ceil function which computes the smallest integer greater than* **tempn_2**.

```
/*                                                    Beginning of comment section
    Input the following information (required)
    alpha: Level of Significance: range 0.0 to 1.0 (e.g., 0.05),
    power: Power: range 0.0 to 1.0 (e.g., 0.80), and
    mu1: Mean in Group 1,
    mu2: Mean in Group 2, and
    sigma: Standard Deviation
*/                                                    End of comment section

cards;                                                Beginning of Raw Data section
0.05 0.90 0.8 0.4 0.3                                 actual observations
0.05 0.90 0.8 0.5 0.3
0.05 0.80 0.8 0.4 0.3
0.05 0.80 0.8 0.5 0.3
run;

proc print;                                           Procedure call.  Print to display
  var alpha beta mu1 mu2 sigma power es n_2;          computed variables.
run;
```

Statistical Inference for $(\mu_1 - \mu_2)$

Summary of SAS Procedures and Functions

The SAS Ttest procedure is used to run a two independent samples test of means. SAS also provides a preliminary test so that the user can determine if the population variances are equal or unequal. The Means Procedure is used to run a one sample test. Here we use the Means procedure to test if the mean difference (in the two dependent samples problem) is significantly different from zero. The procedures and specific options to perform the tests of hypothesis are shown below. Users should refer to the examples in this section for complete descriptions of the procedure and specific options. A general description of the procedure and options is provided in the table below.

Procedure	Sample Procedure Call	Description
proc ttest	proc ttest; class calss_var; var analytic_var;;	Conducts a two independent samples test of equality of means. Class_var is the name of the (dichotomous) variable that distinguishes the two groups and analytic_var is the name of the outcome or analytic variable (continuous).
proc means	proc means *n mean std stderr t prt*; var diff;	Conducts a one sample test of hypothesis (H_0: $\mu_d=0$ vs. H_1: $\mu_d \neq 0$)

The following SAS function was also used in this section in the programs to determine sample size requirements.

Function	Sample Call Statement	Description
Probit	Z_alpha = probit(*alpha*);	Determines the value from the standard normal distribution with lower tail area equal to *alpha*.

only consider case 2 or case 4

7.6   Problems

1.   In the recent years, there have been numerous incidents of mass layoffs, due in part to the changing political and economic climate. Fifteen facilities within the medical field were sampled, the mean number of employees laid off was 138 (s=15.4). Twenty facilities in the government contracting business were sampled, the mean number of employees laid off was 175 (s=20.9). Assume that the medical and contracting facilities are of approximately equal size and that the variations in numbers of layoffs are comparable. Test if there is a difference in the mean number of employee lay offs using a 5% level of significance.

2.   A pediatrician is interested in the long-term effects of an experimental medical treatment designed to improve joint flexibility in children affected with arthritis. Twelve children are randomly selected for the study and a measure of joint flexibility is taken on each child. After using the experimental treatment for 12 months, a second measure of joint flexibility is taken. The mean increase in flexibility is 4.6 units, with a standard deviation of 2.1 units. Construct a 90% confidence interval for the true mean increase in joint flexibility. (Assume that the difference scores are approximately normally distributed.)

3.   A study is conducted to investigate the numbers of hours that graduate students work in addition to a full-time class load. A random sample of 20 male graduate students is selected who work a mean of 16.4 hours per week with a variance of 4.7 hours. A second random sample of 20 female graduate students is selected who work a mean of 13.8 hours per week with a variance of 6.1 hours. Construct a 95% confidence interval for the difference in the mean numbers of hours worked between male and female graduate students. Assume that the variances are equal.

4.   An advertising agency is running a test comparing two forms of advertising for the same product, one involves a television campaign and the other is a print campaign. Market sectors are randomly assigned to receive a particular form of advertising and product sales are recorded during the month following the ad campaigns. Using the data provided below, test if there is a significant difference in product sales by the method of advertisement. Use $\alpha=0.05$.

	number of markets	mean sales	std. dev.
Television:	80	125	24.5
Print:	120	150	46.1

Statistical Inference for $(\mu_1 - \mu_2)$

5. A randomized trial is conducted to evaluate the effectiveness of a newly developed treatment for joint pain in patients with arthritis. The newly developed treatment is compared to an established treatment which has been shown to be effective. Two hundred patients with arthritis agree to participate in the investigation and are randomly assigned to either the newly developed treatment or to the established treatment and monitored. There are several clinical outcomes in the investigation. We focus on a secondary outcome quality of life (QOL). QOL scores (ranging from 0-100 with higher scores indicative of better QOL) are measured on each patient. The following data are observed:

Treatment	No. of Patients	Mean QOL	Std. Dev. QOL
Newly Developed	100	80.2	5.7
Established	100	75.4	6.1

a) Construct a 95% confidence interval for the difference in mean QOL scores between treatments.

b) Based on (a), is there a significant difference in mean QOL scores between treatments? Justify.

6. Following head and/or neck trauma, patients are recommended for long-term physical therapy. An observational study is conducted to assess whether there is a significant difference in the mean numbers of physical therapy sessions attended by male and female patients suffering head and/or neck trauma:

Gender	Number of Patients	Mean Number of Physical Therapy Sessions Attended	Standard Deviation
Male	20	14.6	6.3
Female	15	18.1	5.9

a) Construct a 95% confidence interval for the difference in mean numbers of physical therapy sessions attended between male and female patients suffering head and/or neck trauma.

b) Based on (a), is there a significant difference in the mean numbers of physical therapy sessions attended between male and female patients suffering head and/or neck trauma. Justify briefly.

Statistical Inference for $(\mu_1 - \mu_2)$

7. A randomized trial is run to compare two competing medications for peripheral vascular disease. One of the outcomes is self-reported physical functioning. After taking the assigned medication for 6 weeks, patients provide data on their abilities to perform various physical activities and a score is computed for each individual. The physical functioning scores range from 0 to 100 with higher scores indicative of better functioning. The data are shown below:

Medication	n	$\overline{X}$	s
1	25	70.5	24.3
2	25	76.3	21.6

Generate a 95% confidence interval for the difference in mean physical functioning scores between medications.

8. A clinical trial is conducted to compare an intervention to a control with respect to medication adherence. The intervention treatment consists of personalized scheduling and education regarding medication adherence. The control treatment reflects standard care. For the purposes of the study, medication adherence is measured as the percent of prescribed doses of medication taken over a 3 month period (e.g., a score of 100 indicates that all medications were taken as prescribed, 50 indicates that only half of all medications were taken as prescribed). Use the data given below to test whether medication adherence is significantly higher in the intervention group. Run the appropriate test at $\alpha=0.05$. Assume equal variances between groups.

	Intervention	Control
Number of patients	75	75
Mean Medication Adherence	85	72
Standard Deviation	12.6	14.3

9. A study is conducted comparing two competing medications for asthma. Sixteen subjects are enrolled in the study and patients are randomized to one of the competing medication treatments. The data shown below reflect asthma symptom scores for patients assigned to each treatment. Higher scores are indicative of worse asthma symptoms. Test if there is

Statistical Inference for ($\mu_1 - \mu_2$)

a significant difference in the mean asthma symptom scores between medications. Run the appropriate test at a 5% level of significance.

| Treatment A: | 55 | 60 | 80 | 65 | 72 | 78 | 68 | 71 |
| Treatment B: | 80 | 82 | 86 | 89 | 76 | 81 | 90 | 76 |

10. Suppose we are interested in whether there is a difference between the mean numbers of sick days taken by men and women in a local company. The numbers of sick days taken by men and women are shown below. Use the data to run the appropriate test at a 5% level of significance.

| MEN: | 5 | 10 | 2 | 0 | 6 | 4 | 5 | 15 |
| WOMEN: | 8 | 9 | 3 | 5 | 0 | 4 | 15 | |

11. A study reports that a newly developed medication reduces the systolic blood pressure in patients with hypertension. The medication ultimately makes it to market and a research group in a particular hospital wish to test its effectiveness among their hypertensive patients. A sample of 12 patients with hypertension agree to participate in the investigation. Their systolic blood pressures are measured before initiating treatment (baseline) and after 6 weeks of medication treatment. The mean reduction in systolic blood pressures is 15 units (baseline-post-treatment) with a standard deviation of 12 units. Test if the medication significantly reduces systolic blood pressure. Run the appropriate test at a 5% level of significance.

12. A research group wishes to conduct a study to compare a new medication to a medication they consider standard care in a randomized trial. One hundred patients with hypertension are enrolled and randomized to one of the two comparison treatments. After taking the assigned medication for 6 weeks, each patient has his/her systolic blood pressure (SBP) measured. Summary statistics are given below. Use the data to test if there is a significant difference in systolic blood pressures between medication groups. Run the appropriate test at a 5% level of significance.

Treatment	Number of Patients	Mean SBP	Standard Deviation
New	50	130	12
Standard	50	135	10

Statistical Inference for ($\mu_1 - \mu_2$)

13. The following table was presented in a paper describing a randomized trial comparing an experimental medication to a placebo for treatment of reflux.

Patient Characteristics	Experimental Treatment (n=100)	Placebo (n=100)	P
Mean (SD) Age, years	52 (5.6)	54 (5.2)	0.0090
Gender: % Female	63%	58%	0.4363
Mean (SD) Educational Level, years	13 (3.2)	10 (2.1)	0.0453
Mean (SD) Annual Income, $000s	$42 ($4.5)	$39 ($5.6)	0.0918
Clinical Characteristics			
Mean (SD) Number of Episodes Per Week	8 (3.2)	6 (2.9)	0.0001
Severity of Symptoms: % Minimal	21%	15%	0.0021
% Moderate	45%	21%	
% Severe	24%	64%	

a) Are there any statistically significant differences between the patients assigned to the experimental treatment and placebo? Justify briefly.

b) Consider the comparison of treatment with respect to patient age. Run the appropriate test using the data provided.

c) Consider the comparison of treatment with respect to the number of episodes per week Run the appropriate test using the data provided.

14. We wish to investigate whether there is a significant increase in CD4 cell count following a course of experimental antiretroviral therapy. A sample of 20 patients agree to participate in the investigation. As a comparative measure, each patient's CD4 cell count is measured prior to taking the experimental therapy. After taking the therapy, each patient's CD4 count is again measured. Using the following data, test whether there is a significant increase in CD4 cell counts following the experimental therapy. Run the appropriate test at $\alpha=0.05$.

Statistical Inference for (μ_1 - μ_2)

Statistic	CD4 Count Before Therapy	CD4 Count After Therapy	Increase
n	20	20	20
Mean	474	479	5
Std Dev	8.1	10.2	6.7

15. A study is conducted to assess whether females spend significantly more money out-of-pocket on prescription medications than males. The study is restricted to male and female patients in the same health plan who are 65 years of age or older. Samples of men and women are selected at random and asked the total number of dollars they spent on prescription medications over the last year. Use the following data to run the appropriate test. Use a 5% level of significance.

Gender	Number of Patientss	Mean Dollars Spent	Standard Deviation in Dollars Spent
Male	25	390	57
Female	40	425	65

16. The following table appeared in a journal article describing the results of a research project in which two different pain medications were compared.

Characteristic	Pain Medication 1	Pain Medication 2	p
Age (years)	48	51	0.0854
Educational Level (years)	13.1	12.9	0.3425
Annual Income	$38,456	$41,254	0.4351
No. Visits to MD In Past Year	4.2	5.6	0.0211

Are there any statistically significant differences between groups with respect to the characteristics shown above? Justify (be brief but complete).

Statistical Inference for $(\mu_1 - \mu_2)$

17. The following table summarizes data collected in a study comparing patients between hospitals. The variable summarized below is body mass index (BMI) computed as the ratio of weight in kilograms to height in meters squared.

	Enrollment Site	
BMI	Hospital 1	Hospital 2
N	100	100
Mean	21.6	24.8
Std Dev	2.1	1.8

Test if there is a significant difference in the mean BMI scores between hospitals. Run the appropriate test at a 5% level of significance.

18. The following data reflect body mass index scores for men and women who are considered high risk for coronary heart disease. Construct a 95% confidence interval for the true difference in BMI scores between men and women at high risk for cardiovascular disease.

	High Risk	
BMI	Men	Women
n	20	10
Mean	31.6	28.1
Std Dev	1.7	2.1

19. We wish to design a study to compare two anti-hypertensive medications. Two outcome variables will be considered: systolic and diastolic blood pressure.

a) How many subjects would be required to detect a difference of 20 units in mean systolic blood pressures between groups with 80% power and $\alpha=0.05$? Assume that the standard deviation in systolic blood pressure is 25.

b) How many subjects would be required to detect a difference of 10 units in mean diastolic blood pressures between groups with 80% power and $\alpha=0.05$? Assume that the standard deviation in diastolic blood pressure is 21.

Statistical Inference for $(\mu_1 - \mu_2)$

c)   What is the minimum number of subjects we need to satisfy BOTH (a) and (b)?

d)   Suppose that this study involves enrolling patients and measuring their blood pressures after 6 months – at which time the medications should show an effect. If 10% of the patients drop out of the study by 6 months, how many patients need to be enrolled to satisfy the requirements of (c)?

20.   We wish to design a study to compare an experimental therapy to a standard therapy in patients with diabetes. Subjects will be randomly assigned to one of the two therapy groups and outcomes of treatment will be measured at 6-months and 12-months after initiation of treatment. One of the primary outcomes is blood sugar level which will be measured at 6 months. The other primary outcome is total cholesterol level measured at 12 months. Summary statistics on the two outcome measures from a similar study are as follows: Mean (SD) of blood sugar level = 6.7 (2.1), Mean (SD) of total cholesterol level = 215 (38). We anticipate losing 10% of the subjects enrolled by 6 months and an additional 10% in the period from 6 to 12 months.

a)   How many subjects would we need to enroll in order to detect a difference of 1.8 units in mean blood sugar levels with 80% power (assume $\alpha=0.05$) between treatment groups?

b)   How many subjects would we need to enroll in order to detect a difference of 18 units in mean total cholesterol levels with 80% power (assume $\alpha=0.05$) between treatment groups?

c)   What is the minimum number of subjects we would need to enroll to satisfy both (a) and (b)?

21.   A nutritionist is investigating the effects of a rigorous walking program on systolic blood pressure (SBP) in patients with mild hypertension. Subjects who agree to participate have their systolic blood pressure measured at the start of the study and then after completing the 6-week walking program. The data are shown below:

Statistical Inference for $(\mu_1 - \mu_2)$

Subject	Starting SBP	Ending SBP
1	140	128
2	130	125
3	150	140
4	160	162
5	135	137
6	128	130
7	142	135
8	151	140

Based on the data, is there evidence that SBP is significantly reduced by the walking program? Run the appropriate test at the 5% level of significance.

22. A medical treatment is compared to a surgical treatment for gallstones. The comparison of treatments is based on patient-reported symptom score measured 3 weeks post-treatment. The symptom score ranges from 0 to 40 with higher scores indicative of worse symptoms. Suppose the 20 patients who received the medical treatment reported a mean symptom score of 20 with a standard deviation of 8 and the 20 patients who received the surgical treatment reported a mean symptom score of 14 with a standard deviation of 6. Construct a 95% confidence interval estimate for the difference in mean symptom scores between the medical and surgical treatments.

23. An anti-smoking campaign is being evaluated prior to its implementation in high schools across the state. A pilot study involving 6 volunteers who smoke is conducted. Each volunteer reports the number of cigarettes he/she smoked the day before enrolling in the study. Each then is subjected to the anti-smoking campaign which involves educational material, support groups, formal programs designed to reduce or quit smoking, etc. After 4 weeks, each volunteer again reports the number of cigarettes he/she smoked the day before. Based on the following pilot data, does it appear that the program is effective?

At Enrollment	21	15	8	6	12	20
After Campaign	12	10	10	6	10	20

24. A clinical trial is conducted to compare an intervention to a control with respect to medication adherence. The intervention treatment consists of personalized scheduling and education regarding medication adherence. The control treatment reflects standard

Statistical Inference for $(\mu_1 - \mu_2)$

care. For the purposes of the study, medication adherence is measured as the percent of prescribed doses of medication taken over a 3 month period (e.g., a score of 100 indicates that all medications were taken as prescribed, 50 indicates that only half of all medications were taken as prescribed). Use the data given below to test whether medication adherence is significantly higher in the intervention group. Run the appropriate test at $\alpha=0.05$. Assume equal variances between groups.

	Intervention	Control
Number of patients	75	75
Mean Medication Adherence	85	72
Standard Deviation	12.6	14.3

25. We wish to test the effects of a new diet on weights in animals. Two groups of 20 animals each are to be compared. If the standard deviation in weights are assumed to be 10 pounds in each group, what is the probability that a two independent sample t test will recognize a difference of 1 pound between groups using a 5% level of significance? What is the probability that a two independent sample t test will recognize a difference of 2 pounds between groups using a 5% level of significance?

26. The following table displays the background characteristics of subjects who participated in a trial to compare 2 different medications.

Background Characteristic	Medication 1 (n=100)	Medication 2 (n=100)	p
Mean (SD) Age, years	45 (7.2)	43 (8.1)	.6453
Gender: % Male	65%	48%	.0253
Educational Level			
% Less than High School	8%	6%	.0502
% High School Graduate	12%	14%	
% Some College	36%	38%	
% College Graduate	34%	36%	
% Post-Graduate	10%	6%	
Mean (SD) Annual Income, $	$41,352 ($8754)	$37,459 ($9687)	.2736
Race: % Non-White	36%	39%	.5342

a) Are there any statistically significant differences in background characteristics between patients receiving medications 1 and 2? Justify briefly.

Statistical Inference for ($\mu_1 - \mu_2$)

b) Write out the hypotheses tested and give the formula for the test statistic (no calculations) for comparing ages and incomes of the participants.

c) What is the power of this test to detect a difference in mean ages on the order observed between medication groups ? Use $\alpha=0.05$.

d) What is the power of this test to detect a 10 year difference in mean ages between medication groups ? Use $\alpha=0.05$.

Statistical Inference for $(\mu_1 - \mu_2)$

SAS Problems: Use SAS to solve the following problems.

1. A study is conducted comparing two competing medications for asthma. Sixteen subjects are enrolled in the study and patients are randomized to one of the competing medication treatments. The data shown below reflect asthma symptom scores for patients assigned to each treatment. Higher scores are indicative of worse asthma symptoms. Use SAS Proc Ttest to generate summary statistics on the symptom scores for each treatment and run the appropriate test at a 5% level of significance.

Treatment A:	55	60	80	65	72	78	68	71
Treatment B:	80	82	86	89	76	81	90	76

2. Suppose we are interested in whether there is a difference between the mean numbers of sick days taken by men and women in a local company. The numbers of sick days taken by men and women are shown below. Use SAS Proc Ttest to generate summary statistics on the numers of sick days taken by men and women and run the appropriate test at a 5% level of significance.

MEN:	5	10	2	0	6	4	5	15
WOMEN:	8	9	3	5	0	4	15	

   Use SAS to generate a 95% confidence interval for the difference in the mean number of sick days taken by men and women.

3. We wish to evaluate a program designed to improve health status in older patients with Type II diabetes. Health status is measured on a scale of 0-100 with higher scores indicative f better health. Health status measures are taken on each subject at baseline and then again after participating in the program. Based on the following data, is there evidence that the program significantly improves health status? Use SAS Proc Means to run the appropriate test at a 5% level of significance.

Statistical Inference for $(\mu_1 - \mu_2)$

| Baseline | 80 | 55 | 63 | 76 | 88 | 45 | 65 | 77 |
| Post-Program | 85 | 5 | 65 | 78 | 82 | 55 | 68 | 90 |

4. We wish to design a study in which a two independent samples t test will be run. We wish to consider several scenarios and to estimate the required sample sizes for each. Use SAS to determine the sample sizes required for each scenario. Suppose the mean for group 1 is 80. Consider means for group 2 of 90, 95 and 120; Levels of significance of 0.05 and 0.01; Standard deviations of 7 and 10 (for a total of 3x2x2 = 12 scenarios).

5. An anti-smoking campaign is being evaluated prior to its implementation in high schools across the state. A pilot study involving 6 volunteers who smoke is conducted. Each volunteer reports the number of cigarettes he/she smoked the day before enrolling in the study. Each then is subjected to the anti-smoking campaign which involves educational material, support groups, formal programs designed to reduce or quit smoking, etc. After 4 weeks, each volunteer again reports the number of cigarettes he/she smoked the day before. Based on the following pilot data, does it appear that the program is effective?

| At Enrollment | 21 | 15 | 8 | 6 | 12 | 20 |
| After Campaign | 12 | 10 | 10 | 6 | 10 | 20 |

Use SAS Proc Means to run the appropriate test at a 5% level of significance.

Introductory Applied Biostatistics

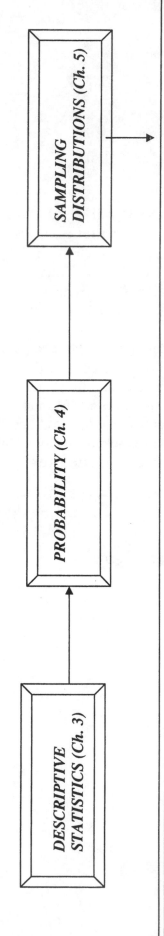

DESCRIPTIVE STATISTICS (Ch. 3) → **PROBABILITY (Ch. 4)** → **SAMPLING DISTRIBUTIONS (Ch. 5)**

STATISTICAL INFERENCE (Chapters 6-14)

Outcome Variable	Grouping Variable(s)/ Predictor(s)	Analysis	Chapter(s)
Continuous	-	Estimate μ, Compare μ to Known, Historical Value	6/13
Continuous	Dichotomous (2 Groups)	Compare Independent Means (Estimate/Test ($\mu_1-\mu_2$)) or the Mean Difference(μ_d)	7/13
Continuous	Discrete (> 2 Groups)	Test the Equality of K Means using Analysis of Variance ($\mu_1=\mu_2=..\mu_k$)	10/13
Continuous	Continuous	Estimate Correlation or Determine Regression Equation	11/13
Continuous	Several Continuous or Dichotomous	Multiple Linear Regression Analysis	11
Dichotomous	-	Estimate p, Compare p to Known, Historical Value	8
Dichotomous	**Dichotomous (2 Groups)**	**Compare Independent Proportions (Estimate/Test (p_1-p_2))**	**8/9**
Dichotomous	**Discrete (>2 Groups)**	**Test the Equality of k Proportions (Chi-Square Test)**	**8**
Dichotomous	Several Continuous or Dichotomous	Multiple Logistic Regression Analysis	12
Discrete	Discrete	Compare Distributions Among k Populations (Ch-Square Test)	8
Time to Event	Several Continuous or Dichotomous	Survival Analysis	14

CHAPTER 8: Categorical Data

8.1 Introduction

In Chapter 3 we defined variables as either *continuous* or *discrete*. Continuous (or measurement) variables assume, in theory, any value between the minimum and maximum value on a given measurement scale. Discrete variables take on a limited number of values, or categories and can be either *ordinal* or *categorical* variables. Ordinal variables take on a limited number of values or categories and the categories are ordered. For example, socioeconomic status (SES) is an ordinal variable with the following response categories: lower SES, lower middle, middle, upper middle, and upper SES. Categorical variables take on a limited number of categories and the categories are unordered. For example, hospital type is a categorical variable with the following response categories: teaching, non-teaching.

Statistical inference techniques applied to continuous (and sometimes ordinal) variables are concerned with means (μ) of those variables. Statistical inference techniques applied to categorical variables are concerned instead with the proportions of subjects in each response category.

In this chapter we present statistical inference techniques for categorical variables. We begin the discussion with the case of dichotomous variables from the binomial distribution (i.e., each observation takes on one of two possible values, usually denoted success and failure), and then consider applications involving variables from multinomial distributions (i.e., each observation takes on one of several (more than two) possible values). We discuss one-sample and two-sample techniques for proportions analogous to the techniques presented in Chapters 6 and 7 relative to means.

In Section 8.2 we present statistical inference techniques for the one sample case in which the analytic variable is dichotomous. In Section 8.3 we discuss cross-tabulation tables and several measures used to compare proportions between two independent populations. In Section 8.4 we discuss the evaluation of diagnostic tests. In Section 8.5 we present statistical inference techniques for the two sample case in which the analytic variable is dichotomous. In Section 8.6 we consider analytic variables from the multinomial distribution and introduce Chi-square tests. In Section 8.7 we discuss power and sample size determination. Key formulas are summarized in Section 8.8 and statistical computing applications are presented in Section 8.9.

8.2   Statistical Inference Concerning p

Recall the binomial distribution in which each observation takes on one of two possible values, called success and failure. Suppose for analytic purposes, successes are coded as 1's and failures are coded as 0's. The parameter of interest is the proportion of successes in the population, or the *population proportion*, denoted p. The population proportion is defined as:

$$p = \frac{\text{number of successes}}{\text{population size}} = \frac{X}{N} \qquad (8.1)$$

There are two areas of statistical inference concerning p, estimation and hypothesis testing. The goal in estimation is to make valid inferences about the population proportion based on a single random sample from the population. There are two types of estimates for the population proportion: *the point estimate* and *the confidence interval estimate*. The point estimate is the "best" single number estimate of the population proportion, and is given by the sample proportion, denoted $\hat{p}$ (shown below). The confidence interval estimate is a range of plausible values for the population proportion. ^: estimate

$$\hat{p} = \frac{\text{Number of successes in the sample}}{n} = \frac{X}{n} \qquad (8.2)$$

Example 8.1: Suppose we wish to estimate the proportion of patients in a particular physician's practice with diagnosed osteoarthritis. A random sample of 200 patients is selected and each patient's medical record is reviewed for the diagnosis of osteoarthritis. Suppose that 38 patients are observed with diagnosed osteoarthritis. A point estimate for the proportion of all patients in this physician's practice with osteoarthritis is given by (8.2): $\hat{p} = 38/200 = 0.19$ or 19%.

Similar to applications concerning means, it is useful to know the standard error in order to assess the variation in the point estimate, in this case the sample proportion. The standard error of the sample proportion is given by the following :

$$s.e.(\hat{p}) = \sqrt{\frac{p(1-p)}{n}} \qquad \text{[handwritten: } p \quad q=1-p \text{]} \qquad (8.3)$$

where s.e. = standard error, and
p = the population proportion.

In most applications, the population proportion, p, is unknown. For large samples the following can be used to estimate the standard error of the sample proportion:

$$s.e.(\hat{p}) = \sqrt{\frac{\hat{p}(1-\hat{p})}{n}} \qquad \text{[handwritten: large samples]} \qquad (8.4)$$

[handwritten: 5 + 5 > 10]

For applications involving binomial variables, a large sample is one with at least 5 successes and 5 failures. A large sample is defined as one that satisfies the following: $\min(n\hat{p}, n(1-\hat{p})) \geq 5$ (i.e., the smaller of $n\hat{p}$ and $n(1-\hat{p})$ must be greater than or equal to 5).

For Example 1, $\min(n\hat{p}, n(1-\hat{p})) = \min(200(0.19), 200(1-0.19)) = \min(38, 162) = 38 \geq 5\checkmark$, therefore the sample is sufficiently large. The standard error of the sample proportion is $\sqrt{0.19*0.81/200} = \sqrt{0.0007695} = 0.028$.

Confidence intervals for the population proportion can be generated using the techniques described in previous chapters. For large samples, we can appeal to the Central Limit Theorem for the derivation of the appropriate confidence interval (and test statistic in the test of hypothesis applications). Table 8.1 contains the formulas for the confidence interval for p and the test statistic for tests concerning p.

Table 8.1 Statistical Inference Concerning p

Attributes	Test Statistic	Confidence Interval
Simple random sample from binomial population,	$$Z = \dfrac{\hat{p} - p_0}{\sqrt{\dfrac{p_0(1-p_0)}{n}}}$$ where p_0 = value specified under H_0	$$\hat{p} \ \pm \ Z_{1-\frac{\alpha}{2}}\sqrt{\dfrac{\hat{p}(1-\hat{p})}{n}}$$ ↑ *table 2A (appendix)*
Large sample[1]	(1) $\min(np_0, n(1-p_0)) \geq 5$	(1) $\min(n\hat{p}, n(1-\hat{p})) \geq 5$

Table 2A in the Appendix contains the values from the standard normal distribution for commonly used confidence levels ($Z_{1-\alpha/2}$). When the sample size is large (i.e., if and only if $\min(n\hat{p}, n(1-\hat{p})) \geq 5$), the confidence interval formula given in Table 8.1 is appropriate. If the sample size is not sufficiently large, alternative formulas are available which are based on the binomial distribution and not the normal approximation which is given here.

Example 8.2: Consider the data from Example 1 and compute a 95% confidence interval for the proportion of all patients in the physician's practice with diagnosed osteoarthritis. The appropriate formula is given in Table 8.1:

$$\hat{p} \ \pm \ Z_{1-\frac{\alpha}{2}}\sqrt{\frac{\hat{p}(1-\hat{p})}{n}}$$

Substituting the sample data and the appropriate value from Table 2A for 95% confidence:

$$0.19 \ \pm \ 1.96\sqrt{\frac{0.19(1-0.19)}{200}}$$

$$0.19 \ \pm \ 1.96\,(0.028)$$

$$0.19 \ \pm \ 0.0549$$

(0.135, 0.245)

Thus, we are 95% confident that the true proportion of patients in this physician's practice with diagnosed osteoarthritis is between 13.5% and 24.5%.

SAS Example 8.2. The following output was generated using SAS Proc Freq which generates a frequency distribution table for a categorical (or ordinal) variable. In this example, we record whether or not each subject has been diagnosed with osteoarthritis (or not). The usual convention is to assign scores of 1 to successes (i.e., diagnosis of osteoarthritis) and scores of 0 to failures (i.e., free of osteoarthritis). The input data consists of designations (0 or 1) for each subject. SAS produces the following. A brief interpretation appears after the output.

SAS Output for Example 8.2

```
                        The FREQ Procedure
                                    Cumulative    Cumulative
    x      Frequency     Percent     Frequency      Percent
    ------------------------------------------------------------
    0         162         81.00          162         81.00
    1          38         19.00          200        100.00
```

Interpretation of SAS Output for Example 8.2

Of interest in the frequency distribution table produced by SAS are the frequencies (i.e., the total numbers) of respondents with 0's and 1's and the percent of respondents with 0's and 1's. There are 38 (out of 200) respondents scored as 1 (success=diagnosis of osteoarthritis). The percent of respondents with diagnosed osteoarthritis is 19%. SAS does not produce a confidence interval for the proportion of successes in a canned procedure. However, the code (i.e., the formula from Table 8.1) can be programmed to generate a confidence interval. We now illustrate the procedure for a test of hypothesis for a population proportion.

Example 8.3. Suppose that a diagnostic test has been shown to be 80% effective in detecting a genetic abnormality in human cells. An investigator modifies the diagnostic testing protocol and wishes to test if the new protocol has a detection rate that is significantly different from 80% in specimens known to possess the abnormality. The new protocol is applied to 300 independent specimens of human cells known to possess the abnormality. The abnormality is detected in 222 specimens. Run the appropriate test at a 5% level of significance.

1. Set up hypotheses.

    H_0: p = 0.80
    H_1: p ≠ 0.80   α = 0.05

2. Select appropriate test statistic.
    First, we check whether or not the sample size is sufficiently large:
    $\min(np_0, n(1-p_0)) = \min(300(0.8), 300(1-0.8)) = \min(240, 60) = 60 \geq 5$ ✓

The appropriate test statistic is given in Table 8.1:

$$Z = \frac{\hat{p} - p_0}{\sqrt{\dfrac{p_0(1-p_0)}{n}}}$$

3. Decision Rule (See Table 2B in the Appendix for the appropriate critical value)

    Reject H_0 if Z ≤ -1.960 or if Z ≥ 1.960
    Do Not Reject H_0 if -1.960 < Z < 1.960

4. Test Statistic
    Substituting the sample data and the value of p specified in H_0 (i.e., $p_0 = 0.80$):

$$\hat{p} = \frac{222}{300} = 0.74$$

$$Z = \frac{\hat{p} - p_0}{\sqrt{\dfrac{p_0(1-p_0)}{n}}}$$

$$Z = \frac{0.74 - 0.80}{\sqrt{\dfrac{0.80(1-0.80)}{300}}} = \frac{-0.06}{0.023} = -2.61$$

5. Conclusion: Reject H_0 since -2.61 $\leq$ -1.960. We have significant evidence, $\alpha = 0.05$, to show that the modified protocol has a significantly different detection rate than 80% (the rate for the original diagnostic test). The effectiveness rate is lower (74%) with the modified protocol as compared to the original. For this example, p=0.010 (see Table 2B).

Example 8.4. In the winter of 1994, fifteen percent of all pediatric outpatient visits at a particular clinic were due to a single strain of flu. An investigator hypothesizes that the proportion of visits due to flu will decrease if patients are provided with flu shots. Suppose that flu shots are given to a random sample of 125 pediatric patients in the fall of 1995. These patients are tracked over the following winter to assess whether or not they come to clinic for flu (visits for other illnesses or injuries are not counted). Twelve percent of these patients are seen in the pediatric clinic in the winter of 1995 for flu. Based on the data, is there evidence of a significant reduction in the proportion of patients seen in clinic for flu after receiving the flu shot ? Use a 5% level of significance.

1. Set up hypotheses.

H_0: p = 0.15
H_1: p < 0.15 $\alpha = 0.05$

2. Select appropriate test statistic.
First, check whether or not the sample size is sufficiently large:
$\min(np_0, n(1-p_0)) = \min(125(0.15), 125(1-0.15)) = \min(18.75, 106.25) = 18.75 \geq 5$ ✓

The appropriate test statistic is given in Table 8.1:

$$Z = \frac{\hat{p} - p_0}{\sqrt{\dfrac{p_0(1-p_0)}{n}}}$$

3. Decision Rule (See Table 2B in the Appendix for the appropriate critical value)

Reject H_0 if $Z \leq -1.645$
Do Not Reject H_0 if $Z > -1.645$

4. Test Statistic

$$\hat{p} = 0.12$$

$$Z = \frac{\hat{p} - p_0}{\sqrt{\dfrac{p_0(1-p_0)}{n}}}$$

Substituting the sample data and the value of p specified in H_0 (i.e., $p_0 = 0.15$):

$$Z = \frac{0.12 - 0.15}{\sqrt{\dfrac{0.15(1-0.15)}{125}}} = \frac{-0.03}{0.032} = -0.938$$

5. Conclusion: Do not reject H_0 since $-0.938 > -1.645$. We do not have significant evidence, $\alpha = 0.05$, to show a significant reduction in the proportion of patients seen in clinic for flu after receiving the vaccine.

Example 8.4 brings up the issue of clinical versus statistical significance. In the formal test of hypothesis we failed to reach statistical significance. However, we may have committed Type II error (e.g., a larger sample size may be required to detect an effect). In any statistical application it is extremely important to look at the direction and magnitude of the effect. In Example 8.4 there is a reduction in the proportion of flu cases seen following the vaccines. A point estimate is 0.12 or 12%. Our test did not indicate that this was statistically significantly lower than 15%, however, there is a reduction and it should be evaluated carefully. Is this reduction clinically important?

8.2   Cross-tabulation Tables

In applications involving discrete variables, *cross-tabulation* tables are often constructed to display the data. Cross-tabulation tables are also called R x C ("R by C") tables where R denotes the number of rows in the table and C denotes the number of columns in the table. A 2 x 2 table is illustrated in Example 8.5.

Example 8.5. A longitudinal study is conducted to evaluate the long term complications in diabetic patients treated under two competing treatment regimens. Complications are measured by incidence of foot disease, eye disease <u>or</u> cardiovascular disease within a 10 year observation period. The following 2 x 2 cross-tabulation table summarizes the data:

	Long Term Complications		
Treatment	YES	NO	Total
Treatment 1	12	88	100
Treatment 2	8	92	100
Total	20	180	200

The estimate of the population proportion of all patients who develop complications under Treatment 1 (p_1) is $\hat{p}_1 = 12/100 = 0.12$, by (8.2). This is equivalent to the estimate of the probability that a single patient develops complications under Treatment 1. The estimate of the probability that a single patient develops complications under Treatment 2 is $\hat{p}_2 = 8/100 = 0.08$.

The probability of success or outcome (in Example 8.5, the outcome of interest is the development of complications) is often called the *risk* of outcome. There are a number of statistics which are used to compare risks of outcomes between populations (or between treatments). These statistics are called *effect measures* and are described in detail in Chapter 9.

SAS Example 8.5. The following output was generated using SAS Proc Freq which generates a contingency table (or cross tabulation) when two variables are specified. SAS produces the following. A brief interpretation appears after the output.

risk difference: $p_1 - p_2$ $(-1, 1)$

relative risk: $\dfrac{p_1}{p_2}$ $(0, \infty)$

odds: $\dfrac{p_1}{1-p_1}$ $\left(\dfrac{\#\ success}{\#\ failure} \right)$

odds ratio: $\dfrac{p_1(1-p_2)}{p_2(1-p_1)}$

Categorical Data

336

SAS Output for Example 8.5

```
The FREQ Procedure
                              Table of trt by compl
                    Frequency|
                    Percent  |
                    Row Pct  |        compl
                    Col Pct  |yes     |z_no    |  Total
                    ---------+--------+--------+
                    trt_1    |     12 |     88 |    100
                             |   6.00 |  44.00 |  50.00
                             |  12.00 |  88.00 |
                             |  60.00 |  48.89 |
             trt    ---------+--------+--------+
                    trt_2    |      8 |     92 |    100
                             |   4.00 |  46.00 |  50.00
                             |   8.00 |  92.00 |
                             |  40.00 |  51.11 |
                    ---------+--------+--------+
                    Total          20      180     200
                                10.00    90.00  100.00

                         Sample Size = 200
```

Interpretation of SAS Output for Example 8.5

SAS generates a contingency table and in each cell of the table displays the Frequency, the Percent, the Row Percent and the Column Percent (See legend in top left corner of table). The Frequency is the number of subjects in each cell. The Percent is the percent of all subjects in each cell. For example, there are 12 patients in the top left cell (i.e., subjects on treatment 1 who also had complications). These patients reflect 6% of the total sample (i.e., 12/200=0.06). The Row Percent is the percent of subjects in the particular row which fall in that cell. For example, there are 100 patients on treatment 1 or 100 patients in the first row of the contingency table. The 12 patients who had complications reflect 12% of all patients on treatment 1 (i.e., 12/100 = 0.12). The Column Percent is the percent of subjects in the particular column which fall in that cell. For example, there are 20 patients who report complications. These 20 patients appear in the first column of the contingency table. The 12 patients in treatment 1 who had complications reflect 60% of all patients who had complications (i.e., 12/20 = 0.60). The row total and column total (called the marginal totals) are displayed to the right and at the bottom of the contingency table, respectively. Both row and column frequencies and percents (of total) are displayed.

Categorical Data

8.4   Diagnostic Tests: Sensitivity and Specificity   *skipped*

A diagnostic test is a tool used to detect outcomes or events which are not directly observable. For example, an individual may have a condition or disease that is not directly observable by a physician. A diagnostic test designed to detect such a condition and it can be used as a tool to assist the physician in detection. Desirable properties in diagnostic tests include the following:

* the diagnostic test will indicate an event when the event is present, and
* the diagnostic test will indicate a non-event when the event is absent.

Example 8.6. A clinical trial is conducted to evaluate a diagnostic screening test designed to detect chromosomal fetal abnormalities. Chromosomal fetal abnormalities are confirmed using amniocentesis. The diagnostic test is performed on a random sample of 200 pregnant women. Each woman later undergoes an amniocentesis. The following 2 x 2 cross-tabulation table summarizes the data:

	Diagnostic Test		
Amniocentesis	Positive	Negative	Total
Abnormal (Disease)	14	6	20
Normal (No Disease)	64	116	180
Total	78	122	200

Based on amniocentesis, the estimate of the population proportion of all women carrying fetuses with chromosomal abnormalities (p) is $\hat{p} = 20/200 = 0.10$, by (8.2).

The following statistics are used to describe diagnostic tests: the *sensitivity* of the test, the *specificity* of the test, the *false positive rate*, and the *false negative rate*. The statistics are defined below:

$$Sensitivity = P(\text{Positive Test}|\text{Disease}) \qquad (8.9)$$

$$Specificity = P(\text{Negative Test}|\text{No Disease})$$

$$False\ Positive\ Rate = P(\text{No Disease}|\text{Positive Test})$$

$$False\ Negative\ Rate = P(\text{Disease}|\text{Negative Test})$$

Categorical Data

In Example 8.6, the estimate of the sensitivity of the test is 14/20 = 0.70. The estimate of the specificity is 116/180 = 0.64. The estimate of the false positive rate is 64/180 = 0.36, and the estimate of the false negative rate is 6/20 = 0.30.

In most cases, higher sensitivities and higher specificities are desirable. There are instances, however, where a better test is determined by only one criterion (e.g., higher sensitivity). The false positive rate is also an important aspect of a test. A test with a high false positive rate needs to be carefully evaluated as patients may be put under unnecessary stress when given an incorrect positive test result.

SAS Example 8.6. The following output was generated using SAS Proc Freq which generates a contingency table (or cross tabulation) when two variables are specified. SAS does not produce the estimates of sensitivity, specificity, false positive rate and false negative rate directly, but these can be extracted from the contingency table. SAS produces the following. The statistics of interest are described after the output.

SAS Output for Example 8.6

```
                       The FREQ Procedure
                    Table of amnio by diagtest
            amnio       diagtest
            Frequency|
            Percent  |
            Row Pct  |
            Col Pct  |positive|negative|  Total
            ---------+--------+--------+
            abnormal |     14 |      6 |     20
                     |   7.00 |   3.00 |  10.00
                     |  70.00 |  30.00 |
                     |  17.95 |   4.92 |
            ---------+--------+--------+
            normal   |     64 |    116 |    180
                     |  32.00 |  58.00 |  90.00
                     |  35.56 |  64.44 |
                     |  82.05 |  95.08 |
            ---------+--------+--------+
            Total          78      122      200
                        39.00    61.00   100.
```

Categorical Data

Interpretation of SAS Output for Example 8.6

The sensitivity is the proportion of abnormal cases correctly classified by the test as positive = 14/20 = 0.70. This is the Row Percent in the top left cell of the table. The specificity is proportion of normal cases that are correctly classified by the test as negative = 116/180 = 0.644. This is the Row Percent of the bottom right cell of the table. The false positive rate is the proportion of normal cases that are incorrectly classified by the test as positive = 64/180 = 0.356. This is the Row Percent of the bottom left cell of the table. The false negative rate is the proportion of abnormal cases incorrectly classified by the test as negative = 6/20 = 0.30. This is the Row Percent of the top right cell of the table.

8.5   Statistical Inference Concerning $(p_1 - p_2)$

We often compare two independent populations with respect to the proportion of successes in each. One parameter of interest is the difference in proportions, the Risk Difference = $(p_1 - p_2)$, where p_1 = the proportion of successes in population 1 and p_2 = the proportion of successes in population 2.

The point estimate for the risk difference, or difference in independent proportions is given by:

$$\hat{p}_1 - \hat{p}_2 \tag{8.10}$$

where $\hat{p}_i$ = the sample proportion in population i (i=1,2).

If samples from both populations are sufficiently large (See criteria in Table 8.3), then the confidence interval formula shown in Table 8.3 can be used to estimate $(p_1 - p_2)$.

Table 8.3 Confidence Interval for (p_1 - p_2)

Attributes	Confidence Interval
Simple random samples from binomial populations, Independent populations, Large samples: $\min(n_1 \hat{p}_1, n_1(1-\hat{p}_1)) \geq 5$ <u>and</u> $\min(n_2 \hat{p}_2, n_2(1-\hat{p}_2)) \geq 5$	$(\hat{p}_1 - \hat{p}_2) \quad \pm \quad Z_{1-\frac{\alpha}{2}} \sqrt{\left(\frac{\hat{p}_1(1-\hat{p}_1)}{n_1} \right) + \left(\frac{\hat{p}_2(1-\hat{p}_2)}{n_2} \right)}$ where $\hat{p}_1 = X_1/n_1$ and $\hat{p}_2 = X_2/n_2$

Table 2A contains the values from the standard normal distribution for commonly used confidence levels. When the sample sizes are adequate (i.e., if and only if $\min(n_1 \hat{p}_1, n_1(1-\hat{p}_1)) \geq 5$ and $\min(n_2 \hat{p}_2, n_2(1-\hat{p}_2)) \geq 5$), the confidence interval formula given in Table 8.3 is appropriate. If either (or both) sample size(s) are not adequate, alternative formulas are available which are based on the binomial distribution and not the normal approximation which is given here.

Example 8.7. We wish to compare Emergency Room (ER) use in asthmatic children who do and do not use steroidal inhalers. We estimate the difference in proportions of children who use the ER for complications of asthma during a 6 month period. A random sample of 375 asthmatic children are selected from a registry. Two hundred and fifty are randomized to the treatment group and are instructed to use steroidal inhalers to manage their asthma. One hundred and twenty five are randomized to the control group and do not use steroidal inhalers. Both groups of children are followed for 6 months and monitored for ER use. Sixty of the children on inhalers used the ER during the 6 months for complications of asthma, while 19 of the children not on inhalers used the ER for complications of asthma during the same period. Construct a 95% confidence interval for the difference in the proportions of asthmatic children on inhalers as compared to those not on inhalers with respect to ER use during the 6 month time period.

The data layout is as follows:

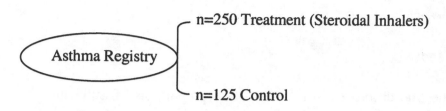

n=250 Treatment (Steroidal Inhalers)   X=60 Use ER

Asthma Registry

n=125 Control   X=19 Use ER

Study Start   6 Months

The sample proportions are shown below:

$$\hat{p}_1 = \frac{60}{250} = 0.24, \quad \hat{p}_2 = \frac{19}{125} = 0.15$$

A point estimate for the difference in proportions is given by (8.10):

$$\hat{p}_1 - \hat{p}_2 = 0.24 - 0.15 = 0.09$$

Now, check whether or not the sample sizes are sufficiently large:

$$\min(n_1 \hat{p}_1, n_1(1-\hat{p}_1)) = \min(250(0.24), 250(1-0.24)) = \min(60, 190) = 60 \geq 5 \checkmark$$
$$\underline{and}$$

$$\min(n_2 \hat{p}_2, n_2(1-\hat{p}_2)) = \min(125(0.15), 125(1-0.15)) = \min(19, 106) = 19 \geq 5 \checkmark$$

The appropriate formula from Table 8.3 is:

$$(\hat{p}_1 - \hat{p}_2) \pm Z_{1-\frac{\alpha}{2}} \sqrt{\left(\frac{\hat{p}_1(1-\hat{p}_1)}{n_1}\right) + \left(\frac{\hat{p}_2(1-\hat{p}_2)}{n_2}\right)}$$

Substituting the sample data and the appropriate value from Table 2A for 95% confidence:

$$0.09 \pm 1.960 \sqrt{\frac{0.24(1-0.24)}{250} + \frac{0.15(1-0.15)}{125}}$$

$$0.09 \pm 1.960 (0.042)$$

$$0.09 \quad \pm \quad 0.082$$

$$(0.008, \ 0.172)$$

Thus, we are 95% confident that the true difference in the population proportions of asthmatic children on inhalers as compared to children not on inhalers who used the ER during a 6 month period is between 0.8% and 17.2%. Based on the confidence interval estimate, can we say that there is a significant difference in the proportions of asthmatic children on inhalers as compared to not on inhalers who used the ER during a 6 month period ? (HINT: Does the confidence interval estimate include 0?)

NOTE: The two sample confidence interval concerning (p_1-p_2) estimates the <u>difference</u> in proportions, as opposed to the value of either proportion (as was the case in the one sample applications).

In some applications, it is of interest to compare two populations on the basis of the proportions of successes in each using a formal test of hypothesis. Table 8.4 contains the test statistic for tests concerning $(p_1 - p_2)$.

Table 8.4 Test Statistic for $(p_1 - p_2)$

Attributes	Test Statistic
Large samples[1], Simple random samples from binomial populations, Independent populations (1) $\min(n_1 \hat{p}_1, n_1(1-\hat{p}_1)) \geq 5$ <u>and</u> $\min(n_2 \hat{p}_2, n_2(1-\hat{p}_2)) \geq 5$	$$Z = \frac{\hat{p}_1 - \hat{p}_2}{\sqrt{\hat{p}(1-\hat{p})\left(\dfrac{1}{n_1} + \dfrac{1}{n_2}\right)}}$$ where $\hat{p}_1 = X_1/n_1$ and $\hat{p}_2 = X_2/n_2$ $$\hat{p} = \frac{(X_1 + X_2)}{(n_1 + n_2)}$$

Example 8.8. A new drug is being compared to an existing drug for its effectiveness in relieving headache pain. One hundred subjects who suffer from chronic headaches are randomly assigned to either Group 1: Existing Drug, or Group 2: New Drug. Subjects do not know which drug they are taking in this experiment. Subjects are provided with a single dose of the assigned drug and instructed to take the full dose as soon as they experience headache pain. They are instructed to record whether or not they experience relief from headache pain within 60 minutes. Among the fifty subjects assigned to Group 1: Existing Drug, 28 reported relief from headache pain within 60 minutes. Among the fifty subject assigned to Group 2: New Drug, 34 reported relief from headache pain within 60 minutes. Based on the data, is the proportion of subjects reporting relief from headache pain within 60 minutes under the New Drug significantly different from the proportion of subjects reporting relief from headache pain within 60 minutes under the Existing Drug ? Use a 5% level of significance.

The data layout is as follows:

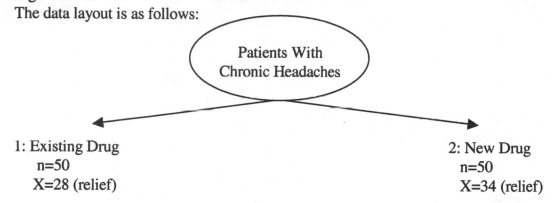

1. Set up hypotheses.

H_0: $p_1 = p_2$
H_1: $p_1 \neq p_2$ $\qquad \alpha = 0.05$

where p_1 = the proportion of patients who experience relief from headache pain using the Existing Drug, and p_2 = the proportion of patients who experience relief from headache pain using the New Drug.

2. Select appropriate test statistic.

The sample proportions are shown below:

$$\hat{p}_1 = \frac{28}{50} = 0.56 , \quad \hat{p}_2 = \frac{34}{50} = 0.68$$

First, check whether or not the sample sizes are sufficiently large:

$$\min(n_1 \hat{p}_1, n_1(1-\hat{p}_1)) = \min(50(0.56), 50(1-0.56)) = \min(28, 22) = 22 \geq 5 \checkmark$$
$$\underline{\text{and}}$$
$$\min(n_2 \hat{p}_2, n_2(1-\hat{p}_2)) = \min(50(0.68), 50(1-0.68)) = \min(34, 16) = 16 \geq 5 \checkmark$$

The appropriate test statistic is given in Table 8.4:

$$z = \frac{\hat{p}_1 - \hat{p}_2}{\sqrt{\hat{p}(1-\hat{p})\left(\frac{1}{n_1} + \frac{1}{n_2}\right)}}$$

3. Decision Rule

Reject H_0 if $Z \leq -1.960$ or if $Z \geq 1.960$
Do Not Reject H_0 if $-1.960 < Z < 1.960$

4. Test Statistic

$$z = \frac{\hat{p}_1 - \hat{p}_2}{\sqrt{\hat{p}(1-\hat{p})\left(\frac{1}{n_1} + \frac{1}{n_2}\right)}}$$

We compute the estimate of the common proportion:

$$\hat{p} = \frac{x_1 + x_2}{n_1 + n_2} = \frac{28 + 34}{50 + 50} = 0.62$$

Now substituting the sample data:

$$z = \frac{0.56 - 0.68}{\sqrt{0.62\,(1-0.62)\left(\dfrac{1}{50} + \dfrac{1}{50}\right)}} = \frac{-0.12}{0.097} = -1.24$$

5.   Conclusion: Do not reject H_0 since $-1.960 < -1.24 < 1.960$. We do not have significant evidence, $\alpha = 0.05$, to show a difference in the proportions of subjects experiencing relief from headache pain with the New Drug within 60 minutes.

8.6   Chi-Square Tests

In the previous sections we focused attention on variables from the binomial distribution (i.e., observations took on one of two possible values, called success and failure). We now consider variables from the multinomial distribution in which each observation takes on one of several (more than 2) possible values. The techniques described in the previous sections can be applied to sample data from a multinomial distribution. However, instead of estimating the proportion of successes in the population, we estimate the proportion of subjects in response category 1, the proportion of subjects in response category 2, and so on.

In this section we describe two tests, called the Goodness of Fit Test and the Test of Independence, which are used for tests of hypotheses in the presence of multinomial data in one-sample and two-or more sample applications, respectively. Both tests involve a test statistic that follows a chi-square distribution (χ^2). The tests and test statistics are described below.

8.6.1.   Goodness of Fit Test

We begin with the case of a categorical variable measured in a single sample. The following example illustrates the Goodness of Fit Test.

Example 8.9. Following coronary artery bypass graft (CABG) surgery, patients are encouraged to participate in a cardiac rehabilitation program. The program lasts approximately 14 weeks and includes exercise training, nutritional information and general lifestyle guidelines. One particular hospital is offering 3 cardiac rehabilitation programs which are identical in content but are offered at three different times. The administration is interested in whether the three time slots are equally popular among (or convenient for) the patients. A total of 100 patients are involved in the in investigation and each patient is asked to select the day and time that is most convenient. The following data are observed:

Time Slot:	Mondays 6:00-7:30 PM	Thursdays 4:00-5:30 PM	Saturdays 8:00-9:30 AM
Number of patients	47	32	21

The day and time variable follows a multinomial distribution with three response categories, denoted k=3 (i.e., Mon 6:00-7:30 PM, Thurs 4:00-5:30 PM, Sat 8:00-9:30 AM). The proportions of patients in the sample selecting each day and time are given by (8.2): $\hat{p}_1 = 47/100 = 0.47$, $\hat{p}_2 = 32/100 = 0.32$, and $\hat{p}_3 = 21/100 = 0.21$.

We wish to use the data to test the hypothesis that the three day and time options are equally popular. Mathematically, this is represented as follows:

1.   Set up hypotheses.

H_0: $p_1 = p_2 = p_3$ (Population proportions are equal)
H_1: H_0 is false   (Population proportions are not all equal)

NOTE: We do not write H_1: $p_1 \neq p_2 \neq p_3$, as we will reject H_0 in favor of H_1 if <u>any</u> of the three proportions are not equal, not only if <u>all</u> three are not equal (e.g., we will reject H_0 if $p_1 \neq p_2$, regardless of whether p_3 is equal to either p_1 or p_2).

$\alpha = 0.05$

Equivalent to the above is the following:

H_0: $p_1 = 0.33$, $p_2 = 0.33$, $p_3 = 0.33$
H_1: H_0 is false

2.   Select appropriate test statistic.

In tests involving multinomial distributions, the test statistic is no longer based on the sample mean but on the *observed frequencies*, or numbers of subjects in each response category. In this example, we observed 32 patients in the first response category, 47 in the second and 21 in the third. If the null hypothesis were true, that is if the true proportions of patients in each response category were equal (i.e., if $p_1 = p_2 = p_3 = 0.33$), then we would have *expected* approximately 33 patients to select each time slot (since n=100).

χ^2 *tests are based on the agreement between expected (under H_0) and observed (sample)* *frequencies.* The χ^2 statistic for testing whether the distribution of a single multinomial variable is as specified under H_0 is given by:

$$\chi^2 = \Sigma \frac{(O-E)^2}{E} \tag{8.11}$$

where Σ indicates summation over the k response categories,
  O = observed frequencies, and
  E = expected frequencies (i.e., if H_0 is true, or under H_0).

The statistic follows a χ^2 distribution and has df = k - 1, where k = the number of response categories. The general form of the χ^2 distribution is shown in Figure 8.1.

Figure 8.1   Chi-Square Distribution

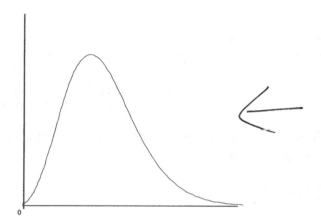

Notice that all of the values in the χ^2 distribution are greater than or equal to zero. In order to test the hypotheses of interest, we need to determine an appropriate critical value. In χ^2 tests, we reject the null hypothesis in favor of the alternative hypothesis if the value of the test statistic is large. The test statistic (8.11) is large when the observed and expected frequencies are not similar. In such a case, we reject H_0.

Categorical Data

3.   Decision Rule
In order to select the appropriate critical value, we first determine the degrees of freedom.
$$df = k - 1 = 3 - 1 = 2.$$
The appropriate critical value of χ^2 is $\chi^2 = 5.99$ from Table 5. The Decision Rule is:

Reject H_0 if $\chi^2 \geq 5.99$
Do Not Reject H_0 if $\chi^2 < 5.99$

4.   Test Statistic
To organize the computations of the test statistic (8.11), the following table is used:

Time Slot:	Mondays 6:00-7:30 PM	Thursdays 4:00-5:30 PM	Saturdays 8:00-9:30 AM	TOTAL
O=observed frequency	47	32	21	100
E=expected frequency	33.3	33.3	33.3	100
(O - E)	13.7	-1.3	-12.3	0
$(O - E)^2/E$	$(13.7)^2/33.3$ =5.64	$(-1.3)^2/33.3$ =0.05	$(-12.3)^2/33.3$ =4.54	10.23

NOTE: The sum of the expected frequencies is equal to the sum of the observed frequencies (n=100).

The test statistic is $\chi^2 = 10.23$.

5.   Conclusion: Reject H_0 since $10.23 \geq 5.99$. We have significant evidence, $\alpha = 0.05$, to show that the three time slots are not equally popular (or convenient) among the patients. In fact, almost half (47%) of the patients selected the Monday 6:00-7:30 PM slot. For this example p=0.01 (See Table 5).

Example 8.10. Volunteers at a teen hotline have been assigned based on the assumption that 40% of all calls are drug-related, 25% are sex-related (e.g., date-rape), 25% are stress-related and 10% concern educational issues. For this investigation, each call is classified into one category based on the primary issue raised by the caller. To test the hypothesis, the following data are collected from

120 randomly selected calls placed to the teen hotline. Based on the data, is the assumption regarding the distribution of topic issues appropriate ?

	Topic Issue			
	Drugs	Sex	Stress	Education
Number of calls	52	38	21	9

1.   Set up hypotheses.

H_0: $p_1 = 0.40$, $p_2 = 0.25$, $p_3 = 0.25$, $p_4 = 0.10$
H_1: H_0 is false
         OR
H_0: Distribution across categories is 0.40, 0.25, 0.25, 0.10
H_1: H_0 is false $\alpha = 0.05$

2.   Select appropriate test statistic.

$$\chi^2 = \Sigma \frac{(O - E)^2}{E}$$

where Σ indicates summation over the k response categories,
        O = observed frequencies, and
        E = expected frequencies (i.e., if H_0 is true, or under H_0).

3.   Decision Rule
     In order to select the appropriate critical value, we first determine the degrees of freedom.
                   df = k - 1 = 4 - 1 = 3.

The appropriate critical value of χ^2 is $\chi^2 = 7.815$ from Table 5. The Decision Rule is:

        Reject H_0 if $\chi^2 \geq 7.815$
        Do Not Reject H_0 if $\chi^2 < 7.815$

Categorical Data

4.   Test Statistic

To organize the computations of the test statistic (8.11), the following table is used:

Topic Issue	Drugs	Sex	Stress	Education	TOTAL
O=observed frequency	52	38	21	9	120
E=expected frequency	48	30	30	12	120
(O - E)	4	8	-9	-3	0
$(O - E)^2/E$	$(4)^2/48 =$ 0.33	$(8)^2/30=$ 2.13	$(-9)^2/30=$ 2.70	$(-3)^2/12=$ 0.75	5.913

NOTE: The expected frequencies are computed assuming that H_0 is true, or under H_0: $120(0.40) = 48$, $120(0.25) = 30$ and $120(0.10) = 12$.

The test statistic is $\chi^2 = 5.913$.

5.   Conclusion: Do not reject H_0 since $5.913 < 7.815$. We do not have significant evidence, $\alpha = 0.05$, to show that the distribution of topic issues in the calls placed to the teen hotline is not as assumed (i.e., 40% drug-related, 25% sex-related, 25% stress-related and 10% education-related).

Example 8.11. Genetic counselors work with pregnant women (usually women at high risk of fetal abnormalities or those who might not be at high risk but screen positive for abnormalities based on standard screening tests) and hypothesize that about ½ of all abnormalities are trisomy 21, 1/3 are trisomy 18 and the remainder are trisomy 13 (trisomy indicates three copies of a particular chromosome, e.g., 21, and reflects a particular abnormality associated with that chromosome). To test the hypothesis, a sample of 200 pregnant women who deliver babies with abnormalities are studied. The specific abnormalities are recorded and are summarized below. Based on the data, is the assumption regarding the distribution of abnormalities appropriate ?

	Trisomy 21	Trisomy 18	Trisomy 13	Total
Number of Women	107	70	23	200

1.   Set up hypotheses.

H_0: $p_1 = 0.50$, $p_2 = 0.33$, $p_3 = 0.17$
H_1: H_0 is false $\alpha = 0.05$

2.   Select appropriate test statistic.

$$\chi^2 = \Sigma \frac{(O - E)^2}{E}$$

where Σ indicates summation over the k response categories,
         O = observed frequencies, and
         E = expected frequencies (i.e., if H_0 is true, or under H_0).

3.   Decision Rule
      In order to select the appropriate critical value, we first determine the degrees of freedom.
                              df = k - 1 = 3 - 1 = 2.

The appropriate critical value of χ^2 is $\chi^2 = 5.99$ from Table 5. The Decision Rule is:
         Reject H_0 if $\chi^2 \geq 5.99$
         Do Not Reject H_0 if $\chi^2 < 5.99$

4.   Test Statistic
      To organize the computations of the test statistic (8.11), the following table is used:

	Trisomy 21	Trisomy 18	Trisomy 13	TOTAL
O=observed frequency	107	70	23	200
E=expected frequency	100	66	34	200
(O - E)	7	4	-11	0
$(O - E)^2/E$	$(7)^2/100 =$ 0.49	$(4)^2/66=$ 0.24	$(-11)^2/34=$ 3.55	4.29

NOTE: The expected frequencies are computed assuming that H_0 is true.

The test statistic is $\chi^2 = 4.29$.

5.   Conclusion: Do not reject H_0 since $4.29 < 5.99$. We do not have significant evidence, $\alpha = 0.05$, to show that the distribution of abnormalities is not as assumed (i.e., 50% trisomy 21, 33% trisomy 18 and 13% trisomy 13).

8.6.2. Tests of Independence

We now consider applications involving two or more samples or two categorical variables, where interest lies in evaluating whether these two categorical variables are related (dependent) or unrelated (independent). The following example illustrates the use of the χ^2 Test of Independence.

Example 8.12. The following data were collected in a multi-site observational study of medical effectiveness in Type II diabetes. Three sites were involved in the study: a health maintenance organization (HMO), a University teaching hospital (UTH), and an independent practice association (IPA). Type II diabetic patients were enrolled in the study from each site and monitored over a three year observation period. The data shown below display the treatment regimens of patients measured at baseline by site.

Site	*Treatment Regimen*			
	Diet & Exercise	Oral Hypoglycemics	Insulin	TOTAL
HMO	294	827	579	1700
UTH	132	288	352	772
IPA	189	516	404	1109
TOTAL	615	1631	1335	3581

The table shown above is a 3X3 *cross-tabulation table* or a *contingency table*. Both site and treatment regimen are categorical variables. Site is called the *row variable* and treatment regimen is called the *column variable*. The number of rows in the table is denoted R and the number of columns is denoted C. In this example, R=3 and C=3. The *row totals* are shown on the right side of the table. The *column totals* are shown at the bottom of the table. The row and column totals are called the *marginal totals*. The 9 combinations of site and treatment regimen are called the cells of the table (e.g., patients in the HMO treated by diet and exercise denote one cell of the table, patients in the HMO treated by oral hypoglycemics denote another, and so on).

We wish to use the data to test the hypothesis that the two variables (site and treatment regimen) are independent (i.e., no difference in treatment regimens across sites). The hypotheses are written as follows:

1. Set up hypotheses.

    H_0: Site and Treatment Regimen are independent   (No relationship between site and treatment regimen)

    H_1: H_0 is false   (Site and treatment regimen are related)

          NOTE: In the Test of Independence, the hypotheses are generally expressed in words as opposed to mathematical symbols.

    $\alpha = 0.05$

2. Select appropriate test statistic.

    The test statistic in the Test of Independence is similar to the test statistic used in the Goodness of Fit test illustrated in the previous section. The test statistic is based on the *observed frequencies*, or numbers of subjects in each cell of the contingency table. In this example, we involved 1700 patients from the HMO, 772 from the UTH and 1109 from the IPA. If the null hypothesis, H_0, is true (i.e., if there is no relationship between site and treatment regimen), we would expect the distribution of patients by treatment regimen to be similar across sites (i.e., the proportions of patients in each treatment regimen would be approximately equal in each of the sites).

    Again, χ^2 tests are based on the agreement between expected (under H_0) and observed (sample) frequencies. Recall from probability theory, that when two events are independent, the probability of their intersection is given by:

$$P (A \text{ and } B) = P(A) \cdot P(B). \tag{8.12}$$

For example, if site and treatment regimen are independent, then the probability that a patient is in the HMO <u>and</u> treated by diet and exercise is given by:

$$P(HMO \text{ and Diet/Exercise}) = P(HMO) \cdot P(Diet/Exercise)$$
$$= (1700/3581)(615/3581) = (0.4747)(0.1717) = 0.0815.$$

Similarly, if site and treatment regimen are independent, then the probability that a patient is in the UTH and treated by insulin is given by:

$$P(\text{UTH and Insulin}) = P(\text{UTH}) \cdot P(\text{Insulin})$$
$$= (772/3581)(1335/3581) = (0.216)(0.373) = 0.0806.$$

Therefore, if site and treatment regimen are independent, the probabilities (or proportions) of patients in each cell of the table can be computed using (8.12). To compute the test statistic, we must compute the <u>expected frequencies</u> (i.e., the <u>expected numbers</u> of patients in each cell of the table if site and treatment regimen are independent). Formula (8.12) yields the proportions of patients in each cell. To convert these proportions to frequencies, we use the following:

$$\text{Expected Cell Frequency} = n \bullet P(\text{cell}) \tag{8.13}$$

This is equivalent to:

$$\text{Expected Cell Frequency} = (\text{Row Total} \bullet \text{Column Total})/n \tag{8.14}$$

For example, if site and treatment regimen are independent, then the expected number of patients in the HMO and treated by diet and exercise is given by:

$$\text{Expected Frequency (HMO and Diet/Exercise)} = 3581(0.0815) = 291.9$$

by (8.13). Equivalent to this is:

$$\text{Expected Frequency (HMO and Diet/Exercise)} = (1700)(615)/3581 = 291.9$$

by (8.14).

The χ^2 statistic for Tests of Independence is given by:

$$\chi^2 = \Sigma \frac{(O - E)^2}{E} \tag{8.15}$$

where Σ indicates summation over all cells of the contingency table,
   O = observed frequencies, and
   E = expected frequencies (i.e., if H_0 is true, or under H_0).

The statistic follows a χ^2 distribution and has df = $(R - 1)(C - 1)$, where R = the number of rows in the contingency table and C = the number of columns in the contingency table.

Similar to the Goodness of Fit tests, we reject the null hypothesis in favor of the alternative hypothesis if the value of the test statistic is large. The test statistic (8.15) is large when the observed and expected frequencies are not similar.

3.   Decision Rule
      In order to select the appropriate critical value, we first determine the degrees of freedom.
$$df = (R - 1)(C - 1) = (3-1)(3-1) = 2(2) = 4.$$

The appropriate critical value of χ^2 is $\chi^2 = 9.49$ from Table 5. The Decision Rule is:
      Reject H_0 if $\chi^2 \geq 9.49$
      Do Not Reject H_0 if $\chi^2 < 9.49$

4. Test Statistic
      To organize the computations of the test statistic (8.15), we use the contingency table given above. The observed frequencies in each cell are shown. We compute the expected frequencies for each cell using (8.14) and display expected frequencies in parentheses to distinguish them from the observed frequencies. The computations are shown in detail for a few sample cells.

Site	Diet & Exercise	Oral Hypoglycemics	Insulin	TOTAL
		Treatment Regimen		
HMO	294 *O* ((1700*615)/3581 = 291.9) *E*	827 ((1700*1631)/3581 = 774.3)	579 ((1700*1335)/3581 = 633.8)	1700
UTH	132 ((772*615/3581 = 132.6)	288 (351.6)	352 (287.8)	772
IPA	189 (190.5)	516 (505.1)	404 (413.4)	1109
TOTAL	615	1631	1335	3581

NOTE: The marginal totals of the expected frequencies = the marginal totals of the observed frequencies. For example, 291.9 + 774.3 + 633.8 = 1700. Similarly, 291.9 + 132.6 + 190.5 = 615.

Using the observed and expected frequencies, we compute the test statistic (8.15):

$$\chi^2 = \Sigma \frac{(O-E)^2}{E} = \frac{(294-291.9)^2}{291.9} + \frac{(827-774.3)^2}{774.3} + \frac{(579-633.8)^2}{633.8}$$

$$+ \frac{(132-132.6)^2}{132.6} + \frac{(288-351.6)^2}{351.6} + \frac{(352-187.8)^2}{287.8}$$

$$+ \frac{(189-190.5)^2}{190.5} + \frac{(516-505.1)^2}{505.1} + \frac{(404-413.3)^2}{413.4}$$

$$\chi^2 = 0.014 + 3.359 + 4.732$$
$$+ 0.003 + 11.509 + 14.320$$
$$+ 0.011 + 0.235 + 0.215 = 34.629$$

5.   Conclusion: Reject H_0 since $34.629 \geq 9.49$. We have significant evidence, $\alpha = 0.05$, to show that site and treatment regimen are not independent (i.e., they are related). For this example, p=0.005 (See Table 5). Notice in the table that there are discrepancies between the observed and expected frequencies, particularly among the University teaching hospital patients.

SAS Example 8.12. The following output was generated using SAS Proc Freq. We requested a chi-square test of independence. In this example, we also requested some additional statistics which are described below. SAS produces the following.

SAS Output for Example 8.12

```
                  The FREQ Procedure
                 Table of site by trt
     site         trt
     Frequency      |
     Expected       |
     Cell Chi-Square|
     Percent        |
     Row Pct        |
     Col Pct        |diet    |insulin |oral    |  Total
     ---------------+--------+--------+--------+
     hmo            |    294 |    579 |    827 |   1700
                    | 291.96 | 633.76 | 774.28 |
                    | 0.0143 | 4.7318 | 3.5895 |
                    |   8.21 |  16.17 |  23.09 |  47.47
                    |  17.29 |  34.06 |  48.65 |
                    |  47.80 |  43.37 |  50.71 |
     ---------------+--------+--------+--------+
     ipa            |    189 |    404 |    516 |   1109
                    | 190.46 | 413.44 |  505.1 |
                    | 0.0112 | 0.2154 |  0.235 |
                    |   5.28 |  11.28 |  14.41 |  30.97
                    |  17.04 |  36.43 |  46.53 |
                    |  30.73 |  30.26 |  31.64 |
     ---------------+--------+--------+--------+
     uth            |    132 |    352 |    288 |    772
                    | 132.58 |  287.8 | 351.61 |
                    | 0.0026 |  14.32 | 11.509 |
                    |   3.69 |   9.83 |   8.04 |  21.56
                    |  17.10 |  45.60 |  37.31 |
                    |  21.46 |  26.37 |  17.66 |
     ---------------+--------+--------+--------+
     Total               615     1335     1631     3581
                       17.17    37.28    45.55   100.00

          Statistics for Table of site by trt
     Statistic                    DF       Value      Prob
     ------------------------------------------------------
     Chi-Square                    4     34.6291    <.0001
     Likelihood Ratio Chi-Square   4     34.4975    <.0001
     Mantel-Haenszel Chi-Square    1     10.5953    0.0011
     Phi Coefficient                      0.0983
     Contingency Coefficient              0.0979
     Cramer's V                           0.0695
                    Sample Size = 3581
```

Categorical Data

SAS generates a contingency table and in each cell of the table displays a number of statistics. We requested some additional statistics here (compare this output with the outputs of SAS Examples 8.5 and 8.6). In each cell, SAS produces the Frequency, the Expected frequency (computed by formula (8.14)), the Cell Chi-Square (computed by formula (8.15) in each cell), the Percent, the Row Percent and the Column Percent (See legend in top left corner of the table).

SAS generates a number of statistics which can be used to assess relationships between variables in a contingency table. We are concerned with the test of H_0: Site and Treatment Regimen are independent vs. H_1: H_0 is false. SAS produces the χ^2 statistic $= 34.629$ and the associated p-value, $p < 0.0001$. Assuming $\alpha = 0.05$ we reject H_0 since $p = 0.0001 < \alpha = 0.05$. We have significant evidence, $\alpha = 0.05$, to show that site and treatment regimen are not independent (i.e., they are related).

To understand the nature of the relationship, we evaluate the Row Percents. For example, among HMO patients 17% are on Diet, 34% on Insulin and 49% on Oral Hypoglycemics. Among the IPA patients, 17% are on Diet, 36% on Insulin and 47% on Oral Hypoglycemics. There is little difference between the treatment regimens of patients in these two sites. The difference appears to be with the UTH patients. Among the UTH patients 17% are on Diet, 46% on Insulin and 37% on Oral Hypoglycemics. A higher proportion of IPA patients are on Insulin as compared to the other sites. This resulted in a statistically significant difference. Is it clinically meaningful?

Example 8.13. Consider the study described in Example 8.12. Suppose an investigator is interested in evaluating whether or not treatment regimens differ by gender. Restricting our analyses to the HMO patients, the following table displays the numbers of male and female patients according to their treatment regimens. Based on the data, is there evidence of a significant relationship between gender and treatment regimen among the HMO patients ?

	Treatment Regimen			
Gender	Diet & Exercise	Oral Hypoglycemics	Insulin	TOTAL
Female	147	392	323	862
Male	147	435	256	838
TOTAL	294	827	579	1700

In the contingency table above, gender is the row variable (R=2) and treatment regimen is the column variable (C=3).

1. Set up hypotheses.

    H_0: Gender and Treatment Regimen are independent   (No relationship between gender and treatment regimen)

    H_1: H_0 is false   (Gender and treatment regimen are related)

    $\alpha = 0.05$

2. Select appropriate test statistic.

$$\chi^2 = \Sigma \frac{(O-E)^2}{E}$$

where Σ indicates summation over all cells of the contingency table, O = observed frequencies, and E = expected frequencies (i.e., if H_0 is true, or under H_0).

3. Decision Rule
    In order to select the appropriate critical value, we first determine the degrees of freedom.
    $$df = (R-1)(C-1) = (2-1)(3-1) = 1(2) = 2.$$

The appropriate critical value of χ^2 is $\chi^2 = 5.99$ from Table 5. The Decision Rule is:
    Reject H_0 if $\chi^2 \geq 5.99$
    Do Not Reject H_0 if $\chi^2 < 5.99$

Categorical Data

4. Test Statistic

To organize the computations of the test statistic (8.15), we use the contingency table given above. The observed frequencies in each cell are shown along with the expected frequencies which are computed by (8.14) and displayed in parentheses.

	Treatment Regimen			
Gender	Diet & Exercise	Oral Hypoglycemics	Insulin	TOTAL
Female	147 (149.1)	392 (419.3)	323 (293.6)	862
Male	147 (144.9)	435 (407.7)	256 (285.4)	838
TOTAL	294	827	579	1700

Using the observed and expected frequencies, we compute the test statistic (8.15):

$$\chi^2 = \Sigma \frac{(O-E)^2}{E} = \frac{(147-149.1)^2}{149.1} + \frac{(392-419.3)^2}{419.3} + \frac{(323-293.6)^2}{293.6}$$

$$+ \frac{(147-144.9)^2}{144.9} + \frac{(435-407.7)^2}{407.7} + \frac{(256-285.4)^2}{285.4}$$

$$\chi^2 = 0.029 + 1.782 + 2.947$$
$$+ 0.030 + 1.833 + 3.031 = 9.652$$

5.   Conclusion: Reject H_0 since $9.652 \geq 5.99$. We have significant evidence, $\alpha = 0.05$, to show that gender and treatment regimen are not independent (i.e., they are related) in the sample of HMO patients. For this example, p=0.01 (See Table 5).

SAS Example 8.13. The following output was generated using SAS Proc Freq. Again we requested a chi-square test of independence. In the following we did not request as many statistics as were requested for SAS Example 8.12. SAS produces the following. A brief interpretation appears after the output.

SAS Output for Example 8.13

```
                        The FREQ Procedure
                        Table of gender by trt
             gender      trt
             Frequency|
             Percent  |
             Row Pct  |
             Col Pct  |diet    |insulin |oral    |  Total
             ---------+--------+--------+--------+
             female   |    147 |    256 |    435 |    838
                      |   8.65 |  15.06 |  25.59 |  49.29
                      |  17.54 |  30.55 |  51.91 |
                      |  50.00 |  44.21 |  52.60 |
             ---------+--------+--------+--------+
             male     |    147 |    323 |    392 |    862
                      |   8.65 |  19.00 |  23.06 |  50.71
                      |  17.05 |  37.47 |  45.48 |
                      |  50.00 |  55.79 |  47.40 |
             ---------+--------+--------+--------+
             Total         294      579      827     1700
                         17.29    34.06    48.65   100.00


            Statistics for Table of gender by trt
        Statistic                     DF       Value      Prob
        ------------------------------------------------------
        Chi-Square                     2      9.6519     0.0080
        Likelihood Ratio Chi-Square    2      9.6684     0.0080
        Mantel-Haenszel Chi-Square     1      2.6751     0.1019
        Phi Coefficient                       0.0753
        Contingency Coefficient               0.0751
        Cramer's V                            0.0753

                      Sample Size = 1700
```

Interpretation of SAS Output for Example 8.13

SAS generates a contingency table and in each cell of the table displays the Frequency, the Percent, the Row Percent and the Column Percent (See legend in top left corner of table). The Frequency is the number of subjects in each cell. The Percent is the percent of all subjects in each cell. For example, there are 147 patients in the top left cell (i.e., females on diet and exercise treatment). These patients reflect 8.65% of the total sample (i.e., 147/1700=0.0865). The Row Percent is the percent of subjects in the particular row which fall in that cell. For example, there are

862 female patients or 862 patients in the first row of the contingency table. The 147 female patients on diet and exercise treatment reflect 17.05% of all female patients (i.e., 147/862 = 0.1705). The Column Percent is the percent of subjects in the particular column which fall in that cell. For example, there are 294 patients following treated by diet and exercise for their diabetes. These 862 patients appear in the first column of the contingency table. The 147 female patients on diet and exercise treatment reflect 50% of all patients treated by diet and exercise (i.e., 147/294 = 0.50). The row total and column total (called the marginal totals) are displayed to the right and at the bottom of the contingency table, respectively. Both row and column frequencies and percents (of total) are displayed.

SAS generates a number of statistics which can be used to assess relationships between variables in a contingency table. We are concerned with the test of H_0: Gender and Treatment Regimen are independent vs. H_1: H_0 is false. SAS produces the χ^2 statistic = 9.652 and the associated p-value, p=0.008. Assuming $\alpha = 0.05$ we reject H_0 since $p = 0.008 < \alpha = 0.05$. We have significant evidence, $\alpha = 0.05$, to show that gender and treatment regimen are not independent (i.e., they are related) in the sample of HMO patients.

NOTE: Chi-square tests are valid when the expected frequencies in each cell (or response category) are greater than or equal to 5. If an expected frequency falls below 5, then alternative techniques should be used.

Example 8.14. Consider the study described in Example 8.11 involving the distribution of abnormalities among women who deliver babies with abnormalities. Suppose we wish to test if there is a difference in the distribution of abnormalities among clinical centers. Based on the data below, is there evidence of a significant relationship between the types of abnormalities and clinical center?

| | Type of Abnormality | | | |
	Trisomy 21	Trisomy 18	Trisomy 13	Total
Center A	107	70	23	200
Center B	65	62	23	150
Center C	60	32	8	100
Total	232	164	54	450

1. Set up hypotheses.
    H_0: Center and Type of Abnormality are independent
    H_1: H_0 is false $\alpha = 0.05$

2. Select appropriate test statistic.

$$\chi^2 = \Sigma \frac{(O - E)^2}{E}$$

3. Decision Rule
    In order to select the appropriate critical value, we first determine the degrees of freedom.
    $$df = (R - 1)(C - 1) = (3-1)(3-1) = 2(2) = 4.$$

The appropriate critical value of χ^2 is $\chi^2 = 9.49$ from Table 5. The Decision Rule is:
        Reject H_0 if $\chi^2 \geq 9.49$
        Do Not Reject H_0 if $\chi^2 < 9.49$

4. Test Statistic
        To organize the computations of the test statistic (8.15), we use the contingency table given above. The observed frequencies in each cell are shown. We compute the expected frequencies for each cell using (8.14) and display expected frequencies in parentheses to distinguish them from the observed frequencies.

	Treatment Regimen			
Centere	Trisomy 21	Trisomy 18	Trisomy 13	TOTAL
Center A	107 (103.1)	70 (72.9)	23 (24)	200
Center B	65 (77.3)	62 (54.7)	23 (18)	150
Center C	60 (51.6)	32 (36.4)	8 (12)	100
TOTAL	232	164	54	450

Using the observed and expected frequencies, we compute the test statistic (8.15). Only the results are shown below.

$$\chi^2 = 0.15 + 0.12 + 0.04$$
$$+ 1.96 + 0.97 + 1.39$$
$$+ 1.37 + 0.53 + 1.33 = 7.86$$

5.   Conclusion: Do not reject H_0 since $7.86 < 9.49$. We do not have significant evidence, $\alpha = 0.05$, to show that there is an association between clinical center and type of trisomy.

8.7   Precision, Power and Sample Size Determination

In Section 8.2 we illustrated estimation techniques for the population proportion p. The following formula is used to determine the sample size requirements to produce an estimate for p with a certain level of precision:

$$n = p(1-p)\left(\frac{Z_{1-\frac{\alpha}{2}}}{E}\right)^2 \qquad (8.16)$$

where $Z_{1-\alpha/2}$ reflects the desired level of confidence (e.g., 95%),
p = the population proportion, and E = margin of error.

Equation (8.16) produces the <u>minimum</u> number of subjects required to ensure a margin of error equal to E in the confidence interval for p with the specified level of confidence. Recall, in estimating the sample size to make an inference about μ, an estimate of σ was required. Several alternatives were suggested to estimate σ. In the binomial case, our goal is to make an inference about p. However, the formula for estimating the sample size (8.16) involves p. An estimate from a previous study or an estimate based on pilot data may be used in (8.16). If such an estimate is not available, it can be shown that p (1-p) is maximized at p=0.50 (See Figure 8.2). Therefore, the most conservative estimate of n is produced by substituting p=0.5 into the formula given above:

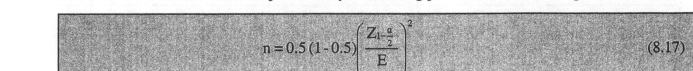

$$n = 0.5(1-0.5)\left(\frac{Z_{1-\frac{\alpha}{2}}}{E}\right)^2 \qquad (8.17)$$

Categorical Data

Which is equivalent to:

$$n = 0.25 \left(\frac{Z_{1-\frac{\alpha}{2}}}{E} \right)^2 \qquad (8.18)$$

where $Z_{1-\alpha/2}$ reflects the desired level of confidence (e.g., 95%),
and E = margin of error.

Figure 8.2 Relationship between p and p(1-p): p(1-p) maximized at p=0.5

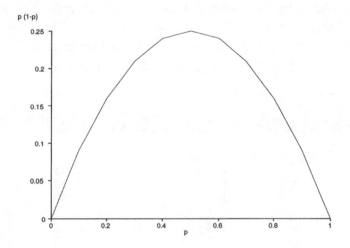

Example 8.15. An investigator wishes to estimate the proportion of patients in a particular health plan in favor of a new policy (e.g., reimbursement for services) and wants the estimate to be within 3% of the true proportion of patients in favor of the new policy. How many subjects would be required to produce such an estimate with 95% confidence ?

Suppose that there is no available data on the issue (i.e., no estimate for p which can be used in (8.16)). Therefore (8.18) is used:

$$n = 0.25 \left(\frac{Z_{1-\frac{\alpha}{2}}}{E} \right)^2$$

Substituting E= 0.03 (the margin of error in the estimate of the proportion) and the appropriate value for 95% confidence:

$$n = 0.25\left(\frac{1.96}{0.03}\right)^2$$

$$n = 1067.1$$

Since (8.18) produces the minimum number of subjects to satisfy our criteria, 1068 patients must be sampled in order to produce a 95% confidence interval estimate for the proportion of patients in favor of the new policy with a margin of error of 3%.

SAS Example 8.15. SAS does not have a procedure to generate sample size requirements. However, we can program formulas (8.16) and (8.18) as we did in previous chapters. In the following we generated sample size requirements for various scenarios, including the one described in Example 8.13. SAS produces the following. A brief interpretation appears after the output.

SAS Output for Example 8.15

Obs	c_level	z	p	e	n
1	0.95	1.95996	0.5	0.03	1068
2	0.95	1.95996	0.1	0.03	385
3	0.95	1.95996	0.2	0.03	683
4	0.95	1.95996	0.3	0.03	897
5	0.95	1.95996	0.4	0.03	1025
6	0.95	1.95996	0.5	0.03	1068
7	0.95	1.95996	0.6	0.03	1025
8	0.95	1.95996	0.7	0.03	897
9	0.95	1.95996	0.8	0.03	683
10	0.95	1.95996	0.9	0.03	385

Interpretation of SAS Output for Example 8.15

We considered several scenarios in the above. The first scenario reflects the situation in Example 8.13 where p is not known and we use the most conservative p=0.5. SAS estimates that n=1068 subjects are required. To illustrate the effect of the estimate of p on the computations, we considered various values of p in scenarios 2-10. Specifically, we considered p=0.10, 0.20, ..., 0.90. Notice that the sample size estimate is largest when p=0.5.

Example 8.16. A study is run to estimate the proportion of high school seniors who smoke. The study involved a random sample of 100 high school seniors and 28 reported that they were smokers. A 95% confidence interval for the true proportion of all high school seniors who smoke was computed as 0.28 ± 0.088. Suppose that we wanted to estimate the true proportion with a margin of error not exceeding 0.05 (the analysis based on the sample of size 100 had a margin of error of 0.088). How many subjects would be required to ensure a margin of error of 0.05 with 95% confidence?

Because we have data from the initial study, we use (8.16)):

$$n = p(1-p)\left(\frac{Z_{1-\frac{\alpha}{2}}}{E}\right)^2 = 0.28(1-0.28)\left(\frac{1.96}{0.05}\right)^2 = 309.8.$$

In order to estimate the true proportion of high school seniors who smoke with a margin of error of 0.05, we need a sample of at least 310 high school seniors.

In Chapter 6 we introduced the concepts of precision (in estimation) and statistical power (in hypothesis testing), respectively. We now present formulas for determining sample size to achieve a specified power in the two sample test for proportions. Recall, power (i.e., Power = 1 - β = P(Reject H_0/H_0 False)) depends on 3 components:

i) n_i = sample sizes (i=1,2)
ii) α = level of significance = P(Type I error)
iii)   ES = the Effect Size = the standardized difference in proportions specified under H_0 and H_1

In the two sample applications the hypotheses are:

H_0: $p_1 = p_2$
H_1: $p_1 \neq p_2$

or equivalently,

H_0: $p_1 - p_2 = 0$
H_1: $p_1 - p_2 \neq 0$.

where p_1 = proportion of successes in population 1 and p_2 = proportion of successes in population 2.

The Effect Size (ES) is defined as the difference in the proportions under H_0 and H_1. In the two sample applications, the parameter of interest is $(p_1 - p_2)$. The ES is defined as follows:

$$ES = |p_2 - p_1| \tag{8.19}$$

In many applications the number of subjects (i.e., n_1 and n_2) that can be involved depends on financial, logistic and/or time constraints. In other cases, the sample size required to ensure a certain level of power can be determined relative to alternative hypotheses of importance. The sample size required to ensure a specific level of power in a <u>two-sided two independent samples test for proportions</u> is given below:

$$n_i = \left(\frac{\sqrt{\overline{p}\,\overline{q}\, 2}\, Z_{1-\alpha/2} + \sqrt{p_1 q_1 + p_2 q_2}\, Z_{1-\beta}}{ES} \right)^2 \tag{8.20}$$

where n_i is the minimum number of subjects required in sample i (i=1,2), $Z_{1-\alpha/2}$ is the value from the standard normal distribution with two-sided tail area equal to $1-\alpha/2$, $Z_{1-\beta}$ is the value from the standard normal distribution with a tail area equal to $1-\beta$, and ES is the effect size (8.19).

In addition, p_1 = proportion of successes in population 1, $q_1=1-p_1$,

p_2 = proportion of successes in population 2, $q_2=1-p_2$, and

$$\overline{p} = \frac{p_1 + p_2}{2}, \overline{q} = 1 - \overline{p}.$$

The following example illustrates the use of formula (8.20).

Example 8.17. Suppose we wish to design a study to compare two treatments with respect to the proportion of successes in each. A two sided test is planned at a 5% level of significance. Based on a review of the literature, $p_1=0.20$. How many subjects would be required per group to detect $p_2=0.10$ with 90% power?

The formula to determine sample sizes is given by (8.20):

$$n_i = \left(\frac{\sqrt{\overline{p}\,\overline{q}\, 2}\, Z_{1-\alpha/2} + \sqrt{p_1 q_1 + p_2 q_2}\, Z_{1-\beta}}{ES} \right)^2$$

Categorical Data

The ES is (8.19):

$$ES = |p_2 - p_1| = |0.10 - 0.20| = 0.10$$

Since $\alpha = 0.05$, $Z_{1-\alpha/2} = Z_{.975} = 1.96$. Similarly, for power=0.90, $\beta = 0.10$, therefore $Z_{1-\beta} = 1.282$.
$$\bar{p} = \frac{0.10 + 0.20}{2} = 0.15, \quad \bar{q} = 1 - 0.15 = 0.85.$$

one side

Substituting:

$$n_i = \left(\frac{\sqrt{0.15(0.85)\,2}\,(1.96) + \sqrt{0.20(0.80) + (0.10)(0.90)}\,(1.282)}{0.10} \right)^2 = 265.9.$$

Thus $n_1 = n_2 = 266$ subjects (532 total) are needed.

SAS Example 8.17. SAS does not have a procedure to generate sample size requirements. However, we can program formulas (8.19) and (8.20) as we did in previous chapters. In the following we generated sample size requirements for various scenarios, including the one described in Example 8.14. SAS produces the following. A brief interpretation appears after the output.

SAS Output for Example 8.17

Obs	alpha	beta	z_alpha2	z_beta	p1	p2	power	es	n 2
1	0.05	0.1	1.95996	1.28155	0.2	0.1	0.9	0.1	266
2	0.05	0.1	1.95996	1.28155	0.2	0.3	0.9	0.1	392
3	0.05	0.2	1.95996	0.84162	0.2	0.1	0.8	0.1	199
4	0.05	0.2	1.95996	0.84162	0.2	0.3	0.8	0.1	294

Interpretation of SAS Output for Example 8.17

We considered several scenarios in the above. The first scenario reflects the situation in Example 8.14. SAS estimates that 266 subjects are required per group to detect the specified difference with 90% power. In scenario 2 we specify $p_2=0.3$. Notice that the effect size is the same as the effect size in scenario 1. However, SAS estimates that 392 subjects are required per group. The values of p_1 and p_2 affect the sample size computation (and not just the difference between them). In the last 2 scenarios we specify power=80%. Notice that fewer cases are required to detect the same differences with lower power.

$$\hat{p} = \frac{\text{Number of successes in the sample}}{n} = \frac{x}{n}$$

$$\text{s.e.}(\hat{p}) = \sqrt{\frac{\hat{p}(1-\hat{p})}{n}} \quad (\text{large size}), \quad \text{otherwise}, \quad \text{s.e.}(\hat{p}) = \sqrt{\frac{p \cdot (1-p)}{n}}$$

(population proportion)

8.8 Key Formulas

$$\min\left(n\hat{p}, \, n(1-\hat{p})\right) \geq 5$$

or $\min\left(np_0, \, n(1-p_0)\right) \geq 5$

APPLICATION	NOTATION/FORMULA	DESCRIPTION
Confidence Interval Estimate for p	$\hat{p} \pm Z_{1-\frac{\alpha}{2}} \sqrt{\dfrac{\hat{p}(1-\hat{p})}{n}}$	See Table 8.1 for necessary conditions (Find Z in Table 2A)
Test H_0: $p = p_0$	$Z = \dfrac{\hat{p} - p_0}{\sqrt{\dfrac{p_0(1-p_0)}{n}}}$	See Table 8.1 for necessary conditions
Confidence Interval Estimate for $(p_1 - p_2)$	$(\hat{p}_1 - \hat{p}_2) \pm Z_{1-\frac{\alpha}{2}} \sqrt{\dfrac{\hat{p}_1(1-\hat{p}_1)}{n_1} + \dfrac{\hat{p}_2(1-\hat{p}_2)}{n_2}}$	See Table 8.3 for necessary conditions (Find Z in Table 2A)
Test H_0: $p_1 = p_2$	$Z = \dfrac{\hat{p}_1 - \hat{p}_2}{\sqrt{\hat{p}(1-\hat{p})\dfrac{1}{n_1} + \dfrac{1}{n_2}}}$ where $\hat{p}_1 = \dfrac{x_1}{n_1}$ $\hat{p}_2 = \dfrac{x_2}{n_2}$ $\hat{p} = \dfrac{x_1 + x_2}{n_1 + n_2}$	See Table 8.4 for necessary conditions and definitions of components of Z
Test H_0: Distribution of responses follows specified pattern	$\chi^2 = \Sigma \dfrac{(O-E)^2}{E}$, df=k-1 1 sample	Chi square goodness of fit test
Test H_0: Two variables are independent	$\chi^2 = \Sigma \dfrac{(O-E)^2}{E}$, df=(r-1)*(c-1) ≥ 2 samples	Chi square test of independence
Find n to Estimate p	$n = p(1-p)\left[\dfrac{Z_{1-\alpha/2}}{E}\right]^2$ 0.5(1-0.5) (if p is not available)	Sample size to estimate p with margin of error E
Find n_1, n_2 to Test Test H_0: $p_1 = p_2$	$n_i = \left(\dfrac{\sqrt{\bar{p}\bar{q}\,2}\,Z_{1-\alpha/2} + \sqrt{p_1 q_1 + p_2 q_2}\,Z_{1-\beta}}{ES}\right)^2$	Sample sizes to detect effect size ES with power 1-β.

Handwritten annotations:
- dichotomous
- $\min\left(n_1\hat{p}_1, \, n_1(1-\hat{p}_1)\right) \geq 5$
- $\min\left(n_2\hat{p}_2, \, n_2(1-\hat{p}_2)\right) \geq 5$
- χ^2 always ≥ 0, so reject is only "≥"

$$ES = |p_2 - p_1|, \quad q_1 = 1-p_1, \quad q_2 = 1-p_2, \quad \bar{p} = \frac{p_1 + p_2}{2},$$

$$\bar{q} = 1 - \bar{p}$$

$Z_{1-\beta}$ is one side (90% power, $Z_{0.1} = 1.282$)

8.9   Statistical Computing

Following are the SAS programs which were used to generate the frequency distribution table, the contingency tables, the chi-square tests of independence and to determine the required sample size to estimate p and to compare to proportions using a test of hypothesis. The SAS procedures used and brief descriptions of their use are noted in the header to each example. Notes are provided to the right of the SAS programs (*in italics*) for orientation purposes and are not part of the programs. In addition, there are blank lines in the programs that follow which are solely to accommodate the notes. Blank lines and spaces can be used throughout SAS programs to enhance readability. A summary of the SAS procedures used in the examples is provided at the end of this section.

Frequency Distribution Table for Single Categorical Variable

SAS EXAMPLE 8.2 Estimate Proportion of Patients with Osteoarthritis

Suppose we wish to estimate the proportion of patients in a particular physician's practice with diagnosed osteoarthritis. A random sample of 200 patients is selected and each patient's medical record is reviewed for the diagnosis of osteoarthritis. Suppose that 38 patients are observed with diagnosed osteoarthritis.   Generate a frequency distribution table and determine a point estimate for the proportion of all patients with diagnosed osteoarthritis using SAS.

Program Code

options ps=62 ls=80;	*Formats the output page to 62 lines in length and 80 columns in width*
data in;	*Beginning of Data Step*
input x count;	*Inputs two variables x and count, here x= 0 of the patient does not have osteoarthritis and x=1 if the patient has osteoarthritis. Count reflects the number of patients in each category.*
cards;	*Beginning of Raw Data section.*
0 162	*actual observations (value of x and*
1 38	*count on each line)*
run;	
proc freq;	*Procedure call. Proc Freq generates a frequency distribution table for a categorical (or ordinal) variable.*
tables x;	*Specification of analytic variable x*
weight count;	*Specification of variable containing the number of subjects in each category, count.*
run;	*End of procedure section.*

Cross-tabulation Table

SAS EXAMPLE 8.5 Cross-tabulation of Treatment by Long Term Complications

A longitudinal study is conducted to evaluate the long term complications in diabetic patients treated under two competing treatment regimens. Complications are measured by incidence of foot disease, eye disease or cardiovascular disease within a 10 year observation period. The following 2 x 2 cross-tabulation table summarizes the data. Estimate the relative risk and odds ratio using SAS.

	Long Term Complications		
Treatment	YES	NO	Total
Treatment 1	12	88	100
Treatment 2	8	92	100
Total	20	180	200

Program Code

```
options ps=62 ls=80;

data in;
  input trt $ compl $ count;
```

Formats the output page to 62 lines in length and 80 columns in width
Beginning of Data Step
*Inputs three variables **trt** (treatment 1 or 2), **compl** (complications yes or no) and **count**, where count reflects the number of patients in each cell of the table. Here we input both trt and compl using character labels, and we indicate this to SAS using the "$" symbol.*

```
cards;
trt_1 z_no 88
trt_1 yes 12
trt_2 z_no 92
trt_2 yes 8
run;
```

Beginning of Raw Data section.
*actual observations (value of **trt**, **compl** and **count** on each line)*

proc freq;	*Procedure call. Proc Freq generates a contingency table for two categorical (or ordinal) variables.*
tables trt*compl;	*Specification of analytic variables **trt** and **compl**.*
weight count;	*Specification of variable containing the number of subjects in each cell of the table, **count**.*
run;	*End of procedure section.*

Contingency Table and Chi-Square Test of Independence

SAS EXAMPLE 8.12   Test if There is a Relationship Between Site and Treatment Regimen

The following data were collected in a multi-site observational study of medical effectiveness in Type II diabetes. Three sites were involved in the study: a health maintenance organization (HMO), a University teaching hospital (UTH), and an independent practice association (IPA). Type II diabetic patients were enrolled in the study from each site and monitored over a three year observation period. The data shown below display the treatment regimens of patients measured at baseline by site. Test if there is a relationship between site and treatment regimen using the chi-square test of independence in SAS.

Site	*Treatment Regimen*			
	Diet & Exercise	Oral Hypoglycemics	Insulin	TOTAL
HMO	294	827	579	1700
UTH	132	288	352	772
IPA	189	516	404	1109
TOTAL	615	1631	1335	3581

Program Code

```
options ps=62 ls=80;

data in;
  input site $ trt $ count;
```

Formats the output page to 62 lines in length and 80 columns in width
Beginning of Data Step
*Inputs three variables **site** (HMO, UTH or IPA), **trt** (diet, oral hypoglycemics or insulin) and **count**, where count reflects the number of patients in each cell of the table. Here we input both site and trt using character labels, and we indicate this to SAS using the "$" symbol.*

Categorical Data

```
cards;
hmo diet 294
hmo oral 827
hmo insulin 579
uth diet 132
uth oral 288
uth insulin 352
ipa diet 189
ipa oral 516
ipa insulin 404
run;


proc freq;



   tables site*trt/expected cellchi2 chisq;











   weight count;




run;
```

Beginning of Raw Data section.
actual observations (value of **site,**
trt *and* **count** *on each line)*

Procedure call. Proc Freq generates a contingency table for two categorical (or ordinal) variables. Specification of analytic variables **site** *and* **trt***. We also request that SAS produce expected frequencies in each cell, the chi-square statistic for each cell and run a chi-square test of independence with the expected, cellchi2 and chisq options, respectively.*
Specification of variable containing the number of subjects in each cell of the table, **count.**
End of procedure section.

Contingency Table and Chi-Square Test of Independence

SAS EXAMPLE 8.13 Test if There is a Relationship Between Gender and Treatment Regimen

Suppose an investigator is interested in evaluating whether or not treatment regimens differ by gender. Restricting our analyses to the HMO patients, the following table displays the numbers of male and female patients according to their treatment regimens. Based on the data, is there evidence of a significant relationship between gender and treatment regimen among the HMO patients ? Test if there is a relationship between site and treatment regimen using the chi-square test of independence in SAS.

	Treatment Regimen			
Gender	Diet & Exercise	Oral Hypoglycemics	Insulin	TOTAL
Female	147	392	323	862
Male	147	435	256	838
TOTAL	294	827	579	1700

Program Code

```
options ps=62 ls=80;

data in;
  input gender $ trt $ count;
```

Formats the output page to 62 lines in length and 80 columns in width
Beginning of Data Step
*Inputs three variablesgender (male or female), **trt** (diet, oral hypoglycemics or insulin) and **count**, where count reflects the number of patients in each cell of the table. Here we input both gender and trt using character labels, and we indicate this to SAS using the "$" symbol.*

```
cards;
male diet 147
male oral 392
male insulin 323
female diet 147
female oral 435
female insulin 256
run;


proc freq;




  tables gender*trt/chisq;







  weight count;




run;
```

Beginning of Raw Data section.
*actual observations (value of **gender**, **trt** and **count** on each line)*

Procedure call. Proc Freq generates a contingency table for two categorical (or ordinal) variables.
*Specification of analytic variables **gender** and **trt**. We request that SAS runs a chi-square test of independence with the chisq option.*
*Specification of variable containing the number of subjects in each cell of the table, **count**.*
End of procedure section.

Determine the Number of Subjects Required to Generate a Confidence Interval for p

SAS EXAMPLE 8.15 Time to Travel Between Hospital Departments

An investigator wishes to estimate the proportion of patients in a particular health plan in favor of a new policy (e.g., reimbursement for services) and wants the estimate to be within 3% of the true proportion of patients in favor of the new policy. How many subjects would be required to produce such an estimate with 95% confidence ?

Program Code

```
options ps=62 ls=80;

data in;
  input c_level e p;
  z=probit((1-c_level)/2);



if p=. then p=0.5;


tempn=(p*(1-p))*(z/e)**2;


n=ceil(tempn);




/*  Input the following information (required)
   c_level: Confidence Level: range 0.0 to 1.0 (e.g., 0.95),
   e: Margin of Error, and
   p  Proportion of Successes: range 0.0 to 1.0
   NOTE: p may not be available, in which case enter . to indicate p is missing
*/
cards;
0.95  0.03  .
0.95  0.03 0.10
0.95  0.03 0.20
0.95  0.03 0.30
0.95  0.03 0.40
0.95  0.03 0.50
```

Formats the output page to 55 lines in length and 80 columns in width
Beginning of Data Step
*Inputs 3 variables **c_level**, **e** and **p**.*
*Determines the value from the standard normal distribution (z) with tail area (1-**c_level**)/2*
If p is not available (input as missing) then SAS recodes p to 0.5 (See (8.17))
*Creates a temporary variable, called **tempn**, determined by formula (8.16).*
*Computes a variable **n** using the ceil function which computes the smallest integer greater than **tempn**.*
Beginning of comment section

End of comment
Beginning of Raw Data section
actual observations
*values of **c_level**, **e** and **p***
on each line

Categorical Data

```
0.95  0.03 0.60
0.95  0.03 0.70
0.95  0.03 0.80
0.95  0.03 0.90
run;

proc print;                    •
  var c_level z p e n;
run;
```

Procedure call. Print to display
input and computed variables.
End of procedure section.

**Determine the Number of Subjects Required to Detect a Specific Effect Size
in a Test of Hypothesis About (p_1-p_2)**

SAS EXAMPLE 8.17 Sample Size Requirements

Suppose we wish to design a study to compare two treatments with respect to the proportion of successes in each. A two sided test is planned at a 5% level of significance. Based on a review of the literature, $p_1=0.20$. How many subjects would be required per group to detect $p_2=0.10$ with 90% power?

Program Code

`options ps=62 ls=80;`	*Formats the output page to 62 lines in length and 80 columns in width*
`data in;`	*Beginning of Data Step.*
`input alpha power p1 p2;`	*Inputs 4 variables **alpha, power, p1**, and **p2**.*
`z_alpha2=probit(1-alpha/2);`	*Determines the value from the standard normal distribution with tail area 1-**alpha**/2.*
`beta=1-power;`	*Computes **beta**.*
`z_beta=probit(1-beta);`	*Determines the value from the standard normal distribution with tail area 1-**beta** (See Z_b above)*
`q1=1-p1;`	*Compute components for (8.20).*
`q2=1-p2;`	
`pbar=(p1+p2)/2;`	
`qbar=1-pbar;`	
`es=abs(p2-p1);`	*Computes the effect size (**es**)*
	Compute (8.20).

`tempn_2=(sqrt(pbar*qbar*2)*z_alpha2+sqrt(p1*q1+p2*q2)*z_beta)**2/(es)**2;`

Categorical Data

`n_2=ceil(tempn_2);`	*Computes a variable **n_2** using the ceil function which computes the smallest integer greater than **tempn_2**.*
`/*`	*Beginning of comment section*
Input the following information (required)	
alpha: Level of Significance: range 0.0 to 1.0 (e.g., 0.05),	
power: Power: range 0.0 to 1.0 (e.g., 0.80),	
p1: Proportion of Successes in Group 1, and	
p2: Proportion of Successes in Group 2.	
`*/`	*End of comment section*
`cards;`	*Beginning of Raw Data section*
`0.05 0.90 0.20 0.10`	*actual observations*
`0.05 0.90 0.20 0.30`	
`0.05 0.80 0.20 0.10`	
`0.05 0.80 0.20 0.30`	
`run;`	
`proc print;`	*Procedure call. Print to display*
`  var alpha beta z_alpha2 z_beta p1 p2 power es n_2;`	*computed variables.*
`run;`	*End of procedure section.*

Summary of SAS Procedures

The SAS Freq Procedure is used to generate a frequency distribution table and to generate contingency tables. Specific options can be requested to produce estimates of relative risks and odds rations, and to run a chi-square test of independence. The specific options are shown in italics below. Users should refer to the examples in this section for complete descriptions of the procedure and specific options. A general description of the procedure and options is provided in the table below.

Procedure	Sample Procedure Call	Description
proc freq	proc freq; tables x; *weight count*;	Generates a frequency distribution table for categorical (or ordinal) variable. The weight option indicates that summary data are input and the variable count contains the numbers of cases in each response category.
	proc freq; tables a*b/*relrisk*;	Generates a contingency table (a by b) and estimates of the relative risk and odds ratio (here b is a dichotomous variable).
	proc freq; tables a*b/expected cellchi2 chisq;	Generates a contingency table and produces the expected frequency in each cell, the value of the chi-square statistic in each cell and runs the chi-square test of independence.

8.10 Problems

1. Recent attention has focused on the health care system, particularly on managed care plans. A study was undertaken within one managed care plan to assess whether patients' reports of satisfaction with the plan were related to their leaving the plan within 1 year. In random sample of 120 patients who reported that they were satisfied with the plan, 30 left that plan within 1 year. In a second random sample of 150 patient who reported that they were not satisfied with the plan, 62 left within one year. Compute a 95% confidence interval for the difference in the proportions of patients who left the plan within one year relative to their satisfaction reports ?

2. An investigator wishes to estimate the proportion of Type II diabetic patients who take insulin to manage their diabetes. A large, national database of Type II diabetic patients is used to generate the estimate. The database involves a random sample of 1200 Type II diabetic patients, 423 of these patients report taking insulin to manage their diabetes.

   a) Compute a point estimate for the proportion of all Type II diabetics who take insulin to manage their diabetes.
   b) Compute the standard error of the point estimate.
   c) Compute a 95% confidence interval for the proportion of all Type II diabetics who take insulin to manage their diabetes.

3. Recent studies have investigated the relationship between gender and career advancement. The following data represent a random sample of mathematicians classified by gender and academic rank. Using the following data, test for a relationship between gender and academic rank:

		Academic Rank		
	Instructor	Assistant Professor	Associate Professor	Full Professor.
Female	12	15	18	7
Male	21	25	30	22

Categorical Data

4. An investigator wishes to test if patients tend to use the middle response in 5-point ordinal scales more frequently than other response options. Scales such as these are used in health status and satisfaction measures. One hinder patients were asked to rate their own health on the following five point ordinal scale. Based on the data, is there evidence that the distribution of responses is 10%, 20%, 40%, 20% 10%, respectively ?

Health Status	:	Excellent	Very Good	Good	Fair	Poor
# Patients	:	12	18	50	10	10

5. A manufacturer has two plants and wants to compare the two on the proportion of defective items produced. In a random sample of 250 items from plant I, 28 were defective. In a random sample of 220 items from plant II, 38 were defective. Is there any significant evidence to support the claim that plant II produces more defective items ? Use a 5% level of significance.

6. As the first step is a study to evaluate highway fatalities, the traffic department wants to evaluate whether the left-hand lane of a 3 lane highway services twice the traffic volume as the other 2 lanes. The following data were collected, reflecting the numbers of vehicles traveling each of the 3 lanes during the rush hour. Use the data to test the claim that the twice as many vehicles travel the left lane as compared to the other 2 lanes ($\alpha = 0.05$).

	Right Lane	Center Lane	Left Lane
# Vehicles	205	220	575

7. A corporation offers 6 different health plans to its employees. Each year the corporation offers its employees an opportunity to switch plans. In 1999, 15% of all employees switched from one plan to another. With all of the emphasis on health plans this year, the corporation thinks that a higher proportion of employees will switch. A random sample of 125 employees is selected and 25 indicate that they will switch plans in 2000. Based on the data is the proportion of employees who switch plans significantly higher in 2000? Run the appropriate test at a 5% level of significance.

Categorical Data

8.10   Problems

1. Recent attention has focused on the health care system, particularly on managed care plans. A study was undertaken within one managed care plan to assess whether patients' reports of satisfaction with the plan were related to their leaving the plan within 1 year. In random sample of 120 patients who reported that they were satisfied with the plan, 30 left that plan within 1 year. In a second random sample of 150 patient who reported that they were not satisfied with the plan, 62 left within one year. Compute a 95% confidence interval for the difference in the proportions of patients who left the plan within one year relative to their satisfaction reports ?

2. An investigator wishes to estimate the proportion of Type II diabetic patients who take insulin to manage their diabetes. A large, national database of Type II diabetic patients is used to generate the estimate. The database involves a random sample of 1200 Type II diabetic patients, 423 of these patients report taking insulin to manage their diabetes.

    a) Compute a point estimate for the proportion of all Type II diabetics who take insulin to manage their diabetes.
    b) Compute the standard error of the point estimate.
    c) Compute a 95% confidence interval for the proportion of all Type II diabetics who take insulin to manage their diabetes.

3. Recent studies have investigated the relationship between gender and career advancement. The following data represent a random sample of mathematicians classified by gender and academic rank. Using the following data, test for a relationship between gender and academic rank:

		Academic Rank		
	Instructor	Assistant Professor	Associate Professor	Full Professor.
Female	12	15	18	7
Male	21	25	30	22

Categorical Data

4. An investigator wishes to test if patients tend to use the middle response in 5-point ordinal scales more frequently than other response options. Scales such as these are used in health status and satisfaction measures. One hundred patients were asked to rate their own health on the following five point ordinal scale. Based on the data, is there evidence that the distribution of responses is 10%, 20%, 40%, 20% 10%, respectively ? Use $\alpha=0.05$.

Health Status	:	Excellent	Very Good	Good	Fair	Poor
# Patients	:	12	18	50	10	10

5. A manufacturer has two plants and wants to compare the two on the proportion of defective items produced. In a random sample of 250 items from plant I, 28 were defective. In a random sample of 220 items from plant II, 38 were defective. Is there any significant evidence to support the claim that plant II produces more defective items ? Use a 5% level of significance.

6. As the first step is a study to evaluate highway fatalities, the traffic department wants to evaluate whether the left-hand lane of a 3 lane highway services twice the traffic volume as the other 2 lanes. The following data were collected, reflecting the numbers of vehicles traveling each of the 3 lanes during the rush hour. Use the data to test the claim that the twice as many vehicles travel the left lane as compared to the other 2 lanes ($\alpha = 0.05$).

	Right Lane	Center Lane	Left Lane
# Vehicles	205	220	575

7. A corporation offers 6 different health plans to its employees. Each year the corporation offers its employees an opportunity to switch plans. In 1999, 15% of all employees switched from one plan to another. With all of the emphasis on health plans this year, the corporation thinks that a higher proportion of employees will switch. A random sample of 125 employees is selected and 25 indicate that they will switch plans in 2000. Based on the data is the proportion of employees who switch plans significantly higher in 2000? Run the appropriate test at a 5% level of significance.

Categorical Data

8. A prospective study is conducted over 10 years to investigate long term complication rates in patients treated with two different therapies. The data are shown below:

| Therapy | Complications | |
	No	Yes
1	911	89
2	873	127

a) Generate a point estimate for the difference in proportions.
b) Compute a 95% confidence interval for the difference in proportions.
c) Based on the above, is there a significant difference in the therapies ?

9. A study is conducted comparing 2 experimental treatments to a control treatment with respect to their effectiveness in reducing joint pain in patients with arthritis. One hundred and fifty patients are randomly assigned to one of the three treatments and the following data are collected, representing the numbers of patients reporting improvement in joint pain after the assigned treatment is administered:

	Control	Experimental 1	Experimental 2
Number of Subjects	50	50	50
Number Reporting Improvement	21	28	34

Is there a significant difference in the proportions of patients reporting improvement in joint pain among the treatments ? Run the appropriate test at the 5% level of significance.

10. Using the data in problem 9,

a) Construct a 95% confidence interval for the difference in the proportions of patients reporting improvement in joint pain between the Control and Experimental 1 treatments ?
b) Based on (a), is there a significant difference in the proportions of patients reporting improvement in joint pain between the Control and Experimental 1 treatments ? Justify (be brief but complete).

Categorical Data

11. An investigator wished to design a study to estimate the proportion of patients in a particular hospital whose primary insurance coverage is Medicaid.

    a) How many subjects would be required to estimate the true proportion within 4% with 95% confidence ?

    b) Suppose a similar study was conducted in 2001 and produced the following 95% confidence interval for the proportion of patients in the same hospital whose primary insurance coverage was Medicaid: $27\% \pm 6\%$. If the point estimate from the 2001 study is used - how does that affect the answer given in (a).

    c) Which answer is more appropriate - (a) or (b) ? Be brief.

12. Prior to being randomized to one of two competing therapies, the severity of participant's migraines are clinically assessed. The following table displays the severity classifications for patients assigned to the medical and non-traditional therapies. Is there a significant association between severity and assigned therapy ? Run the appropriate test at $\alpha=0.05$.

Therapy	Severity Classification			Total
	Minimal	Moderate	Severe	
Medical	90	60	50	200
Non-Traditional	50	60	90	200
Total	140	120	140	400

13. A survey is conducted among current MPH students to assess their knowledge of safe (alcohol) drinking limits. Two hundred students are randomly selected for the investigation and 60% correctly identified safe drinking limits.

    a) Construct a 95% confidence interval for the proportion of all MPH students who could correctly identify safe drinking limits.

    b) It has been reported that 70% of practicing clinicians can correctly identify safe drinking limits when asked in a survey format. One hundred of the MPH students involved in the survey are former or current practicing clinicians and 72% of these individuals correctly identified safe drinking limits in the survey. Is the proportion of current MPH students who are former or current practicing clinicians significantly different from the proportion reported in the literature ? Run the appropriate test at $\alpha=0.05$.

Categorical Data

14. Suppose we wish to design a study to estimate the proportion of patients who received pain therapy following a particular surgical procedure in a local hospital. How many subjects would be required in the study to ensure that the estimate was within 4 percentage points of the true proportion with 95% confidence ?

15. The following table was derived from a study of HIV patients and the data reflect the numbers of subjects classified by their primary HIV risk factor and gender. Test if there is a relationship between HIV risk factor and gender using a 5% level of signficance.

		Gender	
		Male	Female
HIV Risk Factor	IV Drug User	24	40
	Homosexual	32	18
	Other	15	25

16. Under standard care, 10% of all patients suffering their first MI are readmitted to the hospital within 6 weeks. A new protocol for care following the first MI is proposed and evaluated. Two hundred patients receive the new protocol following an MI and 12 are readmitted to the hospital within 6 weeks. Is there a significant reduction in the proportion of patients readmitted to the hospital within 6 weeks under the new protocol ? Run the appropriate test at the 5% level of significance.

17. Suppose that the data in problem 16 are reported and are criticized based on the use of a 10% readmission rate under standard care. A new study is mounted to directly compare standard care to the new protocol using a randomized trial

Treatment	Number of Patients	Number Readmitted within 6 Weeks
Standard Care	125	16
New Protocol	125	11

Based on the following data, is there a significant reduction in the readmission rate under the new protocol ? Run the appropriate test at the 5% level of significance.

Categorical Data

18.   Some investigators suggest that medication adherence exceeding 85% is sufficient to classify a patient as "adherent", while medication adherence below 85% suggests that the patient is "not adherent". Suppose that each patient is classified as either adherent or not based on this definition.

    a)   Construct a 95% confidence interval for the difference in proportions of adherent patients between the intervention and control groups using the data shown below.

	Intervention	Control
Number of patients	75	75
Number with Medication Adherence $\geq$ 85%	47	36

    b)   Based on (a), is there a significant difference in proportions of adherent patients between the intervention and control groups? Justify briefly.

19.   Patients were enrolled into the study described in problem 18 from three clinical centers. Based on the following data, is there a significant difference in the proportions of adherent patients among the clinical centers? Run the appropriate test at $\alpha=0.05$.

	Enrollment Sites		
	Center 1	Center 2	Center 3
Adherent ($> 85\%$)	28	30	25
Not Adherent ($< 85\%$)	21	29	17

20.   Suppose an observational study is conducted to investigate the smoking behaviors of male and female patients with a history of coronary heart disease. Among 220 men surveyed 80 were smokers. Among 190 women surveyed, 95 were smokers.

    a)   Construct a 95% confidence interval for the difference in the proportions of males and females who smoke.
    b)   Based on (a), would you reject H_0: $p_1=p_2$ in favor of H_1: $p_1 \neq p_2$? Justify briefly.

Categorical Data

21.　Suppose we wish to estimate what proportion of patients in an HMO spend more than $500 on prescription medications over 12 months. In a random sample of 150 patients, 34% spent more than $500.

   a) Construct a 95% confidence interval for the true proportion of patients who spend more than $500 on prescription medications per year.
   b) How many subjects would be required to estimate the true proportion who spend more than $500 on prescription medications per year with a margin of error no more than 2% with 95% confidence?

22.　Suppose a cross-sectional study is conducted to investigate cardiovascular risk factors among a sample of patients seeking medical care at one of two local hospitals. A total of 200 patients are enrolled. Construct a 95% confidence interval for the difference in proportions of patients with a family history of cardiovascular disease (CVD) between hospitals.

Family History of CVD	Enrollment Site	
	Hospital 1	Hospital 2
Definite	24	14
No	76	86
Total	100	100

23.　The following table was constructed based on a comparison of various sociodemographic characteristics between men and women enrolled in the study of cardiovascular risk factors.

Characteristic	Men (n=160)	Women (n=140)	p
Mean Age (SD)	45 (7.8)	46 (8.6)	0.7256
% High School Graduate	78	64	0.0245
Mean Annual Income (SD)	47,345 (8,456)	31.987 (9,645)	0.0001
% with No Insurance	8	9	0.9876

   a)　Which, if any, of the characteristics shown above are significantly different between men and women? Assume $\alpha=0.05$. Justify.

Categorical Data

b) Write the hypotheses tested and show the formula of the test statistic used to compare educational levels between men and women.

c) Suppose we wish to test whether the study population had a significantly higher than average proportion of patients with no insurance. The reported proportion is 7%. Write the hypotheses we would test and show the formula of the test statistic we would use to conduct such a test.

24. Investigators want to use a self-reported measure of pain to evaluate the effectiveness of a new medication designed to reduce post-operative pain. Before using the measure they examine its distributional properties in a sample of patients recently undergoing knee surgery. Post-procedure, each patient is asked to rate his or her pain on the following scale: No pain, minimal pain, some pain, moderate pain, or severe pain. The investigators hypothesize that the distribution of responses will be approximately normal, on the order of 1 to 2 to 4 to 2 to 1 across the response categories. Use the following data to test the claim at a 5% level of significance.

	Pain				
	None	Minimal	Some	Moderate	Severe
Number of Patients	15	25	46	36	18

25. Suppose the measure described in problem 24 is used to compare a newly developed pain medication against a standard medication with respect to self-reported pain. Using the data given below, is there a difference in the pain levels of patients on the different medications? Run the appropriate test at a 5% level of significance.

	Pain				
	None	Minimal	Some	Moderate	Severe
New	20	35	41	15	6
Standard	15	25	46	36	18

Categorical Data

26. Consider the application described in problem 25. Suppose the investigators want to compare the new and standard drugs but instead of using the 5-level pain measure they collapse the responses as follows: {None, Minimal and Some = Low Pain} and {Moderate and Severe=High Pain}. Is there a significant difference in the proportions of patients with High Pain between medication groups? Run the appropriate test at a 5% level of significance.

27. A study is conducted comparing three different prenatal care programs among women at high risk for preterm delivery. The programs differ in intensity of medical intervention. Women who meet the criteria for high risk of preterm delivery are asked to participate in the study and are randomly assigned to one of the three prenatal care programs. At the time of delivery, they are classified as preterm or term delivery. Based on the following data, is there a relationship between the prenatal care programs and preterm delivery? Run the appropriate test at α=0.05.

	Program 1: Minimal Intensity	Program 2: Moderate Intensity	Program 3: High Intensity
Preterm	34	18	12
Term	36	52	58

28. A national study reports that 28% of high school students smoke. We wish to test if the smoking is rate is higher than the national rate among freshman at a local university. To run the test, we take a random sample of 200 freshmen and 32% report that they smoke. Run the appropriate test at a 5% level of significance

SAS Problems: Use SAS to solve the following problems.

1. Recent studies have investigated the relationship between gender and career advancement. The following data represent a random sample of mathematicians classified by gender and academic rank.

		Academic Rank		
	Instructor	Assistant Professor	Associate Professor	Full Professor.
Female	12	15	18	7
Male	21	25	30	22

Use SAS Proc Freq to generate a contingency table and run a chi-square test of independence. Use a 5% level of significance.

1. A prospective study is conducted over 10 years to investigate long term complication rates in patients treated with two different therapies. The data are shown below:

	Complications	
Therapy	No	Yes
1	911	89
2	873	127

Use SAS Proc Freq to generate a contingency table and estimate the test if there is a significant difference in the proportions of patients with complications between therapies.

2. Prior to being randomized to one of two competing therapies, the severity of participant's migraines are clinically assessed. The following table displays the severity classifications for patients assigned to the medical and non-traditional therapies.

	Severity Classification			
Therapy	Minimal	Moderate	Severe	Total
Medical	90	60	50	200
Non-Traditional	50	60	90	200
Total	140	120	140	400

Categorical Data

Use SAS Proc Freq to generate a contingency table and run a chi-square test of independence. Use a 5% level of significance.

3. Use SAS to estimate the sample sizes required in each group to test the hypotheses H_0: $p_1=p_2$. Consider a two sided test. Suppose that $p_1=0.4$ and consider the following values for p_2: 0.20, 0.25, 0.30. How many subjects would be required to detect the specified differences with power of 80% at a 5% level of significance?

4. Suppose we wish to estimate what proportion of patients in an HMO spend more than $500 on prescription medications over 12 months. In a random sample of 150 patients, 34% spent more than $500.

   a) Use SAS to determine the number of subjects that would be required to estimate the true proportion who spend more than $500 on prescription medications per year with a margin of error no more than 2% with 95% confidence.

   b) Consider scenarios where the margin of error is 1%, 3% and 5%. What is the effect of changing the margin of error on the number of subjects required.

   c) Suppose that the sample described above was not available. What would the estimate be in (a) without that information?

9.2 Effect Measures

Dichotomous outcome variables in biostatistical applications usually represent events such as the development of a disease, a change in disease severity, or mortality. The parameters of interest are the population proportions, or the conditional probabilities of the event in the two populations. For example, consider the example above in which one group receives treatment 1 and the other receives treatment 2 and the event of interest is stroke. The two parameters are the probability of having a stroke during the five year period conditional on receiving treatment 1 (p_1) and the probability of having a stroke during the five year period conditional on receiving treatment 2 (p_2). The probability of having the outcome of interest is often called the *risk* of that event. As described in Chapter 8, we estimate these probabilities, or risks, using the sample proportions in each treatment group, $\hat{p}_1$ and $\hat{p}_2$.

Example 9.1. A trial of gamma globulin in the treatment of children with Kawasaki syndrome randomized approximately half of the patients to receive gamma globulin. The standard treatment for Kawasaki syndrome was a regimen of aspirin; however, about one quarter of these patients developed coronary abnormalities even under the standard treatment. The outcome of interest was the development of coronary abnormalities (CA) within seven weeks of treatment. The following 2 X 2 cross-tabulation table summarizes the results of the trial.

Treatment Group	Coronary Abnormalities CA = 1	No Coronary Abnormalities CA = 0	Total
Gamma Globulin GG = 1	5	78	83
Aspirin GG = 0	21	63	84
Total	26	141	167

Overall, 26 of the 167 patients developed coronary abnormalities. The estimated conditional probability or risk of developing coronary abnormalities given the standard treatment (aspirin) is $\hat{p}_0$ = 21/84 = .25. The risk of CA given treatment with gamma globulin is $\hat{p}_1$ = 5/83 = .06.

There are a number of statistics that are used to compare conditional probabilities of outcomes between populations (or treatments). Measures of difference in risk are known as effect measures. Three of these measures are the risk difference, the relative risk, and the odds ratio.

The simple difference between risks is called the risk difference and is estimated as follows:

$$\text{Estimate of Risk Difference} = \hat{R}D = \hat{p}_1 - \hat{p}_0 \qquad (9.1)$$

A risk difference of zero indicates no difference in risks, a positive RD indicates higher risk in Group 1, and a negative RD indicates lower risk in Group 1.

In Example 9.1, the risk difference is estimated as $\hat{R}D = 0.06 - 0.25 = -0.19$. The risk of CA in patients treated with gamma globulin is 0.19 lower than the risk in patients on the standard treatment.

Note that you can reverse the roles of the groups and calculate the risk difference as $RD = \hat{p}_0 - \hat{p}_1$. In example 9.1 this would yield an estimate of 0.19 (instead of –0.19), indicating a higher risk in Group 0.

The relative risk is the ratio of the risks and is estimated as follows:

$$\text{Estimate of Relative Risk} = \hat{R}R = \hat{p}_1/\hat{p}_0 \qquad (9.2)$$

A relative risk of one indicates no difference in risks, a RR greater than 1 indicates a higher risk in Group 1 and a RR less than 1 indicates a lower risk in Group 1.

In Eample 9.1, the relative risk is estimated as $\hat{R}R = (0.06/0.25) = 0.24$. The risk of CA in patients treated with gamma globulin is 24% (less than a quarter) of the risk of those on standard treatment.

Again, we can reverse the groups and estimate the relative risk as $\hat{R}R = \hat{p}_0/\hat{p}_1$. However, in many cases, such as in clinical trials, it is easier to interpret a relative risk in which Group 1 is compared to Group 0. For example, in a clinical trial, the relative risk as defined in (9.2) compares the treated group to the control group. In Example 9.1, the estimated $\hat{R}R = 0.24$, which is the risk in the gamma globulin treatment group, relative to the risk in the standard treatment group. Reversing the comparison groups would yield $\hat{R}R = (0.25/0.06) = 4.17$, and is interpreted (awkwardly) as lack of treatment with gamma globulin more than quadruples the risk of coronary abnormalities.

The third measure commonly used to compare risks in two populations is the odds ratio, based on a comparison of the odds of the outcome in the two groups. Suppose that x = the number of outcome events in a trial of size n. We estimate the risk of the event as $\hat{p} = \dfrac{x}{n}$. The odds of the event is defined as $\hat{o} = \dfrac{x}{(n - x)}$, the ratio of events to non-events.

In Example 9.1, the odds of CA in the standard treatment group (GG=0) is estimated at $\hat{o}_0 = 21/63 = 0.330$. The odds of CA in the gamma globulin group is estimated at $\hat{o}_1 = 5/78 = 0.064$.

Comparing Risks in Two Populations

The odds can also be estimated using   the sample proportion as follows:

$$\text{Estimate of Odds} = \hat{o} = \frac{x}{(n-x)} = \frac{x/n}{(n-x)/n} = \frac{x/n}{(1-x/n)} = \frac{\hat{p}}{(1-\hat{p})}$$

The odds ratio is estimated by the ratio of the odds in Group 1 to the odds in Group 2:

$$\text{Estimate of Odds Ratio} = \hat{O}R = \frac{\hat{o}_1}{\hat{o}_0} = \frac{\hat{p}_1/(1-\hat{p}_1)}{\hat{p}_0/(1-\hat{p}_0)} \qquad (9.3)$$

In Example 9.1, the odds ratio is estimated as $\hat{O}R = (5/78)/(21/63) = 0.064/0.330 = 0.192$. The odds of CA in the gamma globulin group is less than a fifth the odds of CA in the standard treatment group.

The interpretation of the odds ratio is not as intuitive as is the relative risk. However, there are some research designs in which the relative risk cannot be estimated and in those situations, the odds ratio can be used to estimate the relative risk.

If the event is rare (i.e., p is small), the odds ratio is a very good estimate of relative risk. This is because when p is small, (1-p) is very close to 1, so the odds o = p/(1-p) is very close to p. For example, if p = 0.1, then (1-p) = 0.9 and o = (0.1/0.9) = 0.11, which is close to p = 0.1. For more common events, the odds is not a good estimate of the probability of that event. For example, if p = 0.5, then (1-p) = 0.5, and o = (0.5/0.5) = 1.0, which is not close to p.

When the prevalence is less than 0.10, the odds of an event generally provides a good estimate of the probability of the event. In example 9.1, the risk of CA is low in the GG group (0.06), and $\hat{o}_1 = 0.064$, very close to $\hat{p}_1 = 0.060$. In the standard treatment group, the risk is higher (0.250) and $\hat{o}_0 = 0.330$ is larger than $\hat{p}_0 = 0.250$. Overall the risk of CA is 0.156. The estimate of the odds ratio $\hat{O}R = 0.192$ is smaller than the estimate of the relative risk, $\hat{R}R = 0.240$.

9.3 Confidence Intervals for Effect Measures

In Chapter 8 we presented techniques for statistical inference concerning the difference in proportions and introduced the Chi-square tests of Goodness of Fit and Independence. In this section we provide formulas for calculating confidence intervals around the three effect measures used to compare risks in two populations. In addition, we introduce the Chi-square test of Homogeneity used to test hypotheses about two population risks.

The general case involving two populations with a dichotomous outcome variable can be represented by the general 2 X 2 table in Table 9.1. This framework facilitates many of the calculations needed for statistical inference.

Comparing Risks in Two Populations

In this framework, the letter *a* represents x_1, the number of successes in group 1, and (a + b) represents n_1, the sample size in group 1. Likewise, the letter c represents x_0, the number of successes in group 0, and (c + d) represents n_0, the sample size in group 0.

Table 9.1 General Form of a 2 x 2 Table

		Event		
		1	0	
Treatment or	1	a	b	a + b
Comparison Group	0	c	d	c + d
		a + c	b + d	N

The convention is to label events as 1 if present and 0 if absent. Similarly, the active or experimental treatment is usually labeled 1 and the control (e.g., placebo) is usually labeled 0. The estimated risks and effect measures can be expressed as follows using the notation of Table 9.1:

$$\hat{p}_1 = \frac{a}{(a + b)} \qquad \hat{o}_1 = a/b$$

$$\hat{p}_0 = \frac{c}{(c + d)} \qquad \hat{o}_0 = c/d$$

$$\text{Estimate of Risk Difference} = \hat{R}D = (\hat{p}_1 - \hat{p}_0) = \frac{a}{(a + b)} - \frac{c}{(c + d)} \qquad (9.4)$$

$$\text{Estimate of Relative Risk} = \hat{R}R = \hat{p}_1/\hat{p}_0 = \frac{a/(a + b)}{c/(c + d)} \qquad (9.5)$$

$$\text{Estimate of Odds Ratio} = \hat{O}R = \frac{\hat{o}_1}{\hat{o}_0} = \frac{a/b}{c/d} \qquad (9.6)$$

The risk difference defined in (9.1) and (9.4) is simply the point estimate for the difference in proportions ($p_1 - p_0$) introduced in Chapter 8. The confidence interval estimate of the risk difference is given in Table 8.3 and again in Table 9.2. Confidence interval formulas for the other two effect measures are also given in Table 9.2. Notice that the formulas for the confidence intervals for RR and OR are given in terms of the notation in Table 9.1.

Table 9.2 Confidence Intervals for RD, RR, and OR

Risk Difference	$\hat{RD} \pm Z_{1-\alpha/2} \sqrt{\left(\dfrac{\hat{p}_0(1-\hat{p}_0)}{n_0}\right) + \left(\dfrac{\hat{p}_1(1-\hat{p}_1)}{n_1}\right)}$
Relative Risk	$\exp\left(\ln(\hat{RR}) \pm Z_{1-\alpha/2} \sqrt{\dfrac{(d/c)}{n_0} + \dfrac{(b/a)}{n_1}} \right)$
Odds Ratio	$\exp\left(\ln(\hat{OR}) \pm Z_{1-\alpha/2} \sqrt{\left(\dfrac{1}{a} + \dfrac{1}{b} + \dfrac{1}{c} + \dfrac{1}{d}\right)} \right)$

As in Table 8.3, the three confidence intervals in Table 9.2 all assume that $\min(n_0\hat{p}_0, n_0(1-\hat{p}_0)) \geq 5$ <u>and</u> $\min(n_1\hat{p}_1, n_1(1-\hat{p}_1)) \geq 5$. This assumption is satisfied if the cell frequencies a, b, c, and d are each at least 5.

Recall Example 9.1 in which we compared gamma globulin to standard treatment with respect to the development of coronary abnormalities. The results are summarized below using the notation of Table 9.1.

Example 9.1 continued

Treatment Group	Coronary Abnormalities CA = 1	No Coronary Abnormalities CA = 0	Total
Gamma Globulin GG = 1	a=5	b=78	83
Aspirin GG = 0	c=21	d=63	84
Total	26	141	167

We now calculate 95% confidence intervals for the RD, RR and OR using the formulas in Table 9.2. First we check the sample size assumption, and note that each of the cell frequencies is at least 5. Next, we substitute the following into the formulas in Table 9.2.

$$\hat{RD} = -0.19, \ \hat{RR} = 0.24, \ \hat{OR} = 0.19$$

$$\hat{p}_0 = 0.25, \hat{p}_1 = 0.06, n_0 = 84, n_1 = 83$$

$$a = 5, b = 78, c = 21, d = 63$$

Comparing Risks in Two Populations

A 95% confidence interval for the <u>risk</u>   <u>difference</u> is given by:

$$\hat{RD} \pm Z_{1-\alpha/2} \sqrt{\left(\frac{\hat{p}_0(1-\hat{p}_0)}{n_0} \right) + \left(\frac{\hat{p}_1(1-\hat{p}_1)}{n_1} \right)}$$

$$-0.19 \pm 1.96 \sqrt{\left(\frac{0.25(0.75)}{84} \right) + \left(\frac{0.06(0.94)}{83} \right)}$$

$$-0.19 \pm 1.96 \, (.054)$$

$$-0.19 \pm 0.106$$

$$(-0.296, \ -0.084)$$

The risk of coronary abnormalities in patients treated with gamma globulin is *lower* than that in patients on the standard treatment by 19%. We are 95% confident that the true difference in risk of coronary abnormalities in patients treated with gamma globulin as compared to patients on the standard treatment is between -29.6% and –8.4% (or is from 8.4% to 29.6% lower).

A 95% confidence interval for the <u>relative risk</u> is given by:

$$\exp\left(\ln(\hat{RR}) \pm Z_{1-\alpha/2} \sqrt{\frac{(d/c)}{n_0} + \frac{(b/a)}{n_1}} \right)$$

$$\exp\left(\ln(0.24) \pm 1.96 \sqrt{\frac{(63/21)}{84} + \frac{(78/5)}{83}} \right)$$

$$\exp\left(-1.423 \pm 1.96(0.473) \right)$$

$$\exp\left(-1.423 \pm 0.927 \right)$$

$$(0.095, 0.609)$$

The risk of coronary abnormalities in patients treated with gamma globulin is less than a quarter the risk in patients on the standard treatment (RR = 0.24). We are 95% confident that the true relative risk of coronary abnormalities in patients treated with gamma globulin as compared to patients on the standard treatment is between 0.095 and 0.609.

A 95% confidence interval for the <u>odds ratio</u> is given by:

Comparing Risks in Two Populations

$$\exp\left(\ln(\hat{OR}) \pm Z_{1-\alpha/2}\sqrt{\left(\frac{1}{a}+\frac{1}{b}+\frac{1}{c}+\frac{1}{d} \right)} \right)$$

$$\exp\left(\ln(0.19) \pm 1.96\sqrt{\left(\frac{1}{5}+\frac{1}{78}+\frac{1}{21}+\frac{1}{63} \right)} \right)$$

$$\exp\left(-1.649 \pm 1.96(0.526) \right)$$

$$\exp\left(-1.649 \pm 1.030 \right)$$

$$(0.069, 0.539)$$

The odds of coronary abnormalities in patients treated with gamma globulin is less than one fifth the odds in patients on the standard treatment (OR = 0.19). We are 95% confident that the true relative risk of coronary abnormalities in patients treated with gamma globulin as compared to patients on the standard treatment is between 0.069 and 0.539.

SAS Example 9.1. The following output was generated for the data in Example 9.1 using SAS Proc Freq, which produces a contingency table (or cross tabulation) when two variables are specified. Notice that the row percentages in the "CA" column are 6.02% and 25% for those in the gamma globulin and standard treatment groups, respectively. These are the estimated risks we calculated as 0.06 and 0.25.

We requested that SAS provide the odds ratio and relative risk and 95% confidence intervals. It is important that the event of interest is reported in the first column as SAS calculates the relative risk and odds ratios assuming that is the case. The relative risks and odds ratios compare the row 1 group to the row 2 group. For that reason we make sure that the GG = 1 group is in the first row so that the effect measures will compare the gamma globulin group to the standard treatment group. A brief interpretation appears after the output.

Comparing Risks in Two Populations

SAS Output for Example 9.1

The FREQ Procedure
Table of group by event

group   event

Frequency Percent Row Pct Col Pct	CA	noCA	Total
GG	5 2.99 6.02 19.23	78 46.71 93.98 55.32	83 49.70
noGG	21 12.57 25.00 80.77	63 37.72 75.00 44.68	84 50.30
Total	26 15.57	141 84.43	167 100.00

Statistics for Table of group by event
Estimates of the Relative Risk (Row1/Row2)

Type of Study	Value	95% Confidence Limits	
Case-Control (Odds Ratio)	0.1923	0.0686	0.5388
Cohort (Col1 Risk)	0.2410	0.0954	0.6089
Cohort (Col2 Risk)	1.2530	1.0948	1.4340

Sample Size = 167

Comparing Risks in Two Populations

We requested that SAS provide the odds ratio and the relative risk. The OR is given in the first row and the RR is given in the next row, labeled Col1 Risk as it is the relative risk of the event in the first column of the 2 X 2 table. Note that the values SAS provides for the OR and RR agree with those we calculated above. The 95% confidence limits are also given for the odds ratio and relative risk.

9.4 The Chi-Square Test of Homogeneity

The Chi-square test was introduced in Chapter 8 as a technique to be used in tests of goodness of fit and independence. The Chi-square test of homogeneity can be used to test hypotheses concerning risks in two populations. The null hypothesis is that the risks are the same in the two populations, or that the two populations are *homogeneous* with respect to risk.

$$H_0: p_0 = p_1$$
$$H_1: p_0 \neq p_1$$

Now consider the relationship between the effect measures and the risks. If the two risks are the same, then the risk difference must be equal to zero and both the relative risk and odds ratios must be equal to one. Thus, the null hypothesis may equivalently be written in terms of

the risk difference, $\quad\quad\quad\quad\quad\quad H_0: RD = 0$
$\quad\quad\quad\quad\quad\quad\quad\quad\quad\quad\quad\quad\quad H_1: RD \neq 0$

the relative risk, $\quad\quad\quad\quad\quad\quad\quad H_0: RR = 1$
$\quad\quad\quad\quad\quad\quad\quad\quad\quad\quad\quad\quad\quad H_1: RR \neq 1$

or the odds ratio, $\quad\quad\quad\quad\quad\quad\quad H_0: OR = 1$
$\quad\quad\quad\quad\quad\quad\quad\quad\quad\quad\quad\quad\quad H_1: OR \neq 1.$

As stated in Chapter 8, χ^2 tests are based on the agreement between observed frequencies and frequencies expected under H_0. We therefore calculate expected frequencies using the fact that the risks in the two groups are equal under the null hypothesis.

Example 9.2. Consider again the example of a clinical trial of a proposed stroke prevention medication. Suppose that 250 participants are randomly assigned to receive either the new treatment ($n_1 = 120$) or a placebo ($n_0 = 130$) and are followed over the course of a 5 year study period. During the 5 year study period, 12 of the 120 subjects in the new treatment group (T=1)

Comparing Risks in Two Populations

and 28 of the 130 subjects in the placebo group (T=0) had strokes. Thus, $\hat{p}_1 = 0.100$ and $\hat{p}_0 = 0.215$.

These results are given in the following table.

		Event		
		Stroke	No Stroke	Total
Treatment	T=1	12	108	120
Group	T=0	28	102	130
		40	210	250

Suppose that we want to test the hypothesis that the risk of stroke is the same in the two groups.

1. Set up hypotheses

H_0: $p_0 = p_1$

H_1: $p_0 \neq p_1$.

Alternatively, the hypotheses can be phrased in terms of the relative risk of stroke in the new treatment group (T=1) as compared to the placebo group (T=0):

H_0: RR = 1

H_1: RR $\neq$ 1.

2. Select appropriate test statistic

We test the null hypothesis of homogeneity using the χ^2 test of homogeneity

$$\chi^2 = \sum \frac{(O - E)^2}{E} \tag{9.7}$$

where Σ indicates summation over all four cells of the table,
O = observed frequency and
E = expected frequency.

The test statistic χ^2 follows a χ^2 distribution with df = 1 as long as the expected frequencies in each cell are at least 5. The critical value (see Table 5) for a test with $\alpha = 0.05$ is $\chi^2 = 3.84$.

2b. Check that the test is valid

We must calculate the expected frequencies and check that each is at least 5 so that we can be sure that the test statistic follows the χ^2 distribution. Combining the two treatment groups, we see that 40 of the 250 subjects had strokes in the 5 year study period. The estimated 5

Comparing Risks in Two Populations

year risk of stroke is thus $\hat{p} = 40/250 = 0.16$.   Assuming that the null hypothesis is true, the risk of stroke is the same in both treatment groups, and we would expect that 16% of the subjects in each group would have a stroke during the study period. Thus, among the120 subjects in the new treatment group (T=1), we expect that 16%, or (.16)(120) = 19.2 subjects would have a stroke. Similarly, we expect that 16% of the 130 subjects in the placebo group (T=0), or (.16)(130) = 20.8 subjects would have a stroke. The following table contains the expected cell frequencies under the null hypothesis. The expected frequencies in the "No Stroke" column are obtained by subtraction (for example, 120 − 19.2 = 100.8). Notice that we do not round the expected frequencies to integers.

		Event		
		Stroke	No Stroke	Total
Treatment	T=1	19.2	100.8	120
Group	T=0	20.8	109.2	130
		40	210	250

Each of the expected frequencies (19.2, 100.8, 20.8, and 109.2) is at least 5 so we can compare the test statistic to the the critical value $\chi^2 = 3.84$.

3. Decision Rule

Reject H_0 if $\chi^2 \geq 3.84$.
Do not reject H_0 if $\chi^2 < 3.84$.

1. Test Statistic

We now calculate the test statistic using equation (9.7) and the expected frequencies.

$$\chi^2 = \frac{(12-19.2)^2}{19.2} + \frac{(108-100.8)^2}{100.8} + \frac{(28-20.8)^2}{20.8} + \frac{(102-109.2)^2}{109.2} = 6.181$$

5. Conclusion: Reject H_0 since $6.181 \geq 3.84$. There is significant evidence that the relative risk is not equal to 1.

A computational formula for the χ^2 statistic in the special situation of a 2 x 2 table is

$$\chi^2 = \frac{(ad-bc)^2(N)}{(a+b)(c+d)(a+c)(b+d)} \tag{9.8}$$

Comparing Risks in Two Populations

This formula does not require the calculation   of expected frequencies in a separate step. However, it is still necessary to check that the cell sizes are sufficient (that each expected frequency is at least 5).

We can use this formula to calculate the test statistic in Example 9.2.

$$\chi^2 = \frac{(ad - bc)^2(N)}{(a+b)(c+d)(a+c)(b+d)}$$

$$\chi^2 = \frac{\{(12)(102) - (108)(28)\}^2(250)}{(120)(130)(40)(210)}$$

$$\chi^2 = \frac{810,000,000}{131,040,000}$$

$$\chi^2 = 6.181$$

SAS Example 9.2 The following output was generated for the data in Example 9.2 using SAS Proc Freq. We requested that SAS provide the expected cell frequencies and the row percents, and the odds ratio, relative risk and 95% confidence intervals. We also requested the Chi-square test of homogeneity. Again, we make sure that the event of interest (stroke) is reported in the first column and the treatment group is in the first row. A brief interpretation appears after the output.

SAS Output for Example 9.2

The FREQ Procedure
Table of group by stroke

```
group        stroke

Frequency  |
Expected   |
Row Pct    | yes      | no       | Total
-----------+----------+----------+--------
Treatment  |      12  |     108  |   120
           |    19.2  |   100.8  |
           |   10.00  |   90.00  |
-----------+----------+----------+--------
Placebo    |      28  |     102  |   130
           |    20.8  |   109.2  |
           |   21.54  |   78.46  |
-----------+----------+----------+--------
Total           40         210       250
```

Comparing Risks in Two Populations

Statistics for Table of group by stroke

Statistic	DF	Value	Prob
Chi-Square	1	6.1813	0.0129
Likelihood Ratio Chi-Square	1	6.3540	0.0117
Continuity Adj. Chi-Square	1	5.3526	0.0207
Mantel-Haenszel Chi-Square	1	6.1566	0.0131
Phi Coefficient		-0.1572	
Contingency Coefficient		0.1553	
Cramer's V		-0.1572	

Statistics for Table of group by stroke
Estimates of the Relative Risk (Row1/Row2)

Type of Study	Value	95% Confidence Limits	
Case-Control (Odds Ratio)	0.4048	0.1954	0.8386
Cohort (Col1 Risk)	0.4643	0.2475	0.8710
Cohort (Col2 Risk)	1.1471	1.0296	1.2779

Sample Size = 250

Interpretation of SAS Output for Example 9.2

The first number in each cell is the observed frequency, the second is the expected frequency and the final number is the row percent. For example, in the "Treatment" – "stroke" cell (the "a" cell), the observed frequency is 12, the expected frequency is 10.2 and the row percent is 10%. The estimated relative risk is $\hat{R}R$ = 0.464 with 95% confidence limits (0.248, 0.871). Thus the risk of stroke in subjects on the new treatment is less than half the risk of stroke in those in the placebo group. Notice that the confidence interval does not include the null value 1.

The requested Chi-square procedure allows us to test the null hypothesis that the risk of stroke is the same in the two groups. Here we will phrase the hypotheses in terms of the relative risk, H_0: RR = 1 and H_1: RR $\neq$ 1. The value of the Chi-square statistic is 6.181, as we calculated above, and the associated p-value is p = 0.0129. Assuming α = 0.05, we reject H_0 since p = 0.0129 < α = 0.05. There is significant evidence (p = 0.0129) that the relative risk of stroke in treated subjects as compared to those on placebo is less than 1.

Comparing Risks in Two Populations

Example 9.3. A study of stroke patients who   had survived six months after the stroke found that 6/45 men and 22/63 women lived in an institution (such as an assisted living or nursing facility). Is there evidence that women are more likely than men to live in an institution after stroke?

The null hypothesis is that men and women stroke survivors are equally likely to live in an institution, or that the relative risk of living in an institution for men compared to women is equal to 1. First, we set up the 2 X 2 table below.

		Institution Yes	No (Home)	Total
Comparison Group: Gender	Men	6	39	45
	Women	22	41	63
		28	80	108

We estimate the probabilities of living in an institution for men and women separately: $\hat{p}_M = 0.133$ and $\hat{p}_w = 0.349$. The relative risk comparing men to women is RR = 0.381; among stroke survivors, men are only 38% as likely as women to live in an institution. Suppose that we want to test the hypothesis that the risk of living in an institution is the same for men and women.

1.   Set up hypotheses
H_0: $p_M = p_W$
H_1: $p_M \neq p_W$.

Again, the hypotheses can be phrased in terms of the relative risk of living in an institution for men as compared to women:

H_0: RR = 1
H_1: RR $\neq$ 1.

2.   Select appropriate test statistic

We test the null hypothesis of homogeneity using the χ^2 test of homogeneity

$$\chi^2 = \Sigma \frac{(O - E)^2}{E},$$

where Σ indicates summation over all four cells of the table,
O = observed frequency and
E = expected frequency.

The test statistic χ^2 follows a χ^2 distribution with df = 1 as long as the expected frequencies in each cell are at least 5.

Check that the test is valid

Before we can use the Chi-square test, we must check that each of the expected frequencies is each at least 5. Under the null hypothesis we assume that the probability of living in an institution is the same for men and women ($p_M = p_W = p$). The best estimate of this probability is obtained by combining men and women and calculating the overall sample proportion. Based on this combined sample, the estimated probability of living in an institution is $\hat{p} = (28/108) = .259$. Thus we expect about 25.9% of men and 25.9% of women to be living in an institution. Under the null hypothesis, in this sample of 45 men, we expect $(.259)(45)$ or 11.67 men to be living in an institution, while the remaining 33.33 live at home. Similarly, among the women, we expect $(.259)(63) = 16.33$ to be living in an institution and 46.67 to be living at home. The expected frequencies are 11.67, 33.33, 16.33, and 46.67, and each is of sufficient size (i.e., each is at least 5), so we can use the Chi-square test to test H_0.

3.   Decision Rule

The decision rule is

Reject H_0 if $\chi^2 \geq 3.84$.
Do not reject H_0 if $\chi^2 < 3.84$.

4.   Test Statistic

We now calculate the test statistic using the computational formula (9.7).

$$\chi^2 = \frac{(ad - bc)^2 (N)}{(a + b)(c + d)(a + c)(b + d)}$$

$$\chi^2 = \frac{\{(6)(41) - (31)(22)\}^2 (108)}{(45)(63)(28)(80)}$$

$$\chi^2 = \frac{20,530,368}{6,350,400}$$

$$\chi^2 = 3.233$$

5.   Conclusion: Do not reject H_0 since $\chi^2 = 3.233 < 3.84$. There is no significant evidence that the probability of living in an institution is different for men and women stroke survivors. Another interpretation is that there is no significant evidence that the relative risk of living in an institution comparing men and women is significantly different from 1.

Comparing Risks in Two Populations

9.5 Fisher's Exact Test

In Chapter 8 and in Sections 9.3 and 9.4 above, we stressed that the Chi-square tests are valid if each of the expected cell frequencies is at least 5. If one of the four expected cell frequencies in a 2 X 2 table is less than 5, the Chi-square test is not valid and we must use another method. One such method is Fisher's Exact test. A discussion of the mechanics of this test is beyond the scope of this text, here we simply present examples using SAS.

SAS Example 9.4. A substudy of the stroke study described above involved a comparison of the functional ability of stroke patients to patients who did not suffer stroke of the same age and gender. As many as possible of the stroke survivors were matched to controls who were the same age and sex. Cases (stroke patients) and controls (persons who did not suffer stroke) were questioned with respect to various measures of disability. Among the 24 stroke cases, only 16 were able to walk unassisted, while 23 of the 24 controls were able to walk unassisted. Using the data below, we test whether the stroke cases were more likely to need assistance walking as compared to age- and sex-matched controls. The analysis is conducted using SAS Proc Freq and a brief interpretation appears after the output.

SAS Output for Example 9.4

Table of group by Walk

group	Walk		
Frequency Expected Row Pct	Needs Assistance	Able to Walk	Total
Stroke Cases	8 4.5 33.33	16 19.5 66.67	24
Controls	1 4.5 4.17	23 19.5 95.83	24
Total	9	39	48

Comparing Risks in Two Populations

Statistics for Table of group by Walk

Statistic	DF	Value	Prob
Chi-Square	1	6.7009	0.0096
Likelihood Ratio Chi-Square	1	7.4609	0.0063
Continuity Adj. Chi-Square	1	4.9231	0.0265
Mantel-Haenszel Chi-Square	1	6.5613	0.0104
Phi Coefficient		0.3736	
Contingency Coefficient		0.3500	
Cramer's V		0.3736	

WARNING: 50% of the cells have expected counts less
than 5. Chi-Square may not be a valid test.
Fisher's Exact Test

Cell (1,1) Frequency (F)	8
Left-sided Pr <= F	0.9992
Right-sided Pr >= F	0.0113
Table Probability (P)	0.0105
Two-sided Pr <= P	0.0226

Sample Size = 48

The analysis is conducted using SAS Proc Freq and a brief interpretation appears after the output.

Interpretation of SAS Output for Example 9.4

The expected cell frequencies in the four cells are 4.5, 4.5, 19.5, and 19.5. Two of these are less than the required 5. Notice that SAS provides a "Warning" statement, indicating that 50% of the cells have expected counts less than 5 and that the Chi-square test may not be valid. Below this, SAS provides results of the Two-sided Fisher's Exact test which is recommended when the expected counts are small (i.e., < 5).

One third (33.3%) of the stroke cases need assistance to walk while only 4.2% of their age and sex-matched controls need assistance. The estimated RR is 8.0 (0.333/0.417). The null hypothesis is H_0: RR = 1 and H_1: RR $\neq$ 1. We use the Two-sided Fisher's Exact test to test the hypothesis. Assuming $\alpha = 0.05$, we reject H_0 since p = 0.0226 < α = 0.05. There is significant evidence (p = 0.0226) to show that stroke cases are more likely to need assistance walking as compared to age- and sex-matched controls.

9.5 Cox-Mantel-Haenszel Method

The Cox-Mantel-Haenszel method is an extension of the chi-square method and applied when interest lies in comparing two groups in terms of a dichotomous outcome over several levels of a third variable. For example, suppose there are a series of 2 X 2 tables, one for each of several strata (maybe the strata reflect different levels of severity of the index condition or different clinical centers), and interest lies in combining the information to take into account the differences in the strata. In this section we will describe a method for used to adjust the estimate of relative risk and the test statistic to account for a third variable in the analysis.

Example 9.5. Suppose we want to compare two medications, call them Drug A and Drug B, with respect to the development of or risk of hypertension (HTN). A total of 200 subjects are involved in the analysis and half are randomized to receive Drug A and half are randomized to receive Drug B. The following table summarizes the results of the study.

		HTN		
		Yes	No	
Medication	Drug A	56	44	100
	Drug B	32	68	100
Total		88	112	200

Overall, 88/200 subjects, or 44%, developed hypertension (HTN). The risk of HTN among subjects taking Drug A is $\hat{p}_A = 32/100 = 0.32$ and the risk of HTN among subjects taking Drug B is $\hat{p}_B = 56/100 = 0.56$ and the estimated relative risk comparing subjects on B to those on Drug A is $\hat{RR} = (0.56/0.32) = 1.75$. A test of the hypothesis H_0: RR =1 yields $\chi^2 = 11.688$ with $p < 0.001$. We reject H_0 and conclude that there is significant evidence that the relative risk of HTN comparing subjects taking Drug B to those taking Drug A is significantly different from 1.

Half of the subjects were randomized to receive Drug A and the other half received Drug B. However, further inspection of the data revealed that this balance was not maintained within age groups. In fact it turns out that 60% of the 100 older subjects (age 65 and older) received Drug A, while only 40% of the 100 younger subjects (age less than 65) received Drug A. We can repeat the analysis within the two age groups (i.e., two separate replications).

Age 65 +					Age < 65			
	HTN					HTN		
	Yes	No				Yes	No	
Drug B	32	8	40		Drug B	24	36	60
Drug A	24	36	60		Drug A	8	32	40
Total	56	44	100		Total	32	68	100

Comparing Risks in Two Populations

$$\hat{R}R = 0.8/0.4 = 2.0 \qquad\qquad \hat{R}R = 0.4/0.2 = 2.0$$

The overall risk of hypertension is 56% among those age 65 and older, while it is only 32% for those less than 65 years of age. However, the relative risk of HTN comparing those taking Drug B to those taking Drug A is the same in the two age groups. The baseline level of HTN is different but the relative impact of the two drugs is the same.

This would appear to indicate that combining the two age groups and reporting results based on the combined data is fine. In fact, since the estimated relative risk is 2.0 in each age group, the relative risk based on the combined data should surely also be 2.0. However, recall that we actually obtained an estimate of $\hat{R}R = 1.75$. This is caused by the combination of different baseline risks in the age groups and the imbalance in the treatment group allocation in the age groups, and is called <u>confounding</u> by age group.

In this section we present the Mantel-Haenszel method which is used to adjust the relative risk and the chi-square statistic to remove the effect of confounding. In Chapter 10 we introduce another method to adjust for confounding.

With a series of 2 X 2 tables, each table reflecting the relationship between group and event for a different strata (e.g., different level of a confounder), the Mantel-Haenszel method can be used to calculate an *adjusted chi-square test statistic and an adjusted relative risk*. The procedure is as follows.

First, create a 2x2 table <u>within</u> each stratum of the stratification variable. As before, the basic notation we use for the 2 X 2 table is as follows.

Stratum 1
Event

		1	0	
Group	1	a	b	a + b
	0	c	d	c + d
		a + c	b + d	N

Next, calculate

$$\chi^2_{MH} = \frac{\left(\sum \dfrac{(ad - bc)}{N}\right)^2}{\sum \dfrac{(a+b)(c+d)(a+c)(b+d)}{(N-1)N^2}} \tag{9.9}$$

Comparing Risks in Two Populations

where the summations are over the strata. This test statistic has a χ^2 distribution with 1 df, so we compare the test statistic to the critical value of 3.84 assuming a 5% level of significance.

Returning to Example 9.5, the strata are based on age and the two 2 X 2 tables are

	Age 65 + HTN					Age < 65 HTN		
	Yes	No				Yes	No	
Drug B	32	8	40		Drug B	24	36	60
Drug A	24	36	60		Drug A	8	32	40
Total	56	44	100		Total	32	68	100

The adjusted test statistic is computed as follows:

$$\chi^2_{MH} = \frac{\left(\sum \frac{(ad - bc)}{N} \right)^2}{\sum \frac{(a+b)(c+d)(a+c)(b+d)}{(N-1)N^2}}$$

$$\chi^2_{MH} = \frac{(\frac{(32)(36) - (8)(24)}{100} + \frac{(24)(32) - (36)(86)}{100})^2}{\frac{(40)(60)(56)(44)}{(99)100^2} + \frac{(60)(40)(32)(68)}{(99)100^2}}$$

$$\chi^2_{MH} = \frac{207.36}{11.248} = 18.435$$

The null hypothesis is that the relative risk of hypertension comparing subjects on Drug B to subjects on Drug A, adjusted for age, is equal to 1. We reject H_0 since $\chi^2_{MH} = 18.435 > 3.84$, and conclude that, after adjusting for age, the risk of HTN in subjects taking Drugs A and B are not equal.

We can also calculate adjusted estimate of the relative risk. The crude relative risk, $\hat{R}R_C$ is the relative risk obtained by combining all the data and estimating RR. In Example 9.5, $\hat{R}R_C$ = (0.56/0.32) = 1.75. The Mantel-Haenszel adjusted RR is estimated by the following.

$$\hat{RR}_{MH} = \frac{\sum \frac{a(c+d)}{N}}{\sum \frac{c(a+b)}{N}}$$

(9.10)

where the summations are over the strata.

In Example 9.5,

$$\hat{RR}_{MH} = \frac{\frac{(32)(60)}{100} + \frac{(24)(40)}{100}}{\frac{(24)(40)}{100} + \frac{(8)(60)}{100}}$$

$$\hat{RR}_{MH} = \frac{28.8}{14.4} = 2.0$$

The relative risk adjusted for age is $\hat{RR}_{MH} = 2.0$, which is the same relative risk we obtained in each of the age groups.

SAS Example 9.5. We can request that SAS perform the Mantel-Haenszel adjusted chi-square test and produce a Mantel-Haenszel adjusted relative risk. The analysis is conducted using SAS Proc Freq and a brief interpretation appears after the output.

SAS Output for Example 9.5

```
                    Table 1 of group by htn
                    Controlling for age=65+

          group        htn

          Frequency|
          Row Pct  |HTN     |noHTN   | Total
          ---------+--------+--------+
          Drug A   |    24  |    36  |    60
                   | 40.00  | 60.00  |
          ---------+--------+--------+
          Drug B   |     8  |    32  |    40
                   | 20.00  | 80.00  |
          ---------+--------+--------+
          Total         32       68     100
```

Comparing Risks in Two Populations

```
                    Table 2 of group by htn
                    Controlling for age=<65

          group      htn

          Frequency|
          Row Pct  |HTN     |noHTN   |  Total
          ---------+--------+--------+
          Drug A   |     32 |      8 |    40
                   |  80.00 |  20.00 |
          ---------+--------+--------+
          Drug B   |     24 |     36 |    60
                   |  40.00 |  60.00 |
          ---------+--------+--------+
          Total         56       44      100
```

Summary Statistics for group by htn

Controlling for age

Cochran-Mantel-Haenszel Statistics (Based on Table Scores)

Statistic	Alternative Hypothesis	DF	Value	Prob
1	Nonzero Correlation	1	18.4345	<.0001
2	Row Mean Scores Differ	1	18.4345	<.0001
3	General Association	1	18.4345	<.0001

Estimates of the Common Relative Risk (Row1/Row2)

Type of Study	Method	Value	95% Confidence Limits	
Case-Control	Mantel-Haenszel	4.0000	2.0791	7.6955
(Odds Ratio)	Logit	4.0000	2.0707	7.7268
Cohort	Mantel-Haenszel	2.0000	1.4427	2.7727
(Col1 Risk)	Logit	2.0000	1.4670	2.7266
Cohort	Mantel-Haenszel	0.5714	0.4351	0.7505
(Col2 Risk)	Logit	0.6722	0.5286	0.8546

Comparing Risks in Two Populations

```
         Breslow-Day Test for
      Homogeneity of the Odds Ratios
      -----------------------------
      Chi-Square               1.4664
      DF                            1
      Pr > ChiSq               0.2259

         Total Sample Size = 200
```

Interpretation of SAS Output for Example 9.5

SAS produces individual 2 X 2 tables for each of the strata (age groups). The Mantel-Haenzel adjusted Chi-square statistic and its associated p-value is given under "Cochran-Mantel-Haenszel Statistics (Based on Table Scores)"; in this situation all three given statistics are identical. The Mantel-Haenszel adjusted relative risk and the 95% confidence interval are given under "Estimates of the Common Relative Risk (Row1/Row2)" in the row labeled Type of Study = "Cohort (Col 1 Risk)" and Method = "Mantel-Haenszel."

This method is only valid if the relative risks are homogeneous across strata. SAS provides the p-value associated with the Breslow-Day test for the hypothesis of homogeneity. If this p-value is > 0.5, we assume that the relative risks are homogeneous and that the Mantel-Haenszel statistics are valid.

In this example, the p-value for the Breslow-Day test is $p = 0.226 > 0.05$, so the Mantel-Haenszel method is valid. From the output, we see that $\hat{RR}_{MH} = 2.0$ with 95% confidence interval (1.44, 2.77). The $\chi^2_{MH} = 18.435$, with $p = < 0.001$.

9.7 Precision, Power and Sample Size Determination

The formulas given in Section 8.7 may be used to calculate the minimum number of subjects needed to ensure a specified level of power to detect a difference in proportions (or risks) between two groups. Equation (8.20) is given below using the terminology of this chapter.

$$n_1 = \left(\frac{\sqrt{2\bar{p}\bar{q}}\,Z_{1-\alpha/2} + \sqrt{p_1 q_1 + p_2 q_2}\,Z_{1-\beta}}{\hat{RD}} \right)^2 \qquad (9.11)$$

where $\bar{p} = (\hat{p}_1 + \hat{p}_2)/2$ and $\bar{q} = 1 - \bar{p}$.

Comparing Risks in Two Populations

9.8 Key Formulas (using 2 X 2 table notation in Table 9.1)

APPLICATION	NOTATION/FORMULA	DESCRIPTION
Risk Difference	$\hat{RD} = \hat{p}_1 - \hat{p}_0 = \dfrac{a}{(a+b)} - \dfrac{c}{(c+d)}$	
Relative Risk	$\hat{RR} = \hat{p}_1/\hat{p}_0 = \dfrac{a/(a+b)}{c/(c+d)}$	
Odds Ratio	$\hat{OR} = \dfrac{\hat{o}_1}{\hat{o}_0} = \dfrac{\hat{p}_1/(1-\hat{p}_1)}{\hat{p}_0/(1-\hat{p}_0)} = \dfrac{a/b}{c/d}$	
Confidence Interval for Risk Difference	$\hat{RD} \pm Z_{1-\alpha/2}\sqrt{\left(\dfrac{\hat{p}_0(1-\hat{p}_0)}{n_0}\right) + \left(\dfrac{\hat{p}_1(1-\hat{p}_1)}{n_1}\right)}$	
Confidence Interval for Relative Risk	$\exp\left(\ln(\hat{RR}) \pm Z_{1-\alpha/2}\sqrt{\dfrac{(d/c)}{n_0} + \dfrac{(b/a)}{n_1}}\right)$	
Confidence Interval for Odds Ratio	$\exp\left(\ln(\hat{OR}) \pm Z_{1-\alpha/2}\sqrt{\left(\dfrac{1}{a} + \dfrac{1}{b} + \dfrac{1}{c} + \dfrac{1}{d}\right)}\right)$	
H_0: RD = 0 or H_0: RR = 1 or H_0: OR = 1	$\chi^2 = \sum \dfrac{(O-E)^2}{E}$, df = 1	Chi Square test of Homogeneity. Assumes all expected cell counts are ≥ 5
H_0: RD = 0 or H_0: RR = 1 or H_0: OR = 1	$\chi^2 = \dfrac{(ad-bc)^2(N)}{(a+b)(c+d)(a+c)(b+d)}$	Chi Square test of Homogeneity: computational formula. Assumes all expected cell counts are ≥ 5
H_0: RD = 0 or H_0: RR = 1 or H_0: OR = 1 Each adjusted for confounder	$\chi^2_{MH} = \dfrac{\left(\sum \dfrac{(ad-bc)}{N}\right)^2}{\sum \dfrac{(a+b)(c+d)(a+c)(b+d)}{(N-1)N^2}}$	Mantel-Haenszel Chi square test of Homogeneity, adjusted for confounder
Eatimate of Relative Risk, adjusted for confounder	$RR_{MH} = \left(\sum \dfrac{a(c+d)}{N}\right)\Big/\left(\sum \dfrac{c(a+b)}{N}\right)$	Mantel-Haenszel estimate of relative risk adjusted for confounder
Sample Size required per group to test H_0: RD = 0 with power 1-β	$n_i = \left(\dfrac{\sqrt{2\overline{pq}}\,Z_{1-\alpha/2} + \sqrt{p_1q_1 + p_2q_2}\,Z_{1-\beta}}{\hat{RD}}\right)^2$	

Comparing Risks in Two Populations

9.9   Statistical Computing

Following are the SAS programs which were used to generate the analyses presented here. The SAS procedures used and brief descriptions of their use are noted in the header to each example. Notes are provided to the right of the SAS programs (*in italics*) for orientation purposes and are not part of the programs. In addition, there are blank lines in the programs that follow which are solely to accommodate the notes. Blank lines and spaces can be used throughout SAS programs to enhance readability. A summary of the SAS procedures used in the examples is provided at the end of this section.

Estimate Relative Risk and Odds Ratio

SAS EXAMPLE 9.1 Compare risk of coronary abnormalities on Gamma Globulin as compared to Aspririn.

A trial of gamma globulin in the treatment of children with Kawasaki syndrome randomized approximately half of the patients to receive gamma globulin plus aspirin; the other half received standard treatment of aspirin. The outcome of interest was the development of coronary abnormalities (CA) within seven weeks of treatment. The following 2 X 2 cross-tabulation table summarizes the results of the trial. Use SAS to estimate the relative risk (and odds ratio) of CA in patients treated with gamma globulin as compared to those on standard treatment.

Treatment Group	Coronary Abnormalities CA = 1	No Coronary Abnormalities CA = 0	Total
Gamma Globulin GG = 1	5	78	83
Aspirin GG = 0	21	63	84
Total	26	141	167

Program Code

options ps=62 ls=80;	*Formats the output page to 62 lines in length and 80 columns in width*
data in;	*Beginning of Data Step*
input group & event $ f;	*Inputs three variables, **group** (noGG or GG), **event** (CA or noCA) and **f**, the cell frequency. **Group** and **event** are defined as character variables using the $.*
cards;	*Beginning of Raw Data section.*
noGG noCA 63	*Actual observations (values of*
noGG CA 21	***group, event** and **f** on each line)*
GG noCA 78	
GG CA 5	
run;	

```
proc freq;
```
Procedure call. Proc freq generates a 2x2 table for two categorical variables.

```
   tables group*event/relrisk;
```
*Specifies the analytic variables to form the rows (**group**) and columns (**event**) of the table. The relrisk option requests that SAS generates estimates of the relative risk and odds ratio.*

```
   weight f;
```
*Specifies the variable that contains the cell counts, **f.***

```
run;
```
End of procedure section.

Comparing Risks in Two Populations

Estimate Relative Risk and Test H_0: RR=1 Versus H_1: RR $\neq$ 1

SAS EXAMPLE 9.2 Estimate the relative risk of stroke comparing a proposed stroke medication to placebo, and test the hypothesis that the relative risk is equal to 1.

Two hundred and fifty participants in a clinical trial are randomly assigned to receive either a proposed stroke prevention medication (T_1) or a placebo (T_0) and are followed over the course of a 5 year study period. These results are given in the following table. Using SAS, estimate the relative risk of stroke comparing those on the medication to those on placebo, and test the hypothesis that the relative risk is equal to 1.

		Event		
		Stroke	No Stroke	Total
Treatment Group	T=1	12	108	120
	T=0	28	102	130
		40	210	250

Program Code

options ps=62 ls=80;	*Formats the output page to 62 lines in length and 80 columns in width*
proc format;	*Proc format defines formats to label the values of the variables.*
value eventf 1='yes' 2='no';	*Variables with values of 1 and 2 will*
value grpf 1='Treatment' 2='Placebo';	*be labeled as 'yes' and 'no' respectively if formatted with the **eventf** format. Variables wuth values of 1 and 2 will be labeled as 'Treatment' and 'Placebo' respectively if formatted with the **grpf** format.*
run;	*End of format procedure.*
data in;	*Beginning of Data Step*
input group stroke f;	*Inputs three variables **group** (1 or 2), **stroke** (1 or 2) and **f** (the cell frequency).*

Comparing Risks in Two Populations

```
format stroke eventf. group grpf.;
the
```

The format statement assigns

*format **eventf** to the variable **stroke**, and the format **grpf** to the variable **group**. Note that the format names are followed by a period to distinguish them from variable names.*

```
cards;
1 1 12
1 2 108
2 1 28
2 2 102
run;
```

Beginning of Raw Data section.
*Actual observations (values of **group**, **stroke** and **f** on each line).*

```
proc freq;
```

Procedure call. Proc freq generates a 2x2 table for two categorical variables.

```
  tables group*stroke/nocol nopercent expected
      relrisk chisq;
```

*Specifies the analytic variables to form the rows (**group**) and columns (**stroke**) of the table. The nocol and nopercent options suppress the printing of the column and overall percents, respectively. The relrisk option requests that SAS generates estimates of the relative risk and odds ratio. The chisq option requests the chi-square tes.t*

```
  weight f;
```

*Specifies the variable that contains the cell counts, **f**.*

```
run;
```

End of procedure section.

Estimate Relative Risk and Test H_0: RR=1 Versus H_1: RR $\neq$ 1

SAS EXAMPLE 9.4 Are stroke cases more likely to need assistance walking as compared to age- and sex-matched controls?

As part of the stroke study described above, as many as possible of the stroke survivors were matched to controls who were the same age and sex. Cases and controls were questioned with respect to various measures of disability. Among the 24 stroke cases, only 16 were able to walk unassisted, while 23 of the 24 controls were able to walk unassisted. Use SAS to test whether stroke cases more likely to need assistance walking as compared to age- and sex-matched controls?

Program Code

Code	Description
options ps=62 ls=80;	*Formats the output page to 62 lines in length and 80 columns in width*
proc format;	*Proc format defines formats to label the values of the variables.*
value eventf 1='Needs Assistance' 2='Able to Walk';	*Variables with values of 1 and 2 will be labeled as 'Needs Assistance' and 'Able to Walks' respectively if formatted with the **eventf** format.*
value grpf 1='Stroke Cases' 2='Controls';	*Variables wuth values of 1 and 2 will be labeled as 'Stroke Cases' and 'Controls' respectively if formatted with the **grpf** format.*
run;	*End of format procedure.*
data in; input group walk f;	*Beginning of Data Step* *Inputs three variables **group** (1 or 2), **walk** (1 or 2) and **f** (the cell frequency).*
format walk eventf. group grpf.;	*The format statement assigns the format **eventf** to the variable **walk**, and the format **grpf** to the variable **group**. Note that the format names are followed by a period to distinguish them from variable names.*

cards;	*Beginning of Raw Data*
section.	
1 1 8	*Actual observations (values of*
1 2 16	***group***, ***walk*** *and **f** on each line).*
2 1 1	
2 2 23	
run;	

proc freq;

Procedure call. Proc freq generates a 2x2 table for two categorical variables.

    tables group*walk/nocol nopercent expected
       relrisk chisq;

*Specifies the analytic variables to form the rows (**group**) and columns (**walk**) of the table. The nocol and nopercent options suppress the printing of the column and overall percents, respectively. The relrisk option requests that SAS generates estimates of the relative risk and odds ratio. The chisq option requests the chi-square tes.t*

  weight f;

*Specifies the variable that contains the cell counts, **f**.*

run;

End of procedure section.

Estimate Mantel-Haenszel adjusted relative risk and perform Mantel-Haenszel Chi-square test of homogeneity.

SAS EXAMPLE 9.5. **Estimate Mantel-Haenszel adjusted relative risk and perform Mantel-Haenszel Chi-square test of homogeneity.**

We want to compare the risk of hypertension (HTN) in two groups of subjects, those treated with a Drug A and those treated with Drug B, adjusted for age (< 65 versus 65+). The assignments to drug treatment and the outcome status (Hypertension yes./no) for each age group is summarized in the tables below. Use SAS to estimate a relative risk adjusted for age and perform the Mantel-Haenszel test of homogeneity adjusting for age.

	Age 65 + HTN Yes	No			Age < 65 HTN Yes	No	
Drug B	32	8	40	Drug B	24	36	60
Drug A	24	36	60	Drug A	8	32	40
Total	56	44	100	Total	32	68	100

Program Code

options ps=62 ls=80;	*Formats the output page to 62 lines in length and 80 columns in width*
proc format;	*Proc format defines formats to label the values of the variables.*
value eventf 1='HTN' 2='noHTN';	*Variables with values of 1 and 2 will be labeled as 'HTN' and 'noHTN' respectively if formatted with the **eventf** format.*
value grpf 1='Drug A' 2='Drug B';	*Variables wuth values of 1 and 2 will be labeled as 'Drug A' and 'Drug B' respectively if formatted with the **grpf** format.*
value agef 1='65+' 2='<65';	*Variables wuth values of 1 and 2 will be labeled as '65+' and '<65' respectively if formatted with the **agef** format.*
run;	*End of format procedure.*

```
data in;
  input group htn age f;



  format htn eventf. group grpf. age agef.;





cards;
1 1 1 24
1 2 1 36
2 1 1 8
2 2 1 32
1 1 2 32
1 2 2 8
2 1 2 24
2 2 2 36
run;


proc freq;




  tables group*htn/nocol nopercent relrisk chisq;













  weight f;

run;
```

Beginning of Data Step
*Inputs four variables **group** (1 or 2),*
***htn** (1 or 2, age (1 or 2) and **f** (the*
cell frequency).
The format statement assigns the
*format **eventf** to the variable **htn**,*
*the format **grpf** to the variable **group***
*and the format **agef** to the variable*
***age**.. Note that the format names are*
followed by a period to distinguish
them from variable names.
Beginning of Raw Data section.

Procedure call. Proc freq generates
a 2x2 table for two categorical
variables.
Specifies the analytic variables to
*form the rows **(group)** and columns*
***(htn)** of the table. The nocol and*
nopercent options suppress the
printing of the column and overall
percents, respectively. The relrisk
option requests that SAS generates
estimates of the crude relative risk
and odds ratio. The chisq option
requests the chi-square tes.t
Specifies the variable that contains
*the cell counts, **f**.*
End of procedure section.

Comparing Risks in Two Populations

```
proc freq;
```
Procedure call. Proc freq generates a 2x2 table for two categorical variables.

```
    tables age*group*htn/cmh nocol nopercent;
```
*Specifies the analytic variables to form the rows (**group**) and columns (**htn**) of the table, stratified by the third variable (**age**). The cmh option requests an estimate of the Mantel-Haenszel relative risk and requests the Mentel-Haenszel test of homogeneity. Again, the nocol and nopercent options suppress the printing of the column and overall percents, respectively*

```
    weight f;
```
*Specifies the variable that contains the cell counts, **f**.*

```
run;
run;
```
End of procedure section.
End of procedure section.

Summary of SAS Procedures

The SAS Freq Procedure is used to generate RxC tables for categorical variables. Specific options can be requested to produce crude and adjusted estimates of relative risks and odds ratios, and to run unadjusted and adjusted chi-square tests of homogeneity. The specific options are shown in italics below. Users should refer to the examples in this section for complete descriptions of the procedure and specific options. A general description of the procedure and options is provided in the table below.

Procedure	Sample Procedure Call	Description
proc freq	proc freq; tables a*b; *weight f*;	Generates a 2x2 contingency table with row variable a (a dichotomous variable) and column variable b (a dichotomous variable). The weight option indicates that summary data are input and the variable f contains the cell frequencies.
	proc freq; tables a*b/*relrisk chisq*;	Generates a 2x2 contingency table (a by b) and requests estimates of the crude or unadjusted relative risk and odds ratio and runs the unadjusted chi-square test of homogeneity.
	proc freq; tables strata*a*b/*cmh*;	Generates 2x2 contingency tables (a by b) stratified by the third variable strata. The cmh option requests estimates of the adjusted relative risk and odds ratio and runs the Mantel-Haenszel adjusted chi-square test of homogeneity..

9.10   Problems

1. A prospective study is conducted over 10 years to investigate long term complication rates in patients treated with two different therapies. The data are shown below:

	Complications	
Therapy	Yes	No
1	89	911
2	127	873

   a)   Estimate the risk difference.
   b)   Estimate the relative risk.
   c)   Estimate the odds ratio.
   d)   Based on the above, how do the therapies compare ?

2. A study is conducted comparing an experimental treatment to a control treatment with respect to their effectiveness in reducing joint pain in patients with arthritis. One hundred patients are randomly assigned to one of the competing treatments and the following data are collected, representing the numbers of patients reporting improvement in joint pain after the assigned treatment is administered:

	Control	Experimental
Number Reporting Improvement	21	28
Total Number of Subjects	50	50

   Is there a significant difference in the proportions of patients reporting improvement in joint pain among the treatments ? Run the appropriate test at the 5% level of significance.

3.   Using the data in problem 2,

   a)   Construct a 95% confidence interval for the risk difference of patients reporting improvement in joint pain between the Control and Experimental treatments ?
   b)   Based on (a), is there a significant difference in the proportions of patients reporting improvement in joint pain between the Control and Experimental treatments ? Justify (be brief but complete).

4. Patients were enrolled in a study from two clinical centers. The objective of the study was to assess medication adherence (i.e., taking medications as prescribed). Some investigators suggest that adherence exceeding 85% is sufficient to classify a patient as adherent, while medication adherence below 85% suggests that the patient is not adherent to the prescribed schedule. Based on the following data, is there a significant difference in the proportions of adherent patients among the clinical centers? Run the appropriate test at $\alpha=0.05$.

	Enrollment Site	
	Center 1	Center 2
Adherent (> 85%)	28	30
Not Adherent (< 85%)	21	29

5. Suppose an observational study is conducted to investigate the smoking behaviors of male and female patients with a history of coronary heart disease. Among 220 men surveyed 80 were smokers. Among 190 women surveyed, 95 were smokers.

a) Estimate the odds ratio of smoking for male versus female patients with a history of coronary heart disease.

b) Construct a 95% confidence interval for the odds ratio.

c) Based on (a), would you reject H_0: OR=1 in favor of H_1: OR $\neq$ 1? Justify briefly.

6. Suppose a cross-sectional study is conducted to investigate cardiovascular risk factors among a sample of patients seeking medical care at one of two local hospitals. A total of 200 patients are enrolled. Construct a 95% confidence interval for the difference in proportions of patients with a family history of cardiovascular disease (CVD) between hospitals.

Family History of CVD	Enrollment Site	
	Hospital 1	Hospital 2
Definite	24	14
No	76	86
Total	100	100

7.   A randomized trial is conducted to compare a newly developed pain medication against a standard medication with respect to self-reported pain. Pain is reported on a 5 point scale. Suppose we are interested in the proportions of patients reporting moderate or severe pain between medications. Organize the data given below into a dichotomous outcome {moderate or severe pain} Versus {none, minimal or some pain}. Estimate the relative risk of moderate or severe pain between treatment groups. Test the hypotheses H_0: RR=1 versus H_1: RR ≠ 1   at a 5% level of significance.

	Pain				
	None	Minimal	Some	Moderate	Severe
New	20	35	41	15	6
Standard	15	25	46	36	18

8.   A randomized trial is conducted comparing two different prenatal care programs among women at high risk for preterm delivery. The programs differ in intensity of medical intervention. Women who meet the criteria for high risk of preterm delivery are asked to participate in the study and are randomly assigned to one of the prenatal care programs. At the time of delivery, they are classified as preterm or term delivery. The results are summarized below.

	Preterm Delivery	Term Delivery
Intensive Prenatal Care	12	43
Standard Prenatal Care	18	47

a)   Estimate the relative risk of preterm delivery for women in the intensive prenatal care program as compared to the standard.

b)   Construct a 95% confidence interval for the relative risk.

c)   Based on (a), would you reject H_0: RR=1 in favor of H_1: RR ≠ 1? Justify briefly.

9.   A clinical trial of a new treatment for migraine headaches was conducted using a sample
      of 200 young adults with documented history of migraine headache. Among the 100
      subjects on the new treatment (coded as 1), 32 experienced at least one migraine
      headache episode (MH=1) during the study period, and 56 of the 100 subjects on the
      standard treatment (coded as 0) experienced at least one migraine headache episode
      during the study period.

      a)   Complete the table of results below.
      b)   Estimate the RR and test H_0: RR=1.

Migraine Headache (MH)

Treatment	Yes (1)	No (0)
New (1)		
Standard (0)		
Total		200

One of the investigators suggested adjustment for age since the risk of MH was higher in older
subjects. The results are given below by age (older versus younger)

Age=Old

Migraine

Trt	1	0	
1	24	36	60
0	32	8	40
Total	56	44	100

Age=Young

Migraine

Trt	1	0	
1	8	32	40
0	24	36	60
Total	32	36	100

      c)   Estimate the RR adjusted for age, and test H_0: RR=1.

Comparing Risks in Two Populations

SAS Problems. Use SAS to solve the following problems.

1. A prospective study is conducted over 10 years to investigate long term complication rates in patients treated with two different therapies. The data are shown below:

	Complications	
Therapy	Yes	No
1	89	911
2	127	873

Use SAS Proc Freq to generate a 2x2 contingency table and estimate the relative risk and odds ratio of complications. Test whether the risk of complications is different between therapies using the chi-square test of homogeneity.

1. A randomized trial is conducted comparing two different prenatal care programs among women at high risk for preterm delivery. The programs differ in intensity of medical intervention. Women who meet the criteria for high risk of preterm delivery are asked to participate in the study and are randomly assigned to one of the prenatal care programs. At the time of delivery, they are classified as preterm or term delivery. The results are summarized below.

	Preterm Delivery	Term Delivery
Intensive Prenatal Care	12	43
Standard Prenatal Care	18	47

Use SAS Proc Freq to generate a 2x2 contingency table and estimate the relative risk and odds ratio of complications. Generate 95% confidence intervals for the relative risk and odds ratio. Test whether the risk of complications is different between therapies using the chi-square test of homogeneity.

Comparing Risks in Two Populations

3.   A clinical trial of a new treatment for migraine headaches was conducted using a sample of 200 young adults with documented history of migraine headache. Among the 100 subjects on the new treatment (coded as 1), 32 experienced at least one migraine headache episode (MH=1) during the study period, and 56 of the 100 subjects on the standard treatment (coded as 0) experienced at least one migraine headache episode during the study period. Use SAS to etimate the RR and test H_0: RR=1.

One of the investigators suggested adjustment for age since the risk of MH was higher in older subjects. The results are given below by age (older versus younger). Use SAS to estimate the RR adjusted for age and test H_0: RR=1 using the following data.

	Age=Old					Age=Young		
	Migraine (MH)					Migraine (MH)		
Trt	1	0			Trt	1	0	
1	24	36	60		1	8	32	40
0	32	8	40		0	24	36	60
Total	56	44	100		Total	32	36	100

CHAPTER 10: Analysis of Variance

10.1   Introduction

Analysis of variance (ANOVA) is one of the most widely used statistical techniques for testing the equality of population means. ANOVA is used to test the equality of more than 2 treatment means (to test the equality of 2 treatment means, we use the techniques described in Chapter 7). Specifically, the hypotheses are:

H_0: $\mu_1 = \mu_2 = \mu_3 = \ldots = \mu_k$
H_1: means not all equal

where k = the number of populations under consideration (k > 2).

NOTE: The alternative hypothesis is not written H_1: $\mu_1 \neq \mu_2 \neq \mu_3 \neq \ldots \neq \mu_k$, as we want to reject the null hypothesis (H_0) if <u>any</u> of the population means are not equal (i.e., if at least one pair of means is not equal).

The assumptions necessary for valid applications of ANOVA are:

(i)   Random samples from each of the k the populations under consideration,
(ii)   Large samples ($n_i \geq 30$, where i=1, 2, ... , k) or normal populations,
(iii)   k independent populations (k > 2),
(iv)   Equal population variances (i.e., $\sigma_1^2 = \sigma_2^2 = \ldots \ldots = \sigma_k^2 = \sigma^2$).

The logic of the analysis of variance technique is presented in Section 10.2. Notation and computations are illustrated through examples in Section 10.3. In Section 10.4 we define fixed and random effects models. In Section 10.5 we present a statistic to assess the magnitude of the treatment effect. In Section 10.6 we define multiple comparison procedures and illustrate the use of two specific procedures. In Section 10.7 we outline simple analysis of variance procedures for dependent samples, called repeated measures analysis of variance. Key formulas are summarized in Section 10.8 and statistical computing applications are presented in Section 10.9.

10.2   Background Logic

Consistent with other tests of hypotheses, we take random samples from each population of interest and evaluate sample statistics as a means of assessing the likelihood that H_0 is true. Consider the following examples which illustrate the logic of the analysis of variance procedure.

Example 10.1. An experiment is conducted in which three treatments are compared with respect to their effectiveness. For the purposes of this example, effectiveness is evaluated in terms of time to relief of symptoms, reported in minutes. We assume that the distribution of times to relief are approximately normal, and the test of interest is the following:

H_0: $\mu_1 = \mu_2 = \mu_3$
H_1: means not all equal.

Fifteen subjects are randomly selected to participate in the investigation. Five subjects are randomly assigned to each treatment and each subject reports the time to relief of symptoms, in minutes, following their assigned treatment. The sample data, and summary statistics, are shown below.

Analysis of Variance

Treatment 1	Treatment 2	Treatment 3
29.0	25.1	20.1
29.2	25.0	20.0
29.1	25.0	19.9
28.9	24.9	19.8
28.8	25.0	20.2
Summary Statistics by Treatment		
$\overline{X}_1 = 29.0$	$\overline{X}_2 = 25.0$	$\overline{X}_3 = 20.0$
$s_1^2 = 0.025$	$s_2^2 = 0.005$	$s_3^2 = 0.025$
$s_1 = 0.158$	$s_2 = 0.071$	$s_3 = 0.158$

From the summary statistics we see that the sample means are numerically different (i.e., 29.0 vs. 25.0 vs. 20.0 minutes). In addition, the variability <u>within</u> each sample is small (i.e., the standard deviations are 0.158, 0.071, 0.158, respectively), which implies that observations within each sample are tightly clustered about their respective sample means.

Suppose that the population means are equal (i.e., H_0: $\mu_1 = \mu_2 = \mu_3$ is true). We can assume then, that the three samples are drawn from the same population and can pool all of the observations together (i.e., n=15). The overall sample mean is $\overline{X}$ = 24.67 minutes, with a sample variance $s^2 = 14.54$, and standard deviation, s = 3.81. The variability in the pooled sample is large. In ANOVA we compare the variation <u>within</u> samples to the variation <u>between</u> samples to assess the equality of the population means. If the observations within a sample are similar in value (i.e., small within-sample variation) and the means are different across samples (large between-sample variation) then a real difference is said to exist in the population means. Figure 10.1 displays the sample means and ranges of observations within each sample in Example 10.1.

Figure 10.1 Variation Between and Within Treatments in Example 10.1

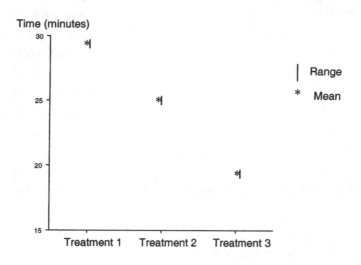

Example 10.2. Consider the investigation described in Example 10.1. Suppose that the following sample data are observed in a similar investigation. Summary statistics for each sample are also shown.

Treatment 1	Treatment 2	Treatment 3
29.0	33.1	15.2
14.2	7.4	39.3
45.1	17.6	14.8
48.9	44.2	25.5
7.8	22.7	5.2
Summary Statistics by Treatment		
$\overline{X}_1 = 29.0$	$\overline{X}_2 = 25.0$	$\overline{X}_3 = 20.0$
$s_1^2 = 330.87$	$s_2^2 = 201.07$	$s_3^2 = 167.96$
$s_1 = 18.19$	$s_2 = 14.18$	$s_3 = 12.96$

Analysis of Variance

From the summary statistics we see that again the sample means are numerically different (i.e., 29.0 vs. 25.0 vs. 20.0 minutes). However the variability within each sample is large. (i.e., the standard deviations are 18.19, 14.18, 12.96, respectively), which implies that the observations within each sample are not tightly clustered about their respective sample means, but widely spread.

Suppose again we pool all of the observations together (n=15) and compute summary statistics. The overall sample mean is $\overline{X}$ = 24.67, the sample variance is s^2 = 214.49, and the standard deviation is s = 14.65. In Example 10.2, the variation between samples is the same as in Example 10.1, however, there is large variation within samples, which suggests that there is no real difference in the population means (See Figure 10.2).

Figure 10.2 Variation Between and Within Treatments in Example 2

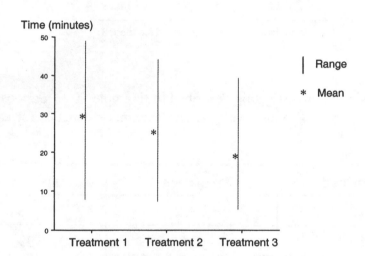

In ANOVA we wish to test the following: H_0: $\mu_1 = \mu_2 = \mu_3 = ... = \mu_k$ vs. H_1: Means are not all equal (or at least one pair is not equal), where k = the number of populations under consideration. To test H_0 we compute two estimates of the common population variance (σ^2). The first estimate is independent of H_0 (i.e., we do not assume that the population means are equal and we treat each sample separately). This estimate is called the estimate of the <u>within treatment variation</u>. The second estimate is based on the assumption that H_0 is true (i.e., population means are equal) and we pool all data together. This second estimate is called the estimate of the <u>between treatment variation</u>. The formulas for the estimates of the between and within variation are illustrated below for the case of equal sample sizes (i.e., $n_1 = n_2 = ... = n_k$). Sample sizes do not

Analysis of Variance

have to be equal in practice. The formulas presented in the next section illustrate the computations for both equal and unequal sample size applications.

Our first estimate is the estimate of the <u>within treatment variation</u>. To compute this estimate we assume nothing about the population means. We do, however, assume that $\sigma_1^2 = \sigma_2^2 = \ldots\ldots = \sigma_k^2 = \sigma^2$. (This assumption is required for appropriate use of ANOVA.) It follows that each sample variance s_j^2 (where $j=1,2,\ldots,k$) is an estimate of σ^2 (the common population variance). Again, to simplify things, we assume that $n_1 = n_2 = \ldots = n_k$. An estimate of σ^2 is obtained by taking the mean of the sample variances (s_j^2) over the k treatments:

$$s_w^2 = \sum_{j=1}^{k} \frac{s_j^2}{k} = \frac{s_1^2 + s_2^2 + \ldots + s_k^2}{k} \tag{9.1}$$

where $s_w^2 = $ denotes the within treatment variance or within variation.

This estimate of σ^2 depends only on the assumptions for ANOVA, specifically on the assumption of equal variances ($\sigma_1^2 = \sigma_2^2 = \ldots\ldots = \sigma_k^2 = \sigma^2$).

The second estimate of σ^2 is called the <u>between variation</u>. This second estimate depends on the assumptions for ANOVA (e.g., $\sigma_1^2 = \sigma_2^2 = \ldots\ldots = \sigma_k^2 = \sigma^2$) and also on the assumption that H_0 is true (i.e., $\mu_1 = \mu_2 = \mu_3 = \ldots = \mu_k = \mu$). If the population means are all equal then each sample mean $\overline{X}_j$ ($j=1,2,\ldots,k$) is an estimate of the common mean, μ. The $\overline{X}_j$'s can be viewed as a simple random sample of size k from a population with mean $\mu_{\overline{x}j} = \mu$ and variance $\sigma^2_{\overline{x}j} = \dfrac{\sigma^2}{n}$ by the Central Limit Theorem. Our goal is to generate an estimate of σ^2. We can estimate $\sigma^2_{\overline{x}j}$ which is equal to $\dfrac{\sigma^2}{n}$, and then use algebra to solve for σ^2.

The variance of the k sample means is:

$$s_{x_j}^2 = \sum_{j=1}^{k} \frac{(\overline{X}_j - \overline{X})^2}{k-1} \tag{9.2}$$

where $\overline{X}$ is the overall mean (i.e., based on all observations pooled together).

Analysis of Variance

$s^2 \overline{x}_j$ is an estimate of $\dfrac{\sigma^2}{n}$. Therefore $n\, s^2 \overline{x}_j$ is an estimate of σ^2:

$$s_b^2 = n\, s_{\overline{x}_j}^2 \qquad\qquad (9.3)$$

where $s_b^2 =$ denotes the between treatment variance or between variation.

The test statistic in ANOVA is the based on the ratio of these two estimates:

$$F = \frac{s_b^2}{s_w^2} \qquad\qquad (9.4)$$

The test statistic follows an F distribution (See Figure 10.3). The F distribution is an asymmetric distribution which takes on values greater than or equal to zero.

Figure 10.3   The F Distribution

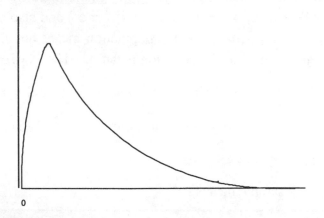

If the two estimates of σ^2 are close in value (i.e., F is approximately equal to 1), then we have no reason to reject H_0. However, if the variation between samples (s_b^2) is large and the variation within samples (s_w^2) is small (i.e., F is large), then we reject H_0. To draw a conclusion

Analysis of Variance

regarding H_0 we need a critical value from the F distribution. In order to select the appropriate value, we need 2 degrees of freedom: the numerator degrees of freedom, denoted $df_1 = k-1$, and the denominator degrees of freedom, denoted $df_2 = k(n-1)$, where k = the number of populations or treatments under consideration and n = the sample size per treatment. (NOTE: $k(n-1) = kn-k$, where n is the common sample size or sample size in each group, we sometimes let N=kn, where N = the total sample size (all groups combined)). Once the appropriate critical value is selected we can construct our decision rule which is of the form: Reject H_0 if $F \geq F_{k-1, N-k}$. Critical values of the F distribution are contained in Table 4a for $\alpha=0.05$ (and in Table 4b for $\alpha=0.025$).

The ANOVA technique is illustrated in the next section through examples. The computations are based on the logic presented here.

10.3   Notation and Examples

We now illustrate the computations using examples. In each example, the test of interest is a test of hypotheses involving more than two populations. Computations are illustrated for both equal and unequal sample size situations.

Example 10.3. An investigator wishes to compare the average time to relief of headache pain under three distinct medications, call them Drugs A, B and C. Fifteen patients who suffer from chronic headaches are randomly selected for the investigation. Five subjects are randomly assigned to each treatment. The following data reflect times to relief (in minutes) after taking the assigned drug:

Drug A	Drug B	Drug C
30	25	15
35	20	20
40	30	25
25	25	20
35	30	20

Notation

X_{ij} denotes the ith observation in the jth treatment
(e.g., $X_{11} = 30$, $X_{42} = 25$, $X_{53} = 20$)

Analysis of Variance

A "." in place of a subscript (either i or j) denotes summation over that index:

(e.g., $X_{.1}$ = sum over observations in treatment 1,

$X_{.1} = 30+35+40+25+35 = 165$,

$X_{1.}$ = sum first observations over treatments,

$X_{1.} = 30+25+15 = 70$.

$\overline{X}_{.1}$ = sample mean in treatment 1.

$\overline{X}_{..}$ = overall sample mean (taken over all observations and treatments).

Using the notation the summary statistics are:

Drug A	Drug B	Drug C
$\overline{X}_{.1} = 33.0$	$\overline{X}_{.2} = 26.0$	$\overline{X}_{.3} = 20.0$
$s_{.1} = 5.7$	$s_{.2} = 4.2$	$s_{.3} = 3.5$

To test whether the true mean times to relief under the three different drugs are equal, we use the same procedure used in other tests of hypotheses.

1. Set up hypotheses.

$H_0: \mu_1 = \mu_2 = \mu_3$

H_1: means not all equal $\alpha = 0.05$

The assumptions for ANOVA are:

(i)   Random samples from each of the 3 the populations under consideration

(ii)   The times to relief are approximately normally distributed.

(iii)   The 3 populations are independent.

(iv)   The population variances are equal (i.e., $\sigma_1^2 = \sigma_2^2 = \sigma_3^2 = \sigma^2$).

2. Select appropriate test statistic.

$$F = \frac{s_b^2}{s_w^2}$$

where s_b^2 = denotes the between variation which is also denoted MS_b = Mean Square between,

s_w^2 = denotes the within variation which is also denoted MS_w = Mean Square within (See Table 10.1 below).

Analysis of Variance

3.   Decision Rule

To select the appropriate critical value from the F distribution (Table 4a), we first compute the numerator degrees of freedom (df_1) and the denominator degrees of freedom (df_2):

$$df_1 = k - 1 = 3 - 1 = 2$$
$$df_2 = k(n - 1) = 3 (5 - 1) = 3(4) = 12$$

In ANOVA, we reject H_0 if the test statistic is larger than the critical value. The critical value of F with 2 and 12 degrees of freedom, relative to a 5% level of significance, is denoted $F_{2,12}$ and can be found in Table 4: $F_{2,12} = 3.89$. The decision rule is:

Reject H_0 if $F \geq 3.89$
Do Not Reject H_0 if $F < 3.89$

4.   Compute the test statistic.

To organize the computations in ANOVA applications, generally an ANOVA table is constructed. The table contains all of the computations, the degrees of freedom which are used to select the appropriate critical value and the test statistic. The general form of the ANOVA table is shown in Table 10.1.

Table 10.1   Analysis of Variance Table

Source of Variation	Sums of Squares SS	Degrees of freedom df	Mean Squares MS	F
Between	$SS_b = \Sigma n_j (\overline{X}_{.j} - \overline{X}_{..})^2$	k-1	$S_b^2 = MS_b = \dfrac{SS_b}{k - 1}$	$F = \dfrac{MS_b}{MS_w}$
Within	$SS_w = \Sigma \Sigma (X_{ij} - \overline{X}_{.j})^2$	N-k	$S_w^2 = MS_w = \dfrac{SS_w}{N - k}$	
Total	$SS_{total} = \Sigma \Sigma (X_{ij} - \overline{X}_{..})^2$	N-1		

where X_{ij} = ith observation in the jth treatment,
    $\overline{X}_{.j}$ = sample mean of jth treatment,
    $\overline{X}_{..}$ = overall sample mean,
    k = # treatments, and
    N = Σn_j = total sample size.

Now, using the data in Example 10.3, we construct an ANOVA table. We first compute the sums of squares. The between treatment sums of squares (also called the sums of square due to treatments), is computed by summing the squared differences between each treatment mean and the overall mean. We first compute the overall mean. In Example 10.3, the sample sizes are equal, therefore the overall mean is the mean of the three treatment means :

$$\overline{X}_{..} = \frac{(33.0 + 26.0 + 20.0)}{3} = 26.3$$

The between treatment sums of squares is:

$$SS_b = \Sigma n_j (\overline{X}_{.j} - \overline{X}_{..})^2 = 5((33-26.3)^2 + (26-26.3)^2 + (20-26.3)^2) = 423.3$$

The within treatment sums of squares (also called the sums of square due to error), is computed by summing the squared differences between each observation and its treatment mean.

To compute the within sums of squares, $SS_w = \Sigma\Sigma (X_{ij} - \overline{X}_{.j})^2$, we construct the following table to organize our computations:

Drug A		Drug B		Drug C	
$(X_{i1} - \overline{X}_{.1})$	$(X_{i1} - \overline{X}_{.1})^2$	$(X_{i2} - \overline{X}_{.2})$	$(X_{i2} - \overline{X}_{.2})^2$	$(X_{i3} - \overline{X}_{.3})$	$(X_{i3} - \overline{X}_{.3})^2$
-3	9	-1	1	-5	25
2	4	-6	36	0	0
7	49	4	16	5	25
-8	64	-1	1	0	0
2	4	4	16	0	0
$\Sigma(X_{i1} - \overline{X}_{.1})$ $=0$	$\Sigma(X_{i1} - \overline{X}_{.1})^2$ $=130$	$\Sigma(X_{i2} - \overline{X}_{.2})$ $=0$	$\Sigma(X_{i2} - \overline{X}_{.2})^2$ $=70$	$\Sigma(X_{i3} - \overline{X}_{.3})$ $=0$	$\Sigma(X_{i3} - \overline{X}_{.3})^2$ $=50$

The within treatment sums of squares is:

$$SS_w = \Sigma\Sigma (X_{ij} - \overline{X}_{.j})^2 = 130 + 70 + 50 = 250$$

The total sums of squares is computed by summing the squared differences between each observation and the overall mean:

Analysis of Variance

$$SS_{total} = \Sigma \Sigma (X_{ij} - \overline{X}_{..})^2 = (30 - 26.3)^2 + \ldots + (20 - 26.3)^2 = 673.3$$

However, since $SS_{total} = SS_b + SS_w$, it is not necessary to compute SS_{total} directly: $SS_{total} = 423.3 + 250 = 673.3$.

We now construct the ANOVA table for Example 10.3.

Source of Variation	Sums of Squares SS	Degrees of freedom df	Mean Squares MS	F
Between	423.3	2	211.67	10.16
Within	250.0	12	20.82	
Total	673.3	14		

5.   Conclusion. Reject H_0, since $10.16 > 3.89$. We have significant evidence, $\alpha = 0.05$, to show that the mean times to relief from headache pain under the three Drugs A, B and C are not all equal. Because we do not have more extensive tables for the F distribution here, we cannot compute a p-value. However, when an ANOVA is run using SAS as exact p-value is generated.

SAS Example 10.3. The following output was generated using SAS Proc ANOVA which runs an analysis of variance test for equality of means. SAS produces the following. A brief interpretation appears after the output.

449

SAS Output for Example 10.3

```
                                 The ANOVA Procedure
Dependent Variable: time
                                      Sum of
Source                      DF        Squares      Mean Square    F Value   Pr >

Model                        2     423.3333333    211.6666667      10.16   0.0026
Error                       12     250.0000000     20.8333333

Corrected Total             14     673.3333333

             R-Square     Coeff Var      Root MSE       time Mean
             0.628713     17.33299       4.564355       26.33333
Source                      DF       ANOVA SS      Mean Square    F Value   Pr > F
trt                          2     423.3333333    211.6666667      10.16   0.0026
```

Interpretation of SAS Output for Example 10.3

SAS has two procedures for analysis of variance applications. The first is the ANOVA procedure which is used when the sample sizes are equal, and the second is the Glm (general linear models) procedure, can be used when the sample sizes are unequal or equal. Since the sample sizes are equal in Example 10.3, we used the ANOVA procedure.

SAS produces an ANOVA table, similar to the table displayed in Table 10.1. SAS uses the term 'Model' to refer to the between treatment variation and 'Error' to refer to the within treatment variation. SAS also presents the degrees of freedom before the Sums of Squares, otherwise the table is identical. In Example 10.3, the test statistic if F = 10.16 with p = 0.0026, which would lead to rejection of H_0 since p = 0.0026 < α = 0.05.

Example 10.4. The following data reflect ages of students at completion of 8th grade. Test if there is a significant difference in the mean age at completion of 8th grade for Rural, Suburban and Urban students using a 5% level of significance. The following data were collected from randomly selected students Rural, Suburban and Urban schools.

Sample Data:
Rural: 14 14 14 14 13 13 13 12
Suburban: 14 14 14 13 13 13 13 13 12 12
Urban: 16 16 15 15 15 14 14 14 13 12

Analysis of Variance

1. Set up hypotheses.

$H_0 : \mu_1 = \mu_2 = \mu_3$
H_1 : means not all equal $\qquad \alpha = 0.05$

2. Select appropriate test statistic.

$$F = \frac{s_b^2}{s_w^2}$$

where s_b^2 = denotes the between variation which is also denoted MS_b = Mean Square between,
s_w^2 = denotes the within variation which is also denoted MS_w = Mean Square within (See Table 10.1 below).

3. Decision Rule

To select the appropriate critical value from the F distribution (Table 4), we first compute the numerator degrees of freedom (df_1) and the denominator degrees of freedom (df_2):

$df_1 = k - 1 = 3 - 1 = 2$
$df_2 = N - k = 28 - 3 = 25$

The critical value of F with 2 and 25 degrees of freedom, relative to a 5% level of significance, is found in Table 4: $F_{2,25} = 3.39$. The decision rule is:

Reject H_0 if $F \geq 3.39$
Do Not Reject H_0 if $F < 3.39$

4. Compute the test statistic.

Again, we will construct an ANOVA table to organize our computations. In this example, some of the formulas used have been modified to accommodate the unequal sample sizes.
The following table contains summary statistics on the age data:

Analysis of Variance

	Rural	Suburban	Urban
n_j	8	10	10
ΣX_{ij}	107	131	144
$\overline{X}_{.j}$	13.4	13.1	14.4

The between sums of squares (also called the sums of square due to treatments), is computed by summing the squared differences between each treatment mean and the overall mean. In Example 10.4, the sample sizes are unequal, therefore the overall mean is computed by summing all of the observations and dividing by N :

$$\overline{X}_{..} = \frac{(107 + 131 + 144)}{28} = 13.7$$

The between sums of squares is:

$$SS_b = \Sigma \, n_j (\overline{X}_{.j} - \overline{X}_{..})^2 = \{8(13.4 - 13.7)^2 + 10(13.1 - 13.7)^2 + 10(14.4 - 13.7)^2\} = 9.22$$

The within sums of squares (also called the sums of square due to error), is computed by summing the squared differences between each observation and its treatment mean.

To compute the within sums of squares, $SS_w = \Sigma \Sigma \, (X_{ij} - \overline{X}_{.j})^2$, we construct the following table:

Analysis of Variance

Rural		Suburban		Urban	
$(X_{i1} - \overline{X}_{.1})$	$(X_{i1} - \overline{X}_{.1})^2$	$(X_{i2} - \overline{X}_{.2})$	$(X_{i2} - \overline{X}_{.2})^2$	$(X_{i3} - \overline{X}_{.3})$	$(X_{i3} - \overline{X}_{.3})^2$
0.6	0.36	0.9	0.81	1.6	2.56
0.6	0.36	0.9	0.81	1.6	2.56
0.6	0.36	0.9	0.81	0.6	0.36
0.6	0.36	-0.1	0.01	0.6	0.36
-0.4	0.16	-0.1	0.01	0.6	0.36
-0.4	0.16	-0.1	0.01	-0.4	0.16
-0.4	0.16	-0.1	0.01	-0.4	0.16
-1.4	1.96	-0.1	0.01	-0.4	0.16
		-1.1	1.21	-1.4	1.96
		-1.1	1.21	-2.4	5.76
$\Sigma(X_{i1} - \overline{X}_{.1})$ $=0$	$\Sigma(X_{i1} - \overline{X}_{.1})^2$ $=3.88$	$\Sigma(X_{i2} - \overline{X}_{.2})$ $=0$	$\Sigma(X_{i2} - \overline{X}_{.2})^2$ $=4.90$	$\Sigma(X_{i3} - \overline{X}_{.3})$ $=0$	$\Sigma(X_{i3} - \overline{X}_{.3})^2 =$ 14.40

The within sums of squares is:

$$SS_w = \Sigma\Sigma(X_{ij} - \overline{X}_{.j})^2 \quad = 3.88 + 4.90 + 14.40 = 23.18$$

The total sums of squares is computed by summing the squared differences between each observation and the overall mean:

$$SS_{total} = \Sigma\Sigma(X_{ij} - \overline{X}_{..})^2 \quad = (14 - 13.7)^2 + ... + (12 - 13.7)^2 = 32.4$$

Again, since $SS_{total} = SS_b + SS_w$, it is not necessary to compute SS_{total} directly: $SS_{total} = 9.22 + 23.18 = 32.40$.

We now construct the ANOVA table for Example 10.4.

Analysis of Variance

Source of Variation	Sums of Squares SS	Degrees of freedom df	Mean Squares MS	F
Between	9.22	2	4.63	4.99
Within	23.18	25	0.93	
Total	32.40	27		

5.   Conclusion. Reject H_0, since 4.99 > 3.39. We have significant evidence, $\alpha = 0.05$, to show that the mean ages at completion of 8th grade are not equal for Rural, Suburban and Urban students.

SAS Example 10.4. The following output was generated using SAS Proc Glm which runs an analysis of variance test for equality of means. Glm is used when the sample sizes are unequal. SAS produces the following. A brief interpretation appears after the output.

SAS Output for Example 10.4

```
                               The GLM Procedure
Dependent Variable: age
                                    Sum of
Source                   DF        Squares      Mean Square    F Value    Pr > F

Model                     2     9.25357143      4.62678571       4.99    0.0150
Error                    25    23.17500000      0.92700000

Corrected Total          27    32.42857143

            R-Square      Coeff Var      Root MSE      age Mean
            0.285352      7.057234       0.962808      13.64286

Source                   DF      Type I SS      Mean Square    F Value    Pr > F
school                    2     9.25357143      4.62678571       4.99    0.0150

Source                   DF     Type III SS     Mean Square    F Value    Pr > F
school                    2     9.25357143      4.62678571       4.99    0.0150
```

Analysis of Variance

Interpretation of SAS Output for Example 10.4

Here we used the Glm procedure to run the ANOVA because the sample sizes are unequal. Similar to the ANOVA procedure, SAS produces an ANOVA table, similar to the table displayed in Table 10.1. Again, SAS uses the term 'Model' to refer to the between treatment variation and 'Error' to refer to the within treatment variation. SAS also presents the degrees of freedom before the Sums of Squares, otherwise the table is identical. In Example 10.4, the test statistic if $F = 4.99$ with $p = 0.0150$, which would lead to rejection of H_0 since $p = 0.0150 < \alpha = 0.05$.

Example 10.5. A study is developed to examine the effects of vitamins and milk supplements on infant weight gain. Four diet plans are considered: Diet A involves a regular diet plus the vitamin supplement, Diet B involves a regular diet plus the special milk formula, Diet C is our control diet (i.e., no restrictions or special considerations), and Diet D involves a regular diet plus the vitamin and the special milk formula. Twenty infants are selected for the investigation and each is randomized to one of the four competing diet programs. The table below displays weight gains, measured in pounds, after 1 month on the respective diet.

Diet A	Diet B	Diet C	Diet D
2.0	1.6	1.5	2.1
1.5	1.9	2.0	2.4
2.4	2.1	1.8	1.9
1.9	1.1	1.3	1.8
2.6	1.7	1.2	2.2

1. Set up hypotheses.

   $H_0 : \mu_1 = \mu_2 = \mu_3 = \mu_4$

   $H_1 :$ means not all equal $\qquad \alpha = 0.05$

2. Select appropriate test statistic.

$$F = \frac{s_b^2}{s_w^2}$$

   where s_b^2 = denotes the between variation which is also denoted MS_b = Mean Square between,

Analysis of Variance

s_w^2 = denotes the within variation which is also denoted MS_w = Mean Square within (See Table 10.1).

3.   Decision Rule

To select the appropriate critical value from the F distribution (Table 4), we first compute the numerator degrees of freedom (df_1) and the denominator degrees of freedom (df_2):

$$df_1 = k - 1 = 4 - 1 = 3$$
$$df_2 = k(n - 1) = 4 (5 - 1) = 4(4) = 16$$

The critical value of F with 3 and 16 degrees of freedom, relative to a 5% level of significance, is found in Table 4: $F_{3,16} = 3.24$. The decision rule is:

Reject H_0 if $F \geq 3.24$
Do Not Reject H_0 if $F < 3.24$

4.   Compute the test statistic.

Again, we will construct an ANOVA table to organize our computations. The following table displays summary statistics on the infant weights:

	Diet A	Diet B	Diet C	Diet D
n_j	5	5	5	5
ΣX_{ij}	10.4	8.4	7.8	10.4
$\overline{X}_{.j}$	2.1	1.7	1.6	2.1

The between sums of squares (also called the sums of square due to treatments), is computed by summing the squared differences between each treatment mean and the overall mean. Since the sample sizes are equal, the overall mean can be found by computing the mean of the 4 treatment means:

$$\overline{X}_{..} = \frac{(2.1 + 1.7 + 1.6 + 2.1)}{4} = 1.88$$

The between sums of squares is:

$$SS_b = \Sigma \, n_j \, (\overline{X}_j - \overline{X}_{..})^2 \;\; = 5((2.1-1.88)^2 + (1.7-1.88)^2 + (1.6-1.88)^2 + (2.1-1.88)^2) = 1.04.$$

The within sums of squares (also called the sums of square due to error), is computed by summing the squared differences between each observation and its treatment mean.

To compute the within sums of squares, $SS_w = \Sigma\Sigma \, (X_{ij} - \overline{X}_j)^2$, we construct a table similar to those presented in Examples 10.3 and 10.4. Here, we present only the results.

The within sums of squares is:

$$SS_w = \Sigma\Sigma \, (X_{ij} - \overline{X}_j)^2 \;\; = 0.75 + 0.57 + 0.46 + 0.23 = 2.01.$$

The total sums of squares is computed by adding the between and within sums of squares:

$$SS_{total} = SS_b + SS_w = 1.04 + 2.01 = 3.05$$

We now construct the ANOVA table for Example 10.5.

Source of Variation	Sums of Squares SS	Degrees of freedom df	Mean Squares MS	F
Between	1.04	3	0.35	2.69
Within	2.01	16	0.13	
Total	3.05	19		

5.   Conclusion. Do not reject H_0, since $2.69 < 3.24$. We do not have significant evidence, $\alpha = 0.05$, to show that the mean weight gains under the 4 different diets are not equal.

SAS Example 10.5. The following output was generated using SAS Proc Glm which runs an analysis of variance test for equality of means. Glm can accommodate applications where the sample sizes are equal or unequal. SAS produces the following. A brief interpretation appears after the output.

SAS Output for Example 10.5

The GLM Procedure

Dependent Variable: gain

Source	DF	Sum of Squares	Mean Square	F Value	Pr > F
Model	3	1.09400000	0.36466667	2.92	0.0659
Error	16	1.99600000	0.12475000		
Corrected Total	19	3.09000000			

R-Square	Coeff Var	Root MSE	gain Mean
0.354045	19.09187	0.353200	1.850000

Source	DF	Type I SS	Mean Square	F Value	Pr > F
diet	3	1.09400000	0.36466667	2.92	0.0659

Source	DF	Type III SS	Mean Square	F Value	Pr > F
diet	3	1.09400000	0.36466667	2.92	0.0659

Interpretation of SAS Output for Example 10.5

Here we used the Glm procedure to run the ANOVA. In Example 10.5, the test statistic is $F = 2.92$ with $p = 0.0659$. We do not reject H_0 since $p = 0.0659 > \alpha = 0.05$. In this example, the SAS calculations are slightly different from our hand calculations. This is due to rounding. SAS carries many more decimal places in computations and produces a more exact solution. The conclusion of the tests (performed by hand and by SAS) are the same. It should be noted that the test is marginally significant. Although we did not reach the specified significance level of 0.05, these results should be evaluated carefully as $p=0.0659$, indicating that there are some differences among the diets. If we evaluate the mean weight gains, Diets A and D produce the largest gains, Diet B is not much different than the control diet, Diet C.

10.4 Fixed versus Random Effects Models

There are two types of analysis of variance applications; fixed effects models and random effects models. In fixed effects models the treatment groups under study (e.g., the three headache medications (Drugs A, B and C), the four infant diets (A, B, C and D)) represent all treatments of interest. In our concluding statement we say there is (or is not) significant evidence of a difference

Analysis of Variance

in means among the treatments studied (e.g., there is a difference in the mean times to relief of headache pain under Drugs A, B and C).

In random effects models we randomly select k treatments for the investigation from a larger pool of available treatments. For example, suppose there are 10 competing treatments for headache pain or 12 well-known infant diets and we randomly select 3 or 4 to study (e.g., Drugs A, B and C or Diets A, B, C and D). In our concluding statement we say there is (or is not) significant evidence of a difference in means among ALL treatments (e.g., there is a difference in weight gain across all 12 well known infant diets), though we studied only a subset.

Basically, in random effects models we can generalize our results to the pool of all treatments since we randomly selected a subset for the investigation. In fixed effects models, our conclusions apply only to the treatments studied. The logic and the computations presented in the preceding sections are appropriate only for fixed effects models. Modifications are necessary for random effects models (See Cobb GW (1998). *Introduction to Design and Analysis of Experiments*, Springer-Verlag, New York, Inc.)

10.5   Evaluating Treatment Effects   *skipped*

If an ANOVA is performed and it has been established that a difference in means exists (i.e., we reject H_0), we want to assess the magnitude of the effect due to the treatments. That is, we want to address the question, How much variation in the data is due to the treatments?

The following statistic, called "eta-squared," is the ratio of variation due to the treatments (SS_b) to the total variation:

$$\eta^2 = \frac{SS_b}{SS_{total}} \tag{9.5}$$

where $0 \le \eta^2 \le 1$.

Values of η^2 that are closer to 1 imply that more variation in the data is attributable to the treatments.

In Example 10.3 comparing the times to relief among the three headache medications:

$$\eta^2 = \frac{423.3}{673.3} = 0.629$$

Thus, 62.9% of the variation in the times to relief is due to the medications (i.e., Drug A, B or C). In Example 10.4 comparing students' ages:

Analysis of Variance

$$\eta^2 = \frac{9.25}{32.43} = 0.285$$

Thus only 28.5% of the variation in the students ages is due to the location of the school (i.e., rural, suburban, urban).

10.6   Multiple Comparisons Procedures

Once we reject $H_0 : \mu_1 = \mu_2 = \ldots \mu_k$ in an ANOVA application, we say there is a significant difference among all of the treatment means (or at least one pair of means are not equal). It is often of interest to then test specific hypotheses comparing certain treatments. For example, in Example 10.3 we concluded that the mean times to relief for the three headache medications were significantly different. Suppose we are particularly interested in comparing only the first two medications (i.e., H_0: $\mu_1 = \mu_2$) or the first and the third (i.e., H_0: $\mu_1 = \mu_3$). Tests of this type are called pairwise comparisons, since they involve pairs of treatment means. It is also possible to construct more complicated comparisons. For example, it may be of interest to compare the mean time to relief for patients assigned to either Drug A or B to the mean time to relief for patients assigned to Drug C. The hypotheses would be denoted as follows: $H_0 : (\mu_1 + \mu_2)/2 = \mu_3$. Both pairwise (two-at-a-time) and more complicated comparisons are generally called contrasts.

There are a number of statistical procedures for handling these applications, which are called multiple comparison procedures, or MCP's. Different procedures are recommended for different applications. The procedures differ according to the types of comparisons of interest (e.g., pairwise comparisons), the number of comparisons, and the number of treatments in the ANOVA application. The procedures also differ with respect to their treatment of Type I errors (i.e., P(Reject H_0/H_0 true)). Techniques such as the two independent samples t test can be used to test pairs of treatment means but the Type I error rate over all comparisons of interest is not controlled. In applications involving k treatments, there are as many as k(k-1)/2 possible pairwise comparisons. In the worst possible case, the Type I error can be as large as $\alpha\{k(k-1)/2\}$. For example, if three pairwise comparisons are performed the Type I error rate could be 15%, or as large as 50% if 10 pairwise comparisons are performed. In most every application, these levels would be unacceptable.

Consider the following definitions:

Error rate per comparison (ER_PC)	(9.6)
= P(Type I error) on any one test or comparison	

Analysis of Variance

In general, the error rate per comparison is 0.05.

Error rate per experiment (ER_PE)   (9.7)

= the number of Type I errors we expect to make in any experiment under H_0.

For example, suppose we collect data on 100 different variables (e.g., systolic blood pressure, diastolic blood pressure, total cholesterol, age) from a random sample of females and from a random sample of males and wish to test for mean differences between males and females on each variable. The experiment involves 100 two independent samples t-tests, each performed at the $\alpha = 0.05$ level. The error rate over the entire experiment (ER_PE) is $100(0.05) = 5$. That is, we expect 5 tests to be significant (i.e., we reject H_0 in 5 tests) solely by chance (i.e., we expect to make 5 Type I errors in the experiment). Notice that the error rate per experiment is a frequency and not a probability.

Familywise error rate (FW_ER)   (9.8)

= P(at least 1 Type I error) in experiment.

The FW_ER is computed by : $1 - (1-\alpha_i)^c$, where α_i is the error rate per comparison (ER_PC) and c represents the number of contrasts, or comparisons, in the experiment. The formula is illustrated in Example 10.6.

Example 10.6. Suppose we test the equality of 5 treatment means using ANOVA and the null hypothesis is rejected at $a = 0.05$. Suppose that it is of interest to perform all pairwise comparisons. There are $k(k-1)/2 = 5(5-1)/2 = 10$ distinct pairwise comparisons (e.g., $H_0: \mu_1 = \mu_2$ vs. $H_1: \mu_1 \neq \mu_2$, $H_0: \mu_1 = \mu_3$ vs. $H_1: \mu_1 \neq \mu_4$, $H_0: \mu_1 = \mu_4$ vs. $H_1: \mu_1 \neq \mu_4$, . . . , $H_0: \mu_4 = \mu_5$ vs. $H_1: \mu_4 \neq \mu_5$). Suppose we wish to conduct each comparison at a 5% level of significance. It should be noted that one should perform only tests that are of substantive interest and not just all possible tests.

The error rate per comparison, ER_PC, = 0.05.
The error rate per experiment, ER_PE = $10(0.05) = 0.5$.
The familywise error rate, FW_ER = $1 - (1 - 0.05)^{10} = 0.401$.

In this example, we expect to make one Type I error solely by chance, and the probability of at least one Type I error is large (40.1% chance).

Analysis of Variance

Again, there are a number of multiple comparison procedures which differ according to their treatment of the error rates per experiment and family-wise error rates. Other MCPs (not discussed here) include the Duncan procedure (also called the multiple range test), Fisher's Least Significant Difference, the Newman-Keuls test, and Dunnett's test (used to compare a control to several active treatments). For further details, see D'Agostino RB, Massaro J, Kwan H, Cabral H (1993). Strategies for Dealing with Multiple Treatment Comparisons in Confirmatory Clinical Trials, *Drug Information Journal*; 27: 625-641. We now illustrate the use of two popular multiple comparison procedures, the Scheffe and Tukey procedures.

10.6.1 The Scheffe Procedure

The Scheffe procedure is a MCP which controls the familywise error rate. That is, the P(Type I error) is controlled (and equal to α) over the family of all comparisons. For example, if 6 or 8 or 20 comparisons are performed within a particular experiment, the familywise error rate is 5%. The Scheffe procedure is most desirable in applications involving more than a few contrasts, however, it is a conservative procedure (i.e., has lower statistical power) as compared to competing procedures.

The Scheffe procedure is outlined below, for the case of pairwise comparisons in Example 10.7. More complicated contrasts are considered in Example 10.8.

1. Set up hypotheses.

   $H_0: \mu_i = \mu_j$
   $H_1: \mu_i \neq \mu_j$

   where μ_i and μ_j are two of k treatment means which were found to be significantly different based on ANOVA.

2. Select appropriate test statistic.
   The test statistic for pairwise comparisons takes the following form:

$$F = \frac{(\overline{X}_{.i} - \overline{X}_{.j})^2}{s_w^2 \left(\dfrac{1}{n_i} + \dfrac{1}{n_j} \right)}$$

Analysis of Variance

where s_w^2 is the estimate of the within variation, and equal to the Mean Square Within or Mean Square Error (from ANOVA table).

3.   Decision Rule.

The test statistic follows an F distribution, therefore a critical value is selected from the F distribution table (Table 4). The critical value for the Scheffe procedure (pairwise as well as other contrasts) is the product of (k-1) and the critical value from the ANOVA:

Reject H_0 if $F \geq (k-1) F_{k-1, N-k}$
Do Not Reject H_0 if $F < (k-1) F_{k-1, N-k}$

Example 10.7. Consider Example 10.3 in which we compared the mean times to relief from headache pain under three competing medications. An ANOVA was performed at the 5% level of significance and we rejected $H_0: \mu_1 = \mu_2 = \mu_3$. Suppose we now wish to compare the medications taken two-at-a-time (i.e., pairwise comparisons). The summary statistics are shown below:

Drug A	Drug B	Drug C
$\overline{X}_{.1} = 33.0$	$\overline{X}_{.2} = 26.0$	$\overline{X}_{.3} = 20.0$
$s_{.1} = 5.7$	$s_{.2} = 4.2$	$s_{.3} = 3.5$

Drug A Vs. Drug B

1.   Set up hypotheses.
$H_0 : \mu_1 = \mu_2$
$H_1 : \mu_1 \neq \mu_2$ $\alpha = 0.05$

2.   Select appropriate test statistic.

$$F = \frac{(\overline{X}_{.1} - \overline{X}_{.2})^2}{S_w^2 \left(\dfrac{1}{n_1} + \dfrac{1}{n_2}\right)}$$

3.   Decision Rule.
Reject H_0 if $F \geq (k-1) F_{2,12} = 2(3.89) = 7.78$
(where $F_{2,12} = 3.89$ was used in Example 10.3 in the ANOVA)

4.   Compute the test statistic.

$$F = \frac{(33.0 - 26.0)^2}{20.82\,(\frac{1}{5} + \frac{1}{5})} = 5.88$$

5.   Conclusion. Do not reject H_0 since $5.88 < 7.78$. We do not have significant evidence, $\alpha = 0.05$, to show that $\mu_A \neq \mu_B$.

Drug A Vs. Drug C

1.   Set up hypotheses.
        $H_0 : \mu_1 = \mu_3$
        $H_1 : \mu_1 \neq \mu_3$ $\alpha = 0.05$

2.   Select appropriate test statistic.

$$F = \frac{(\overline{x}_{.1} - \overline{x}_{.3})^2}{S_w^2\,(\dfrac{1}{n_1} + \dfrac{1}{n_3})}$$

3.   Decision Rule.
        Reject H_0 if $F \geq (k-1)\,F_{2,12} = 2(3.89) = 7.78$

4.   Compute the test statistic.

$$F = \frac{(33.0 - 20.0)^2}{20.82\,(\frac{1}{5} + \frac{1}{5})} = 20.28$$

5.   Conclusion. Reject H_0 since $20.28 \geq 7.78$. We have significant evidence, $\alpha = 0.05$, to show that $\mu_A \neq \mu_C$.

Drug B Vs. Drug C

1.   Set up hypotheses.
        $H_0 : \mu_2 = \mu_3$
        $H_1 : \mu_2 \neq \mu_3$ $\alpha = 0.05$

Analysis of Variance

2. Select appropriate test statistic.

$$F = \frac{(\bar{x}_{.2} - \bar{x}_{.3})^2}{S_w^2 \left(\frac{1}{n_2} + \frac{1}{n_3}\right)}$$

3. Set up Decision Rule.
    Reject H_0 if $F \geq (k-1) F_{2,12} = 2(3.89) = 7.78$

4. Compute the test statistic.

$$F = \frac{(26.0 - 20.0)^2}{20.82 \left(\frac{1}{5} + \frac{1}{5}\right)} = 4.32$$

5. Conclusion. Do not reject H_0 since $4.32 < 7.78$. We do not have significant evidence, $\alpha = 0.05$, to show that $\mu_B \neq \mu_C$.

SAS Example 10.7. The following output was generated using SAS Proc ANOVA. We specified an option to conduct all pairwise comparisons using the Scheffe MCP. SAS produces the following. A brief interpretation appears after the output.

SAS Output for Example 10.7

```
                          The ANOVA Procedure
Dependent Variable: time
                                    Sum of
Source                     DF       Squares    Mean Square   F Value   Pr > F
Model                       2   423.3333333   211.6666667     10.16   0.0026
Error                      12   250.0000000    20.8333333
Corrected Total            14   673.3333333

              R-Square     Coeff Var      Root MSE     time Mean
              0.628713     17.33299       4.564355      26.33333

Source                     DF      ANOVA SS    Mean Square   F Value   Pr > F
trt                         2   423.3333333   211.6666667     10.16   0.0026
```

Analysis of Variance

```
                    Scheffe's Test for time
NOTE: This test controls the Type I experimentwise error rate.
          Alpha                               0.05
          Error Degrees of Freedom              12
          Error Mean Square                20.83333
          Critical Value of F               3.88529
          Minimum Significant Difference     8.047

Means with the same letter are not significantly different.
     Scheffe Grouping         Mean       N     trt
                      A      33.000       5      A
              B       A      26.000       5      B
              B              20.000       5      C
```

Interpretation of SAS Output for Example 10.7

Again, SAS produces an ANOVA table, similar to the table displayed in Table 10.1. In Example 10.3, we had a test statistic if $F = 10.16$ with $p = 0.0026$, which would led to rejection of H_0 since $p = 0.0026 < \alpha = 0.05$.

The next section of the output displays the results of the Scheffe procedure for pairwise comparisons. Along with the results of the Scheffe pairwise comparisons, SAS prints a note that the Scheffe procedure is one which controls the Type I error rate but has a higher Type II error rate than alternative procedures. SAS also copies the Mean Square Error (MSE) from the ANOVA table above, which is used in the test statistics. The pairwise comparisons are performed internally and SAS indicates which means are significantly different from others by assigning letters (Scheffe Grouping) to each treatment. If the same letters are assigned to different treatments, the treatment means are not significantly different. However, if different letters are assigned to the treatments, the treatment means are significantly different. In the output above, the treatments (or Drugs) A and B are assigned the letter "A" indicating that the means of Drugs A and B are not significantly different. Drug C is assigned the letter "B." Therefore the means of Drugs A and C are significantly different. Drug B is also assigned the letter "B" indicating that the means of Drugs B and C are not significantly different.

Example 10.8. Consider Example 10.4 above comparing the ages of students at the completion of 8th grade from Rural, Suburban and Urban schools. A one-way ANOVA was performed in which $H_0: \mu_1 = \mu_2 = \mu_3$ was rejected at $\alpha=0.05$. Suppose we wish to compare the Urban students to the Rural and Suburban students, combined. The summary statistics on the student's ages are given below:

Analysis of Variance

	Rural	Suburban	Urban
n_j	8	10	10
ΣX_{ij}	107	131	144
$\overline{X}_{.j}$	13.4	13.1	14.4

1. Set up hypotheses.
    $H_0 : 1/2 \, (\mu_1 + \mu_2) = \mu_3$
    $H_1 : 1/2 \, (\mu_1 + \mu_2) \neq \mu_3$ $\qquad \alpha = 0.05$

    NOTE: In the null hypothesis we state that the mean age for Rural and Suburban students combined is equal to the mean age for Urban students.

2. Select appropriate test statistic.

    $$F = \frac{\left(\dfrac{\overline{X}_{.1} + \overline{X}_{.2}}{2} - \overline{X}_{.3} \right)^2}{s_w^{\,2} \left[\dfrac{1}{2^2}\left(\dfrac{1}{n_1}\right) + \dfrac{1}{2^2}\left(\dfrac{1}{n_2}\right) + \dfrac{1}{n_3} \right]}$$

    In this example, we consider a more complicated contrast. Notice that the test statistic is modified accordingly. The numerator of the test statistic compares point estimates and the denominator involves the reciprocal of each sample size, weighted by the coefficient associated with that population mean squared. For example, in H_0 we weight both μ_1 and μ_2 by 1/2, which is squared in the denominator of the statistic.

3. Decision Rule.

    Reject H_0 if $F \geq (k-1) \, F_{2,25} = 2(3.39) = 6.78$
    (where $F_{2,25} = 3.39$ was used in Example 4 in the ANOVA)

Analysis of Variance

467

4.   Compute the test statistic.

$$F = \frac{\left(\frac{(13.4+13.1)}{2} - 14.4\right)^2}{0.927\left[\frac{1}{4}\left(\frac{1}{8}\right) + \frac{1}{4}\left(\frac{1}{10}\right) + \frac{1}{10}\right]} = 9.13$$

5.   Conclusion. Reject H_0 since $9.13 > 6.78$. We have significant evidence, $\alpha = 0.05$, to show that the mean age for Rural and Suburban students combined is not equal to the mean age of Urban students.

10.6.2 The Tukey Procedure

The Tukey procedure, also called the Studentized Range test, is a MCP which also controls the familywise error rate and is a popular, widely applied MCP. The Tukey procedure is appropriate for pairwise comparisons. It does not handle general contrasts. The Tukey procedure is a less conservative procedure (i.e., has better statistical power) than the Scheffe procedure when there are a large number of pairwise comparisons.

The Tukey procedure is outlined below and illustrated in Example 10.9. The procedure involves several steps. In the first step, treatments are ordered according to the magnitude of their respective sample means. Let $\overline{X}_{1'}$ denote the largest sample mean, $\overline{X}_{2'}$ denote the second largest sample mean, and so on, to $\overline{X}_{k'}$ which denotes the smallest sample mean. In the Tukey test, pairwise comparisons are made in a specific order. The first comparison involves a comparison of the treatments with the largest and smallest sample means. If this test is significant, then a test comparing the treatment with the largest sample mean to the treatment with the next-to-smallest sample mean is performed. If this test is significant then one proceeds to compare the treatment with the largest sample mean to the treatment with the third-to-smallest sample mean, and so on. The specifics of each test are outlined below. The example refers to the first test in the sequence of tests which involves the treatments with the largest and smallest sample means.

1.   Set up hypotheses.
        H_0: $\mu_{1'} = \mu_{k'}$
        H_1: $\mu_{1'} \neq \mu_{k'}$

Analysis of Variance

where $\mu_{1'}$ and $\mu_{k'}$ are the means of the treatments with the largest and smallest sample means, respectively.

2.   Select appropriate test statistic.

$$q_k = \frac{(\overline{X}_{1'} - \overline{X}_{k'})}{\sqrt{\dfrac{s_w^2}{n}}}$$

where s_w^2 is the estimate of the within variation, and equal to the Mean Square Within or Mean Square Error (from ANOVA table).

3.   Decision Rule.

The critical value for the Tukey test can be found in Table 6: Critical Values of the Studentized Range Distribution. The critical value depends upon the level of significance, α, the number of treatments involved in the analysis, k, and the error degrees of freedom from the ANOVA Table. Table 6 contains critical values for $\alpha=0.05$. The same critical value is used for all pairwise comparisons, and the decision rule is of the form:

Reject H_0 if $q_k \geq q_\alpha$ (k, df$_{error}$)
Do Not Reject H_0 if $q_k < q_\alpha$ (k, df$_{error}$)

Example 10.9. Consider Example 10.3 in which we compared the mean times to relief from headache pain under three competing medications. An ANOVA was performed at the 5% level of significance and we rejected H_0: $\mu_1 = \mu_2 = \mu_3$. Suppose we now wish to compare the medications taken two-at-a-time (i.e., pairwise comparisons) using the Tukey procedure. The summary statistics are shown below:

Drug A	Drug B	Drug C
$\overline{X}_{.1} = 33.0$	$\overline{X}_{.2} = 26.0$	$\overline{X}_{.3} = 20.0$
$s_{.1} = 5.7$	$s_{.2} = 4.2$	$s_{.3} = 3.5$

The first step is to order the treatments according to the magnitude of their respective sample means: $\overline{X}_{1'} = 33.0$, $\overline{X}_{2'} = 26.0$, and $\overline{X}_{3'} = 20.0$. (NOTE: Only coincidentally, the sample means array from largest to smallest as presented.) The first test in the Tukey procedure involves a comparison of Drugs A and C (largest Vs. smallest).

Analysis of Variance

Drug A Vs. Drug C

1. Set up hypotheses.

   $H_0 : \mu_1 = \mu_3$

   $H_1 : \mu_1 \neq \mu_3$ $\alpha = 0.05$

2. Select appropriate test statistic.

$$q_k = \frac{(\overline{X}_1 - \overline{X}_3)}{\sqrt{\dfrac{s_w^2}{n}}}$$

   where s_w^2 is the estimate of the within variation, and equal to the Mean Square Within or Mean Square Error (from ANOVA table).

3. Decision Rule.

   The appropriate critical value from Table 6 for k=3 and $df_{error} = 12$ is 3.77.

   Reject H_0 if $q_k \geq 3.77$

   Do Not Reject H_0 if $q_k < 3.77$

4. Compute the test statistic.

$$q_3 = \frac{(33.0 - 20.0)}{\sqrt{\dfrac{20.82}{5}}} = 6.37$$

5. Conclusion. Reject H_0 since 6.37 < 3.77. We have significant evidence, $\alpha = 0.05$, to show that $\mu_A \neq \mu_C$.

Because the first test was significant, we proceed to test the equality of treatments whose sample means reflect the next largest difference.

Drug A Vs. Drug B

1. Set up hypotheses.

   $H_0 : \mu_1 = \mu_2$

   $H_1 : \mu_1 \neq \mu_2$ $\qquad \alpha = 0.05$

2. Select appropriate test statistic.

$$q_k = \frac{(\overline{X}_1 - \overline{X}_2)}{\sqrt{\frac{s_w^2}{n}}}$$

where s_w^2 is the estimate of the within variation, and equal to the Mean Square Within or Mean Square Error (from ANOVA table).

3. Decision Rule.

   Reject H_0 if $q_k \geq 3.77$

   Do Not Reject H_0 if $q_k < 3.77$

4. Compute the test statistic.

$$q_3 = \frac{(33.0 - 26.0)}{\sqrt{\frac{20.82}{5}}} = 3.43$$

5. Conclusion. Do not reject H_0 since $3.43 < 3.77$. We do not have significant evidence, $\alpha = 0.05$, to show that $\mu_A \neq \mu_B$. Because this test is not significant, we do not go on to test $H_0 : \mu_2 = \mu_3$ Vs.$H_1 : \mu_2 \neq \mu_3$.

SAS Example 10.9. The following output was generated using SAS Proc ANOVA. We specified an option to conduct all pairwise comparisons using the Tukey MCP. SAS produces the following. A brief interpretation appears after the output.

SAS Output for Example 10.9

```
Analysis of Variance Procedure

Dependent Variable: TIME
                            Sum of          Mean
Source              DF      Squares         Square       F Value    Pr > F
Model                2      423.33333333    211.66666667  10.16     0.0026
Error               12      250.00000000     20.83333333
Corrected Total     14      673.33333333

            R-Square              C.V.        Root MSE           TIME Mean
            0.628713           17.33299       4.5643546          26.333333

Source              DF      ANOVA SS      Mean Square    F Value    Pr > F
TRT                  2      423.33333333  211.66666667    10.16     0.0026
```

```
              Tukey's Studentized Range (HSD) Test for time
NOTE: This test controls the Type I experimentwise error rate, but it
generally has a higher Type II error rate than REGWQ.

            Alpha                                    0.05
            Error Degrees of Freedom                   12
            Error Mean Square                    20.83333
            Critical Value of Studentized Range   3.77278
            Minimum Significant Difference         7.7012

    Means with the same letter are not significantly different.
          Tukey Grouping            Mean      N     trt
                        A          33.000      5     A
                  B     A          26.000      5     B
                  B                20.000      5     C
```

Interpretation of SAS Output for Example 10.9

In the second section of the output, along with the results of the Tukey pairwise comparisons, SAS prints a note that the Tukey procedure is one which controls the Type I error rate but has a higher Type II error rate than alternative procedures. SAS also prints the level of significance, $\alpha=0.05$, the error degrees of freedom, df=12, and the Mean Square Error (MSE) from the ANOVA table (MSE=20.83333) above, which is used in the test statistics. The treatments are ordered according to the magnitude of their respective sample means (largest to smallest). The pairwise comparisons are performed internally and SAS indicates which means are significantly different from others by assigning letters (Tukey Grouping) to each treatment. If the same letters

Analysis of Variance

are assigned to different treatments, the treatment means are not significantly different. However, if different letters are assigned to the treatments, the treatment means are significantly different. Recall, in the Tukey procedure, one first compares the treatments with the largest and smallest sample means. In this application Drug A has the largest mean and Drug C has the smallest. Drug A is assigned the letter "A" while Drug C is assigned the letter "B". Since the Tukey groupings (assigned letters) are different, the means of Drugs A and C are significantly different. The next test is the test comparing the means of Drugs A and B. In the output above, Drugs A and B are assigned the letter "A" indicating that the means of Drugs A and B are not significantly different. Although SAS allows for the test comparing Drugs B and C, the Tukey procedure should be terminated following the non significant result in the second test.

9.7 Repeated Measures Analysis of Variance

Skipped

In some applications, it is of interest to assess changes in a particular measure over time. For example, suppose an intervention is designed to improve patient's medication adherence, which is a particularly important issue in the management of many chronic diseases. The intervention is administered at a point in time and measurements are taken at predetermined intervals to assess adherence. A hypothesis might be that the intervention has a decreasing effect over time. The data layout is as follows:

Subject	Time 1	Time 2	. . .	Time k
1	X_{11}	X_{12}		X_{1k}
2	X_{21}	X_{22}		X_{2k}
.				
.				
n	X_{n1}	X_{n2}		X_{nk}

where X_{sj} represents the measurement (X) on the sth subject (s=1,2, …, n) on the jth occasion (j=1,2, …, k).

In these applications we have one sample of n subjects and take multiple, or repeated, measurements on each subject (k measurements in total). Without the subject column, the data layout are identical to the layout for analysis of variance procedures described in previous sections (assuming equal sample sizes). It is important to note that in these applications, there is one sample and repeated measures are taken on these subjects introducing a dependency among the measurements. Because of this dependency, the test statistic needs modification.

For reference, recall the two independent and two dependent sample applications described in Chapter 7. In the two dependent samples applications we focused on difference scores. Because there are now more than two measurements (k>2) the appropriate analysis is analysis of variance. The test is outlined below.

1. Set up hypotheses.

H_0: $\mu_1 = \mu_2 = \ldots = \mu_k$
H_1: means not all equal

2. Select appropriate test statistic.

$$F = \frac{s_b^2}{s_w^2}$$

where s_b^2 = denotes the between variation which is also denoted MS_b = Mean Square between,
s_w^2 = denotes the within variation which is also denoted MS_w = Mean Square within.

This test statistic is identical in form to the test statistic used above. The analysis of variance table is different, however, and includes an additional source of variation – variation due to the subjects. Because there are now multiple measurements taken on each subject, we can measure variation within given subjects (in the applications described in previous sections we had only a single measurement on each subject and could not measure variation within a given subject). The analysis of variable table for repeated measures analysis of variance procedures is given below in Table 10.2

Analysis of Variance

Table 10.2   Repeated Measures Analysis of Variance Table

Source of Variation	Sums of Squares SS	Degrees of freedom df	Mean Squares MS	F
Between Subjects	$SS_{subj} = \Sigma k (\overline{X}_{s.} - \overline{X}_{..})^2$	n-1		
Between Treatments	$SS_b = \Sigma n \ (\overline{X}_{.j} - \overline{X}_{..})^2$	k-1	$S_b^2 = MS_b = \dfrac{SS_b}{k-1}$	$F = \dfrac{MS_b}{MS_w}$
Within	$SS_w = SS_{total} - SS_{subj} - SS_b$	(n-1)(k-1)	$S_w^2 = MS_w = \dfrac{SS_w}{(n-1)(k-1)}$	
Total	$SS_{total} = \Sigma \Sigma (X_{sj} - \overline{X}_{..})^2$	nk-1		

where X_{sj} = measurement on the sth subject in the jth treatment (or at the jth time point),

$\overline{X}$ s. = sample mean of sth subject,

$\overline{X}$ .j = sample mean of jth treatment (or j^{th} time point),

$\overline{X}$ .. = overall sample mean,

k = # measurements per subject, and

n = number of subjects.

3.   Decision Rule

To select the appropriate critical value from the F distribution (Table 4), we use tha appropriate numerator and denominator degrees of freedom, df_1 and df_2, respectively (df_1 = k-1, and df_2 = (n-1)(k-1)). Again, in ANOVA, we reject H_0 if the test statistic is larger than the critical value. We now illustrate the use of these formulas with an example.

Example 10.10. An investigator is interested in comparing the cardiovascular fitness of elite runners on three different training courses, each of which covers 10 miles. The courses differ in terms of terrain, Course 1 is flat, Course 2 has graded inclines and Course 3 includes steep inclines. Each runner has his/her heart rate monitored at mile 5 of the run on each course. Ten runners are involved and their heart rates measured on each course are shown below.

Runner Number	Course 1	Course 2	Course 3
1	132	135	138
2	143	148	148
3	135	138	141
4	128	131	139
5	141	141	150
6	150	156	161
7	131	134	138
8	150	156	162
9	142	145	151
10	139	165	160

Is there a significant difference in the mean heart rates of runners on the three courses? Run the appropriate test at a 5% level of significance.

Because we have a single sample of 10 subjects, and 3 measures taken on each, the appropriate analysis is a repeated measures analysis of variance. The test is carried out below.

1. Set up hypotheses.

$H_0 : \mu_1 = \mu_2 = \mu_3$
H_1 : means not all equal $\qquad \alpha = 0.05$

2. Select appropriate test statistic.

$$F = \frac{s_b^2}{s_w^2}$$

3. Decision Rule

To select the appropriate critical value from the F distribution (Table 4), we first compute the numerator degrees of freedom (df_1) and the denominator degrees of freedom (df_2):

$$df_1 = k - 1 = 3 - 1 = 2 \qquad\qquad df_2 = (n-1)(k-1) = 9*2 = 18$$

The critical value of F with 2 and 18 degrees of freedom, relative to a 5% level of significance, is found in Table 4: $F_{2,18} = 3.55$. The decision rule is:

Reject H_0 if $F \geq 3.55$
Do Not Reject H_0 if $F < 3.55$

4. Compute the test statistic.

  Again, we will construct an ANOVA table to organize our computations. We will first compute the between subjects sums of squares (necessary to account for the repeated measurements taken on each subject). The between subjects sums of squares is computed by summing the squared differences between each subject's mean heart rate and the overall mean heart rate. The overall mean is computed by summing all of the measurements and dividing by the total number of measurements $(nk=10(3)=30)$:

$$\overline{X}_{..} = \frac{4328}{30} = 144.3$$

Runner Number	Course 1	Course 2	Course 3	Subject Mean $\overline{X}_{s.}$	$(\overline{X}_{s.} - \overline{X}_{..})^2$
1	132	135	138	135.0	86.5
2	143	148	148	146.3	4.0
3	135	138	141	138.0	39.7
4	128	131	139	132.7	134.6
5	141	141	150	144.0	0.1
6	150	156	161	155.7	130.0
7	131	134	138	134.3	100.0
8	150	156	162	156.0	136.9
9	142	145	151	146.0	2.9
10	139	165	160	154.7	108.2
Course Means $\overline{X}_{.j}$	139.1	144.9	148.8	$\overline{X}_{..}=144.3$	742.9

The between subjects sums of squares is:

$$SS_{subj} = \Sigma\, k\, (\overline{X}_{s.} - \overline{X}_{..})^2 = 3(742.9) = 2228.7.$$

  The between treatments sums of squares is computed by summing the squared differences between each treatment mean and the overall mean. The treatment (or course) means are shown along the bottom of the table above. The between treatments sums of squares is:

Analysis of Variance

$$SS_b = \Sigma n (\overline{X}_{.j} - \overline{X}_{..})^2 = 10 \{(139.1 - 144.3)^2 + (144.9 - 144.3)^2 + (148.8 - 144.3^2\} = 477$$

The total sums of squares is computed by summing the squared differences between each observation and the overall mean. This is equivalent to the numerator of the sample variance. Recall from Chapter 3 the shortcut formula for the sample variance and the following alternative formula for the total sums of squares:

$$SS_{total} = \Sigma \Sigma (\overline{X}_{ij} - \overline{X}_{..})^2 = \Sigma X_{ij}^2 - \frac{(\Sigma X_{ij})^2}{nk} .$$

It is tedious to compute the sum of each observation squared, but for this example, $\Sigma X_{ij}^2 = 627,362$. The sum of the observations is $\Sigma X_{ij} = 4328$, and the total sums of squares is:

$$SS_{total} = \Sigma X_{ij}^2 - \frac{(\Sigma X_{ij})^2}{nk} = 627,362 - \frac{(4328)^2}{30} = 2975.9.$$

The within sums of squares is computed by subtraction: $SS_{within} = SS_{total} - SS_{subj} - SS_b$.

$$SS_{within} = 2975.9 - 2228.7 - 477 = 270.2.$$

We now construct the ANOVA table for Example 10.10.

Source of Variation	Sums of Squares SS	Degrees of freedom df	Mean Squares MS	F
Between Subjects	2228.7	9		
Between Treatments	477	2	238.5	15.9
Within	270.2	18	15.0	
Total	2975.9	29		

5.   Conclusion. Reject H_0, since $15.9 > 3.55$. We have significant evidence, $\alpha = 0.05$, to show that there is a difference in mean heart rates of runners on the three courses.

Analysis of Variance

SAS Example 10.10. The following output was generated using SAS Proc Glm which runs a repeated measures analysis of variance test for equality of means. The repeated option (see SAS program code in section 10.9 for the details) is used to indicate that the measurements are dependent. SAS produces the following. A brief interpretation appears after the output.

SAS Output for Example 10.10

```
                      The GLM Procedure
            Repeated Measures Analysis of Variance
             Repeated Measures Level Information
          Dependent Variable     course1  course2  course3
             Level of course         1        2        3

                      The GLM Procedure
            Repeated Measures Analysis of Variance
     Univariate Tests of Hypotheses for Within Subject Effects
```

Source	DF	Type III SS	Mean Square	F Value	Pr > F
course	2	476.4666667	238.2333333	15.60	0.0001
Error(course)	18	274.8666667	15.2703704		

Interpretation of SAS Output for Example 10.10

The Glm procedure generates more output than is shown above for a repeated measures analysis of variance. The section shown above is the most relevant for performing the test of interest. SAS generates an abbreviated ANOVA table, similar to the table displayed in Table 10.2. Again, SAS presents the degrees of freedom before the Sums of Squares. For the repeated measures applications, only the between treatment variation (labeled "course" in the output) and the within treatment variation (labeled "error(course)" in the output) rows of the AVOVA table are shown. These are the most relevant for computing the test statistic which is F=15.60 with p = 0.0001. We computed F=15.9 by hand; the difference is due to rounding. With p=0.0001, we would reject H_0 and conclude that the mean heart rates are different for the runners on the three courses.

Analysis of Variance

9.8   Key Formulas

APPLICATION	NOTATION/FORMULA	DESCRIPTION
Test $H_0: \mu_1 = \mu_2 = \ldots \mu_k$	$F = \dfrac{s_b^{\;2}}{s_w^{\;2}}$	See Table 10.1 for ANOVA computations (Table 10.2 for repeated measures ANOVA)
Variation explained by treatments	$\eta^2 = \dfrac{SS_b}{SS_{total}}$	See (10.5)
Test $H_0: \mu_i = \mu_j$ using Scheffe MCP	$F = \dfrac{(\overline{X}_{.i} - \overline{X}_{.j})^2}{s_w^2 \left(\dfrac{1}{n_i} + \dfrac{1}{n_j} \right)}$	MCP which controls familywise error rate
Test $H_0: \mu_i = \mu_j$ using Tukey MCP	$q_k = \dfrac{(\overline{X}_{1'} - \overline{X}_{k'})}{\sqrt{\dfrac{s_w^2}{n}}}$	MCP which controls familywise error rate

reject if $F \geq (k-1)F_c$

if compare combined with one: (P466)

10.9   Statistical Computing

Following are the SAS programs which were used to conduct tests of equality of k means (k>2) using ANOVA. Also shown are programs to run Scheffe and Tukey multiple comparisons (pairwise comparisons). The SAS procedures used and brief descriptions of their use are noted in the header to each example. Notes are provided to the right of the SAS programs (*in italics*) for orientation purposes and are not part of the programs. In addition, there are blank lines in the programs that follow which are solely to accommodate the notes. Blank lines and spaces can be used throughout SAS programs to enhance readability. A summary of the SAS procedures used in the examples is provided at the end of this section.

Analysis of Variance (ANOVA): Equal Sample Sizes

SAS EXAMPLE 10.3 Compare Mean Times to Relief Among 3 Treatments

An investigator wishes to compare the average time to relief of headache pain under three distinct medications, call them Drugs A, B and C. Fifteen patients who suffer from chronic headaches are randomly selected for the investigation. Five subjects are randomly assigned to each treatment. The following data reflect times to relief (in minutes) after taking the assigned drug: Run an ANOVA using SAS.

Drug A	Drug B	Drug C
30	25	15
35	20	20
40	30	25
25	25	20
35	30	20

Program Code

```
options ps=62 ls=80;


data in;
  input trt $ time;




cards;
A 30
A 35
A 40
A 25
A 35
B 25
B 20
```

Formats the output page to 62 lines in length and 80 columns in width
Beginning of Data Step
*Inputs two variables **trt** and **time**, where **trt** is a character variable (A, B or C).*

Beginning of Raw Data section.
*actual observations (value of **trt** and **time** on each line)*

Analysis of Variance

```
B 30
B 25
B 30
C 15
C 20
C 25
C 20
C 20
run;

proc ANOVA;


  class trt;;


  model time=trt;


run;
```

Procedure call. Proc ANOVA tests the equality of k treatment means when sample sizes are equal.
Specification of grouping variable (i.e., variable that defines k comparison groups).
*Specification of outcome variable (**time**). SAS requires a model statement relating the outcome to the grouping variable (**trt**).*
End of procedure section.

Analysis of Variance (ANOVA): Unequal Sample Sizes

SAS EXAMPLE 10.4 Compare Mean Ages Among 3 Groups of Students

The following data reflect ages of students at completion of 8th grade. Test if there is a significant difference in the mean age at completion of 8th grade for Rural, Suburban and Urban students using SAS and a 5% level of significance. The following data were collected from randomly selected students Rural, Suburban and Urban schools.

Rural: 14 14 14 14 13 13 13 12
Suburban: 14 14 14 13 13 13 13 13 12 12
Urban: 16 16 15 15 15 14 14 14 13 12

Program Code

options ps=62 ls=80;	*Formats the output page to 62 lines in length and 80 columns in width*
data in;	*Beginning of Data Step*
input school $ age;	*Inputs two variables **school** and **age**, where **school** = Rural, Suburban or Urban.*
cards;	*Beginning of Raw Data section.*
rural 14	*actual observations (value of **school** and **age** on each line)*
rural 14	
rural 14	
rural 14	
rural 13	
rural 13	
rural 13	
rural 12	
suburban 14	
suburban 14	
suburban 14	
suburban 13	
suburban 13	
suburban 13	
suburban 13	

Analysis of Variance

```
suburban 13
suburban 12
suburban 12
urban 16
urban 16
urban 15
urban 15
urban 15
urban 14
urban 14
urban 14
urban 13
urban 12
run;
```

```
proc glm;
```
Procedure call. Proc Glm tests the equality of k treatment means when sample sizes are unequal.

```
  class school;
```
Specification of grouping variable (i.e., variable that defines k comparison groups).

```
  model age=school;
```
*Specification of outcome variable (**age**). SAS requires a model statement relating the outcome to the grouping variable (**school**).*

```
run;
```
End of procedure section.

Analysis of Variance

Analysis of Variance (ANOVA)

SAS EXAMPLE 10.5 Compare Mean Weight Gains Among 4 Diets

A study is developed to examine the effects of vitamins and milk supplements on infant weight gain. Four diet plans are considered: Diet A involves a regular diet plus the vitamin supplement, Diet B involves a regular diet plus the special milk formula, Diet C is our control diet (i.e., no restrictions or special considerations), and Diet D involves a regular diet plus the vitamin and the special milk formula. Twenty infants are selected for the investigation and each is randomized to one of the four competing diet programs. The table below displays weight gains, measured in pounds, after 1 month on the respective diet. Run an ANOVA using SAS.

Diet A	Diet B	Diet C	Diet D
2.0	1.6	1.5	2.1
1.5	1.9	2.0	2.4
2.4	2.1	1.8	1.9
1.9	1.1	1.3	1.8
2.6	1.7	1.2	2.2

Program Code

options ps=62 ls=80;	*Formats the output page to 62 lines in length and 80 columns in width*
data in;	*Beginning of Data Step*
input diet $ gain;	*Inputs two variables **diet** and **gain**, where **diet** = A, B, C or D.*
cards;	*Beginning of Raw Data section.*
A 2.0	*actual observations (value of **diet***
A 1.5	*and **gain** on each line)*
A 2.4	
A 1.9	
A 2.6	
B 1.6	

Analysis of Variance

```
B 1.9
B 2.1
B 1.1
B 1.7
C 1.5
C 2.0
C 1.8
C 1.3
C 1.2
D 2.1
D 2.4
D 1.9
D 1.8
D 2.2
run;

proc glm;


  class diet;



  model gain=diet;




run;
```

Procedure call. Proc Glm tests the equality of k treatment means when sample sizes are equal or unequal.
Specification of grouping variable (i.e., variable that defines k comparison groups).
*Specification of outcome variable (**gain**). SAS requires a model statement relating the outcome to the grouping variable (**diet**).*
End of procedure section.

Analysis of Variance (ANOVA): Pairwise Comparisons Using Scheffe and Tukey

SAS EXAMPLES 10.7 & 10.9   **Pairwise Multiple Comparisons**

In SAS EXAMPLE 10.3 we ran an ANOVA and rejected H_0. Run pairwise comparisons using the Scheffe and Tukey procedures to determine which means are different from others.

Program Code

options ps=62 ls=80;	*Formats the output page to 62 lines in length and 80 columns in width*
data in;	*Beginning of Data Step*
input trt $ time;	*Inputs two variables **trt** and **time**, where **trt** = A, B or C.*
cards;	*Beginning of Raw Data section.*
A 30	*actual observations (value of **trt** and*
A 35	***time** on each line)*
A 40	
A 25	
A 35	
B 25	
B 20	
B 30	
B 25	
B 30	
C 15	
C 20	
C 25	
C 20	
C 20	
run;	

Analysis of Variance

proc ANOVA;	*Procedure call. Proc ANOVA tests the equality of k treatment means when sample sizes are equal.*
class trt;;	*Specification of grouping variable (i.e., variable that defines k comparison groups).*
model time=trt;	*Specification of outcome variable (**time**). SAS requires a model statement relating the outcome to the grouping variable (**trt**).*
means trt/scheffe;	*Means option requires a comparison of means among the comparison groups defined by **trt**, using the Scheffe MCP. (SAS EXAMPLE 10.7)*
means trt/tukey;	*Means option requires a comparison of means among the comparison groups defined by **trt**, using the Tukey MCP. (SAS EXAMPLE 10.9)*
run;	*End of procedure section.*

Analysis of Variance

Repeated Measures Analysis of Variance (ANOVA)

SAS EXAMPLE 10.10   **Compare Mean Heart Rates on Three Training Courses**

      An investigator is interested in comparing the cardiovascular fitness of elite runners on three different training courses, each of which covers 10 miles. The courses differ in terms of terrain, one includes steep inclines, the other more graded inclines and the third is flat. Each runner has his/her heart rate monitored at mile 5 of the run on each course. Ten runners are involved and their heart rates measured on each course are shown below. Run a repeated measures ANOVA using SAS.

Runner Number	Course 1	Course 2	Course 3
1	132	135	138
2	143	148	148
3	135	138	141
4	128	131	139
5	141	141	150
6	150	156	161
7	131	134	138
8	150	156	162
9	142	145	151
10	139	165	160

Program Code
options ps=62 ls=80;

data in;
  input runnerid course1 course2 course3;

    subjmean=mean(course1,course2,course3);

cards;
1   132   135   138
2   143   148   148
3   135   138   141
4   128   131   139

Formats the output page to 62 lines in length and 80 columns in width
Beginning of Data Step
Inputs four variables **runnerid** *(subject identifier) and 3 measurements on each subject –* **course1**, **course2** *and* **course3**.
Computes a new variable, **subjmean**, *which is the mean heart rate for each subject.*
Beginning of Raw Data section.
actual observations (value of **runnerid, course1, course2** *and* **course3** *on each line)*

Analysis of Variance

```
5    141    141    150
6    150    156    161
7    131    134    138
8    150    156    162
9    142    145    151
10   139    165    160
run;
proc print;

   var course1 course2 course3 subjmean;
run;


proc means;



   var course1 course2 course3 subjmean;




run;
proc glm;

   model course1 course2 course3=/nouni;











   repeated course;









run;
```

Procedure call. Proc Print lists the values of the specified variables. Specification of variables for listing.

*Procedure call. Proc Means generates summary statistics on specified variables. Specification of variables (we are most interested in the means for each course. The mean of **subjmean** is the overall mean).*

Procedure call. Proc Glm is used for analysis of variance applications. Specification of variables for analysis. In repeated measures analysis, the variables listed after the model statement are those measured on each subject. The nouni option indicates that we do not want univariate tests (which test if the mean for each variable is zero or not).

*The repeated option indicates a repeated measures ANOVA. After the repeated statement, the user specifies a label for the repeated measurements which appears on the output (**course**).*

Analysis of Variance

Summary of SAS Procedures

The SAS ANOVA Procedure is used to run and ANOVA when the sample sizes are equal in all comparison groups. The SAS Glm Procedure is used to run and ANOVA when the sample sizes are either equal or unequal. Specific options can .be requested in these procedures to run pairwise multiple comparisons procedures (e.g., Scheffe or Tukey). The options are shown in italics below. Users should refer to the examples in this section for complete descriptions of the procedure and specific options. A general description of the procedure and options is provided in the table below.

Procedure	Sample Procedure Call	Description
proc ANOVA	proc ANOVA; class group; model outcome=group;	Runs an ANOVA comparing the mean outcome scores among groups, when sample sizes are equal in all comparison groups..
proc glm	proc glm; class group; model outcome=group;	Runs an ANOVA comparing the mean outcome scores among groups.
	proc ANOVA; class group; model outcome=group; means group/*scheffe;*	Runs an ANOVA comparing the mean outcome scores among groups, and generates pairwise comparisons using the Scheffe procedure. Here we assume equal sample sizes.
	proc glm; class group; model outcome=group; means group/*tukey;*	Runs an ANOVA comparing the mean outcome scores among groups, and generates pairwise comparisons using the Tukey procedure.
	proc glm; model outcome1 outcome2 … outcomek=/nouni; repeated label*;*	Runs a repeated measures ANOVA comparing the mean outcome scores over time. The nouni option suppresses univariate tests and the user specifies a label describing the nature of the repeated assessments.

10.10 Problems

1. A pharmaceutical company is interested in the effectiveness of a new preparation designed to relieve arthritis pain. Three variations of the compound have been prepared for investigation. The three preparations differ according to the proportion of the active ingredients: T15 contains 15% active ingredients, T40 contains 40% active ingredients, and T50 contains 50% active ingredients. A sample of 20 patients are selected to participate in a study comparing the three variations of the compound. A control compound, which is currently available over the counter, is also included in the investigation. Patients are randomly assigned to one of the 4 treatments (control, T15, T40, T50) and the time (in minutes) until pain relief is recorded on each subject. The data are given below:

Control	12	15	18	16	20
T15	20	21	22	19	20
T40	17	16	19	15	19
T50	14	13	12	14	11

a) Test if there is a difference in the mean time to relief among the 4 treatments. Use a 5% level of significance.

b) Using Scheffe multiple comparisons, test if there is a significant difference in the mean time to relief between the control and each of the experimental treatments (i.e., T15, T40, and T50), considered separately.

2. A study is performed to compare mean numbers of primary care visits over three years among 4 different health maintenance organizations (HMO). Fifty patients are randomly sampled from each HMO.

a) Write the hypotheses to be tested.

b) Complete the following ANOVA table.

Source of Variation	Sums of Squares	Degrees of Freedom	Mean Square	F
Between	574.3			
Within				
Total	2759.8			

c) Is there a significant difference in mean numbers of primary care visits among the 4 HMOs ? Use a 5% level of significance.

3. Six different doses of a particular drug are compared in an effectiveness study. The study involves 30 subjects and equal numbers of subjects are randomly assigned to each dose group.

a) What are the null and alternative hypotheses in the comparison ?
b) Complete the following ANOVA table.

Source of Variation	Sums of Squares	df	Mean Squares	F
Between	189.85			
Within				
Total	352.57			

c) Is there a significant difference in the effectiveness among the doses ? Use $\alpha = 0.05$.

d) Suppose we wish to compare dose groups 1 and 2, and the following are available: $\overline{X}_{.1} = 21.4$, $\overline{X}_{.2} = 27.6$. Run the appropriate test to assess whether there is a significant difference in effectiveness between dose groups 1 and 2 at the 5% level of significance.

Analysis of Variance

4.   The following data were collected as part of a study comparing a control treatment to an active treatment. Three doses of the active treatment were considered in the study. The following table displays summary statistics on the ages of participants enrolled in the study classified by treatment group:

Treatment	# Participants	Mean Age	Std. Dev.
Control	8	29.5	3.74
Low Dose	8	34.5	2.88
Moderate Dose	8	15.9	3.72
High Dose	8	44.0	6.65

a)   Test if there is a significant difference in the mean ages of participants across treatment groups. Complete the following table and show all parts of the test. Use $\alpha=0.05$.

Source	Sums of Squares	Degrees of Freedom	Mean Squares	F
Between				
Within				
Total	3860.97			

b)   Is there a significant difference in the mean ages of participants in the Control and Low Dose groups? Run the appropriate test at $\alpha=0.05$.

Analysis of Variance

5.	An investigation is performed to evaluate two new experimental treatments for allergies. Eighteen subjects who suffer from allergies are enrolled in the study and randomly assigned to one of three treatments: Control Treatment, Experimental Treatment 1 or Experimental Treatment 2. Each subject is instructed to take the assigned treatment and symptoms of allergies are recorded on a scale of 1 to 20 with higher scores indicating worse symptoms. The data are shown below.

Control Treatment	Experimental Treatment 1	Experimental Treatment 2
19	12	5
18	14	6
16	13	3
15	10	2
12	9	3
17	10	4
$\overline{X}_1 = 16.2$	$\overline{X}_2 = 11.3$	$\overline{X}_3 = 3.8$

a)	Test if there is a significant difference in mean symptom scores among the three treatments under investigation. Use a 5% level of significance. HINT: SS $_{total}$ = 524.4.

b)	Which treatment is the "best" of those investigated ? (Justify briefly)

6.	 A randomized trial is conducted to compare 4 treatment regimens for HIV. Each regimen is based on a combination of medications, each medication involving a specific time schedule. Twenty patients are involved in the investigation and are randomly assigned to one of four treatments: A, B, C or D (5 subjects per treatment group). An interim outcome is self-reported symptoms. An overall symptom score, ranging from 0 to 20, is computed for each individual by aggregating their reports of the frequency and severity of an array of symptoms. Higher symptom scores indicate more frequent and/or severe symptoms. Summary statistics are shown below.

Treatment Regimen	Symptom Score
A	7.9
B	5.4
C	12.4
D	5.9

Analysis of Variance

Test whether the mean symptom scores are equal among treatment regimens. Run the appropriate test at the 5% level of significance. HINT: $SS_w = 135.4$.

7. A study is conducted among college students who smoke to assess whether there is a relationship between the number of cigarettes they smoke per day and the smoking status of their parents. For the purposes of the study, smoking status is defined as follows: Never smoked, Former smokers or Current smokers. The number of cigarettes smoked per day by each student is recorded and data are shown below:

Parents Smoking Status	Number of students	Mean Number of Cigarettes Smoked/Day	Standard Deviation
Never Smoked	30	12.6	3.1
Former Smokers	34	14.1	3.7
Current Smokers	53	18.3	4.3
ALL	117	15.6	6.5

Based on the data, is there a significant difference in the mean numbers of cigarettes smoked per day according to the parents smoking status? Run the appropriate test at $\alpha=0.05$. HINT: $SS_{total} = 3428.4$.

8. A study is conducted comparing birth weights (in pounds) of infants born to mothers of various ages. The data are shown below.

Mother's Age		
< 20 Years	20-29 Years	30+ Years
8.4	7.5	6.9
7.3	6.3	7.1
9.1	6.9	5.7
7.8	5.4	6.5
8.4	7.1	6.6

Is there a significant difference in the mean birth weights for mother's of different ages? Run the appropriate test at $\alpha=0.05$.

Analysis of Variance

9. Consider a continuous measure of medication adherence (scores from 0-100), and use the following data to test if there is a significant difference in the mean medication adherence scores in patients of different age groups. Use the following data to run the appropriate test at $\alpha=0.05$. HINT: $\Sigma\Sigma (X_{ij}-\overline{X})^2 = 29{,}159.26$

	Age Group			
	20-29	30-39	40-49	50-59
Number of patients	48	53	37	12
Mean Medication Adherence	69.2	83.3	82.1	83.4
Standard Deviation	11.9	13.2	12.8	11.3

10. A clinical trial is performed comparing a test (newly developed) drug to an active control (a drug already proven effective) and to a placebo. Persons with diagnosed hypertension who are at least 18 years of age are eligible for the trial. Persons agreeing to participate are randomly assigned to one of the competing drugs. The outcome measure is systolic blood pressure (SBP) which is measured 4 weeks post-randomization. Summary statistics are given below.

Drug	n	Mean (SD) SBP
Test	15	125 (25)
Control	15	135 (21)
Placebo	15	160 (22)

a) Is there a significant difference in mean SBPs 4 weeks post-randomization? Run the appropriate test at a 5% level of significance. HINT: $SS_{total} = 31{,}450$.

b) Is there a significant difference in mean SBPs 4 weeks post-randomization between the Test Drug and Placebo? Run the appropriate test at a 5% level of significance.

c) Is there a significant difference in mean SBPs 4 weeks post-randomization between the Test and Control Drugs Placebo? Run the appropriate test at a 5% level of significance.

Analysis of Variance

11.   The following output was generated from SAS:

```
                    Analysis of Variance Procedure
                        Class Level Information
                    Class       Levels     Values
                    GROUP          3        a b c
               Number of observations in data set = 24
```

Dependent Variable: SCORE

Source	DF	Sum of Squares	Mean Square	F Value	Pr > F
Model	2	839.58333333	419.79166667	43.16	0.0001
Error	21	204.25000000	9.72619048		
Corrected Total	23	1043.83333333			

a)   Test if there is a significant difference in mean scores between treatments. Write the hypotheses and justify your conclusion (use the SAS output).

b)   How much of the variation in scores is explained by the groups?

12.   The following table summarizes data collected in a multi-center study. The variable summarized below is body mass index (BMI) computed as the ratio of weight in kilograms to height in meters squared.

		Enrollment Site		
BMI	Overall	Hospital 1	Hospital 2	Hospital 3
n	300	100	100	100
Mean	24.8	21.6	24.8	27.9
Std Dev	2.5	2.1	1.8	1.3

Test if there is a significant difference in the mean BMI scores among hospitals. Show all parts of the test and use a 5% level of significance. (HINT: $MS_w = 3.1$).

Analysis of Variance

13. A research group wishes to conduct a larger study to compare a newly approved medication, an experimental medication (one not yet available on market) and a medication they consider standard care in a randomized trial. One hundred and fifty patients with hypertension are enrolled and randomized to one of the three comparison treatments. After taking the assigned medication for 6 weeks, each patient has his/her systolic blood pressure (SBP) measured. Summary statistics are given below. Use the data to test if there is a significant difference in systolic blood pressures among medication groups. Run the appropriate test at a 5% level of significance.

Treatment	Number of Patients	Mean SBP	Standard Deviation
New	50	130	12
Standard	50	135	10
Experimental	50	115	17

Is there a difference in systolic blood pressures among the treatments? Run the appropriate test using a 5% level of significance. HINT: $SS_{within} = 26,117$ and $SS_{total} = 36,951$.

14. The following data represent birth weights of siblings born to six different mothers:

Mother	Birth weights			
	Child 1	Child 2	Child 3	Child 4
1	6.4	6.9	6.7	7.1
2	8.5	7.8	7.8	8.3
3	7.6	8.7	9.9	8.2
4	5.3	6.7	7.5	6.4
5	6.2	5.6	6.4	5.5
6	7.0	7.8	8.6	6.6

Is there a significant difference in birth weights of siblings? Run the appropriate test at $\alpha=0.05$.

Analysis of Variance

15. Suppose in a clinical trial , systolic blood pressures are measured at 4, 8 and 12 weeks post-randomization in a subgroup of patients receiving a test drug. We wish to test if there are significant differences in systolic blood pressures over time.

Subject	Systolic Blood Pressure		
	4 weeks	8 weeks	12 weeks
1	120	125	130
2	110	115	118
3	105	110	100
4	140	130	140
5	150	145	140

Test if there is a significant difference in mean systolic blood pressures over time using a 5% level of significance.

16. Suppose the data in the previous problem were collected from 3 independent random samples of subjects (i.e. $n_1=5$, $n_2=5$, $n_3=5$, total sample size =15). Test if there is a significant difference in mean systolic blood pressures over time using a 5% level of significance. HINT: You do not need to recalculate the sums of squares - compare formulas first!

SAS Problems: Use SAS to solve the following problems.

1. A pharmaceutical company is interested in the effectiveness of a new preparation designed to relieve arthritis pain. Three variations of the compound have been prepared for investigation. The three preparations differ according to the proportion of the active ingredients: T15 contains 15% active ingredients, T40 contains 40% active ingredients, and T50 contains 50% active ingredients. A sample of 20 patients are selected to participate in a study comparing the three variations of the compound. A control compound, which is currently available over the counter, is also included in the investigation. Patients are randomly assigned to one of the 4 treatments (control, T15, T40, T50) and the time (in minutes) until pain relief is recorded on each subject. The data are given below:

Control	12	15	18	16	20
T15	20	21	22	19	20
T40	17	16	19	15	19
T50	14	13	12	14	11

Use SAS Proc ANOVA to test if the equality of means. In addition, run all pairwise comparisons using the Scheffe procedure. Use a 5% level of significance.

2. An investigation is performed to evaluate two new experimental treatments for allergies. Eighteen subjects who suffer from allergies are enrolled in the study and randomly assigned to one of three treatments: Control Treatment, Experimental Treatment 1 or Experimental Treatment 2. Each subject is instructed to take the assigned treatment and symptoms of allergies are recorded on a scale of 1 to 20 with higher scores indicating worse symptoms. The data are shown below.

Control Treatment	Experimental Trt 1	Experimental Trt 2
19	12	5
18	14	6
16	13	3
15	10	2
12	9	3
17	10	4

Analysis of Variance

Use SAS Proc ANOVA to test if the equality of means. In addition, run all pairwise comparisons using the Tukey procedure. Use a 5% level of significance.

3. A study is conducted comparing birth weights (in pounds) of infants born to mothers of various ages. The data are shown below.

Mother's Age		
< 20 Years	20-29 Years	30+ Years
8.4	7.5	6.9
7.3	6.3	7.1
9.1	6.9	5.7
7.8	5.4	6.5
8.4	7.1	6.6

Use SAS Proc Glm to test if there a significant difference in the mean birth weights for mother's of different ages? Run the appropriate test at $\alpha=0.05$.

4. The following data represent birth weights of siblings born to six different mothers:

Mother	Birth weights			
	Child 1	Child 2	Child 3	Child 4
1	6.4	6.9	6.7	7.1
2	8.5	7.8	7.8	8.3
3	7.6	8.7	9.9	8.2
4	5.3	6.7	7.5	6.4
5	6.2	5.6	6.4	5.5
6	7.0	7.8	8.6	6.6

Use SAS Proc Glm (with a repeated option) to test if there a significant difference in birth weights of siblings? Run the appropriate test at $\alpha=0.05$.

Analysis of Variance

5. Suppose in a clinical trial , systolic blood pressures are measured at 4, 8 and 12 weeks post-randomization in a subgroup of patients receiving a test drug. We wish to test if there are significant differences in systolic blood pressures over time. The following data were collected at 4 weeks, 8 weeks and 12 weeks post-randomization in sample of 5 patients receiving the test drug.

Subject	Systolic Blood Pressure		
	4 weeks	8 weeks	12 weeks
1	120	125	130
2	110	115	118
3	105	110	100
4	140	130	140
5	150	145	140

Use SAS Proc Glm (with a repeated option) to test if there a significant difference in birth weights of siblings? Run the appropriate test at $\alpha=0.05$.

Analysis of Variance

Introductory Applied Biostatistics

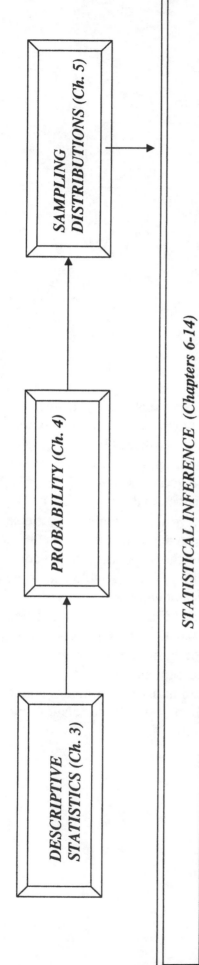

DESCRIPTIVE STATISTICS (Ch. 3) → PROBABILITY (Ch. 4) → SAMPLING DISTRIBUTIONS (Ch. 5)

STATISTICAL INFERENCE (Chapters 6-14)

Outcome Variable	Grouping Variable(s)/ Predictor(s)	Analysis	Chapter(s)
Continuous	-	Estimate μ, Compare μ to Known, Historical Value	6/13
Continuous	Dichotomous (2 Groups)	Compare Independent Means (Estimate/Test ($\mu_1-\mu_2$)) or the Mean Difference(μ_d)	7/13
Continuous	Discrete (> 2 Groups)	Test the Equality of K Means using Analysis of Variance ($\mu_1=\mu_2=..\mu_k$)	10/13
Continuous	**Continuous**	**Estimate Correlation or Determine Regression Equation**	**11/13**
Continuous	**Several Continuous or Dichotomous**	**Multiple Linear Regression Analysis**	**11**
Dichotomous	-	Estimate p, Compare p to Known, Historical Value	8
Dichotomous	Dichotomous (2 Groups)	Compare Independent Proportions (Estimate/Test (p_1-p_2))	8/9
Dichotomous	Discrete (>2 Groups)	Test the Equality of k Proportions (Chi-Square Test)	8
Dichotomous	Several Continuous or Dichotomous	Multiple Logistic Regression Analysis	12
Discrete	Discrete	Compare Distributions Among k Populations (Ch-Square Test)	8
Time to Event	Several Continuous or Dichotomous	Survival Analysis	14

CHAPTER 11: Correlation and Regression

11.1 Introduction

In correlation and regression applications, we consider the relationship between 2 continuous (or measurement) variables. For example, we might be interested in the relationship between systolic blood pressure and age, or the relationship between the number of hours of aerobic exercise per week and percent body fat. In correlation and regression applications, we measure both variables on each of n randomly selected subjects. We denote the variables X and Y. (In prior applications, we used the variable name X to denote the single characteristic of interest.) The variable name X is used to represent the *independent, or predictor, variable,* while the variable name Y is used to represent the *dependent, or response, variable.* In many applications it will be clear which is the independent or predictor variable and which is the dependent or outcome variable. In some applications the investigator must specify explicitly which variable should be considered the independent and which should be considered the dependent variable.

In Section 11.2 we discuss correlation analysis. We present and illustrate the computation of the sample correlation coefficient and then present techniques for statistical inference concerning the correlation coefficient. In Section 11.3 we present and illustrate simple linear regression analysis. In Section 11.4 we provide an introduction to multiple regression analysis and in Section 11.5 we introduce logistic regression analysis. In Section 11.6 we summarize key formulas and in Section 11.7 we display the statistical computing programs used to generate correlation and regression analyses presented in this chapter.

11.2   Correlation Analysis

The goal of correlation analysis is understand the nature and strength of the association between the two measurement variables, denoted X and Y. A first step in understanding the relationship between two variables is through a *scatter diagram.* A scatter diagram is a plot of the (X,Y) pairs recorded on each of the n subjects. The X (independent or predictor) variable is plotted on the horizontal axis and the Y (dependent or outcome) variable is plotted on the vertical axis. The scatter diagram shown in Figure 11.1 indicates a positive or direct linear relationship between X and Y.

Figure 11.1   Direct Linear Relationship Between X and Y

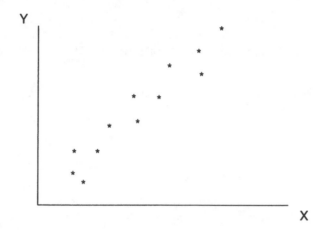

Figure 11.2 displays an negative or inverse linear association between 2 variables X and Y.

Figure 11.2   Inverse Linear Relationship Between X and Y

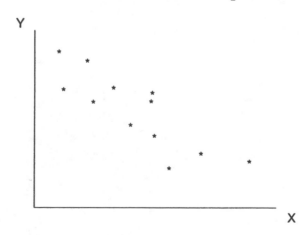

Correlation and Regression

The population correlation coefficient, ρ (rho), is a statistic which quantifies the nature and strength of the linear relationship between X and Y. The population correlation coefficient takes on values in the range of -1 to 1 ($-1 \leq \rho \leq +1$). The sign of the correlation coefficient indicates the nature of the relationship between X and Y (i.e., positive or direct, negative or inverse), while the magnitude of the correlation coefficient indicates the strength of the linear association between the two variables. Figures 11.3a - 11.3d display scatter diagrams corresponding to the following values of the correlation coefficient: $\rho = -1$, $\rho = -0.5$, $\rho = 0.5$, and $\rho = 1$. These values represent a perfect inverse relationship between X and Y, a moderate inverse relationship between X and Y, a moderate direct relationship between X and Y, and a perfect direct relationship between X and Y, respectively.

Figure 11.3   The Correlation Coefficient

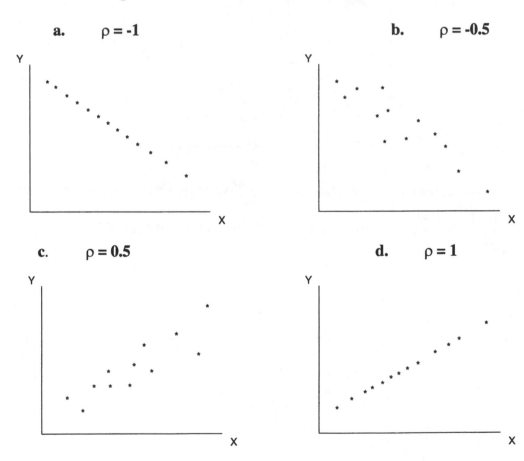

The correlation coefficient is a measure of the linear association between X and Y. One must be cautious in interpreting the value of the correlation coefficient. It is always useful to generate a scatter diagram to assist in the interpretation of the correlation coefficient. For example, the data displayed in both Figures 11.4a and 11.4b produce correlation coefficients of zero. In Figure 11.4a there is no association between X and Y. However, in Figure 11.4b there is an association between X and Y, though not a linear one.

Figure 11.4   Correlation Coefficient = 0

a. No relationship between X and Y   **b. Nonlinear relationship between X and Y**

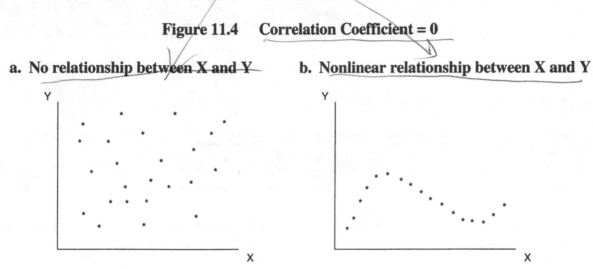

The correlation coefficient can be affected by truncation. For example, suppose we analyze the relationship between SAT scores measured during the senior year of high school and grade point averages (GPAs) measured at the completion of freshman year in college. Because individuals who do poorly on the SAT are less likely to attend college, the correlations might be distorted due to the fact that these individuals are not included.

In other situations, a correlation between two variables, X and Y, may turn out to be zero due to a confounding variable (i.e., a variable which affects either X or Y or both). For example, suppose we investigate the relationship between the size of a home and its selling price. Selling prices might be different each year depending on the market and the economy, but if we pool data from several years there may appear to be no linear association. In other applications, a correlation may be large due to a confounding variable. For example, say we investigate the association between age of first job and starting salary. Observed salaries may be higher for individuals who start working later. This might be due to the fact that these individuals completed a college education, thus delaying their entry into the workforce.

11.2.1 The Sample Correlation Coefficient r

As with other applications, we will have data measured on a sample of subjects. The sample correlation coefficient, denoted r, is computed as follows:

$$r = \frac{Cov(X,Y)}{\sqrt{Var(X)\ Var(Y)}} \tag{11.1}$$

where Cov(X,Y) is the covariance of X and Y (defined below), Var(X) and Var(Y) are the sample variances of X and Y, respectively. Recall $Var(X) = \frac{\Sigma(X-\overline{X})^2}{n-1}$, and $Var(Y) = \frac{\Sigma(Y-\overline{Y})^2}{n-1}$.

$$Cov(X,Y) = \frac{\Sigma(X-\overline{X})(Y-\overline{Y})}{n-1} \tag{11.2}$$

We now illustrate the estimation of the population correlation between two continuous variables through examples.

Example 11.1. Suppose we are interested in the relationship between body mass index (computed as the ratio of weight in kilograms to height in meters squared) and systolic blood pressure in males 50 years of age. A random sample of 10 males 50 years of age is selected and their weights, heights and systolic blood pressures are measured. Their weights and heights are transformed into body mass index scores and are given below. In this analysis, the independent (or predictor) variable is body mass index and the dependent (or response) variable is systolic blood pressure.

X = Body Mass Index	Y = Systolic Blood Pressure
18.4	120
20.1	110
22.4	120
25.9	135
26.5	140
28.9	115
30.1	150
32.9	165
33.0	160
34.7	180

The first step in assessing the relationship between body mass index and systolic blood pressure is through a scatter diagram (Figure 11.5). Notice that the independent variable (body mass index) is displayed on the horizontal axis and the dependent variable (systolic blood pressure) is displayed on the vertical axis.

Figure 11.5   Scatter Diagram Relating Body Mass Index and Systolic Blood Pressure

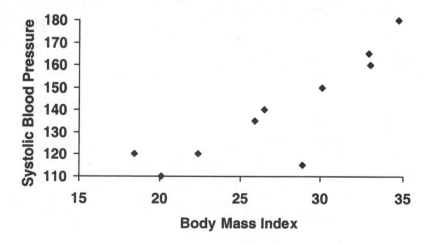

Correlation and Regression

From the scatter diagram, it appears that there is a positive (direct) association between body mass index and systolic blood pressure. We now compute the sample correlation coefficient, r, to quantify the degree of linear association between body mass index and systolic blood pressure. In order to compute r, we must first compute the variances of X and Y as well as the covariance between X and Y.

We first compute the variance of X: $Var(X) = \dfrac{\Sigma(X - \overline{X})^2}{n-1}$. The following table summarizes the computations:

X = Body Mass Index	$(X - \overline{X})$	$(X - \overline{X})^2$
18.4	-8.89	79.032
20.1	-7.19	51.696
22.4	-4.89	23.912
25.9	-1.39	1.932
26.5	-0.79	0.624
28.9	1.61	2.592
30.1	2.81	7.896
32.9	5.61	31.472
33.0	5.71	32.604
34.7	7.41	54.908
$\Sigma X = 272.9$, $\overline{X} = 27.29$	0	286.669

So, $Var(X) = \dfrac{\Sigma(X - \overline{X})^2}{n-1} = 286.669/9 = 31.852.$

Correlation and Regression

We now compute the variance of Y: $Var(Y) = \dfrac{\Sigma(Y-\overline{Y})^2}{n-1}$. The following table summarizes the computations:

Y = Systolic Blood Pressure	$(Y - \overline{Y})$	$(Y - \overline{Y})^2$
120	-19.5	380.25
110	-29.5	870.25
120	-19.5	380.25
135	-4.5	20.25
140	0.5	0.25
115	-24.5	600.25
150	10.5	110.25
165	25.5	650.25
160	20.5	420.25
180	40.5	1640.25
$\Sigma Y = 1395.0$, $\overline{Y} = 139.5$	0	5072.50

So, $Var(Y) = \dfrac{\Sigma(Y-\overline{Y})^2}{n-1} = 5072.50/9 = 563.611$.

Finally, we compute the covariance of X and Y: $Cov(X,Y) = \dfrac{\Sigma(X-\overline{X})(Y-\overline{Y})}{n-1}$. The following table summarizes the computations:

Correlation and Regression

$(X - \overline{X})$	$(Y - \overline{Y})$	$(X - \overline{X})(Y - \overline{Y})$
-8.89	-19.5	173.355
-7.19	-29.5	212.105
-4.89	-19.5	95.355
-1.39	-4.5	6.255
-0.79	0.5	-0.395
1.61	-24.5	-39.455
2.81	10.5	29.505
5.61	25.5	143.055
5.71	20.5	117.055
7.41	40.5	300.105
		1036.95

So, $\mathrm{Cov(X,Y)} = \dfrac{\Sigma (X - \overline{X})(Y - \overline{Y})}{n - 1} = 1036.95/9 = 115.22.$

Substituting into (11.1), the sample correlation coefficient is:

$$r = \frac{115.22}{\sqrt{(31.852)(563.611)}} = 0.859$$

Based on the sign and the magnitude of r, there is a strong, positive association between body mass index and systolic blood pressure. In fact this correlation is artificially large (the data are hypothetical). In practice correlations on the order of 0.3 or larger in absolute value are usually indicative of a meaningful or important relationship.

Correlation and Regression

11.2.2 Statistical Inference Concerning ρ

It is often of interest to draw inferences about the correlation between 2 variables in the population through a formal test of hypothesis. The sample correlation coefficient, r, is a point estimate for the population correlation coefficient, ρ. In general, tests of hypothesis concerning ρ address whether there is a linear association in the population ($\rho \neq 0$) or not ($\rho = 0$). The hypotheses are of the form:

$$H_0: \rho = 0 \quad \text{(No linear association)}$$
$$H_1: \rho \neq 0 \quad \text{(Linear association)}$$

The test statistic follows a t distribution with n-2 degrees of freedom:

$$t = r\sqrt{\frac{n-2}{1-r^2}}, df=n-2 \tag{11.3}$$

NOTE: The t statistic can be used regardless of the sample size (even when n is large), as the critical value of t (Table 3) reflects the exact sample size. For example, when n is large, the two-sided critical value of t for $\alpha=0.05$ is 1.96.

We now illustrate the test using the data from Example 11.1. The issue at hand is whether there is a significant correlation between systolic blood pressure and body mass index among all males 50 years of age. We have a random sample of 10 males 50 years of age and observed a sample correlation coefficient of 0.859.

1. Set up hypotheses.

$H_0: \rho = 0$ $\alpha = 0.05$
$H_1: \rho \neq 0$

2. Select appropriate test statistic.

$$t = r\sqrt{\frac{n-2}{1-r^2}}$$

514

3.   Decision Rule (two sided test, $\alpha = 0.05$).

To determine the appropriate value from the t distribution table, we fist compute the degrees of freedom. For Example 11.1, df = n - 2 = 10 - 2 = 8. The critical value is t = 2.306. The decision rule is given below.

Reject H_0 if $t \geq 2.306$ or if $t \leq -2.306$

Do not reject Ho if $-2.306 < t < 2.306$

4.   Test statistic.

$$t = r\sqrt{\frac{n-2}{1-r^2}} = 0.859\sqrt{\frac{10-2}{1-0.859^2}} = 4.70$$

5.   Conclusion: Reject H_0 since $4.70 \geq 2.306$. We have significant evidence, $\alpha = 0.05$, to show that $\rho \neq 0$. For this example, p=0.01 (See Table 3). Therefore, there is evidence of a significant linear association between systolic blood pressure and body mass index among all males 50 years of age.

Example 11.2 Consider the application described in Example 11.1. Suppose for the same sample, we also recorded the number of hours of vigorous exercise in a typical week. We wish to investigate the relationship between the number of hours of exercise per week and systolic blood pressure in males 50 years of age.   In this analysis, the independent (or predictor) variable is number of hours of exercise per week and the dependent (or response) variable is systolic blood pressure. A scatter diagram is shown below (Figure 11.6)

Figure 11.6   Scatter Diagram Relating Hours of Exercise and Systolic Blood Pressure

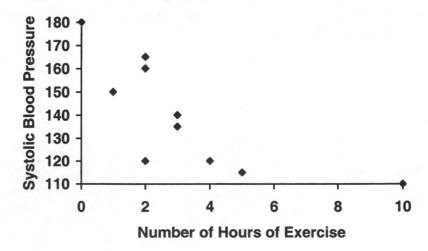

Correlation and Regression

X = Number of Hours of Exercise Per Week	Y = Systolic Blood Pressure
4	120
10	110
2	120
3	135
3	140
5	115
1	150
2	165
2	160
0	180

In order to compute r, we must first compute the variances of X and Y as well as the covariance between X and Y.

We first compute the variance of X: $Var(X) = \dfrac{\Sigma (X - \overline{X})^2}{n - 1}$. The following table summarizes the computations:

X = Number of Hours of Exercise Per Week	$(X - \overline{X})$	$(X - \overline{X})^2$
4	0.8	0.64
10	6.8	46.24
2	-1.2	1.44
3	-0.2	0.04
3	-0.2	0.04
5	1.8	3.24
1	-2.2	4.84
2	-1.2	1.44
2	-1.2	1.44
0	-3.2	10.24
$\Sigma X = 32$, $\overline{X} = 3.2$	0	69.60

So, $Var(X) = \dfrac{\Sigma (X - \overline{X})^2}{n-1} = 69.60/9 = 7.73$.

We computed the variance of Y in Example 11.1, $Var(Y) = \dfrac{\Sigma (Y - \overline{Y})^2}{n-1} = 5072.50/9 = 563.611$. We now need to compute the covariance of X and Y: $Cov(X,Y) = \dfrac{\Sigma (X - \overline{X})(Y - \overline{Y})}{n-1}$. The following table summarizes the computations:

$(X - \overline{X})$	$(Y - \overline{Y})$	$(X - \overline{X})(Y - \overline{Y})$
0.8	-19.5	-15.6
6.8	-29.5	-200.6
-1.2	-19.5	23.4
-0.2	-4.5	0.9
-0.2	0.5	-0.1
1.8	-24.5	-44.1
-2.2	10.5	-23.1
-1.2	25.5	-30.6
-1.2	20.5	-24.6
-3.2	40.5	-129.6
		-444.0

So, $\text{Cov}(X, Y) = \dfrac{\Sigma (X - \overline{X})(Y - \overline{Y})}{n - 1} = -444.0/9 = -49.33$.

Substituting into (11.1), the sample correlation coefficient is:

$$r = \frac{-49.33}{\sqrt{(7.73)(563.611)}} = -0.75$$

Based on the sign and the magnitude of r, there is a strong, negative association between the number of hours of exercise per week and systolic blood pressure (i.e., more hours of exercise per week are associated with lower systolic blood pressures).

Using the same data (Example 11.2) we now test whether there is a significant correlation between systolic blood pressure and the number of hours of exercise per week among all males 50 years of age. We have a random sample of 10 males 50 years of age and observed a sample correlation coefficient of –0.75.

1.   Set up hypotheses.

   H_0: $\rho = 0$ $\alpha = 0.05$
   H_1: $\rho \neq 0$

2.   Select appropriate test statistic.

$$t = r \sqrt{\frac{n-2}{1-r^2}}$$

3.   Decision Rule (two sided test, $\alpha = 0.05$).

To determine the appropriate value from the t distribution table, we fist compute the degrees of freedom. For Example 1, df = n - 2 = 10 - 2 = 8. The critical value is t = 2.306. The decision rule is given below.

   Reject H_0 if t $\geq$ 2.306 or if t $\leq$ -2.306
   Do not reject Ho if -2.306 < t < 2.306

4.   Test statistic.

$$t = r \sqrt{\frac{n-2}{1-r^2}} = -0.75 \sqrt{\frac{10-2}{1-(-0.75)^2}} = -3.18$$

5.   Conclusion: Reject H_0 since $-3.18 \leq$ -2.306. We have significant evidence, $\alpha = 0.05$, to show that $\rho \neq 0$. For this example, p=0.02 (See Table 3). Therefore, there is evidence of a significant linear association between systolic blood pressure and the number of hours of exercise per week among all males 50 years of age.

11.3   Simple Linear Regression

In regression analysis, we develop the mathematical equation that best describes the relationship between two variables, X and Y. In correlation analysis, it is actually not necessary to specify which of the two variables is the independent and which is the dependent variable. In contrast, in regression analysis the independent and dependent variables must be specified. In regression analysis we address the following issues:

i) What mathematical form describes the relationship between X and Y (e.g., a line or a curve of some form) ?

ii) How do we best estimate the equation that describes the relationship between X and Y ?

iii) Is the form specified in (i) appropriate ?

We begin with the simplest situation, the one in which the relationship between X and Y is linear. The equation of the line relating Y to X is called the *simple linear regression equation* and is given below:

$$Y = \beta_0 + \beta_1 X + \varepsilon \qquad (11.4)$$

where Y is the dependent variable,
X is the independent variable,
β_0 is the Y-intercept (i.e., the value of Y when X = 0),
β_1 is the slope (i.e., the expected change in Y relative to one unit change in X), and
ε is the random error.

The estimates of β_0 and β_1, denoted $\hat{\beta}_0$ and $\hat{\beta}_1$, respectively, are determined using the following equations:

$$\hat{\beta}_1 = r \sqrt{\frac{Var(Y)}{Var(X)}}$$

$$\hat{\beta}_0 = \overline{Y} - \hat{\beta}_1 \overline{X} \qquad (11.5)$$

These estimates (11.5) are called the least squares estimates of the slope and intercept. The formulas shown above are those which minimize the squared errors (i.e., minimize $\Sigma \varepsilon^2$). The derivations of the formulas are not shown here.

The estimate of the simple linear regression equation is given by substituting the least squares estimates (11.5) into equation (11.4):

$$\hat{Y} = \hat{\beta}_0 + \hat{\beta}_1 X \qquad\qquad (10.6)$$

where $\hat{Y}$ is the expected value of Y for a given value of X.

Example 11.1 (continued). We now estimate the simple linear regression equation for the data given in Example 11.1. The dependent variable (Y) is systolic blood pressure, the independent variable, X, is body mass index and the correlation was estimated at 0.859. We first estimate the slope by substituting the appropriate statistics into (11.5):

$$\hat{\beta}_1 = r \sqrt{\frac{\text{Var}(Y)}{\text{Var}(X)}} = 0.859 \sqrt{\frac{563.611}{31.852}} = 3.61.$$

Now, substituting the above, we compute the Y-intercept:

$$\hat{\beta}_0 = \overline{Y} - \hat{\beta}_1 \overline{X} = 139.5 - (3.61)(27.29) = 40.98.$$

The simple linear regression equation for Example 11.1 is given below:

$$\hat{Y} = 40.98 + 3.61\ X.$$

The Y-intercept is 40.98 which indicates the systolic blood pressure (Y) when body mass index (X) is zero. In this example, the Y-intercept is not meaningful since it is not possible to observe a body mass index equal to zero. However, the estimated slope, $\hat{\beta}_1 = 3.61$, indicates that a one unit increase in body mass index is associated with an increase of 3.61 units in systolic blood pressure. If we compare two males 50 years of age, the first with a body mass index 10 units higher than the second, we would expect the first subject's systolic blood pressure to be $10(3.61) = 36.1$ units higher than the second's.

We can also can use the equation to generate an estimate of systolic blood pressure (Y) for a person with a specific body mass index (X). For example, suppose we wish to estimate the systolic blood pressure of a male age 50 whose body mass index is 20. Using the simple linear regression equation:

$$\hat{Y} = 40.98 + 3.61\ X = 40.98 + 3.61\ (20) = 113.81.$$

The above can be interpreted in two ways. It is the expected systolic blood pressure of a male age 50 whose body mass index is 20. It is also the expected systolic blood pressure for all males age 50 with body mass index of 20.

SAS Example 11.1. The following output was generated using SAS Proc Plot, SAS Proc Corr and SAS Proc Reg which generate a scatter diagram and perform correlation and regression analyses, respectively. SAS produces the following. A brief interpretation appears after the output.

SAS Output for Example 11.1

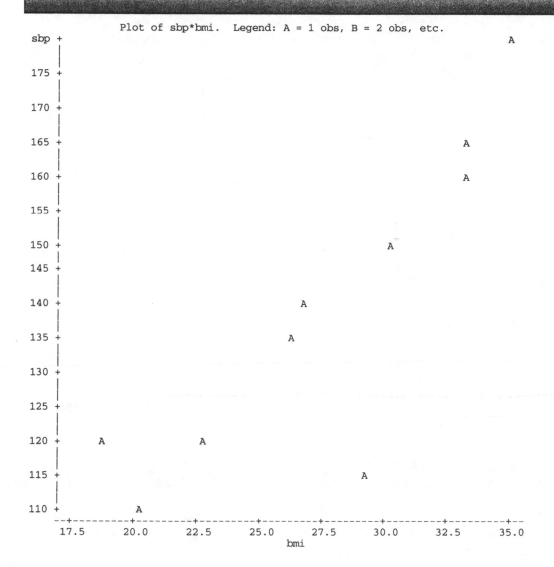

Correlation and Regression

```
                     The CORR Procedure
              2  Variables:     bmi      sbp

                 Covariance Matrix, DF = 9
                           bmi                   sbp
         bmi         31.8521111          115.2166667
         sbp        115.2166667          563.6111111

                      Simple Statistics
Variable      N       Mean      Std Dev        Sum      Minimum      Maximum
bmi          10    27.29000      5.64377    272.90000    18.40000     34.70000
sbp          10   139.50000     23.74050       1395     110.00000    180.00000

          Pearson Correlation Coefficients, N = 10
                Prob > |r| under H0: Rho=0

                          bmi               sbp
         bmi          1.00000           0.85992
                                        0.0014
         sbp          0.85992           1.00000
                      0.0014
```

```
                     The REG Procedure
                       Model: MODEL1
                  Dependent Variable: sbp

                     Analysis of Variance
                              Sum of             Mean
Source               DF      Squares            Square      F Value      Pr > F
Model                 1    3750.89494        3750.89494      22.71       0.0014
Error                 8    1321.60506         165.20063
Corrected Total       9    5072.50000

         Root MSE               12.85304      R-Square      0.7395
         Dependent Mean        139.50000      Adj R-Sq      0.7069
         Coeff Var               9.21365

                      Parameter Estimates
                     Parameter        Standard
Variable     DF       Estimate           Error      t Value      Pr > |t|
Intercept     1       40.78558        21.11158        1.93        0.0895
bmi           1        3.61724         0.75913        4.76        0.0014
```

Interpretation of SAS Output for Example 11.1

SAS can produce high-level graphical displays, however, these require some programming to implement. The Plot Procedure is easy to use and generates a scatter diagram for two continuous variables. SAS displays points in the scatter diagram with letters of the alphabet. The letter "A"

Correlation and Regression

appears when there is a single point in a particular location, "B" indicates that two points fall in the same location, and so on. Although the scatter diagram is somewhat crude, it clearly conveys the nature of the relationship between the two variables under investigation. In Example 11.1 there is a strong positive association between BMI and SBP in this sample.

The SAS Corr procedure generates a correlation analysis. In this example, we requested that SAS produce a Covariance Matrix, which appears at the top of the correlation output. The Covariance Matrix contains covariances between the variables in the rows and columns of the matrix. The covariance between a variable and itself is identical to the variance of that variable. For example, the Cov(SBP,SBP)=Var(SBP) = 563.61. The Cov(SBP, BMI) = Cov(BMI,SBP) = 115.22, and Cov(BMI, BMI) = Var(BMI) = 31.85. By default, the Corr procedure also generates summary statistics for each variable. The last section of the output contains the Correlation Matrix. The Correlation Matrix contains sample correlations (r) between variables in the rows and columns of the matrix. Underneath each sample correlation is a two-sided p-value for testing H_0: $\rho=0$ against H_1: $\rho\neq0$. The correlation between a variable and itself is always 1.00 (and no test is performed). Of interest are the correlations between distinct variables. Specifically, Corr(SBP, BMI) = Corr(BMI,SBP) = 0.85992, and the correlation between SBP and BMI is significant, $p=0.0014 < 0.05$.

The SAS Reg procedure generates an ANOVA table which is used to test whether the regression is significant or not, and then provides estimates of the parameters of the regression equation (e.g., estimates of the intercept and slope). The ANOVA table is set up exactly like the ANOVA tables we used in analysis of variance applications to test the equality of k treatment means. In regression analysis, the total sum of squares represents the variation in the dependent variable Y. The total variation is partitioned into two components, labeled Model and Error. The Model sum of squares is also referred to as the regression sums of squares and reflects the variation in Y accounted for by the regression equation. The Error sum of squares is also referred to as the residual sum of squares and reflects the variation in Y which is not accounted for by the regression. The p value (0.0014) is used to test whether the regression is significant. When there is a single independent variable (i.e., simple linear regression), this p-value is identical to the p-value in the correlation analysis and also identical to the p-value used to test if the slope is significant (described below). The next part of the regression output includes estimates of the regression parameters. The estimate of the intercept is 40.79 and the estimate of the slope is 3.62. SAS produces two-sided p-values to test if the intercept and slope are significantly different from zero. It is usually not of interest to test if the intercept is significantly different from zero. It is of interest to test if the slope is different from zero. Here p=0.0014, indicating that the slope (reflecting the change in SBP associated with a one unit change in BMI) is significant.

Correlation and Regression

If a regression is run and it has been established that the regression is significant, we often want to quantify how much variation in the dependent variable is "explained" by the independent variable (or set of independent variables, in the case of multiple regression analysis).

The *coefficient of determination*, denoted R^2, "R-squared," is the ratio of the regression (or Model) sums of squares to the total sums of squares (from the ANOVA table).

$$R^2 = \frac{SS_{regression}}{SS_{total}} \qquad (10.7)$$

where $0 \leq R^2 \leq 1$.

Higher values of R^2 imply that more variation in the dependent variable is "explained" by the independent variable(s).

In Example 11.1:

$$R^2 = \frac{3750.89}{5072.50} = 0.7395$$

Thus, 73.9% of the variation in systolic blood pressures is explained by body mass index. Because this example is artificial, the value of R^2 is very high. In practice, values of $R^2 > 0.3$ are generally considered clinically significant.

Example 11.2 (continued). We now estimate the simple linear regression equation for the data given in Example 11.2. The dependent variable (Y) is systolic blood pressure, the independent variable (X) is the number of hours of exercise per week and the correlation was estimated at –0.75. We first estimate the slope by substituting the appropriate statistics into (11.5):

$$\hat{\beta}_1 = r\sqrt{\frac{Var(Y)}{Var(X)}} = -0.75\sqrt{\frac{563.611}{7.73}} = -6.38$$

Now, substituting the above, we compute the Y-intercept:

$$\hat{\beta}_0 = \overline{Y} - \hat{\beta}_1\overline{X} = 139.5 - (-6.38)(3.2) = 159.9$$

The simple linear regression equation for Example 11.2 is given below:

$$\hat{Y} = 159.9 - 6.38\,X$$

The Y-intercept is 159.9 which is the expected systolic blood pressure (Y) when the number of hours of exercise (X) is zero. In this example, the Y-intercept is meaningful because it is possible to observe X = 0 (i.e., no exercise). The estimated slope, $\hat{\beta}_1$ = -6.38, indicates that each additional hour of exercise per week (a one unit increase) is associated with a decrease of 6.38 units in systolic blood pressure. If we compare two males 50 years of age, the first exercises 5 hours in a typical week and the second exercises 4 hours in a typical week, we would expect the first subject's systolic blood pressure to be 6.38 units lower than the second's.

We can also use the equation to generate estimates of values of systolic blood pressure (Y) for a person who exercises a specific number of hours per week (X). For example, suppose we wish to estimate the systolic blood pressure of a male age 50 who exercises 3 hours per week. Using the simple linear regression equation:

$$\hat{Y} = 159.9 - 6.38\,(3) = 159.9 - 19.14 = 140.76.$$

SAS Example 11.2. The following output was generated using SAS Proc Corr and SAS Proc Reg which perform correlation and regression analyses, respectively. SAS produces the following. A brief interpretation appears after the output.

SAS Output for Example 11.2

```
                      The CORR Procedure
             2  Variables:     exercise sbp

                    Simple Statistics
Variable      N       Mean      Std Dev        Sum      Minimum      Maximum
exercise     10    3.20000      2.78089    32.00000            0     10.00000
sbp          10  139.50000     23.74050        1395    110.00000    180.00000

            Pearson Correlation Coefficients, N = 10
                 Prob > |r| under H0: Rho=0

                        exercise           sbp
          exercise       1.00000      -0.74725
                                        0.0130
          sbp           -0.74725       1.00000
                         0.0130
```

Correlation and Regression

```
                      The REG Procedure
                      Model: MODEL1
                   Dependent Variable: sbp

                        Analysis of Variance
                                 Sum of            Mean
Source                  DF       Squares          Square    F Value    Pr > F
Model                   1      2832.41379      2832.41379     10.12    0.0130
Error                   8      2240.08621       280.01078
Corrected Total         9      5072.50000


             Root MSE             16.73352    R-Square     0.5584
             Dependent Mean      139.50000    Adj R-Sq     0.5032
             Coeff Var            11.99536


                          Parameter Estimates
                        Parameter      Standard
      Variable    DF     Estimate        Error    t Value    Pr > |t|
      Intercept   1     159.91379       8.31854     19.22     <.0001
      exercise    1      -6.37931       2.00578     -3.18     0.0130
```

Interpretation of SAS Output for Example 11.2

The SAS Corr procedure generates a correlation analysis. In this example, we did not request that SAS produce a Covariance Matrix (as we had requested in Example 11.1). The first part of the output contains summary statistics for each variable. The next section of the output includes the Correlation Matrix. The Correlation Matrix contains sample correlations (r) between variables showing in the rows and columns of the matrix. Underneath each sample correlation is a two-sided p-value for testing H_0: $\rho=0$ against H_1: $\rho \neq 0$. The correlation between a variable and itself is always 1.00. Of interest are the correlations between distinct variables. Specifically, Corr(SBP, EXERCISE) = Corr(EXERCISE,SBP) = -0.74725, and this correlation between SBP and EXERCISE is significant, p=0.0130 < 0.05.

The SAS Reg procedure generates an ANOVA table which is used to test whether the regression is significant or not, and then provides estimates of the parameters of the regression equation (e.g., estimates of the intercept and slope). In regression analysis, the total sums of squares represents the variation in the dependent variable Y. The total variation is partitioned into two components, labeled Model and Error. The Model sums of squares is also referred to as the regression sums of squares and reflect the variation in Y accounted for by the regression equation. The Error sums of squares is also referred to as the residual sums of squares and reflects the variation in Y which is not accounted for by the regression. The p value (0.0130) is used to test whether the regression is significant. When there is a single independent variable (i.e., simple linear regression), this p-value is identical to the p-value in the correlation analysis and also identical to the p-value used to test if the slope is significant (described below). The next part of the regression output includes estimates of the regression parameters. The estimate of the intercept is 159.91 and the estimate of the slope is -6.38. SAS produces two-sided p-values to test if the intercept and slope are significantly different from zero. It is usually not of interest to test if the intercept is significantly different from zero. It is of interest to test if the slope is different from zero. Here p=0.0130, indicating that the slope (reflecting the change in SBP associated with a one unit change in the number of hours of exercise per week) is significant. In this example, $R^2 = 0.56$ suggesting that 56% of the variation in systolic blood pressures is explained by the number of hours of exercise per week.

11.4   Multiple Regression Analysis

In Examples 11.1 and 11.2 we found that body mass index and the number of hours of exercise per week were significantly associated with systolic blood pressure (when considered separately). There are many characteristics that might be related to systolic blood pressure, such as age, gender, diet, family history, race, smoking status and so on. In many applications, we wish to assess which of a set of candidate variables are related to a particular dependent variable, and investigate their relative importance.

In multiple regression analysis we consider applications involving a single continuous dependent variable, Y, and multiple independent variables, denoted X_1, X_2, and so on. The form of the multiple linear regression equation is given below:

$$Y = \beta_0 + \beta_1 X_1 + \beta_2 X_2 + \ldots + \beta_p X_p + \varepsilon \qquad (11.8)$$

where   Y is the dependent variable,
X_1 to X_p are the independent variables,
β_0 is the intercept (i.e., the value of Y when $X_1 = X_2 = \ldots = X_p = 0$),
β_i (i=1, ... p) are the slope coefficients, also called the regression parameters
(i.e., the expected change in Y relative to one unit change in X_i), and
ε is the random error.

NOTE: The dependent variable, Y, is a continuous variable. The independent variables, X_i, can be continuous variables or dichotomous (sometimes called indicator) variables. In the example below we consider both continuous independent variables (e.g., age) and indicator variables (e.g., smoking status, where 1=smoker, 0=non-smoker).

The formulas to estimate the intercept and slope coefficients are computationally complex. We therefore restrict our attention to interpreting a multiple regression analysis performed in SAS. Interested readers can refer to *Applied Regression Analysis and Other Multivariable Methods 2nd Edition* (by Kleinbaum DG, Kupper LL and Muller KE (1988), PWS-Kent Publishing Company, Boston, MA) for a more complete discussion of multiple regression analysis. Here we present only a single example as a means of illustrating the conceptual framework of the application in a basic sense. Consider the following example.

SAS Example 11.3. In the following we used SAS to analyze a large dataset (n=5078). The dependent variable is systolic blood pressure (SBP) and we consider three independent variables: AGE (a continuous variable, measured in years), MALE (an indicator variable, coded 1 for males and 0 for females) and SMOKER (an indicator variable, coded 1 for smokers and 0 for non-smokers). In the following we estimated a multiple regression equation using SAS. A brief interpretation appears after the output.

SAS Output for Example 11.3

```
                      The REG Procedure
                       Model: MODEL1
                   Dependent Variable: SBP

                       Analysis of Variance
                            Sum of        Mean
Source              DF     Squares       Square    F Value   Pr > F
Model                3      446821       148940     493.03   <.0001
Error             5074     1532804    302.08991
Corrected Total   5077     1979625

          Root MSE            17.38073   R-Square   0.2257
          Dependent Mean     129.60339   Adj R-Sq   0.2253
          Coeff Var           13.41071
```

Parameter Estimates

Variable	DF	Estimate	Error	t Value	DF	Pr > |t|
Intercept	1	90.57147	1.12386	80.59	1	<.0001
age	1	0.77398	0.02094	36.96	1	<.0001
male	1	4.39800	0.49022	8.97	1	<.0001
smoker	1	-1.84484	0.50366	-3.66	1	0.0003

Interpretation of SAS Output for Example 11.3

The SAS Reg procedure generates an ANOVA table which is used to test whether the regression is significant or not, and then provides estimates of the regression parameters (e.g., estimates of the intercept and regression slopes). The ANOVA table is set up exactly like the ANOVA tables we used in analysis of variance applications to test the equality of k treatment means. In regression analysis, the total sums of squares represents the variation in the dependent variable Y. The total variation is partitioned into two components, labeled Model and Error. The Model sums of squares are also referred to as the regression sums of squares and reflects the variation in Y accounted for by the regression equation. In multiple regression analysis, the ANOVA table is used to perform a global test. In particular, to test if the collection of variables are significant. The p value (<.0001) is used to test whether the set of independent variables considered is significant, here whether AGE, MALE and SMOKER, considered simultaneously, are significant in explaining variation in SBP. The next part of the regression output includes estimates of the regression parameters. The estimate of the intercept is 90.57 and is the expected systolic blood pressure if all of the independent variables are zero (i.e., if AGE=0 (which is unreasonable), MALE=0 (which is the code for female) and SMOKER=0 (which is the code for a non-smoker)). The estimate of the regression parameter associated with AGE is 0.77. A one year increase in AGE is associated with a 0.77 unit increase in SBP. The estimate of the regression parameter associated with MALE is 4.40. Because MALE is an indicator variable, the effect is interpreted as follows. On average, males (MALE=1) have SBPs which are 4.40 units higher than females (MALE=0). Smoking status is interpreted in a similar fashion. The estimate of the regression parameter associated with SMOKER is −1.84. On average, smokers (SMOKER=1) have SBPs which are 1.84 units lower than non-smokers (SMOKER=0). We should be cautious interpreting this effect of smoking as smoking status may be related to other behaviors and smoking, per se, may not be associated with lower systolic blood pressure. Each of these parameter estimates is adjusted for the other independent variables in the model. They reflect the impact of each independent variable on SBP after considering the other variables in the model. SAS produces two-sided p-values to test if the intercept and regression coefficients are significantly different from zero. Again, it is usually not of interest to test if the intercept is significantly different from zero. It is of interest to test if the other coefficients are different from zero. Here p=0.0001 for each independent variable indicating that each of the variables is highly significant. In this example the sample size is very large which accounts, in part, for the highly significant results. It is often of interest to determine the relative importance of variables in a multiple regression analysis. This is determined by the magnitude of the t statistics (or associated p-values). One cannot determine relative importance based on the magnitude of the parameter estimates as the estimates are affected by the scales on which variables are measured. In this example, AGE is the most important (most significant) independent variable, followed by MALE and then SMOKER. These three variables account for, or explain, 22.6% of the variation in the dependent variable SBP.

11.5   Logistic Regression Analysis   *Skipped*

In logistic regression analysis we consider applications involving a single dependent variable, Y, which is dichotomous (i.e., Success Vs. Failure). The technique is discussed in detail in Chapter 12, here we present a brief overview. Suppose in Example 11.4 the dependent variable was not systolic blood pressure (a continuous variable), but instead diagnosis of hypertension (hypertensive or not, a dichotomous variable). Logistic regression applications can involve one or several independent variables, denoted X_1, X_2, and so on. The form of the logistic regression equation is given below:

$$\ln\left\{\frac{Y}{(1-Y)}\right\} = \beta_0 + \beta_1 X_1 + \beta_2 X_2 + \ldots + \beta_p X_p + \varepsilon \qquad (11.9)$$

where   Y is the dichotomous dependent variable,
X_1 to X_p are the independent variables,
β_0 is the intercept,
β_i (i=1, ... p) are the slope coefficients, also called the regression parameters, and
ε is the random error.

NOTES: The left hand side of the equation displays the dependent or outcome variable in a specific form. The quantity Y/(1 - Y) is the "odds" of possessing the event (Y) to not possessing the event (1 – Y). The expression log{Y/(1 - Y)} is called the log odds or the "logit" of Y. The slope coefficients, β_i (i=1, ... p), reflect the change in the logit or log odds of Y relative to a one unit change in X_i.

The dependent variable, Y, is a dichotomous variable. The independent variables, X_i, can be continuous variables or dichotomous (sometimes called indicator) variables. In the example below we consider both continuous independent variables (e.g., age) and indicator variables (e.g., smoking status, where 1=smoker, 0=non-smoker).

The formulas to estimate the intercept and slope coefficients are computationally complex. We again restrict our attention to interpreting a logistic regression analysis performed in SAS.

Readers are referred elsewhere for a more complete discussion of logistic regression analysis. Here we present only a single example as a means of illustrating the conceptual framework of the application in a basic sense. Consider the following example.

SAS Example 11.4. In the following we used SAS to analyze a large dataset (n=5078). The dependent variable is a dichotomous variable (HTN, coded 1 for patients classified as hypertensive and 0 otherwise) and we consider three independent variables: AGE (a continuous variable, measured in years), MALE (an indicator variable, coded 1 for males and 0 for females) and SMOKER (an indicator variable, coded 1 for smokers and 0 for non-smokers). In the following we generated estimated a logistic regression equation using SAS. A brief interpretation appears after the output.

SAS Output for Example 11.4

```
                        The LOGISTIC Procedure
                          Response Profile
            Ordered                          Total
             Value            HTN          Frequency
               1               1              349
               2               0             4729

                 Probability modeled is HTN=1.
                    Model Convergence Status
          Convergence criterion (GCONV=1E-8) satisfied.
                     Model Fit Statistics
                          Intercept      Intercept and
           Criterion        Only           Covariates
           AIC             2544.410         2246.882
           SC              2550.943         2273.013
           -2 Log L        2542.410         2238.882

             Testing Global Null Hypothesis: BETA=0
     Test                Chi-Square      DF       Pr > ChiSq
     Likelihood Ratio     303.5282        3         <.0001
     Score                291.3059        3         <.0001
     Wald                 259.4685        3         <.0001
                        The LOGISTIC Procedure

           Analysis of Maximum Likelihood Estimates
                              Standard        Wald
     Parameter   DF   Estimate    Error    Chi-Square   Pr > ChiSq
     Intercept    1   -7.1813    0.3371     453.9220      <.0001
     age          1    0.0654    0.00534    149.6551      <.0001
     male         1    1.2628    0.1254     101.4887      <.0001
     smoker       1    0.9520    0.1187      64.3556      <.0001
```

```
            Odds Ratio Estimates
                 Point          95% Wald
       Effect   Estimate    Confidence Limits
       age       1.068       1.056      1.079
       male      3.535       2.765      4.520
       smoker    2.591       2.053      3.269


  Association of Predicted Probabilities and Observed Responses
     Percent Concordant      77.0    Somers' D   0.550
     Percent Discordant      22.1    Gamma       0.555
     Percent Tied             0.9    Tau-a       0.070
     Pairs                1650421    c           0.775
```

Interpretation of SAS Output for Example 11.4

The SAS Logistic procedure first displays the numbers of subjects classified in each of the two response categories. In this example, 349 subjects are hypertensive and 4729 are not. SAS then indicates that the model is set up to predict the probability that a person is hypertensive (HTN=1). The next sections of the output are entitled "Model Fit Statistics" and "Testing Global Null Hypothesis: BETA=0". The statistics presented here are useful for testing whether the collection of variables considered are significant and for comparing models. In this application we considered three independent variables. The statistic labeled -2 log L for model containing intercept only (2542.410) represents a measure of the overall variation in the dependent variable. This quantity is similar to the total sum of squares in a multiple regression analysis. The Chi-Square for Covariates (303.528) is computed by taking the difference between -2 log L for model containing intercept only and -2 log L for model with intercept and covariates. This statistic is used to perform a global test of significance of the set of independent variables considered. In Example 11.4 the p value =0.0001, so AGE, MALE and SMOKER, considered simultaneously, are significant in explaining variation in hypertensive status. The next part of the regression output includes estimates of the regression parameters. The estimate of the intercept is –7.1813. The estimate of the regression parameter associated with AGE is 0.0654. A one year increase in AGE is associated with a 0.0654 unit increase in the logit or log odds of Y. SAS also produces odds ratios for each independent variable (computed by $\exp(\beta_i)$). The odds ratio for AGE is estimated as 1.068. A one year increase in age is associated with a 1.068 increase in the odds of having hypertension. The estimate of the regression parameter associated with MALE is 1.2628. Because MALE is an indicator variable, the effect is interpreted as follows. On average, the logit or log odds of Y is 1.2628 times higher for males (MALE=1) as compared to females (MALE=0). Smoking status is interpreted in a similar fashion. The estimate of the regression parameter associated with SMOKER is –0.9520. On average, smokers (SMOKER=1) are less likely to be hypertensive than non-smokers. The odds ratio for MALE is 3.535, so males are much more likely to have hypertension than females (odds of being hypertensive are 3.535 times higher for males as compared to females). The odds ratio for SMOKER is 0.386, so smokers are less likely to have hypertension than non-smokers (odds of being hypertensive are 0.386 (to 1) for smokers as compared to no-smokers). Again, we should be cautious interpreting this effect of smoking as smoking status may be related to other behaviors and smoking, per se, may not be associated with hypertensive status. Each of these parameter estimates and estimates of odds ratios is adjusted for the other independent variables in the model. They reflect the impact of each independent variable on hypertensive status after considering the other variables in the model. SAS produces two-sided p-values to test if the intercept and regression coefficients are significantly different from zero. Again, it is usually not of interest to test if the intercept is significantly different from zero. It is of interest to test if the other coefficients are different from zero. Here p=0.0001 for each independent variable indicating that each of the variables is highly significant. In this example the sample size is very large which accounts, in part, for the highly significant results. It is often of interest to determine the relative importance of variables in a multiple regression analysis. This is determined by the magnitude of the Wald Chi-Square statistics (or associated p-values). One cannot determine relative importance based on the magnitude of the parameter estimates as the estimates are affected by the scales on which variables are measured. In this example, AGE is the most important (most significant) independent variable, followed by MALE and then SMOKER. It is not possible to compute R^2 (11.7) for a logistic regression to describe how much variation in the dependent variable is explained by the independent variables. In logistic regression, we use the c statistic for a similar purpose. The c statistic in Example 11.4 is 0.775 and represents the extent to which the actual values of the dependent variable and the predicted values (generated by the estimated model) agree. Values exceeding 0.7 are generally considered adequate.

A more detailed discussion of logistic regression analysis is contained in Chapter 12.

11.6   Key Formulas

$$Var(X) = \frac{\Sigma(x-\bar{x})^2}{n-1} = \frac{\Sigma x^2 - (\Sigma x)^2/n}{n-1}$$

$$Var(Y) = \underline{\quad\quad} = \underline{\quad\quad}$$

APPLICATION	NOTATION/FORMULA	DESCRIPTION
Estimate Sample Correlation Coefficient	$r = \dfrac{Cov(X,Y)}{\sqrt{Var(X)\ Var(Y)}}$ $Cov(X,Y) = \dfrac{\Sigma(x-\bar{x})(Y-\bar{Y})}{n-1}$	Quantifies the nature and extent of linear association between X and Y
Test $H_0: \rho = 0$	$t = r\sqrt{\dfrac{n-2}{1-r^2}}$, $df = n-2$ $\left(\begin{array}{c}\text{only } t \\ \text{no size} \\ \text{difference}\end{array}\right)$ $\rightarrow \hat{Y} = \hat{\beta}_0 + \hat{\beta}_1 X$	Test for significant correlation between X and Y
Simple Linear Regression Equation	$Y = \beta_0 + \beta_1 X + \varepsilon$ $\hat{\beta}_1 = r\sqrt{\dfrac{Var(Y)}{Var(X)}}$, $\hat{\beta}_0 = \bar{Y} - \hat{\beta}_1\bar{X}$	See (11.5) for least squares estimates of regression parameters
Estimate R^2	$R^2 = \dfrac{SS_{regression}}{SS_{total}}$ $0 \le R^2 \le 1$ $\left(\text{?\% of the variance in "Y" is explained by "X"}\right)$	Proportion of variation in dependent variable explained by independent variable(s)
Multiple Linear Regression Equation	$Y = \beta_0 + \beta_1 X_1 + \beta_2 X_2 + \ldots + \beta_p X_p + \varepsilon$	Continuous dependent variable Y, continuous or dichotomous independent variables X_i.
Multiple Logistic Regression Equation	$\ln\left\{\dfrac{Y}{(1-Y)}\right\} = \beta_0 + \beta_1 X_1 + \beta_2 X_2 + \ldots + \beta_p X_p + \varepsilon$	Dichotomous dependent variable Y, continuous or dichotomous independent variables X_i (See Chapter 12)

Correlation and Regression

11.7   Statistical Computing

Following are the SAS programs which were used to generate a scatter diagram, to estimate correlations between variables, to estimate a simple linear regression equation, a multiple regression equation and a logistic regression equation. The SAS procedures used and brief descriptions of their use are noted in the header to each example. Notes are provided to the right of the SAS programs (*in italics*) for orientation purposes and are not part of the programs. In addition, there are blank lines in the programs that follow which are solely to accommodate the notes. Blank lines and spaces can be used throughout SAS programs to enhance readability. A summary of the SAS procedures used in the examples is provided at the end of this section.

Scatter Diagram, Correlation and Simple Linear Regression Analysis

SAS EXAMPLE 11.1 Assess the Relationship Between Body Mass Index and Systolic Blood Pressure

Suppose we are interested in the relationship between body mass index (computed as the ratio of weight in kilograms to height in meters squared) and systolic blood pressure in males 50 years of age. A random sample of 10 males 50 years of age is selected and their weights, heights and systolic blood pressures are measured. Their weights and heights are transformed into body mass index scores and are given below. In this analysis, the independent (or predictor) variable is body mass index and the dependent (or response) variable is systolic blood pressure. Generate a scatter diagram to assess the relationship between body mass index and systolic blood pressure, using SAS. In addition, estimate the correlation between body mass index and systolic blood pressure and the regression equation relating body mass index to systolic blood pressure using SAS.

X = Body Mass Index	Y = Systolic Blood Pressure
18.4	120
20.1	110
22.4	120
25.9	135
26.5	140
28.9	115
30.1	150
32.9	165
33.0	160
34.7	180

Correlation and Regression

Program Code

```
options ps=62 ls=80;
```
Formats the output page to 62 lines in length and 80 columns in width

```
data in;
```
Beginning of Data Step

```
  input bmi sbp;
```
Inputs two variables **bmi** and **sbp**, (both continuous variables).

```
cards;
```
Beginning of Raw Data section.

```
18.4 120
```
actual observations (value of **bmi** and **sbp** on each line)

```
20.1 110
22.4 120
25.9 135
26.5 140
28.9 115
30.1 150
32.9 165
33.0 160
34.7 180
run;
proc plot;
```
Procedure call. Proc Plot generates a scatter diagram for two continuous variables.

```
  plot sbp*bmi;
```
Specification of analytic variables. First variable specified is plotted on the vertical axis (Y).

```
run;
```
End of procedure section.

```
proc corr cov;
```
Procedure call. Proc Corr estimates correlation coefficients and tests their significance. The Cov option produces a covariance matrix.

```
  var bmi sbp;
```
Specification of analytic variables.

```
run;
```
End of procedure section.

```
proc reg;
```
Procedure call. Proc Reg estimates regression parameters and tests significance.

```
  model sbp=bmi;
```
Specification of regression model. The format is dependent (**sbp**) = independent variable(s) (**bmi**).

```
run;
```
End of procedure section.

Correlation and Regression

Correlation and Simple Linear Regression Analysis

SAS EXAMPLE 11.2 Assess the Relationship Between Number of Hours of Exercise Per Week and Systolic Blood Pressure

Consider the application described in Example 11.1. Suppose for the same sample, we also recorded the number of hours of vigorous exercise in a typical week. We wish to investigate the relationship between the number of hours of exercise per week and systolic blood pressure in males 50 years of age.   In this analysis, the independent (or predictor) variable is number of hours of exercise per week and the dependent (or response) variable is systolic blood pressure. Estimate the correlation between the number of hours of exercise and systolic blood pressure and the regression equation relating the number of hours of exercise to systolic blood pressure using SAS.

X = Number of Hours of Exercise Per Week	Y = Systolic Blood Pressure
4	120
10	110
2	120
3	135
3	140
5	115
1	150
2	165
2	160
0	180

Program Code

```
options ps=62 ls=80;
```
Formats the output page to 62 lines in length and 80 columns in width

```
data in;
```
Beginning of Data Step

```
  input exercise sbp;
```
*Inputs two variables **exercise** and **sbp**, (both continuous variables).*

```
cards;
```
Beginning of Raw Data section.

```
4 120
10 110
```
*actual observations (value of **exercise** and **sbp** on each line)*

```
2 120
3 135
3 140
5 115
1 150
2 165
2 160
0 180
run;
```

```
proc corr;
```
Procedure call. Proc Corr estimates correlation coefficients and tests their significance.

```
  var exercise sbp;
```
Specification of analytic variables.

```
run;
```
End of procedure section.

```
proc reg;
```
Procedure call. Proc Reg estimates regression parameters and tests their significance.

```
  model sbp=exercise;
```
*Specification of regression model. The format is dependent variable (**sbp**) = independent variable(s) (**exercise**).*

```
run;
```
End of procedure section.

Multiple Regression Analysis

SAS EXAMPLE 11.3 Assess the Relationship Between Age, Gender and Smoking Status, considered simultaneously, and Systolic Blood Pressure

In the following we use SAS to analyze a large dataset (n=5078). The dependent variable is systolic blood pressure (SBP) and we consider three independent variables: AGE (a continuous variable, measured in years), MALE (an indicator variable, coded 1 for males and 0 for females) and SMOKER (an indicator variable, coded 1 for smokers and 0 for non-smokers).   In the following we estimated a multiple regression equation using SAS – the data is abbreviated below.

Program Code

options ps=62 ls=80;	*Formats the output page to 62 lines in length and 80 columns in width*
data in;	*Beginning of Data Step*
input sbp age male smoker;	*Inputs four variables **sbp, age, male** and **smoker**.*
cards;	*Beginning of Raw Data section.*
120 55 1 0	*actual observations (value of **sbp,***
110 40 0 0	***age, male,** and **smoker** on each line)*
.	
.	
.	
140 35 1 1	
run;	
proc reg;	*Procedure call. Proc Reg estimates regression parameters and tests their significance.*
model sbp=age male smoker;	*Specification of regression model. The format is dependent variable (**sbp**) = independent variable(s) (**age, male** and **smoker**).*
run;	*End of procedure section.*

Logistic Regression Analysis

SAS EXAMPLE 11.4 Assess the Relationship Between Age, Gender and Smoking Status, considered simultaneously, and Hypertensive Status

In the following we use SAS to analyze a large dataset (n=5078). The dependent variable is a dichotomous variable (HTN, coded 1 for patients classified as hypertensive and 0 otherwise) and we consider three independent variables: AGE (a continuous variable, measured in years), MALE (an indicator variable, coded 1 for males and 0 for females) and SMOKES (an indicator variable, coded 1 for smokers and 0 for non-smokers). In the following we generated estimated a logistic regression equation using SAS – the data is abbreviated below.

Program Code	
Program Code	
options ps=62 ls=80;	*Formats the output page to 62 lines in length and 80 columns in width*
data in;	*Beginning of Data Step*
input htn age male smoker;	*Inputs four variables **htn, age, male** and **smoker**.*
cards;	*Beginning of Raw Data section.*
0 55 1 0	*actual observations (value of **htn**,*
0 40 0 0	***age, male**, and **smoker** on each line)*
.	
.	
1 35 1 1	
run;	
proc logistic descending;	*Procedure call. Proc Logistic estimates logistic regression parameters and tests their significance. The descending option indicates that the value 1 is the outcome of interest (the value 0 is the comparison).*
model htn=age male smoker;	*Specification of regression model. The format is dependent variable (**htn**) = independent variable(s) (**age**, **male** and **smoker**).*
run;	*End of procedure section.*

Correlation and Regression

Summary of SAS Procedures

The SAS Procedures for correlation and regression analysis are summarized below. Specific options can .be requested in these procedures to produce specific results. The options are shown in italics below. Users should refer to the examples in this section for complete descriptions of the procedure and specific options. A general description of the procedure and options is provided in the table below.

Procedure	Sample Procedure Call	Description
proc plot	proc plot; plot y*x;	Generates a scatter diagram for two continuous variables with Y on the vertical axis and X on the horizontal axis.
proc corr	proc corr *cov*; var y x1 x2 x3 x4 ;	Produces a correlation matrix for specified variables. Matrix includes estimates of correlations between variables and associated significance tests. Cov option generates a covariance matrix.
proc reg	proc reg; model y=x1 x2 x3 x4;	Estimates a linear regression model relating dependent variable Y to independent variables X1 X2 X3 X4.
proc logistic	proc logistic *descending*; model y = x1 x2 x3 x4;	Estimates a logistic regression model relating dichotomous dependent variable Y to independent variables X1 X2 X3 X4. The descending option considers the response with the higher numerical code the outcome of interest (e..g, 1=outcome of interest, 0=comparison).

11.8 Problems

not turn in

1. Data were collected from a random sample of 8 patients currently undergoing treatment for hypertension. Each subject reported the average number of cigarettes smoked per day and each subject was assigned a numerical value reflecting risk of cardiovascular disease (CVD). The risk assessments were computed by physicians and based on blood pressure, cholesterol level and exercise status. The risk assessments ranged from 0 to 100, higher values indicated increased risk.

Number of Cigarettes:	0	2	6	8	12	0	2	20
Risk of CVD :	12	20	50	68	75	8	10	80

a) Compute the correlation between Number of Cigarettes and Risk of CVD

b) Based on this sample, is there evidence of a significant correlation between Number of Cigarettes and Risk of CVD among all hypertensive?

c) Compute the equation of the line that best fits the data to predict Risk of CVD from the Number of cigarettes (i.e., risk of CVD = dependent variable).

2. Consider the following data reflecting lengths of stay in the hospital (recorded in days) and the total charge (in $000's) for 6 patients undergoing a minor surgical procedure :

Length of Stay :	5	7	9	10	12	15
Total Charge :	6	5	7.2	8	9.4	7.9

Consider length of stay as the independent and total charge as the dependent variable.

a) Describe the relationship between length of stay and total charge using a scatter diagram.

b) Compute the sample correlation coefficient

c) Compute the regression equation

d) Estimate the total charge for an individual who stays 11 days in the hospital.

e) Suppose we compare two patients, one stays 3 days longer in the hospital than the other. What is the expected difference in total charges between these patients ?

Correlation and Regression

3. Consider a study assessing the quality of life (QOL) of patients with arthritis. The following figure describes the relationship between QOL and the duration of arthritis, measured in years.

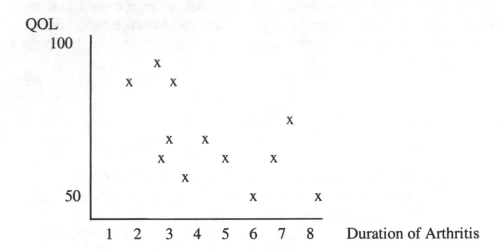

The mean QOL score is 75.8 with a standard deviation of 6.3. The mean duration of arthritis is 4 years with a standard deviation of 0.7 years. The covariance between QOL and duration is -2.9.

    a)   What is the nature of the relationship between duration of arthritis and QOL based on the figure ? Briefly.

    b)   Estimate the sample correlation coefficient.

    c)   Compute the equation of the line that best describes the relationship between duration of arthritis and QOL.

    d)   Interpret the estimated slope.

4. Consider the study described in problem 3, suppose that we investigate the relationship between age and the severity of arthritis (measured on a scale from 0-100 with higher scores indicative of more severe disease). The following summary statistics are available on the total sample (n=175):

Characteristic	Mean	Standard Deviation
Age	57.7	3.8
Severity Score	59.8	19.5

a) The point estimate for the correlation between age and severity score is 0.36 (p=0.0001). Is there a evidence of a significant correlation between age and severity score among all patients ?

b) Estimate the equation of the line that best describes the relationship between age and severity score (consider severity score as the dependent variable).

c) What is the expected severity score for an individual 60 years of age ?

5. An analysis is performed to investigate the relationship between cardiovascular exercise and percent body fat. A sample of 50 subjects are involved and the number of hours of cardiovascular exercise over the previous week and percent body fat are measured on each subject. Summary statistics are shown below:

Variable	n	mean	std. dev.
# hours of cardiovascular exercise	50	3.7	1.9
Percent body fat	50	28.5	5.8

a) Suppose the sample correlation coefficient is r =-0.48. Is there evidence of a significant correlation ? Run the appropriate test at α= 0.05.

b) Suppose that the investigation involved 20 men and 30 women and separate correlation analyses were performed which produced the following results: r_{men}= -0.17 (p=0.4748), r_{women} = -0.59 (p=0.0006). How would one interpret these results ?

c) Summary statistics for the 20 men and 30 women are shown below:

Variable	Men mean (std. dev.)	Women mean (std. dev.)
# hours of cardiovascular exercise	4.7 (2.2)	2.5 (1.4)
Percent body fat	20.1 (3.1)	31.6 (8.1)

d) Estimate the regression equations relating the number of hours of exercise to percent body fat for men and women (separately). (Let Y=percent body fat.)

e) What is the expected percent body fat for a male who exercises 2 hours per week?

f) What is the expected percent body fat for a female who exercises 2 hours per week?

Correlation and Regression

6.   An investigation is performed to understand the relationship between age and satisfaction with medical care. Each individual in the investigation is asked to rate their satisfaction with medical care on a scale of 0 to 100, with 100 denoting complete satisfaction. Other data, including sociodemographic characteristics (age, gender, race), are also recorded on each individual. A total of 200 individuals are involved in the investigation, and the following results were obtained: $r=0.43$ ($p=0.0001$), satisfaction$=45.2 + 0.9$ age.

   a)   Based on the correlation, are older or younger individuals more likely to be more satisfied with medical care ? (Be brief).
   b)   Is the correlation between age and satisfaction significant ? Justify.
   c)   Estimate the satisfaction rating of a 50 year old individual.

7.   Anecdotal evidence suggests that there is an inverse relationship between alcohol consumption and medication adherence. A study is run involving $n=50$ subjects and data are measured on each subject reflecting the number of alcoholic drinks consumed in the week prior to study enrollment. The sample correlation coefficient between the number of alcoholic drinks consumed and medication adherence is -0.47.

   a)   Is there evidence of a significant correlation between the number of alcoholic drinks consumed and medication adherence? Run the appropriate test at $\alpha=0.05$.
   b)   Using the following information, estimate the equation of the line that best describes the relationship between the number of alcoholic drinks consumed and medication adherence. The mean number of alcoholic drinks consumed (per week) is 14 with a standard deviation of 4.5. The mean medication adherence score is 78.5 with a standard deviation of 13.5. Consider medication adherence as the dependent variable.
   c)   What is the expected medication adherence score for a person who does not drink alchohol?

8.   A study was conducted on a random sample of 15 undergraduates to assess whether there is a relationship between stress and GPA. Stress was measured on a scale of 0 to 100 with higher scores indicative of more stress. Descriptive statistics were generated in SAS and are given below:

Correlation and Regression

```
                    Correlation Analysis
               2 'VAR' Variables:  STRESS   GPA
                  Covariance Matrix      DF = 14
                              STRESS              GPA
          STRESS          1075.666667       -11.328571
          GPA              -11.328571         0.226857

                      Simple Statistics
Variable      N      Mean      Std Dev       Sum      Minimum     Maximum
STRESS        15   54.66667    32.79736   820.00000   5.00000    100.00000
GPA           15    2.76000     0.47630    41.40000   2.00000      3.90000
```

a)   Compute the sample correlation coefficient.

b)   Is there evidence of a significant correlation between stress and GPA? Run the appropriate test at $\alpha=0.05$.

c)   Compute the equation of the line that best describes the relationship between stress and GPA. Assume that GPA is the dependent variable.

d)   What is the expected GPA for a person with a stress score of 80?

SAS Problems: Use SAS to solve the following problems.

1. Data were collected from a random sample of 8 patients currently undergoing treatment for hypertension. Each subject reported the average number of cigarettes smoked per day and each subject was assigned a numerical value reflecting risk of cardiovascular disease (CVD). The risk assessments were computed by physicians and based on blood pressure, cholesterol level and exercise status. The risk assessments ranged from 0 to 100, higher values indicated increased risk.

Number of Cigarettes:	0	2	6	8	12	0	2	20
Risk of CVD :	12	20	50	68	75	8	10	80

Use SAS to generate a scatter diagram, to estimate the correlation between variables and test if the correlation is significant. In addition, estimate a simple linear regression equation and compute R^2.

2. Consider the following data reflecting lengths of stay in the hospital (recorded in days) and the total charge (in $00's) for 6 patients undergoing a minor surgical procedure :

Length of Stay :	5	7	9	10	12	15
Total Charge :	6	5	7.2	8	9.4	7.9

Consider length of stay as the independent and total charge as the dependent variable and use SAS to generate a scatter diagram, to estimate the correlation between variables and test if the correlation is significant. In addition, estimate a simple linear regression equation and compute R^2.

Introductory Applied Biostatistics

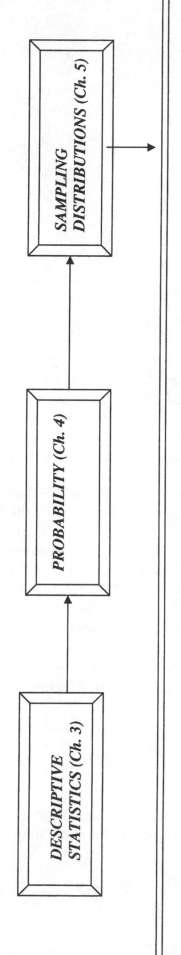

DESCRIPTIVE STATISTICS (Ch. 3) → **PROBABILITY (Ch. 4)** → **SAMPLING DISTRIBUTIONS (Ch. 5)**

STATISTICAL INFERENCE (Chapters 6-14)

Outcome Variable	Grouping Variable(s)/ Predictor(s)	Analysis	Chapter(s)
Continuous	–	Estimate μ, Compare μ to Known, Historical Value	6/13
Continuous	Dichotomous (2 Groups)	Compare Independent Means (Estimate/Test ($\mu_1-\mu_2$)) or the Mean Difference(μ_d)	7/13
Continuous	Discrete (> 2 Groups)	Test the Equality of K Means using Analysis of Variance ($\mu_1=\mu_2=\ldots\mu_k$)	10/13
Continuous	Continuous	Estimate Correlation or Determine Regression Equation	11/13
Continuous	Several Continuous or Dichotomous	Multiple Linear Regression Analysis	11
Dichotomous	–	Estimate p, Compare p to Known, Historical Value	8
Dichotomous	Dichotomous (2 Groups)	Compare Independent Proportions (Estimate/Test (p_1-p_2))	8/9
Dichotomous	Discrete (>2 Groups)	Test the Equality of k Proportions (Chi-Square Test)	8
Dichotomous	**Several Continuous or Dichotomous**	**Multiple Logistic Regression Analysis**	**12**
Discrete	Discrete	Compare Distributions Among k Populations (Ch-Square Test)	8
Time to Event	Several Continuous or Dichotomous	Survival Analysis	14

CHAPTER 12: Logistic Regression Analysis

12.1 Introduction
12.2 The Logistic Model
12.3 Statistical Inference for Simple Logistic Regression
12.4 Multiple Logistic Regression
12.5 ROC Area
12.6 Key Formulas
12.7 Statistical Computing
12.8 Problems

12.1 Introduction

In Chapter 9 we introduced several effect measures used to compare risks in two populations. We presented estimation and statistical inference procedures for crude comparisons and a technique for estimation and statistical inference adjusting for one categorical confounder. In Chapter 10 we presented simple linear regression techniques used to estimate the linear equation describing the relationship between two continuous variables. We also introduced multiple linear regression techniques which are used when the continuous outcome variable is linearly related to a combination of independent variables.

In this chapter we introduce logistic regression which can be used to model the effect of independent variables on the risk of a dichotomous outcome. In Section 12.2 we introduce the logistic model and in Section 12.3 we focus on statistical inference for situations with a single independent variable. In Section 12.4 we extend our discussion to the multiple logisitic model in which a single dichotomous outcome variable is a function of a linear combination of independent variables. We briefly describe the use of the Receiver Operating Charatertsitic (ROC) curve as a measure of the goodness- of-fit of logistic regression models in Section 12.5, and we summarize key formulas and present statistical computing applications in Sections 12.6 and 12.7, respectively.

As in linear regression, our goal is to estimate the regression coefficients in a model, given a sample of (X, Y) pairs. In the case of logistic regression, the X's can be continuous or dichotomous, but the Y's are all equal to 0 (for those who do not have the event) or 1 (for those who do have the event).

12.2 The Logistic Model

The simple logistic model is based on a linear relationship between the natural logarithm (ln) of the odds of an event, and a continuous independent variable. The form of this relationship is as follows:

$$L = \beta_0 + \beta_1 X + \varepsilon \qquad (12.1)$$

where Y is the dichotomous event of interest,

   $L = \ln(\text{odds of event } Y)$,

   X is the independent variable,

   β_0 is the intercept,

   β_1 is the regression coefficient, and

   ε is the random error

The relationship between X and L is linear:

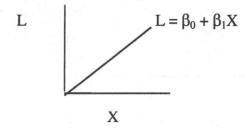

Recall from Chapter 9 that if p is the probability of the event, then the odds of the event is

$$\text{Odds} = o = \frac{p}{(1-p)}$$

We defined $L = \ln(\text{odds of event } Y)$, sometimes called the "log odds" of Y. We can write L in terms of p, Probability $(Y = 1)$, as follows:

$$L = \ln(o) = \ln\left(\frac{p}{(1-p)}\right).$$

We then use the laws of exponents and logs and some algebra to express p in terms of L:

$$e^L = o$$

$$e^L = \frac{p}{(1-p)}$$

$$p = e^L(1-p)$$

$$p = e^L - pe^L$$

$$p + pe^L = e^L$$

Logistic Regression Analysis

$$p(1+e^L)=e^L$$

$$p=\frac{e^L}{(1+e^L)}$$

This is called the logistic function and its graph is as follows. Notice that *p*, the probability of the event, increases from 0 to 1.

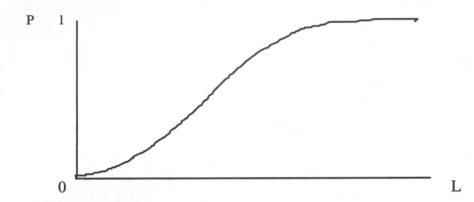

| Logistic function | $p=\dfrac{e^L}{(1+e^L)}$ | (12.2) |

12.3 Statistical Inference for Simple Logistic Regression

The logistic regression model (12.1) may be written in terms of p, the risk of event Y, using equations (12.1) and (12.2). We assume that L is a linear function of X, and we substitute (12.1) into the logistic funtion (12.2).

$$p=\frac{e^{\beta_0+\beta_1 X+\varepsilon}}{(1+e^{\beta_0+\beta_1 X+\varepsilon})} \qquad (12.3)$$

As in linear regression, our goal is to estimate the regression coefficients β_0 and β_1 given a sample of (X, Y) pairs. Here, as in the linear regression framework, the X's can be continuous or dichotomous. However, the Y's are not continuous as in linear regression; rather, they are coded 0 (for those who do not have the event) or 1 (for those who do have the event) . Unfortunately, the techniques used in linear regression to estimate the regression coefficients cannot be applied to the logistic regression case. The iterative maximum likelihood estimation process used to estimate coefficients is computationally complex and is beyond the scope of this book. Here we restrict our attention to interpretation of the logistic model estimated by SAS.

Logistic Regression Analysis

SAS Example 12.1. Recall the clinical trial described in Example 9.1 in which patients with Kawasaki syndrome were treated with gamma globulin (GG=1) or with the standard treatment (GG=0) and the primary outcome was the development of coronary abnormalities (CA). A possible explanation for the effect of treatment with gamma globulin on the development of CA is that it reduces the elevated white blood count (WBC) early in the course of the disease, and that WBC is related to CA. This example addresses the relationship between WBC and CA. The data set contains a total of n = 167 observations, each participant has a continuous value for WBC and a dichotomous value (0 or 1) for CA. The following is a subset of the output generated by SAS Proc Logistic. A brief interpretation follows the output.

SAS Output for Example 12.1

```
                     The LOGISTIC Procedure
                       Model Information

            Data Set                   WORK.ONE
            Response Variable          CA
            Number of Response Levels  2
            Number of Observations     156
            Model                      binary logit
            Optimization Technique     Fisher's scoring


                        Response Profile

            Ordered                        Total
             Value            CA         Frequency

               1              1              25
               2              0             131


           Probability modeled is CA=1.
```

NOTE: 12 observations were deleted due to missing values for the response or explanatory variables.

```
                   The LOGISTIC Procedure

           Analysis of Maximum Likelihood Estimates

                              Standard      Wald
      Parameter    DF   Estimate   Error   Chi-Square   Pr > ChiSq

      Intercept     1   -3.3848   0.5838    33.6120      <.0001
      WBC           1    0.1253   0.0362    11.9835      0.0005
```

Logistic Regression Analysis

Interpretation of SAS Output for Example 12.1

We coded the variable CA to represent the development of CA and specified that the event of interest is CA=1. SAS reports this under "Probability modeled is CA=1." Among the 167 patients, 12 were missing the WBC measure and SAS notes that 12 observations were deleted due to missing values. Thus the analysis is based on n = 156. The results of the logistic regression estimation procedure are given under "Analysis of Maximum Likelihood Estimates." The estimated coefficients are in the column labeled "Estimate," and are $\hat{\beta}_0 = -3.385$ and $\hat{\beta}_1 = 0.125$.

SAS also provides the p-value associated with the test H_0: $\beta_1 = 0$ in the column labeled "Pr > ChiSq." In this case, p < 0.001, so we reject H_0 and conclude that there is significant evidence that WBC is linearly related to the log odds of CA.

Now that the coefficients have been estimated, we can estimate the risk, $\hat{p}$, of the event Y, given a specific value of X by substituting estimates of β_0 and β_1 into equation (12.3) as follows:

$$\hat{p} = \frac{e^{\hat{\beta}_0 + \hat{\beta}_1 X}}{(1 + e^{\hat{\beta}_0 + \hat{\beta}_1 X})} \qquad (12.4)$$

Consider a person with X = x_a. We estimate the risk of Y given X = x_a as

$$\hat{p}_a = \frac{e^{\hat{\beta}_0 + \hat{\beta}_1 x_a}}{1 + e^{\hat{\beta}_0 + \hat{\beta}_1 x_a}}.$$

For example, in SAS Example 12.1,

$$\hat{p} = \frac{e^{-3.385+0.125(WBC)}}{(1 + e^{-3.385+0.125(WBC)})}$$

The risk of CA for a patient with WBC = 10 is

$$\hat{p} = \frac{e^{-3.385+0.125(10)}}{(1 + e^{-3.385+0.125(10)})}$$

$$\hat{p} = \frac{e^{-2.135}}{(1 + e^{-2.135})}$$

$$\hat{p} = 0.106$$

We can also use the estimate of β_1 to estimate the odds of the event for specified values of X, and the odds ratio comparing the odds of the event for people with different values of X. Recall that the logistic model is based on the linear relationship between X and the log odds: $\ln(o) = L = \beta_0 + \beta_1 X$. Thus the estimated odds are

$$\hat{o} = e^{\hat{L}} = e^{\hat{\beta}_0 + \hat{\beta}_1 X}$$

$$\hat{o} = e^{\hat{L}} = e^{\hat{\beta}_0 + \hat{\beta}_1 X} \tag{12.5}$$

Consider two people with different values of X, one with $X = x_a$ and one with $X = x_b$. The odds of the event given $X = x_a$ are $\hat{o}_a = e^{\beta_0 + \beta_1 x_a}$ and, similarly, for $X = x_b$, the odds are $\hat{o}_b = e^{\hat{\beta}_0 + \hat{\beta}_1 x_b}$.

For example, the odds of CA given the WBC are

$$\hat{o} = e^{\hat{L}} = e^{\hat{\beta}_0 + \hat{\beta}_1 X} = e^{-3.385+0.125(WBC)}$$

For a patient with WBC = 10, the odds of CA are

$$\hat{o} = e^{-3.385+0.125(10)} = e^{-2.135} = 0.118.$$

The odds ratio comparing the odds of an event for two people, one with $X = x_a$ and one with $X = x_b$ is:

$$\hat{O}R = \frac{e^{\hat{\beta}_0 + \hat{\beta}_1 x_a}}{e^{\hat{\beta}_0 + \hat{\beta}_1 x_b}}.$$

Recall some of the properties of exponents:

$$e^{A+B} = e^A e^B$$

$$e^{A-B} = e^A / e^B \qquad (12.6)$$

$$e^{AB} = (e^A)^B$$

Then,

$$\hat{O}R = \frac{e^{\hat{\beta}_0 + \hat{\beta}_1 x_a}}{e^{\hat{\beta}_0 + \hat{\beta}_1 x_b}}.$$

This can we written as

$$\hat{O}R = e^{(\hat{\beta}_0 + \hat{\beta}_1 x_a) - (\hat{\beta}_0 + \hat{\beta}_1 x_a)},$$

which is equivalent to

$$\hat{O}R = e^{\hat{\beta}_1 (x_a - x_b)}.$$

The odds ratio comparing the odds of the event for a person $X = x_a$ compared to one with $X = x_b$ is

$$\hat{O}R = e^{\hat{\beta}_1 (x_a - x_b)} \qquad (12.7)$$

In SAS Example 12.1, the odds ratio comparing a person with WBC=15 to a person with WBC=10 is:

$$OR = e^{\hat{\beta}_1 (x_a - x_b)} = e^{0.1253(15-10)} = e^{0.1253(5)} = e^{.6265} = 1.87$$

Notice that this depends only on the difference in the X values, $(x_a - x_b)$, not on their actual values. Thus the odds ratio comparing one person with WBC=10 to another person with WBC=5 is also equal to 1.87, as is the odds ratio comparing any two people whose WBC values are 5 units apart.

The formula for a 95% confidence interval around the odds ratio is

$$\left(e^{(\hat{\beta}_1-1.96se(\hat{\beta}))(x_a-x_b)},e^{(\hat{\beta}_1+1.96se(\hat{\beta}))(x_a-x_b)}\right) \tag{12.8}$$

where $se(\hat{\beta}_1)$ is the standard error of $(\hat{\beta}_1)$. The calculation of $se(\hat{\beta}_1)$ is beyond the scope of this book and we use the value provided by SAS to demonstrate the calculation of confidence intervals around the odds ratio.

In SAS Example 12.1, the standard error of $\hat{\beta}_1$, $se(\hat{\beta}_1)$ = 0.0362. Thus, using (12.8), the 95% confidence interval for the odds ratio comparing the odds of CA in one child with Kawasaki syndrome whose WBC is 5 units higher than another child is as follows.

$$\left(e^{(\hat{\beta}_1-1.96se(\hat{\beta}))(x_a-x_b)},e^{(\hat{\beta}_1+1.96se(\hat{\beta}))(x_a-x_b)}\right)=\left(e^{(0.1253-1.96(0.0362))(5)},e^{(0.1253+1.96(0.0362))(5)}\right)$$

$$=\left(e^{(0.2717)},e^{(0.9813)}\right)$$

$$=(1.31,2.67)$$

Thus, we conclude that the odds ratio comparing the odds of CA in one child with Kawasaki syndrome whose WBC is 5 units higher than another child is 1.87 and that the 95% confidence interval for the odds is (1.31,2.67).

In SAS Example 12.1, the odds ratio comparing two people with WBC values one unit apart is $OR=e^{\hat{\beta}_1}=e^{0.125}=1.13$. The 95% confidence interval is

$$\left(e^{(\hat{\beta}_1-1.96se(\hat{\beta}))(x_a-x_b)},e^{(\hat{\beta}_1+1.96se(\hat{\beta}))(x_a-x_b)}\right)=\left(e^{(0.1253-1.96(0.0362))(1)},e^{(0.1253+1.96(0.0362))(1)}\right)$$

$$=\left(e^{(0.0543)},e^{(0.1963)}\right)$$

$$=(1.056,1.217)$$

SAS automatically provides the odds ratio (and the associated 95% confidence interval) comparing the odds of the event for two people with X values one unit apart ($x_a-x_b=1$). This is $OR=e^{\hat{\beta}_1(x_a-x_b)}=e^{\hat{\beta}_1(1)}=e^{\hat{\beta}_1}$. The label is simply "Odds Ratio Estimates," but it is important to remember that this is the estimate of the odds ratio comparing people exactly one unit apart.

SAS Output for Example 12.1 (Continued)

```
                    The LOGISTIC Procedure

              Analysis of Maximum Likelihood Estimates

                               Standard        Wald
    Parameter    DF    Estimate    Error    Chi-Square    Pr > ChiSq

    Intercept    1     -3.3848     0.5838     33.6120       <.0001
    WBC          1      0.1253     0.0362     11.9835       0.0005



                     Odds Ratio Estimates

                      Point          95% Wald
         Effect      Estimate     Confidence Limits

          WBC         1.133       1.056      1.217
```

In clinical trials, X is usually a dichotomous variable representing membership in one treatment group or another. For example, X=1 could denote the treatment group, and X=0 could denote the control group. Suppose that X is dichotomous (0 or 1) and we use (12.7) to estimate the odds ratio. The estimate of the odds ratio comparing those with X = 1 to those with X = 0 reduces to:

$$\hat{OR} = e^{\hat{\beta}_1(1-0)} = e^{\hat{\beta}_1(1)} = e^{\hat{\beta}_1}.$$

In this situation, where X is dichotomous, the null hypothesis $H_0: \beta_1 = 0$ can be phrased, equivalently, as $H_0: e^{\beta} = 1$, or $H_0: OR = 1$.

SAS Example 12.2. We now use SAS Proc Logist to esimate the probabilities and odds of the development of coronary abnormalities (CA) in children with Kawasaki syndrome treated with gamma globulin (GG=1) as compared to the standard treatment (GG=0).

SAS Output for Example 12.2

```
                     Model Information

        Data Set                    WORK.ONE
        Response Variable           CA
        Number of Response Levels   2
        Number of Observations      167
        Model                       binary logit
        Optimization Technique      Fisher's scoring
```

Logistic Regression Analysis

Response Profile

Ordered Value	CA	Total Frequency
1	1	26
2	0	141

Probability modeled is CA=1.

NOTE: 1 observation was deleted due to missing values for the response or explanatory variables.

The LOGISTIC Procedure

Analysis of Maximum Likelihood Estimates

Parameter	DF	Estimate	Standard Error	Wald Chi-Square	Pr > ChiSq
Intercept	1	-1.0986	0.2520	19.0094	<.0001
gg	1	-1.6487	0.5257	9.8370	0.0017

Odds Ratio Estimates

Effect	Point Estimate	95% Wald Confidence Limits	
gg	0.192	0.069	0.539

Interpretation of SAS Output for Example 12.2

The estimated coefficients are in the column labeled "Estimate," and are $\hat{\beta}_0 = -1.099$ and $\hat{\beta}_1 = -1.649$. The p-value associated with the test H_0: $\beta_1 = 0$ is p = 0.002, so we reject H_0 and conclude that there is significant evidence that treatment (gamma globulin or standard care) is associated with the development of CA.

The odds ratio comparing children with Kawasaki syndrome treated with gamma globulin (GG=1) to those treated with the standard treatment (GG=0) with respect to the development of coronary abnormalities (CA) is given here as 0.19, with 95% confidence interval (0.069, 0.539). Note that the p-value associated with the test H_0: $\beta_1 = 0$ is also the p-value associated with the test H_0: OR = 1. Here, this p-value is p = 0.002, so we reject H_0 and conclude that there is significant evidence that the odds ratio is not equal to 1.

Recall that we used these data in Example 9.1 and estimated the odds ratio using equation (9.6). This estimate will be identical to that obtained using equation (12.8). In fact, we obtained the same result as we do here: $\hat{OR} = 0.19$. In Chapter 9, we used the formula given in Table 9.2 to estimate the 95% confidence interval for the odds ratio, while in this chapter we provide an estimate of the confidence interval which assumes the logistic model. The difference in

Logistic Regression Analysis

estimated confidence intervals will be very slight. Using Table 9.2, we estimated the 95% confidence interval as (0.069, 0.539), identical to the estimate provided by SAS.

SAS does not automatically provide an estimate of the relative risk as it does the odds ratio. To estimate the relative risk, RR, comparing a patient with $X = x_a$ to one with $X = x_b$, we must first estimate the individual probabilities using (12.4).

$$\hat{R}R = \hat{p}_a/\hat{p}_b = \frac{\left(\dfrac{e^{\hat{\beta}_0+\hat{\beta}_1 x_a}}{1+e^{\hat{\beta}_0+\hat{\beta}_1 x_a}}\right)}{\left(\dfrac{e^{\hat{\beta}_0+\hat{\beta}_1 x_b}}{1+e^{\hat{\beta}_0+\hat{\beta}_1 x_b}}\right)} \qquad (12.9)$$

In SAS Example 12.1 the estimated regression coefficients were $\hat{\beta}_0 = -3.385$ and $\hat{\beta}_1 = 0.125$ and the estimated risk of CA given WBC is $\hat{p} = \dfrac{e^{-3.385+0.125(WBC)}}{(1+e^{-3.385+0.125(WBC)})}$. We estimated the risk of CA for a patient with WBC = 10 to be $\hat{p}_{10} = 0.106$. Similarly the estimate of the risk of CA for a patient with WBC = 5 is $\hat{p}_5 = 0.060$. Thus, the estimated relative risk of CA for a patient with WBC = 10 as compared to a patient with WBC = 5 is $\hat{R}R = (\hat{p}_{10}/\hat{p}_5) = (0.106/0.060) = 1.77$.

Remember that the odds ratio comparing a patient with $X = x_a$ to one with with $X = x_b$ depends only on the difference $(x_a - x_b)$, not on the actual values x_a and x_b. The relative risk comparing those with $X = x_a$ to those with $X = x_b$, however, does depend on the actual values x_a and x_b.

In SAS Example 12.1, the estimated risk of CA for a patient with WBC = 15 is $\hat{p}_{15} = 0.181$. Therefore, the relative risk of CA comparing patients with WBC = 15 to those with WBC = 10 is $\hat{R}R = (\hat{p}_{15}/\hat{p}_{10}) = (0.181/0.106) = 1.71$. An increase of 5 units in WBC is associated with a relative risk $\hat{R}R = 1.77$ if the WBC values are 5 and 10, but is associated with a relative risk $\hat{R}R = 1.71$ if the WBC values are 10 and 15.

Now consider the situation in SAS Example 12.2 in which the comparison is of patients in two treatment groups, of those with $X = 1$ to those with $X = 0$. We can modify (12.9) slightly to estimate the relative risk of the event in patients with $X = 1$ as compared to those with $X = 0$.

$$\hat{R}R = \hat{p}_1/\hat{p}_0 = \frac{\left(\dfrac{e^{\hat{\beta}_0+\hat{\beta}_1}}{1+e^{\hat{\beta}_0+\hat{\beta}_1}}\right)}{\left(\dfrac{e^{\hat{\beta}_0}}{1+e^{\hat{\beta}_0}}\right)} \qquad (12.10)$$

Logistic Regression Analysis

In SAS Example 12.2, the relative risk of CA for patients treated with gamma globulin as compared to those on the standard treatment is estimated as follows from the SAS output.

$$\hat{RR} = \hat{p}_1/\hat{p}_0 = \frac{\left(\dfrac{e^{\hat{\beta}_0+\hat{\beta}_1}}{1+e^{\hat{\beta}_0+\hat{\beta}_1}}\right)}{\left(\dfrac{e^{\hat{\beta}_0}}{1+e^{\hat{\beta}_0}}\right)} = \frac{\left(\dfrac{e^{-1.099-1.649}}{1+e^{-1.099-1.649}}\right)}{\left(\dfrac{e^{-1.099}}{1+e^{-1.099}}\right)}$$

$$= (0.060/0.250)$$
$$= 0.24$$

The estimated relative risk of CA for patients treated with gamma globulin as compared to those on the standard treatment is $\hat{RR} = 0.24$. This may be interpreted as the risk of CA among patients treated with gamma globulin in less than a quarter the risk for those on the statndard treated. Recall that this is the estimate we obtained in chapter 9 using (9.5).

As discussed in Chapter 9, the odds ratio may be used to estimate the relative risk if the prevalence of the disease is low (the overall risk is small). Here we see that it is much easier to obtain the estimate of the odds ratio from the results of a logistic regression than it is to obtain the estimate of the relative risk.

12.4 Multiple Logistic Regression

As in the case of linear regression, logistic regression techniques may be generalized to models with more than one independent variable. The methods used to estimate coefficients, standard errors, and p-values are beyond the scope of this book, but we present some examples of multiple logistic regression in this section using SAS. The general multiple logistic regression model is

$$p = \frac{e^{\beta_o+\beta_1 X_1+\hat{\beta}_2 X_2+...+\beta_k X_k+\varepsilon}}{(1+e^{\beta_o+\beta_1 X_1+\hat{\beta}_2 X_2+...+\beta_k X_k+\varepsilon})} \tag{12.11}$$

where p is the probability of the event and $X_1, X_2,......,X_k$ are independent variables. The independent variables may be dichotomous or continuous.

SAS Example 12.3. Suppose we want to use the sample described used in SAS Example 12.2 to assess the effect of treatment with gamma globulin on CA but we are concerned that white blood count is a potential confounder. Although patients were randomized to one of the two treatment groups, it turns out that the mean WBC in patients treated with gamma globulin was significantly lower than the mean WBC in patients on the standard treatment. Thus the investigators were concerned that the patients treated with gamma globulin were less sick than those on the standard treatment, and that the observed impact of the gamma globulin treatment might not be real. The following output is a multiple logistic regression in which the outcome is the dichotomous variable CA and the two independent variables are the dichotomous treatment variable, GG, and the continuous variable WBC.

SAS Output for Example 12.3

```
                        The LOGISTIC Procedure

                 Analysis of Maximum Likelihood Estimates

                                 Standard        Wald
    Parameter    DF    Estimate     Error    Chi-Square    Pr > ChiSq

    Intercept     1     -2.6352    0.6562      16.1283       <.0001
    gg            1     -1.3152    0.5496       5.7272       0.0167
    WBC           1      0.1048    0.0383       7.4781       0.0062


                        Odds Ratio Estimates

                        Point          95% Wald
            Effect    Estimate     Confidence Limits

            gg          0.268      0.091      0.788
            WBC         1.111      1.030      1.197
```

Interpretation of SAS Output for Example 12.3

In SAS Example 12.2, we estimated the *crude* odds ratio comparing patients treated with gamma globulin to patients on standard treatment. The estimated OR is $\hat{OR} = 0.19$ with p = 0.002. The odds ratio estimated here is the odds ratio comparing patients treated with gamma globulin to patients on standard treatment, *adjusted* for white blood count. The estimated OR is $\hat{OR} = 0.27$ with p = 0.017. As suspected, part of the observed effect of gamma globulin may have been due to the lower initial WBC in patients treated with gamma globulin. Thus we see the estimated OR is closer to 1 after adjustment for WBC. The crude estimate indicates that gamma globulin reduces the odds by just over 80% (the crude $\hat{OR} = 0.19$), while the adjusted $\hat{OR}$ of 0.27 indicates that gamma globulin reduces the odds by 73%. Although the adjustment

Logistic Regression Analysis

appears to reduce the impact of gamma globulin, the effect is still quite important and the hypothesis of no effect is rejected with p = 0.017.

Another application of multiple logistic regression is in model prediction. The multiple logistic regression model can be used to predict probabilities of event for subjects with specific characteristics (i.e., values of the independent variables included in the model).

SAS Example 12.4. Suppose we want to use the sample described used in SAS Example 12.1 to predict the probability of developing CA as a function of both white blood count and hemoglobin. The following output is based on a multiple logistic regression in which the outcome is the dichotomous variable CA and the two independent variables are the continuous variables WBC and HEM.

SAS Output for Example 12.4

```
                    The LOGISTIC Procedure

                      Model Information

      Data Set                      WORK.ONE
      Response Variable             CA
      Number of Response Levels     2
      Number of Observations        151
      Model                         binary logit
      Optimization Technique        Fisher's scoring


                      Response Profile

          Ordered                         Total
           Value           CA           Frequency

             1              1               25
             2              0              126


          Probability modeled is CA=1.
```

NOTE: 17 observations were deleted due to missing values for the response or explanatory variables.

```
                  Model Convergence Status

        Convergence criterion (GCONV=1E-8) satisfied.

                  Model Fit Statistics

                                            Intercept
                           Intercept           and
          Criterion          Only           Covariates

          AIC               137.532           124.420
          SC                140.549           133.472
          -2 Log L          135.532           118.420
```

Logistic Regression Analysis

```
                 Testing Global Null Hypothesis: BETA=0

        Test                  Chi-Square       DF       Pr > ChiSq

        Likelihood Ratio        17.1120         2         0.0002
        Score                   19.1517         2        <.0001
        Wald                    14.1978         2         0.0008
                        The LOGISTIC Procedure

              Analysis of Maximum Likelihood Estimates

                                  Standard        Wald
     Parameter    DF    Estimate    Error    Chi-Square    Pr > ChiSq

     Intercept     1      1.2859    2.2545      0.3253        0.5684
     HEM           1     -0.4272    0.2078      4.2294        0.0397
     WBC           1      0.1135    0.0376      9.0948        0.0026


                       Odds Ratio Estimates

                           Point          95% Wald
            Effect       Estimate     Confidence Limits

            HEM            0.652       0.434       0.980
            WBC            1.120       1.041       1.206


      Association of Predicted Probabilities and Observed Responses

          Percent Concordant     75.0     Somers' D    0.504
          Percent Discordant     24.6     Gamma        0.506
          Percent Tied            0.5     Tau-a        0.140
          Pairs                  3150     c            0.752
```

Interpretation of SAS Output for Example 12.4

We first consider the overall null hypothesis that all the β coefficients in (12.11) are zero. SAS provides three tests of this hypothesis; we recomment the *likelihood ratio test*. In this example, the likelihood ratio test $\chi^2 = 17.11$, with 2 df and the associated $p < 0.001$. Therefore we reject H_0 and conclude that at least one of the coefficients is not equal to zero. We then look at the individual tests $H_0: \beta_1 = 0$ and $H_0: \beta_2 = 0$, where β_1 and β_2 are the coefficients of HEM and WBC, respectively. We reject $H_0: \beta_1 = 0$, with $p = 0.040$, and conclude that hemoglobin is significantly associated with the development of CA, even after adjusting for white blood count. We also reject $H_0: \beta_2 = 0$, with $p = 0.003$, and conclude that white blood count is significantly associated with the development of CA, even after adjusting for hemoglobin. In fact we conclude that hemoglobin and white blood count each make an independent contribution to the risk of CA.

We now turn to the prediction of the risk of CA given hemoglobin and white blood count.

Logistic Regression Analysis

In SAS Example 12.1, we used logistic regression to estimate the probability of CA given WBC. The estimated probability was

$$\hat{p} = \frac{e^{-3.385+0.125(\text{WBC})}}{(1+e^{-3.385+0.125(\text{WBC})})}.$$

We used this to estimate the risk of CA for a patient with WBC = 10 and obtained the estimate $\hat{p} = 0.106$.

The output for SAS Example 12.4 allows us to estimate the probability of CA given both white blood count and hemoglobin.

$$\hat{p} = \frac{e^{1.286-0.427(\text{HEM})+0.114(\text{WBC})}}{(1+e^{1.286-0.427(\text{HEM})+0.114(\text{WBC})})}$$

Now, suppose WBC = 10 and HEM = 10. We estimate the risk of CA by substituting these values for WBC and HEM into the above equation.

$$\hat{p} = \frac{e^{1.286-0.427(10)+0.114(10)}}{(1+e^{1.286-0.427(10)+0.114(10)})}$$

$$\hat{p} = \frac{e^{-1.844}}{(1+e^{-1.844})}$$

$$\hat{p} = 0.137.$$

Similarly, we can estimate the risk of CA given WBC = 10 and HEM = 12, and obtain the estimated risk $\hat{p} = 0.063$. Notice that as HEM increases, the risk of CA decreases. This can be seen from the negative sign of the coefficient, β_1, of HEM. In constrast, β_2 is positive, indicating that the risk of CA increases as WBC increases.

The odds ratios describing the impact of one unit increase in the independent variable on the risk of the event, adjusting for the other independent variables, is automatically displayed. In this example, an increase of one unit in HEM is associated with a 35% decrease in odds of CA ($\hat{\text{OR}} = 0.65$). An increase of one unit in WBC is associated with a 12% increase in odds of CA ($\hat{\text{OR}} = 1.12$).

12.5 ROC Area

We use maximum likelihood estimation to estimate the coefficients in logistic regression models. One application of logistic regression analysis is the comparison of risks between groups. We use the estimated coefficients to estimate odds ratios and relative risks. Another application of logistic regression is in prediction. The estimated coefficients may be used to estimate or predict probabilities of the event for subjects with specified characteristics (i.e., values on the independent variables included in the model). In the latter application, we may also want to assess the extent to which the set of independent variables is associated with the event.

It is relatively easy to compare pairs of observed and predicted values in the linear regression framework as they are both measured on the same scale. For example, pairs $(Y, \hat{Y})$ can be compared by simple subtraction, $(Y - \hat{Y})$ or by squaring the difference, $(Y - \hat{Y})^2$. A summary over the Y values in the sample can be made by summing the squares of the differences (SSE), or by calculating $R^2 = \dfrac{\sum(\hat{Y} - \bar{Y})^2}{\sum(Y - \bar{Y})^2}$. In linear regression modeling, R^2 may be interpreted as the percent of variability in Y which can be explained by variability in the set of independent variables. This gives an idea of how well the selected model fits, or how close the predicted Y values ($\hat{Y}$) are to the observed Y values.

Unfortunately, this statistic has no obvious interpretation in the logistic regression setting, and depends on the overall event prevalence. Consider pairs of observed and predicted values, $(Y, \hat{p})$. The Y values are all equal to either 0 or 1 and the $\hat{p}$ values are all between 0 and 1. Clearly, if $\hat{p}$ is equal to or close to 1 for an individual who had the event $(Y = 1)$, we would say that the model was successful in predicting that individual's risk. However, what if $\hat{p} = 0.8$? 0.25? Suppose that the model only yields the following predicted probabilities: 0, 0.06, 0.25, and 1.00. Suppose also that all the individuals who had the event had $\hat{p} = 0.25$ and those who did not have the event had $\hat{p} = 0.06$. The model seems to distinguish those who did and did not have the event even though the individual $\hat{p}$ values are not particularly close to the observed Y values. We would like a summary measure of the goodness-of-fit which takes into account whether the model distinguishes those who do and do not have the outcome.

Recall in SAS Example 12.2, that the risk of CA in patients treated with gamma globulin was estimated using logistic regression to be $\hat{p} = 0.25$, and the estimated risk of CA in patients on the standard treatment was $\hat{p} = 0.06$. Suppose we decide to classify individuals in terms of predicting event status by using a cut-off in predicted probability. We predict that individuals with $\hat{p} > 0.06$ will have the event and we assign them the predicted value $\hat{Y} = 1$. We predict that individuals with $\hat{p} \leq 0.06$ will not have the event and assign them the predicted value $\hat{Y} = 0$.

Five of the 26 patients who did develop CA (Y = 1) had been treated with gamma globulin and have $\hat{p}$ = 0.06, while the other 21 were on standard treatment and have $\hat{p}$ = 0.25. Thus, if we apply a cut-off of 0.06 and decide that patients with $\hat{p}$ > 0.06 are predicted to have CA, then we have correctly predicted event status in 21 of the 26 patients who actually did have the event. This is called the true positive rate, or sensitivity. In this example, the sensitivity is 21/26 = 0.808.

Conversely, 78 of the 141 patients who did not develop CA (Y = 0) had been on statndard treatment and have $\hat{p}$ = 0.25. If we again apply a cut-off of 0.06 (assigning patients with $\hat{p}$ > .06 to have $\hat{Y}$ = 1, and those with with $\hat{p}$ ≤ 0.06 to have $\hat{Y}$ = 0), then we have correctly predicted event status in 78 of the 141 patients who actually did not have the event. This is called the true negative rate, or specificity. In this example, the specificity is 78/141 = 0.553.

Now consider the cut-off of 0 (individuals with $\hat{p}$ > 0 will be assigned $\hat{Y}$ = 1). Since all the $\hat{p}$ values are either = 0.06 or 0.25, all the patients have $\hat{p}$ > 0 and thus all the patients will be assigned $\hat{Y}$ = 1. The sensitivity = 26/26 = 1, and specificity = 0/141 = 0. The other extreme is the situation in which the cut-off is 1, all the patients are assigned $\hat{Y}$ = 0, the sensitivity is 0 and the specificity is 1. Any other cut-offs will produce the same sensitivity-specificity combination as one of the three above. For example, any cut-off between 0 and 0.06 will have sensitivity and specificity 1 and 0, respectively. Cut-offs between 0.06 and 0.25 will have sensitivity and specificity 0.808 and 0.553, respectively, and cut-offs between 0.25 and 1 will have sensitivity and specificity 0 and 1, respectively.

Models with only one dichotomous variable produce only two possible predicted probabilities: individuals with X = 0 have predicted probability $\hat{p}_0$ and those with X = 1 have predicted probability $\hat{p}_1$. For example, the only possible predicted values in the example were $\hat{p}_0$ = 0.25 and $\hat{p}_1$ = 0.06. As a result, there is only one cut-off (in addition to cut-offs of 0 and 1) that produces distinct values of sensitivity and specificity. Models with one continuous independent variable or more than one independent variable all have more possible values for $\hat{p}$, and there may be many cut-offs that produce distinct values of sensitivity and specificity.

A measure of goodness-of-fit often used to evaluate the fit of a logistic regression model is based on the simultaneous measure of sensitivity and specificity for all possible cut-off points. First, calculate sensitivity and specificity pairs for each possible cut-off point and plot sensitivity on the y-axis by (1-specificity) on the x-axis. This curve is called the Receiver Operating Characteristic (ROC) curve, and the area under the curve, the *area under the ROC curve,* is a summary of the sensitivities and specificities over all possible cut-off points. It ranges from 0.5 to 1.0 with larger values indicative of better fit.

The figure below shows an application in which the area under the ROC curve (the are to the right under the curve) is equal to 0.68.

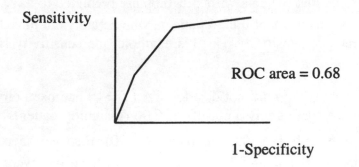

1-Specificity

The following figure shows an application in which the area under the ROC curve is 0.5. When the area under the ROC curve is 0.5, the logistic model is said to classify events and non-events simply by chance or at random.

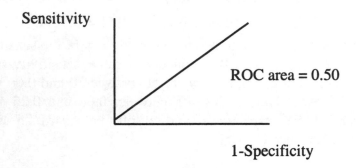

1-Specificity

SAS example 12.5. Proc Logistic produces the "c" statistic automatically in the output of a logistic regression; this is equal to the area under the ROC curve. We can compare the models used in SAS Examples 12.2, 12.3, and 12.5, each of which was used to predict the development of CA. Recall that the independent variables in SAS Examples 12.2, 12.3, and 12.4, were {gamma globulin alone}, {gamma globulin plus white blood count}, and {white blood count plus hemoglobin}, respectively.

SAS Output for Example 12.2 (Continued)

The LOGISTIC Procedure

Analysis of Maximum Likelihood Estimates

Parameter	DF	Estimate	Standard Error	Wald Chi-Square	Pr > ChiSq
Intercept	1	-1.0986	0.2520	19.0094	<.0001
gg	1	-1.6487	0.5257	9.8370	0.0017

Association of Predicted Probabilities and Observed Responses

Percent Concordant	44.7	Somers' D	0.361
Percent Discordant	8.6	Gamma	0.677
Percent Tied	46.7	Tau-a	0.095
Pairs	3666	c	0.680

SAS Output for Example 12.3 (Continued)

The LOGISTIC Procedure

Analysis of Maximum Likelihood Estimates

Parameter	DF	Estimate	Standard Error	Wald Chi-Square	Pr > ChiSq
Intercept	1	-2.6352	0.6562	16.1283	<.0001
gg	1	-1.3152	0.5496	5.7272	0.0167
WBC	1	0.1048	0.0383	7.4781	0.0062

Odds Ratio Estimates

Effect	Point Estimate	95% Wald Confidence Limits	
gg	0.268	0.091	0.788
WBC	1.111	1.030	1.197

Association of Predicted Probabilities and Observed Responses

Percent Concordant	75.5	Somers' D	0.518
Percent Discordant	23.8	Gamma	0.521
Percent Tied	0.7	Tau-a	0.140
Pairs	3275	c	0.759

Logistic Regression Analysis

SAS Output for Example 12.4 (Continued)

The LOGISTIC Procedure

Analysis of Maximum Likelihood Estimates

Parameter	DF	Estimate	Standard Error	Wald Chi-Square	Pr > ChiSq
Intercept	1	1.2859	2.2545	0.3253	0.5684
HEM	1	-0.4272	0.2078	4.2294	0.0397
WBC	1	0.1135	0.0376	9.0948	0.0026

Association of Predicted Probabilities and Observed Responses

Percent Concordant	75.0	Somers' D	0.504
Percent Discordant	24.6	Gamma	0.506
Percent Tied	0.5	Tau-a	0.140
Pairs	3150	c	0.752

Interpretation of SAS Output for Example 12.4 (Continued)

The logistic regression model used to predict the development of CA based on treatment with gamma globulin alone (SAS Example 12.2) has $c = 0.68$. The addition of white blood count to the model increases c to 0.76 (SAS Example 12.3). Thus the model which includes white blood count fits better, as measured by c, the area under the ROC curve.

This model in SAS Example 12.4, which predicts the development of CA on the combination of white blood count and hemoglobin has $c = 0.75$.

Logistic Regression Analysis

12.6 Key Formulas

APPLICATION	NOTATION/FORMULA	DESCRIPTION
Logistic function	$$L = \beta_0 + \beta_1 X + \varepsilon$$ $$p = \frac{e^L}{(1 + e^L)}$$	$L = \ln$(odds of event Y), Y is the dischotomous event of interest X is the independent variable, β_0 is the intercept, β_1 is the coefficient, and ε is the random error
Logistic Regression Model	$$p = \frac{e^{\beta_0 + \beta_1 X + \varepsilon}}{(1 + e^{\beta_0 + \beta_1 X + \varepsilon})}$$	
Estimated Logistic Regression Equation	$$\hat{p} = \frac{e^{\hat{\beta}_0 + \hat{\beta}_1 X}}{(1 + e^{\hat{\beta}_0 + \hat{\beta}_1 X})}$$	
Estimate of the Odds given X	$$\hat{o} = e^{\hat{\beta}_0 + \hat{\beta}_1 X}$$	
Estimate of the Odds Ratio	$$\hat{OR} = e^{\hat{\beta}_1(x_a - x_b)}$$	Compares odds of the event for those with $X = x_a$ to those with $X = x_b$
95% Confidence Interval for OR	$$\left(e^{(\hat{\beta}_1 - 1.96 se(\hat{\beta}))(x_a - x_b)}, e^{(\hat{\beta}_1 + 1.96 se(\hat{\beta}))(x_a - x_b)} \right)$$	
Estimate of the OR in the special case where $x_a = 1$ and $x_b = 0$	$$\hat{OR} = e^{\hat{\beta}_1(1-0)} = e^{\hat{\beta}_1(1)} = e^{\hat{\beta}_1}$$	
Estimate of the Relative Risk	$$\hat{RR} = \hat{p}_a / \hat{p}_b = \frac{\left(\dfrac{e^{\hat{\beta}_0 + \hat{\beta}_1 x_a}}{1 + e^{\hat{\beta}_0 + \hat{\beta}_1 x_a}} \right)}{\left(\dfrac{e^{\hat{\beta}_0 + \hat{\beta}_1 x_b}}{1 + e^{\hat{\beta}_0 + \hat{\beta}_1 x_b}} \right)}$$	Compares the risks of the event for those with $X = x_a$ to those with $X = x_b$
Estimate of the RR in the special case where $x_a = 1$ and $x_b = 0$	$$\hat{RR} = \hat{p}_1 / \hat{p}_0 = \frac{\left(\dfrac{e^{\hat{\beta}_0 + \hat{\beta}_1}}{1 + e^{\hat{\beta}_0 + \hat{\beta}_1}} \right)}{\left(\dfrac{e^{\hat{\beta}_0}}{1 + e^{\hat{\beta}_0}} \right)}$$	

12.7 Statistical Computing

Following are the SAS programs which were used to perform logistic regression analysis. The SAS procedures used and brief descriptions of their use are noted in the header to each example. Notes are provided to the right of the SAS programs (*in italics*) for orientation purposes and are not part of the programs. In addition, there are blank lines in the programs that follow which are solely to accommodate the notes. Blank lines and spaces can be used throughout SAS programs to enhance readability. A summary of the SAS procedures used in the examples is provided at the end of this section.

Estimate Relative Risk and Odds Ratios in Logistic Regression Model

SAS EXAMPLE 12.1 Estimate relative risk and odds ratio of CA comparing patients with Kawasaki syndrome who have different white blood counts.

A trial of gamma globulin in the treatment of children with Kawasaki syndrome randomized approximately half of the patients to receive gamma globulin plus aspirin; the other half received standard treatment of aspirin. The outcome of interest was the development of coronary abnormalities (CA) within seven weeks of treatment. A possible explanation for the effect of treatment with gamma globulin on CA is that it reduces the elevated white blood count (WBC) early in the course of the disease, and that WBC is related to CA. This example addresses the relationship between WBC and CA.

Program Code

```
options ps=62 ls=80;

data in;
   input gg wbc ca;



cards;
0 9.4 0
0 20.4 1
0 8.7 0
.
1 5.7 0
run;
proc logistic descending;
  model ca=wbc
```

Formats the output page to 62 lines in length and 80 columns in width
Beginning of Data Step
*Inputs three variables, **gg** (noGG or GG), **wbc** (white blood count), and **ca** (CA or noCA)*
Beginning of Raw Data section.
*actual observations (value of **group**, **wbc** and **event** on each line)*

Procedure call. Proc logist provides estimates of the coefficients in a logistic regression model. The procedure is set up for outcomes with values of 1 or 2 instead of 0 and 1. The option "descending" ensures that outcomes of 1 are counted as events and outcomes of 0 are counted as non-events. Proc logist also provides estimates of the odds ratio and 95% confidence limits comparing the odds of the event (ca) for two people with a difference of one unit (in wbc).

```
run;
```
End of procedure section

Logistic Regression Analysis

Estimate Relative Risk and Odds Ratios in Logistic Regression Model

SAS EXAMPLE 12.2 Estimate relative risk and odds ratio of CA comparing GG to no GG.

A trial of gamma globulin in the treatment of children with Kawasaki syndrome randomized approximately half of the patients to receive gamma globulin plus aspirin; the other half received standard treatment of aspirin. The outcome of interest was the development of coronary abnormalities (CA) within seven weeks of treatment.

Program Code

options ps=62 ls=80;	*Formats the output page to 62 lines in length and 80 columns in width*
data in;	*Beginning of Data Step*
input gg wbc ca;	*Inputs three variables, **gg** (noGG or GG), **wbc** (white blood count), and **ca** (CA or noCA*
cards;	*Beginning of Raw Data section.*
0 9.4 0	*actual observations (value of **group**,*
0 20.4 1	***wbc** and **event** on each line)*
0 8.7 0	
.	
.	
.	
1 5.7 0	
run;	
proc logistic descending;	*Procedure call. Proc logist provides*
model ca=gg;	*estimates of the coefficients in a logistic regression model. The procedure is set up for outcomes with values of 1 or 2 instead of 0 and The option "descending" ensures that outcomes of 1 are counted as events and outcomes of 0 are counted as non-events. Proc logist also provides estimates of the odds ratio and 95% confidence limits comparing the odds of the event (ca) in those with gg=1 as compared to those with gg=0*
run;	*End of procedure section*

Estimate Adjusted Relative Risk and Odds Ratios in Logistic Regression Model

SAS EXAMPLE 12.3. Estimate the Odds Ratio comparing patients on gamma globulin to those on standard treatment with respect to CA, adjusted for WBC.

Suppose we want to use the sample described used in SAS Example 12.2 to assess the effect of treatment with gamma globulin on CA but we are concerned that white blood count is a potential confounder. We perform a multiple logistic regression in which the outcome is the dichotomous variable CA and the two independent variables are the dichotomous grouping variable, GG, and the continuous variable WBC.

Program Code

options ps=62 ls=80;	*Formats the output page to 62 lines in length and 80 columns in width*
data in;	*Beginning of Data Step*
input gg wbc ca;	*Inputs three variables, **gg** (noGG or GG), **wbc** (white blood count), and **ca** (CA or noCA)*
cards;	*Beginning of Raw Data section.*
0 9.4 0	*actual observations (value of **group**,*
0 20.4 1	***wbc** and **event** on each line)*
0 8.7 0	
.	
.	
.	
1 5.7 0	
run;	
proc logistic descending;	*Procedure call. Proc logist provides*
model ca=gg wbc;	*estimates of the coefficients in a logistic regression model. Proc Logist also provides estimates of the odds ratio and 95% confidence limits comparing the odds of the event (ca) in those with gg=1 as compared to those with gg=0, adjusted for wbc*
run;	*End of procedure section*

Estimate a Multiple Logistic Regression Model

SAS EXAMPLE 12.4. **Perform a multiple logistic regression, predicting the risk of developing CA given both hemoglobin and white blood count.**

Suppose we want to use the sample described used in SAS Example 12.2 to predict the probability of developing CA as a function of both white blood count and hemoglobin. We perform a multiple logistic regression in which the outcome is the dichotomous variable CA and the two independent variables are the continuous variables WBC and HEM.

Program Code	
options ps=62 ls=80;	*Formats the output page to 62 lines in length and 80 columns in width*
data in;	*Beginning of Data Step*
input ca wbc hem;	*Inputs three variables, **ca** (CA or noCA **wbc** (white blood count),and **hem** (hemoglobin).*
cards;	*Beginning of Raw Data section.*
0 9.4 11.0	*actual observations (value of **group**,*
1 20.4 10.1	***wbc** and **event** on each line)*
0 8.7 10.0	
.	
.	
0 5.7 10.2	
run;	
proc logistic descending;	*Procedure call. Proc logist provides*
model ca=wbc hem;	*estimates of the coefficients in a multiple logistic regression model.*
run;	*End of procedure section.*

Summary of SAS Procedures

The SAS Procedures for correlation and regression analysis are summarized below. Specific options can .be requested in these procedures to produce specific results. The options are shown in italics below. Users should refer to the examples in this section for complete descriptions of the procedure and specific options. A general description of the procedure and options is provided in the table below.

Procedure	Sample Procedure Call	Description
proc logistic	proc logistic *descending*; model y = x1 x2 x3 x4;	Estimates a logistic regression model relating dichotomous dependent variable Y to independent variables X1 X2 X3 X4. The descending option considers the response with the higher numerical code the outcome of interest (e..g, 1=outcome of interest, 0=comparison).

12.8 Problems

1. A placebo-controlled clinical trial of a new treatment for chronic pain was performed on 200 patients in a pain clinic. 100 patients were gven the new treatment and a pain questionnaire was administered to all 200 patients one week later. Based on the questionnaire, patients were classified as having moderate to severe pain (pain =1) or mild or no pain (pain=0).

   a) Use the following output from a logistic regression to estimate the probability of pain in patients on placebo (newtreat=0).
   b) Use the following output from a logistic regression to estimate the probability of pain in patients on the new treatment (newtreat=1).
   c) Estimate the relative risk of pain comparing those on the new treatment to those on placebo.
   d) Compare this to the estimated odds ratio provided in the output.
   e) Does the odds ratio provide a reasonable estimate of the relative risk?

The LOGISTIC Procedure

Analysis of Maximum Likelihood Estimates

Parameter	DF	Estimate	Standard Error	Wald Chi-Square	Pr > ChiSq
Intercept	1	-1.2657	0.2414	27.4888	<.0001
newtreat	1	-0.8251	0.4005	4.2435	0.0394

Odds Ratio Estimates

Effect	Point Estimate	95% Wald Confidence Limits	
newtreat	0.438	0.200	0.961

Logistic Regression Analysis

Problems 2-4.   Suppose you have 30-day follow-up data on 350 ischemic stroke patients and you want to investigate whether the risk of recurrent stroke and/or death (RD) depends on the type of stroke. You classify patients according to initial stroke type - having a cerebral embolism (CE=1) or not (CE=0) - and perform a series of logistic regression analyses.

2.   First, suppose you consider the crude relationship between CE and RD.   The following is output from a logistic regression.

The LOGISTIC Procedure

Analysis of Maximum Likelihood Estimates

Parameter	DF	Estimate	Standard Error	Wald Chi-Square	Pr > ChiSq
Intercept	1	-2.8034	0.5149	29.6390	<.0001
ce	1	1.8651	0.6479	8.2874	0.0040

Odds Ratio Estimates

Effect	Point Estimate	95% Wald Confidence Limits	
ce	6.457	1.814	22.986

a)   Use the output to estimate the odds ratio, provide a 95% confidence interval, and test H_0: OR = 1.
b)   What is the risk of RD for CE patients?
c)   What is the risk of RD for non-CE patients?
d)   Is the risk of RD the same for CE and non-CE patients? Estimate the relative risk of RD.

3.   Next, consider the relationship between age and RD.   The following is output from a logistic regression.

The LOGISTIC Procedure

Analysis of Maximum Likelihood Estimates

Parameter	DF	Estimate	Standard Error	Wald Chi-Square	Pr > ChiSq
Intercept	1	-8.0798	8.1749	0.9769	0.3230
age	1	0.0869	0.1149	0.5714	0.4497

Logistic Regression Analysis

Odds Ratio Estimates

Effect	Point Estimate	95% Wald Confidence Limits	
age	1.091	0.871	1.366

a) What is the odds ratio comparing the odds of RD in patients whose ages are one year apart? Provide a 95% confidence interval and test H_0: OR = 1.
b) What is the risk of RD for a patient aged 65?
c) What is the risk of RD for a patient aged 66?
d) Estimate the relative risk of RD comparing patients who are 65 and 66.
e) How does this compare with the odds ratio given in the output?
f) What is the risk of RD for a patient aged 75?
g) Estimate the relative risk of RD comparing patients who are 75 to those who are 65.

4. Finally, consider the relationship between CE and RD, adjusted for age.

The LOGISTIC Procedure

Analysis of Maximum Likelihood Estimates

Parameter	DF	Estimate	Standard Error	Wald Chi-Square	Pr > ChiSq
Intercept	1	-15.3151	9.4983	2.5999	0.1069
ce	1	2.0684	0.6822	9.1919	0.0024
age	1	0.1751	0.1318	1.7654	0.1840

Odds Ratio Estimates

Effect	Point Estimate	95% Wald Confidence Limits	
ce	7.912	2.078	30.129
age	1.191	0.920	1.542

a) Use the output to estimate the odds ratio comparing the odds of RD in patients with CE to those without CE, adjusted for age. Provide a 95% confidence interval, and test H_0: OR = 1.
b) How do these adjusted results differ from the crude results in problem 2a?

Logistic Regression Analysis

5.   A clinical trial was designed to investigate the effect of a new treatment on the course of disease among pediatric patients with a particular infectious disease. Fifty boys and fifty girls were randomized to receive the new treatment (newtreat=1) or the standard treatment (newtreat=0), and the outcome was hospitalization (hospital =1 or 0).

a)   What is the odds ratio comparing patients treated with the new treatment to patients receiving standard care with respect to the odds of requiring hospitalization? Test H_0: OR = 1.

The LOGISTIC Procedure

Analysis of Maximum Likelihood Estimates

Parameter	DF	Estimate	Standard Error	Wald Chi-Square	Pr > ChiSq
Intercept	1	-0.5754	0.2083	7.6273	0.0057
newtreat	1	-0.9410	0.3334	7.9660	0.0048

Odds Ratio Estimates

Effect	Point Estimate	95% Wald Confidence Limits	
newtreat	0.390	0.203	0.750

b)   What is the odds ratio comparing boys treated with the new treatment to boys receiving standard care with respect to the odds of requiring hospitalization? Test H_0: OR = 1.

------------------------------ gender=boy ----------------------------------

Analysis of Maximum Likelihood Estimates

Parameter	DF	Estimate	Standard Error	Wald Chi-Square	Pr > ChiSq
Intercept	1	-0.0800	0.2831	0.0800	0.7774
newtreat	1	-1.0726	0.4356	6.0626	0.0138

Odds Ratio Estimates

Effect	Point Estimate	95% Wald Confidence Limits	
newtreat	0.342	0.146	0.803

Logistic Regression Analysis

c) What is the odds ratio comparing girls treated with the new treatment to girls receiving standard care with respect to the odds of requiring hospitalization? Test H_0: OR = 1.

```
------------------------------ gender=girl ------------------------------
             Analysis of Maximum Likelihood Estimates

                                   Standard          Wald
Parameter    DF     Estimate          Error    Chi-Square    Pr > ChiSq

Intercept     1      -1.1527         0.3311       12.1175        0.0005
newtreat      1      -0.8397         0.5468        2.3581        0.1246


                       Odds Ratio Estimates

                      Point           95% Wald
         Effect    Estimate    Confidence Limits

         newtreat     0.432      0.148       1.261
```

d) Does the new treatment have the same effect on girls and boys?

Logistic Regression Analysis

SAS Problems: Use SAS to solve the following problems.

1. A clinical trial was conducted to compare a new therapy for HIV to a standard therapy with respect to CD4 cell counts and incident opportunistic infections. A total of 30 subjects with diagnosed HIV were enrolled in the trial, randomized to receive standard or new therapy, and studied for three months. The following are data on age and OI (presence of opportunistic infection during the study period).

Standard Therapy Group

Age	23	34	25	29	33	35	31	26	29	20	31	37	28	45	49
OI	0	1	0	0	0	1	1	1	1	1	1	1	0	1	1

New Therapy Group

Age	30	32	40	37	38	28	47	49	50	43	41	38	37	36	34
OI	0	0	0	0	0	0	0	1	1	0	1	1	0	0	0

a) Use logistic regression to estimate the odds ratio comparing the therapies with respect to risk of opportunistic infection.
b) Provide a 95% confidence interval for the odds ratio.
c) Test the appropriate hypothesis.
d) Use your output to provide estimates of the risk of opportunistic infection in each of the therapy groups.
e) Estimate the relative risk of opportunistic infection comparing the new therapy to the standard therapy.
f) Perform a separate logistic regression to estimate the odds ratio comparing the effect of therapy on the risk of opportunistic infection, adjusted for age. Test the appropriate hypothesis.

Introductory Applied Biostatistics

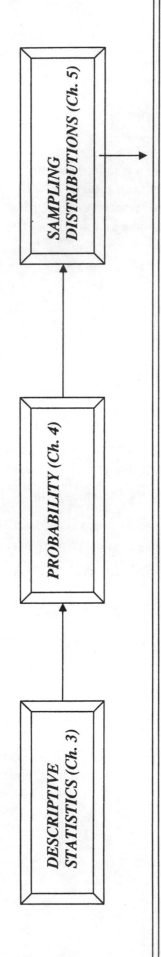

DESCRIPTIVE STATISTICS (Ch. 3) → PROBABILITY (Ch. 4) → SAMPLING DISTRIBUTIONS (Ch. 5)

STATISTICAL INFERENCE (Chapters 6-14)

Outcome Variable	Grouping Variable(s)/ Predictor(s)	Analysis	Chapter(s)
Continuous	-	**Estimate μ, Compare μ to Known, Historical Value**	**6/13**
Continuous	**Dichotomous (2 Groups)**	**Compare Independent Means (Estimate/Test ($\mu_1-\mu_2$)) or the Mean Difference(μ_d)**	**7/13**
Continuous	**Discrete (> 2 Groups)**	**Test the Equality of K Means using Analysis of Variance ($\mu_1=\mu_2=\ldots\mu_k$)**	**10/13**
Continuous	**Continuous**	**Estimate Correlation or Determine Regression Equation**	**11/13**
Continuous	Several Continuous or Dichotomous	Multiple Linear Regression Analysis	11
Dichotomous	-	Estimate p, Compare p to Known, Historical Value	8
Dichotomous	Dichotomous (2 Groups)	Compare Independent Proportions (Estimate/Test (p_1-p_2))	8/9
Dichotomous	Discrete (>2 Groups)	Test the Equality of k Proportions (Chi-Square Test)	8
Dichotomous	Several Continuous or Dichotomous	Multiple Logistic Regression Analysis	12
Discrete	Discrete	Compare Distributions Among k Populations (Ch-Square Test)	8
Time to Event	Several Continuous or Dichotomous	Survival Analysis	14

CHAPTER 13: Nonparametric Tests

13.1   Introduction

In Chapters 6, 7 and 9 we presented techniques for tests of hypothesis concerning one, two (independent and dependent, matched or paired) and more than 2 population means, respectively. Valid application of these tests and procedures requires specific assumptions. These tests assume that the characteristic under investigation is approximately normally distributed or rely on large samples for the application of the Central Limit Theorem. If the samples are small and the analytic variable is clearly not normally distributed we violate the assumptions required for valid application of the procedures. In such cases, alternative methods are required. The techniques we presented in Chapters 6, 7 and 9 are called parametric procedures as they are based on assumptions regarding the distributional form of the analytic variable (e.g., that the analytic variable follows a normal distribution). Methods which do not require such assumptions are called nonparametric procedures. In this chapter we present four nonparametric procedures analogous to the two dependent samples test, the two independent samples test, the k>2 independent samples test and correlation analysis. The hypotheses tested with nonparametric procedures are more general than hypotheses we tested with parametric procedures. For example, the parametric test for two independent means is of the form H_0: $\mu_1=\mu_2$. The nonparametric analog tests for equality of medians in the two distributions.

In Sections 13.2 and 13.3 we present nonparametric tests for two dependent samples. In Section 13.4 we present a technique for the two independent samples test. In Section 13.5 we present a technique for the k>2 independent samples test and in Section 13.6 we present a nonparametric approach to correlation analysis. We summarize key formulas in Section 13.7 and in Section 13.8 we display the statistical computing programs used to generate the nonparametric analyses presented in this chapter.

13.2   The Sign Test (Two Dependent Samples Test)

Recall the two dependent (also called matched or paired) samples procedures we presented in Chapter 7. In the examples we discussed, a single random sample of subjects was selected from the population of interest, and two measurements were taken on each subject. Because the measurements were dependent, we calculated differences between measurements and performed a test on the mean difference (eg., H_0: $\mu_d=0$). When the sample is small and/or if the differences are not normally distributed, we consider a nonparametric test. One popular nonparametric test for two dependent samples is the Sign test. The null hypothesis in the Sign test is that the median difference is zero. We illustrate the test below using an example.

Example 13.1. Suppose we are investigating a new drug hypothesized to lower systolic blood pressure. We select a random sample of subjects and record their resting systolic blood pressure at baseline (i.e., prior to the investigation). Each subject is then given the new drug and resting systolic blood pressure is measured again after 4 weeks of drug treatment. The data are shown below. Based on the data, does it appear that the new drug lowers systolic blood pressure?

Subject Identification Number	Baseline Systolic Blood Pressure	Post-Treatment Systolic Blood Pressure
1	166	138
2	135	120
3	189	176
4	180	180
5	156	160
6	142	150
7	176	152
8	156	140
9	164	160
10	142	130

In the Sign Test, similar to the parametric test we presented in Chapter 7, we focus on differences between measurements. In this application, we are investigating whether there is a significant reduction in systolic blood pressure following drug treatment. In the following table we display the differences between systolic blood pressures (baseline – post-treatment blood pressures). In the rightmost column of the table below we record the sign of the difference (+ or –, or 0 if there is no difference).

Subject Identification Number	Baseline Systolic Blood Pressure	Post-Treatment Systolic Blood Pressure	Difference (Baseline – Post-Treatment)	Sign of Difference
1	166	138	28	+
2	135	120	15	+
3	189	176	13	+
4	180	180	0	0
5	156	160	-4	-
6	142	150	-8	-
7	176	152	24	+
8	156	140	16	+
9	164	160	4	+
10	142	130	12	+

The Sign Test is based on the binomial distribution. There are a total of 10 difference scores (one for each subject in the sample). If the drug has no effect on systolic blood pressure, we expect the post-treatment systolic blood pressures to be the same as the baseline systolic blood pressures. Because there are many factors which affect blood pressure, expecting identical results is unrealistic. In terms of differences, if the drug has no effect, we expect some subjects' post-treatment systolic blood pressures to be higher than their baseline pressures (negative differences) and some subjects' post-treatment systolic blood pressures to be lower than their baseline pressures (positive differences). If the drug has no effect, we expect these differences to be small, and with respect to the signs of the differences, we expect to see about 5 + and 5 − signs (or 50% in each category).

In Example 13.1 we observed 7 (out of 10) + signs. Is a total of 7 + signs indicative of a drug effect (here a + sign indicates a difference in the hypothesized direction, i.e., a lowering of systolic blood pressure)? The exact significance of the test is determined from the binomial distribution (discussed in Chapter 4). The null hypothesis is that, for each person, the probability of a "+" sign is 0.5. We therefore consider a binomial distribution with n=10 (because there are 10 subjects, or trials) and p=0.5. The following values are found in Table 1 in the Appendix.

Number of Successes, x	$P(X=x \mid n=10, p=0.5)$
0	0.0010
1	0.0098
2	0.0439
3	0.1172
4	0.2051
5	0.2461
6	0.2051
7	0.1172
8	0.0439
9	0.0098
10	0.0010

The question of interest is, *How likely is it to observe as many as 7 or more successes out of 10 when the probability of success is 0.5?* The p-value for the test statistic (7 successes or 7 + signs out of 10) is the probability of observing 7 or more successes. (Recall the p-value is defined as the probability of observing a test statistic as or more extreme.) In Example 13.1, the p-value is $P(X \geq 7)$ = $P(X=7)+P(X=8)+P(X=9)+P(X=10) = 0.1172+0.0439+0.0098+0.0010 = 0.1719$. We reject H_0 if the p-value is small (i.e., 0.05 or less). In Example 13.1, we do not reject H_0. We do not have significant evidence, $p=0.1719$, to show that the drug lowers systolic blood pressure.

There are several options for handling observations in which the difference is zero. We ignored the zero (i.e., did not assign a sign). Another alternative is to count a zero difference as 0.5+ and 0.5-. Had we used this strategy we would have counted 7.5 + signs in Example 13.1. The p-value for the test would then have been $P(X \geq 7.5) = P(X=8)+P(X=9)+P(X=10) = 0.0439+0.0098+0.0010 = 0.0547$, which is marginally significant.

Note that we performed a one sided test with the alternative hypothesis that the drug lowers systolic blood pressure. The p-value corresponding to a two sided test with alternative hypothesis that the drug either raises or lowers systolic blood pressure is simply twice the one sided p-value. In our example, the two sided p-value is $P(X \geq 7 \text{ or } X \leq 3) = 2 P(X \geq 7) = 2 (0.1719) = 0.3438$.

The analysis of the data presented in Example 13.1 was based solely on the signs (+ or -) of the differences in blood pressures from baseline to post-treatment. In the parametric test, we also computed differences but focused on the magnitude of those differences. Recall, our test statistic was based on $\overline{X}_d$, the mean of the difference scores. In the Sign Test we do not capture the magnitude of the differences, only the direction. For example if a particular patient has a baseline systolic blood pressure of 150 and a post-treatment blood pressure of 149, we count that observation as evidence in favor of the alternative hypothesis (lowering of blood pressure) because the difference is positive (+). Suppose a second patient has a baseline systolic blood pressure of 150 and a post-treatment blood pressure of 129, we again count this observation as evidence in favor of the alternative hypothesis. In the Sign test, there is no distinction between these subjects, when in fact the second has a much more substantial reduction in blood pressure. Because we are basing the Sign test on limited information (only the sign of the difference in measurements), this nonparametric test often has lower power than competing procedures. A second nonparametric procedure for two dependent samples is described below. The next test incorporates the magnitude of the differences in measurements.

13.3 The Wilcoxon Signed-Rank Test (Two Dependent Samples Test)

The Wilcoxon Signed-Rank test is a second nonparametric alternative for two dependent, matched or paired samples. To perform the Wilcoxon Signed-Rank test we compute differences in measurements as we did to perform the Sign test. However, here we also we take into account the relative magnitude of these differences. The null hypothesis is that the median difference is zero. We illustrate the test below using the data we presented in Example 13.1.

Example 13.2. Consider the data presented in Example 13.1. Again, the question of interest is whether there is significant evidence that the new drug lowers systolic blood pressure. The data are shown below along with the difference scores computed by subtracting the post-treatment systolic blood pressures from the baseline systolic blood pressures.

Subject Identification Number	Baseline Systolic Blood Pressure	Post-Treatment Systolic Blood Pressure	Difference (Baseline – Post-Treatment)
1	166	138	28
2	135	120	15
3	189	176	13
4	180	180	0
5	156	160	-4
6	142	150	-8
7	176	152	24
8	156	140	16
9	164	160	4
10	142	130	12

In the Wilcoxon Signed-Rank test we assign ranks to the *absolute values* of the difference scores (see the third column of the table below). We assign a 1 to the smallest absolute difference, 2 to the next smallest and so on, up to n. For now, we will ignore the case with no difference, although there are alternative methods for handling these cases. If there are ties in the absolute values of the differences, then the mean rank is assigned to both. For example, subjects 5 and 9 had absolute differences of 4 units. These are the first and second smallest differences, so we assign a rank of 1.5 to each. The next (third) smallest absolute difference is measured in subject 6 who is assigned a rank of 3. The ranking continues until all non-zero differences are ranked. Once the ranks are assigned to the absolute differences, we then re-attach the signs of the differences (+ or -) to the ranks (see the rightmost column of the table below).

Subject Identification Number	Difference (Baseline – Post-Treatment)	Ranks of Absolute Values of Differences	Signed Ranks
1	28	9	+9
2	15	6	+6
3	13	5	+5
4	0	-	-
5	-4	1.5	-1.5
6	-8	3	-3
7	24	8	+8
8	16	7	+7
9	4	1.5	+1.5
10	12	4	+4

If the drug has no effect on systolic blood pressures, we expect about half of the post-treatment systolic blood pressures to be higher than the baseline systolic blood pressures and half of the post-treatment systolic blood pressures to be lower than the baseline systolic blood pressures. In addition, we expect the magnitudes of the increases to be about the same as the magnitudes of the decreases. In terms of the signed ranks, if the drug has no effect, we expect the sum of the positive ranks to be approximately equal to the sum of the negative ranks.

The test statistic in the Wilcoxon Signed-Rank test is the sum of the positive ranks, called T. If all the differences are negative, then all the signs will be negative so all the signed ranks will be negative and T will equal 0. The other extreme is all positive differences, which will yield all positive signed ranks and T will therefore be the sum $1+2+3+4+...+n = n(n+1)/2$ (where n represents the number of ranked observations, i.e., the number of observations with non-zero differences). Thus the sum of positive ranks, T, ranges from 0 to $n(n+1)/2$. The median value is $n(n+1)/4$. In Example 13.2, the observed sum of the positive ranks is T=40.5. The maximum possible value of the sum of positive ranks is 9(9+1)/2 = 45 and the median is 9(9+1)/4=22.5. The observed test statistic, T=40.5, is very close to the theoretical maximum (i.e., it falls in the tail of the distribution). To draw a conclusion in the test, we compare the observed test statistic to an appropriate critical value. Because the observed test statistic is large, we would expect a very small p-value and therefore would reject H_0. SAS runs the Wilcoxon Signed-Rank test and produces a p-value based on a normal approximation to the distribution of T. Specifically, SAS converts the sum of positive ranks, T, to a Z score, and then produces a p-value from the standard normal distribution. The Z score (13.1) is computed by subtracting the median n(n+1)/4 and dividing by

the standard error which is $\sqrt{\dfrac{n(n+1)(2n+1)}{24}}$.

$$Z = \frac{T - \dfrac{n(n+1)}{4}}{\sqrt{\dfrac{n(n+1)(2n+1)}{24}}} \tag{13.1}$$

In Example 13.2, $Z = \dfrac{40.5 - \dfrac{9(9+1)}{4}}{\sqrt{\dfrac{9(9+1)(2(9)+1)}{24}}} = \dfrac{18}{8.44} = 2.13.$

The one sided p-value is $P(Z \geq 2.13) = 1 - 0.9834 = 0.0166$. Based on the Wilcoxon Signed-Rank test results, p=0.0166 < 0.05, we reject H_0: The medians are equal in favor of the alternative H_1: The median systolic blood pressure is lower post-treatment as compared to baseline.

SAS Example 13.2. SAS performs the Wilcoxon Signed-Rank test in its Proc Univariate. We used Proc Univariate in Chapter 3 to generate summary statistics for continuous variables. The following output was generated by SAS on the difference variable (i.e., differences between baseline and post-treatment systolic blood pressures). A brief interpretation appears after the output.

SAS Output for Example 13.2

The UNIVARIATE Procedure
Variable: diff
Moments

N	10	Sum Weights	10
Mean	10	Sum Observations	100
Std Deviation	11.785113	Variance	138.888889
Skewness	-0.0712764	Kurtosis	-0.9438912
Uncorrected SS	2250	Corrected SS	1250
Coeff Variation	117.85113	Std Error Mean	3.72677996

Basic Statistical Measures

Location		Variability	
Mean	10.00000	Std Deviation	11.78511
Median	12.50000	Variance	138.88889
Mode	.	Range	36.00000
		Interquartile Range	16.00000

Tests for Location: Mu0=0

Test	-Statistic-		-----p Value------	
Student's t	t	2.683282	Pr > |t|	0.0251
Sign	M	2.5	Pr >= |M|	0.1797
Signed Rank	S	18	Pr >= |S|	0.0313

Tests for Normality

Test	--Statistic---		-----p Value------		
Shapiro-Wilk	W	0.962454	Pr < W		0.8135
Kolmogorov-Smirnov	D	0.167379	Pr > D		>0.1500
Cramer-von Mises	W-Sq	0.035094	Pr > W-Sq		>0.2500
Anderson-Darling	A-Sq	0.209068	Pr > A-Sq		>0.2500

Quantiles (Definition 5)

Quantile	Estimate
100% Max	28.0
99%	28.0
95%	28.0
90%	26.0
75% Q3	16.0
50% Median	12.5
25% Q1	0.0

Nonparametric Tests

```
The UNIVARIATE Procedure
     Variable:  diff
Quantiles (Definition 5)
   Quantile        Estimate
   10%                 -6.0
   5%                  -8.0
   1%                  -8.0
   0% Min              -8.0


       Extreme Observations
----Lowest----          ----Highest---
Value        Obs        Value        Obs
   -8          6           13          3
   -4          5           15          2
    0          4           16          8
    4          9           24          7
   12         10           28          1
```

Interpretation of SAS Output for Example 13.2

The output for the Wilcoxon Signed-Rank test is in the third section of the output. SAS produces the numerator of the test statistic in (13.1). SAS gives "Signed Rank S = 18" which is computed by subtracting the median=22.5 from T=40.5 (S = 40.5-22.5 = 18). SAS then gives a two sided p-value "Pr>=|S| = 0.0313" which is based on the normal approximation. (SAS applies a correction factor to formula (13.1) – See also SAS Example 13.3.) If a one sided test is desired (as is the case in Example 13.2), we divide p=0.0313 / 2 = 0.016. Based on the observed p-value, p=0.016 (< 0.05), we reject H_0: The medians are equal in favor of the alternative H_1: The median systolic blood pressure is lower post-treatment as compared to baseline.

The Signed-Rank test uses more information in the data than the Sign test in that it incorporates the *relative* magnitude of the values through ranks. Ranks are particularly useful in the presence of outliers. However, ranks do not capture the *absolute* magnitude of the differences (in the two dependent samples case). Note that the Signed-Rank test rejected the two-sided null hypothesis, with p = 0.031, while the Sign test did not reject the same null hypothesis, with p = 0.344.

In the next section (just below the results of the Signed-Rank test), SAS produces several tests for normality, we use the Shapiro-Wilk test. In this example, SAS is testing whether the differences follow a normal distribution. The null hypothesis is H_0: Differences follow a normal distribution. The test statistic is W = 0.962454 and the p-value = 0.8135. We would not reject H_0 based on the observed p-value, therefore we do not have significant evidence to show that the data do not follow a normal distribution. This test can be useful to assess whether a parametric test (which assumes that the analytic variable follows a normal distribution) or a nonparametric test should be applied.

13.4 The Wilcoxon Rank Sum Test (Two Independent Samples Test)

We now present a nonparametric test for two independent samples. In the these applications, we have two independent populations which are defined based on a specific attribute of the subjects under study (e.g., gender) or based on the study design (e.g., some patients are assigned to receive Drug A while others receive Drug B). The parametric procedure tests the equality of population means and assumes large samples ($n_i > 30$, i=1, 2) or, if the sample sizes are small, that the analytic variable under investigation is approximately normally distributed. If these assumptions are not met, a nonparametric test might be appropriate. A popular nonparametric test comparing two independent samples is the Wilcoxon Rank Sum test in which the null hypothesis is the equality of medians. This test is equivalent to the nonparametric Mann-Whitney U test. We illustrate the test below using an example.

Example 13.3. Suppose we wish to compare two competing treatments for the management of adult onset diabetes. A total of 8 subjects agree to participate in the investigation and subjects are randomly assigned to one of the competing treatments. After following the prescribed treatment regimen for 6 weeks, we assess the patients' self-reported health status. Patients are asked to rate various aspects of their current health using the following response options: Excellent, Very Good, Good, Fair or Poor. Each attribute rating is ordinal in nature. Investigators often assign numerical values to ordinal responses and analyze these values as if they were continuous. The assignment of numerical values can be made in a variety of ways. Suppose we assign increasing numerical values to the response options as follows: Poor=0, Fair=5, Good=10, Very Good=15 and Excellent=20. Here, higher values indicate better self-reported health status. Each subject's health status score is the mean of his/her responses over all attributes. The test of interest is a two sided test (i.e., Are self reported health status scores different between treatment groups?). The data are shown below.

Self-Reported Health Status	
Treatment 1	Treatment 2
0	8
7	10
11	12
16	15

Nonparametric Tests

In this application we have small samples and the analytic variable is ordinal. Therefore, we consider a nonparametric test. To perform the Wilcoxon Rank Sum test we pool the data from the two groups, and order the values from lowest to highest (i.e., from lowest self-reported health status to highest). We then assign ranks to the values in increasing order. We assign a 1 to the lowest value, 2 to the next lowest and so on, up to $N=n_1+n_2$. When there are ties, the mean ranks are assigned (similar to the procedure we followed with the Signed-Rank test). The table below contains the ranks of the values:

Ranks	
Treatment 1	Treatment 2
1	3
2	4
5	6
8	7

If there is no difference between treatments, we expect to see some low ranks and some high ranks in each group. If there is a difference between treatments, we expect to see clustering of lower ranks in one group and higher ranks in the other.

The Wilcoxon Rank Sum test statistic is S, the smaller of the sums of the ranks in the groups. The sum of all ranks is $N(N+1)/2$. For Example 13.3, the sum of all ranks is $8(8+1)/2 = 36$. If there is no difference between treatments we expect the sums of the ranks to be about 18 in each group. In the most extreme situation (for N=8), the four smallest values would fall in one group producing ranks in that group of 1, 2, 3, and 4, and the smaller sum of ranks would be equal to S=10 (S=1+2+3+4). In Example 13.3, the sum of ranks in Treatment 1 is 16 and the sum of ranks in Treatment 2 is 20. The smaller sum is S=16. Is that significantly different from the expected sum (if there were no difference between treatments) of 18? To draw a conclusion in the test, we compare the observed sum to an appropriate critical value. Again, we will use SAS to run the test. SAS runs the Wilcoxon Rank Sum test and produces a p-value based on a normal approximation. Specifically, SAS converts the smaller sum of ranks, S, to a Z score, and then produces a p-value from the standard normal distribution. The Z score (13.2) is computed by subtracting the mean $n_1(n_1+n_2+1)/2$ and dividing by the standard error which is $\sqrt{\dfrac{n_1 n_2(n_1 + n_2 +1)}{12}}$.

$$Z = \frac{S - \dfrac{n_1(n_1 + n_2 +1)}{2}}{\sqrt{\dfrac{n_1 n_2(n_1 + n_2 +1)}{12}}} \qquad (13.2)$$

Nonparametric Tests

In Example 13.3, $Z = \dfrac{16 - \dfrac{4\ (4+4+1)}{2}}{\sqrt{\dfrac{(4)4(4+4+1)}{12}}} = \dfrac{-2}{3.46} = -0.58.$

The one sided p-value is $P(Z \le -0.58) = 0.2810$. The two sided p-value = $2(0.2810) = 0.5620$. Based on the Wilcoxon Rank Sum test, we would not reject H_0. We do not have significant evidence to show a difference between treatments with respect to self-reported health status because p=0.5620>0.05.

SAS Example 13.3. SAS performs the Wilcoxon Rank Sum test in its Proc Nparlway. The following output was generated by SAS. A brief interpretation appears after the output.

SAS Output for Example 13.3

```
                    The NPAR1WAY Procedure
           Wilcoxon Scores (Rank Sums) for Variable hlthstat
                    Classified by Variable trt
                    Sum of      Expected      Std Dev        Mean
     trt     N      Scores      Under H0      Under H0       Score
     ------------------------------------------------------------------
     1       4      16.0        18.0          3.464102       4.0
     2       4      20.0        18.0          3.464102       5.0

                    Wilcoxon Two-Sample Test
              Statistic              16.0000
              Normal Approximation
              Z                      -0.4330
              One-Sided Pr <  Z       0.3325
              Two-Sided Pr > |Z|      0.6650

              t Approximation
              One-Sided Pr <  Z       0.3390
              Two-Sided Pr > |Z|      0.6780

          Z includes a continuity correction of 0.5.

                    Kruskal-Wallis Test
              Chi-Square              0.3333
              DF                           1
              Pr > Chi-Square         0.5637
```

Nonparametric Tests

Interpretation of SAS Output for Example 13.3

For the Wilcoxon Rank Sum test, SAS produces the sum of the ranks (labeled "Sum of Scores") in each treatment group. SAS then gives the expected sums, the standard deviation and mean (assuming no difference between treatments). SAS then shows the sum of the smaller ranks (S=16) and the corresponding Z score. SAS applies a correction of ½ to formula (13.2) in the standardization as follows: $Z = \dfrac{S - \dfrac{n_1(n_1 + n_2 + 1)}{2} - \dfrac{1}{2}}{\sqrt{\dfrac{n_1 n_2 (n_1 + n_2 + 1)}{12}}}$. For Example 13.3, SAS computes Z = - 0.4330 (with the correction), and gives a two sided p-value of 0.6650. Based on the Wilcoxon Rank Sum test, we would not reject H_0 because p=0.6650 > 0.05 We do not have significant evidence to show a difference between treatments with respect to self-reported health status.

The Rank Sum test provides a comparison of the relative magnitude of values in the two groups but does not use the actual observed values – as does the two independent samples test. Just below the results of the Rank Sum test, SAS provides an approximate p-value for the two independent samples t test for means (Proc Ttest in SAS). For Example 13.3, SAS gives "t Approximation Pr > |Z| = 0.6780." If we had run a two independent samples test for equality of means we would not have rejected H_0: $\mu_1 = \mu_2$ because p=0.6780 > 0.05.

13.5   The Kruskal-Wallis Test (k Independent Samples Test)

We now present a nonparametric procedure for the k independent samples test. This procedure is the nonparametric analog to analysis of variance (ANOVA) we discussed in Chapter 9. A popular nonparametric test for the k independent samples test is the Kruskal-Wallis test. Similar to the Wilcoxon Signed Rank and Rank Sum tests we presented in previous sections, the Kruskal-Wallis test is based on assigning ranks to the observed values and then comparing the observed sums of ranks to what would be expected if there were no difference among groups. The computations involved in the Kruskal-Wallis test are similar to those in the Wilcoxon Ranked Sum test. The computations are complicated by the fact that there are more groups involved. We illustrate the application of the Kruskal-Wallis test using SAS applied to the sample data given in the next example.

Example 13.4. Suppose we wish to compare four treatments for seasonal allergies. A total of 20 subjects agree to participate in the investigation and subjects are randomly assigned to one of the four competing treatments. After following the prescribed treatment regimen for 2 weeks, we assess the subjects' status. The outcome variable in this application is an index score based on three distinct symptoms. At the end of treatment, subjects are asked if they are currently experiencing any (or all) of the following symptoms: Scratchy throat, Itchy eyes, Runny nose. Each subject responds "yes" or "no" to each of the three symptoms. This index score is the sum of affirmative responses. The range of scores is 0 (no symptoms) to 3 (all three symptoms). The data are shown below.

Treatment 1	Treatment 2	Treatment 3	Treatment 4
0	2	0	2
0	2	0	3
1	3	1	2
2	3	1	3
3	2	1	3

Is there is a significant difference among the treatments with respect to symptom scores? Here again we have small samples and an analytic variable with limited response options. Therefore we consider a nonparametric test.

SAS Example 13.4. SAS performs the Kruskal-Wallis test in its Proc Npar1way. The following output was generated by SAS. A brief interpretation appears after the output.

SAS Output for Example 13.4

```
                 The NPAR1WAY Procedure
      Wilcoxon Scores (Rank Sums) for Variable symptoms
                 Classified by Variable trt
                  Sum of      Expected      Std Dev        Mean
  trt    N        Scores      Under H0      Under H0       Score
  ----------------------------------------------------------------
  1      5        40.50       52.50         11.062026      8.10
  2      5        69.50       52.50         11.062026      13.90
  3      5        24.50       52.50         11.062026      4.90
  4      5        75.50       52.50         11.062026      15.10

          Average scores were used for ties.
               Kruskal-Wallis Test
            Chi-Square          10.7013
            DF                        3
            Pr > Chi-Square      0.0135
```

For the Kruskal-Wallis test, SAS produces the sum of the ranks (labeled "Sum of Scores") in each treatment group. SAS then gives the expected sums, the standard deviation and mean (assuming no difference among treatments). SAS then gives a test statistic, in this case a chi-square statistic, and a corresponding p-value. For Example 13.4, SAS computes $\chi^2 = 10.7013$, which has 3 degrees of freedom (number of groups-1), and the p-value=0.0135. Based on the Kruskal-Wallis test, we reject H_0 because p=0.0135 < 0.05. We have significant evidence of a difference among the treatments with respect to median symptom scores.

13.6   Spearman Correlation (Correlation Between Variables)

In Chapter 10 we presented a formula for the sample correlation coefficient, r ($r = \dfrac{Cov(X, Y)}{\sqrt{Var(X)Var(Y)}}$). The correlation coefficient quantifies the nature and strength of the linear association between X and Y. Extreme values can have a substantial impact on the value of the sample correlation coefficient. When data are subject to extremes, an alternative measure of correlation between variables is based on ranks. The correlation based on ranks is called the Spearman correlation. We illustrate the computation of the Spearman correlation below using an example.

Example 13.5. Suppose we wish to assess the relationship between the number of cigarettes smoked per day and the number of hours of aerobic exercise per week. A total of 12 subjects agree to participate in the investigation and we measure the typical number of cigarettes smoked per day and the number of hours of exercise in a typical week on each subject. The data are shown below.

Raw Scores	
X=Number of Cigarettes Per Day	Y=Number of Hours of Exercise Per Week
20	0
0	0
20	1
10	2
5	3
4	5
3	5
5	6
0	3
0	4
0	7
0	8

The first step in the analysis is to replace the raw X and Y scores with ranks. The ranks are assigned for each variable, considered separately. The following table contains the ranks for each variable. In this example, there are several instances of ties in raw scores. When scores are tied, the mean rank is applied to each of the tied values.

Ranks	
Rx=Rank of Number of Cigarettes Per Day	Ry=Rank of Number of Hours of Exercise Per Week
11.5	1.5
3	1.5
11.5	3
10	4
8.5	5.5
7	8.5
6	8.5
8.5	10
3	5.5
3	7
3	11
3	12

The Spearman correlation is computed using the same formula we used in Chapter 10 based on the ranks. The formula is given below, where Rx and Ry are the ranks of X and Y, respectively:

$$r_s = \frac{Cov(Rx, Ry)}{\sqrt{Var(Rx)Var(Ry)}} \tag{13.3}$$

The now calculate the components for r_s.

We first compute the variance of Rx: $Var(Rx) = \frac{\Sigma(Rx - \overline{Rx})^2}{n-1}$. The following table summarizes the computations:

Rx=Rank of Number of Cigarettes Per Day	$(Rx - \overline{R}x)$	$(Rx - \overline{R}x)^2$
11.5	5.0	25.00
3	-3.5	12.25
11.5	5.0	25.00
10	3.5	12.25
8.5	2.0	4.00
7	0.5	0.25
6	-0.5	0.25
8.5	2.0	4.00
3	-3.5	12.25
3	-3.5	12.25
3	-3.5	12.25
3	-3.5	12.25
$\overline{R}x = 6.5$	0	132.5

So, $Var(Rx) = \dfrac{\Sigma (Rx - \overline{Rx})^2}{n-1} = 132.5/11 = 12.05$.

We now compute the variance of Ry: $Var(Ry) = \dfrac{\Sigma (Ry - \overline{Ry})^2}{n-1}$. The following table summarizes the computations:

Ry=Rank of Number of Hours of Exercise Per Week	$(Ry - \overline{R}y)$	$(Ry - \overline{R}y)^2$
1.5	-5.0	25.00
1.5	-5.0	25.00
3	-3.5	12.25
4	-2.5	6.25
5.5	-1.0	1.00
8.5	2.0	4.00
8.5	2.0	4.00
10	3.5	12.25
5.5	-1.0	1.00
7	0.5	0.25
11	4.5	20.25
12	5.5	30.25
$\overline{R}y = 6.5$	0	141.50

So, $Var(Ry) = \dfrac{\Sigma (Ry - \overline{Ry})^2}{n-1} = 141.50/11 = 12.86$.

Finally, we compute the covariance of Rx and Ry: $Cov(Rx, Ry) = \frac{\Sigma (Rx - \overline{Rx})(Ry - \overline{Ry})}{n-1}$. The following table summarizes the computations:

$(Rx - \overline{R}x)$	$(Ry - \overline{R}y)$	$(Rx - \overline{R}x)(Ry - \overline{R}y)$
5.0	-5.0	-25.00
-3.5	-5.0	17.50
5.0	-3.5	-17.50
3.5	-2.5	-8.75
2.0	-1.0	-2.00
0.5	2.0	1.00
-0.5	2.0	-1.00
2.0	3.5	7.00
-3.5	-1.0	3.50
-3.5	0.5	-1.75
-3.5	4.5	-15.75
-3.5	5.5	-19.25
		-62.0

So, $Cov(Rx, Ry) = \frac{\Sigma (Rx - \overline{Rx})(Ry - \overline{Ry})}{n-1} = -62.0/11 = -5.64.$

Substituting into (13.3):

$$r_s = \frac{Cov(Rx, Ry)}{\sqrt{Var(Rx)Var(Ry)}} = \frac{-5.64}{\sqrt{(12.05)(12.86)}} = -0.453.$$

Based on the sign and the magnitude of r_s, there is a strong, inverse association between the number of cigarettes smoked per day and the number of hours of aerobic exercise per week.

Nonparametric Tests

SAS Example 13.5. Users can request that SAS compute a correlation based on ranks by specifying the Spearman option in the call to Proc Corr. Before computing the rank correlation, we generated a scatter diagram using Proc Plot. The following output was generated by SAS. A brief interpretation appears after the output.

SAS Output for Example 13.5

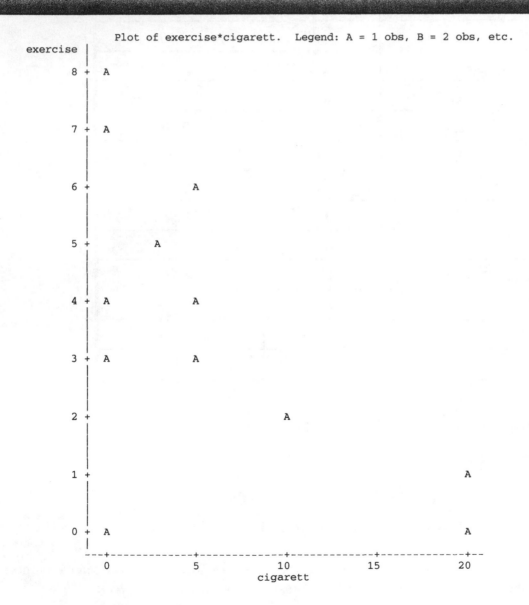

```
              Plot of exercise*cigarett.   Legend: A = 1 obs, B = 2 obs, etc.
  exercise |
           |
       8 + A
           |
           |
           |
       7 + A
           |
           |
           |
       6 +               A
           |
           |
           |
       5 +          A
           |
           |
           |
       4 + A         A
           |
           |
           |
       3 + A         A
           |
           |
           |
       2 +                    A
           |
           |
           |
       1 +                                        A
           |
           |
           |
       0 + A                                      A
           |
         --+-------------+-------------+-------------+-------------+--
           0             5            10            15            20
                                    cigarett
```

Nonparametric Tests

```
                        The CORR Procedure
                 2  Variables:     cigarett exercise
                        Simple Statistics
Variable       N       Mean      Std Dev     Median     Minimum      Maximum
cigarett      12     5.58333     7.39113     3.50000          0     20.00000
exercise      12     3.66667     2.64002     3.50000          0      8.00000


                Spearman Correlation Coefficients, N = 12
                     Prob > |r| under H0: Rho=0
                             cigarett       exercise
                cigarett     1.00000        -0.45366
                                             0.1385
                exercise    -0.45366         1.00000
                             0.1385
```

Interpretation of SAS Output for Example 13.5

The scatter plot shows the inverse relationship between the number of hours of exercise per week and the number of cigarettes smoked per day. The Corr Procedure first generates summary statistics on the raw scores of the variables specified. For example, the mean number of hours of exercise per week in the sample is 3.7 and the mean number of cigarettes smoked per day is 5.7. Because we requested a correlation based on ranks (Spearman correlation) SAS makes this note in the header to the correlation matrix. The Spearman correlation is estimated at –0.45 and the two sided p-value = 0.1385. The correlation is not significantly different from zero, p=0.1385>0.05.

13.7   Key Formulas

APPLICATION	NOTATION/FORMULA	DESCRIPTION
Two Dependent Samples Test	$$Z = \frac{T - \frac{n(n+1)}{4}}{\sqrt{\frac{n(n+1)(2n+1)}{24}}},$$ where T=sum of positive ranks	Normal Approximation for Signed Rank Test
Two Independent Samples Test	$$Z = \frac{S - \frac{n_1(n_1 + n_2 + 1)}{2}}{\sqrt{\frac{n_1 n_2 (n_1 + n_2 + 1)}{12}}},$$ where S=smaller sum of ranks	Normal Approximation for Rank Sum Test
Rank Correlation	$$r_s = \frac{Cov(Rx, Ry)}{\sqrt{Var(Rx)Var(Ry)}}$$	Spearman Correlation Based on Ranks

Guidelines for Determining When to Use a Nonparametric Procedure

There are no "rules" for determining when it is appropriate to use a nonparametric procedure instead of a parametric one. Substantial work has been done to show the robustness of many parametric procedures in the presence of violations of the normality assumption. The following are guidelines that may be useful in determining when it might be appropriate to employ a nonparameteric test.

- When substantive knowledge of the analytic variable suggests non-normality. Some variables clearly do not follow a normal distribution (e.g., ordinal variables with few response options) and analyses focused on these variables are candidates for nonparametric procedures.

- Observed non-normality in the sample data. Visual inspection of the distribution of a variable might suggest that the variable is highly skewed. Comparison of measures of central tendency, such as the mean and median, which are equal in symmetric distributions, can help determine the extent of skewness. Situations in which the standard deviation is much larger than the mean also suggest non normality. A formal test for normality (such as the tests available in proc Univariate) might indicate that a variable does not follow a normal distribution. (These tests have been shown to be a very sensitive test and might suggest non-normality when in fact the data are not substantively different from normal, so they should be interpreted with caution.) In all of these cases, nonparametric procedures might be appropriate.

13.8   Statistical Computing

Following are the SAS programs which were used to conduct the nonparametric tests for two dependent samples, two independent samples, and for k independent samples. Also included is the SAS program used to estimate a correlation based on ranks. The SAS procedures used and brief descriptions of their use are noted in the header to each example. Notes are provided to the right of the SAS programs (*in italics*) for orientation purposes and are not part of the programs. In addition, there are blank lines in the programs that follow which are solely to accommodate the notes. Blank lines and spaces can be used throughout SAS programs to enhance readability. A summary of the SAS procedures used in the examples is provided at the end of this section.

Wilcoxon Signed Rank Test: Two Dependent Samples

SAS EXAMPLE 13.2   **Test to Determine if New Drug Lowers Systolic Blood Pressure**

Suppose we are investigating a new drug hypothesized to lower systolic blood pressure. We select a random sample of subjects and record their resting systolic blood pressure at baseline (i.e., prior to the investigation). Each subject is then given the new drug and resting systolic blood pressure is again measured after 4 weeks of drug treatment. The data are shown below. Based on the data, does it appear that the new drug lowers systolic blood pressure? Run a Wilcoxon Signed Rank test using SAS.

Subject Identification Number	Baseline Systolic Blood Pressure	Post-Treatment Systolic Blood Pressure
1	166	138
2	135	120
3	189	176
4	180	180
5	156	160
6	142	150
7	176	152
8	156	140
9	164	160
10	142	130

Program Code

options ps=62 ls=80;

data in;
  input baseline post_trt;

  diff=baseline-post_trt

Formats the output page to 62 lines in length and 80 columns in width
Beginning of Data Step
*Inputs two variables **baseline** and **post_trt** for each subject.*
*Compute the difference between blood pressures (**diff**).*

Nonparametric Tests

```
cards;
166 138
135 120
189 176
180 180
156 160
142 150
176 152
156 140
164 160
142 130
run;

proc univariate normal;



 var diff;


run;
```

Beginning of Raw Data section.
actual observations

Procedure call. Proc Univariate generates summary statistics and the Wilcoxon Signed Rank test. The normal option requests a test for normality.
*Specification of analytic variable (**diff**).*
End of procedure section.

Nonparametric Tests

Wilcoxon Rank Sum Test: Two Independent Samples

SAS EXAMPLE 13.3   **Test for Difference in Treatments With Respect to Health Status**

We wish to compare two competing treatments for the management of adult onset diabetes. A total of 8 subjects agree to participate in the investigation and subjects are randomly assigned to one of the competing treatments. After subjects follow the prescribed treatment regimen for 6 weeks, we assess their self-reported health status. Subjects are asked to rate various aspects of their current health using the following response options: Excellent, Very Good, Good, Fair or Poor. Each attribute is ordinal in nature. Investigators often assign numerical values to ordinal responses and analyze these values as if they were continuous. The assignment of numerical values can be made in a variety of ways. Suppose we assign increasing numerical values to the response options as follows: Poor=0, Fair=5, Good=10, Very Good=15 and Excellent=20. Here, higher values indicate better self-reported health status. Each subject's health status score is the mean of his/her responses over all attributes. Are self reported health status scores different between treatment groups? Run a Wilcoxon Signed Rank test using SAS.

| Self-Reported Health Status ||
Treatment 1	Treatment 2
0	8
7	10
11	12
16	15

Program Code

```
options ps=62 ls=80;
```
Formats the output page to 62 lines in length and 80 columns in width

```
data in;
  input trt hlthstat;
```
Beginning of Data Step

*Inputs two variables **trt** and **hlthstat** for each subject.*

609

cards; 1 0 1 7 1 11 1 16 2 8 2 10 2 12 2 15 run;	*Beginning of Raw Data section.* *actual observations*
proc npar1way wilcoxon;	*Procedure call. Proc Npar1way runs nonparametric tests. We request a Wilcoxon Rank Sum test by specifying the wilcoxon option.*
class trt;	*Specification of grouping variable (trt).*
var hlthstat;	*Specification of analytic variable (hlthstat).*
run;	*End of procedure section.*

Nonparametric Tests

Kruskal-Wallis Test: k Independent Samples

SAS EXAMPLE 13.4 **Test for Difference Among Treatments With Respect to Symptom Scores**

Suppose we wish to compare four treatments for seasonal allergies. A total of 20 subjects agree to participate in the investigation and subjects are randomly assigned to one of the four competing treatments. After subjects follow the prescribed treatment regimen for 2 weeks, we assess their status. The outcome variable in this application is an index score based on three distinct symptoms. At the end of treatment, subjects are asked if they are currently experiencing any (or all) of the following symptoms: Scratchy throat, Itchy eyes, Runny nose. Each subject responds "yes" or "no" to each of the three symptoms. This index score is the sum of affirmative responses. The range of scores is 0 (no symptoms) to 3 (all three symptoms). Is there is a significant difference among the treatments with respect to symptom scores? Run a Kruskal-Wallis test using SAS.

Treatment 1	Treatment 2	Treatment 3	Treatment 4
0	2	0	2
0	2	0	3
1	3	1	2
2	3	1	3
3	2	1	3

Program Code

options ps=62 ls=80;

data in;
 input trt symptoms;

Formats the output page to 62 lines in length and 80 columns in width
Beginning of Data Step
*Inputs two variables **trt** and **symptoms** for each subject.*

Nonparametric Tests

```
cards;
1 0
1 0
1 1
1 2
1 3
2 2
2 2
2 3
2 3
2 2
3 0
3 0
3 1
3 1
3 1
4 2
4 3
4 2
4 3
4 3
run;


proc npar1way wilcoxon;




   class trt;

   var symptoms;

run;
```

Beginning of Raw Data section.
actual observations

Procedure call. Proc Npar1way runs nonparametric tests. When there are more than two comparison groups the Kruskal-Wallis test is run.
*Specification of grouping variable (**trt**).*
*Specification of analytic variable (**symptoms**).*
End of procedure section.

Spearman (Rank) Correlation

SAS EXAMPLE 13.5   **Assess Relationship Between Number of Cigarettes Smoked per Day and Number of Hours of Exercise per Week**

Suppose we wish to assess the relationship between the number of cigarettes smoked per day and the number of hours of aerobic exercise per week. A total of 12 subjects agree to participate in the investigation and we measure the average number of cigarettes smoked per day and the number of hours of exercise in a typical week on each subject. Estimate the Spearman correlation using SAS.

Raw Scores	
X=Number of Cigarettes Per Day	Y=Number of Hours of Exercise Per Week
20	0
0	0
20	1
10	2
5	3
4	5
3	5
5	6
0	3
0	4
0	7
0	8

Program Code

options ps=62 ls=80;   *Formats the output page to 62 lines in length and 80 columns in width*

data in;   *Beginning of Data Step*
 input cigarett exercise;   *Inputs two variables **cigarett** and **exercise** for each subject.*

```
cards;
20 0
0 0
20 1
10 2
5 3
4 4
3 5
5 6
0 3
0 4
0 7
0 8
run;
```

Beginning of Raw Data section.
actual observations

```
proc plot;
  plot exercise*cigarett;



run;
```

Procedure call. Proc Plot generates a scatter diagram.
*Specification of analytic variables, **exercise** is plotted on the vertical axis and **cigarett** on the horizontal.*
End of procedure section.

```
proc corr spearman;




  var exercise cigarett;
run;
```

Procedure call. Proc Corr runs a correlation analysis. The Spearman option requests that the correlation is based on ranks.
Specification of analytic variables.
End of procedure section.

Summary of SAS Procedures

The SAS procedures for nonparametric analysis are outlined below. Specific options can be requested in these procedures to generate specific tests or analyses. The options are shown in italics below. Users should refer to the examples in this section for complete descriptions of the procedure and specific options. A general description of the procedure and options is provided in the table below.

Procedure	Sample Procedure Call	Description
proc univariate	proc univariate *normal;* var diff;	Runs a Wilcoxon Signed-Rank test. The normal option produces a test for normality. The analytic variable is the difference between measures in dependent samples.
proc npar1way	proc npar1way *wilcoxon*; class group; var x;	Runs nonparametric tests for equality of medians among independent groups. The wilcoxon option requests the Wilcoxon (ranks based) test. When the group variable is dichotomous, a Wilcoxon Rank Sum test is run, when the group variable has k levels a Krusal-Wallis test is run
proc corr	proc corr spearman; var x y;	Runs a correlation analysis. The Spearman option requests that the correlation is based on ranks..

13.9   Problems

1.   A study is conducted comparing two competing medications for asthma. Sixteen subjects are involved in the investigation. The data shown below reflect asthma symptom scores for patients randomly assigned to each treatment. Higher scores are indicative of worse asthma symptoms. Test the null hypothesis that there is no difference in asthma symptoms between medications. Run the Wilcoxon Rank Sum test using the normal approximation at a 5% level of significance.

Treatment A:	55	60	80	65	72	78	68	71
Treatment B:	80	82	86	89	76	81	90	76

2.   We wish to evaluate a program designed to improve quality of life in older patients with coronary heart disease. Quality of life is measured on a scale of 0-100 with higher scores indicative of better quality of life. Quality of life measures are taken on each subject at baseline and then again after participating in the program. Based on the following data, is there evidence that the program significantly improves quality of life? Run the Wilcoxon Signed-Rank test using the normal approximation at a 5% level of significance.

Baseline	80	55	63	76	88	45	65	77
Post-Program	85	5	65	78	82	55	68	90

3.   Data were collected from a random sample of 8 patients currently undergoing treatment for hypertension. Each subject reported the average number of cigarettes smoked per day and each subject was assigned a numerical value reflecting risk of cardiovascular disease (CVD). The risk assessments were computed by physicians and based on blood pressure, cholesterol level and exercise status. The risk assessments ranged from 0 to 100, higher values indicated increased risk. Compute the Spearman correlation coefficient.

Number of Cigarettes:	0	2	6	8	12	0	2	20
Risk of CVD :	12	20	50	68	75	8	10	80

4. Suppose we are interested in whether there is a difference between the mean numbers of sick days taken by men and women in a local company. The numbers of sick days taken by men and women are shown below. Use the data below to test the null hypothesis that there is no difference in sick days between men and women. Run the Wilcoxon Rank Sum test using the normal approximation at a 5% level of significance.

MEN:	5	10	2	0	6	4	5	15
WOMEN:	8	9	3	5	0	4	15	

5. A nutritionist is investigating the effects of a rigorous walking program on systolic blood pressure (SBP) in patients with mild hypertension. Subjects who agree to participate have their systolic blood pressure measured at the start of the study and then after completing the 6-week walking program. The data are shown below:

Subject	Starting SBP	Ending SBP
1	140	128
2	130	125
3	150	140
4	160	162
5	135	137
6	128	130
7	142	135
8	151	140

Based on the following data, is there evidence that SBP is significantly reduced by the walking program? Run the Wilcoxon Signed-Rank test using the normal approximation at a 5% level of significance.

6. An anti-smoking campaign is being evaluated prior to its implementation in high schools across the state. A pilot study involving 6 volunteers who smoke is conducted. Each volunteer reports the number of cigarettes he/she smoked the day before enrolling in the study. Each then is subjected to the anti-smoking campaign which involves educational material, support groups, formal programs designed to reduce or quit smoking, etc. After 4 weeks, each volunteer again reports the number of cigarettes he/she smoked the day before. Based on the following pilot data, does it appear that the program is effective?

| At Enrollment | 21 | 15 | 8 | 6 | 12 | 20 |
| After Campaign | 12 | 10 | 10 | 6 | 10 | 20 |

Run the Wilcoxon Signed-Rank test using the normal approximation at a 5% level of significance.

SAS Problems: Use SAS to solve the following problems.

1. A study is conducted comparing two competing medications for asthma. Sixteen subjects are involved in the investigation. The data shown below reflect asthma symptom scores for patients randomly assigned to each treatment. Higher scores are indicative of worse asthma symptoms. Test the null hypothesis that there is no difference in asthma symptoms between medications.

Treatment A:	55	60	80	65	72	78	68	71
Treatment B:	80	82	86	89	76	81	90	76

Use SAS Proc Univariate to run the Wilcoxon Signed-Rank test. Include the option to perform a test for normality.

2. We wish to evaluate a program designed to improve quality of life in older patients with coronary heart disease. Quality of life is measured on a scale of 0-100 with higher scores indicative of better quality of life. Quality of life measures are taken on each subject at baseline and then again after participating in the program. Based on the following data, is there evidence that the program significantly improves quality of life?

Baseline	80	55	63	76	88	45	65	77
Post-Program	85	5	65	78	82	55	68	90

Use SAS to perform a Wilcoxon Rank Sum test. Perform the test at a 5% level of significance.

3. A pharmaceutical company is interested in the effectiveness of a new preparation designed to relieve arthritis pain. Three variations of the compound have been prepared for investigation. The three preparations differ according to the proportion of the active ingredients: T15 contains 15% active ingredients, T40 contains 40% active ingredients, and T50 contains 50% active ingredients. A sample of 20 patients are selected to participate in a study comparing the three variations of the compound. A control compound, which is currently available over the counter, is also included in the investigation. Patients are

randomly assigned to one of the 4 treatments (control, T15, T40, T50) and the time (in minutes) until pain relief is recorded on each subject. The data are given below:

Control	12	15	18	16	20
T15	20	21	22	19	20
T40	17	16	19	15	19
T50	14	13	12	14	11

Use SAS to perform a Kruskal-Wallis test. Perform the test at a 5% level of significance.

4.   Data were collected from a random sample of 8 patients currently undergoing treatment for hypertension. Each subject reported the average number of cigarettes smoked per day and each subject was assigned a numerical value reflecting risk of cardiovascular disease (CVD). The risk assessments were computed by physicians and based on blood pressure, cholesterol level and exercise status. The risk assessments ranged from 0 to 100, higher values indicated increased risk.

Number of Cigarettes:	0	2	6	8	12	0	2	20
Risk of CVD :	12	20	50	68	75	8	10	80

Use SAS to estimate the Spearman correlation coefficient. Test if the correlation is significant using a 5% level of significance.

Introductory Applied Biostatistics

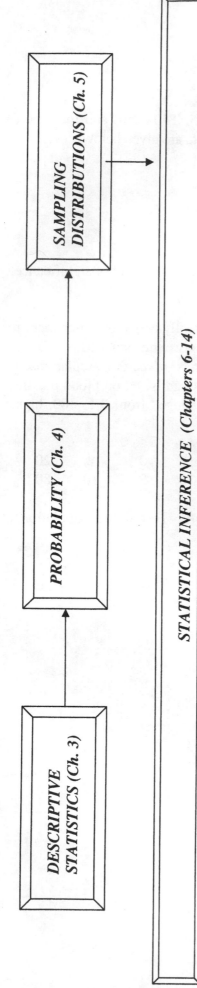

DESCRIPTIVE STATISTICS (Ch. 3) → PROBABILITY (Ch. 4) → SAMPLING DISTRIBUTIONS (Ch. 5)

STATISTICAL INFERENCE (Chapters 6-14)

Outcome Variable	Grouping Variable(s)/ Predictor(s)	Analysis	Chapter(s)
Continuous	-	Estimate μ, Compare μ to Known, Historical Value	6/13
Continuous	Dichotomous (2 Groups)	Compare Independent Means (Estimate/Test $(\mu_1-\mu_2)$) or the Mean Difference(μ_d)	7/13
Continuous	Discrete (> 2 Groups)	Test the Equality of K Means using Analysis of Variance ($\mu_1=\mu_2=...\mu_k$)	10/13
Continuous	Continuous	Estimate Correlation or Determine Regression Equation	11/13
Continuous	Several Continuous or Dichotomous	Multiple Linear Regression Analysis	11
Dichotomous	-	Estimate p, Compare p to Known, Historical Value	8
Dichotomous	Dichotomous (2 Groups)	Compare Independent Proportions (Estimate/Test (p_1-p_2))	8/9
Dichotomous	Discrete (>2 Groups)	Test the Equality of k Proportions (Chi-Square Test)	8
Dichotomous	Several Continuous or Dichotomous	Multiple Logistic Regression Analysis	12
Discrete	Discrete	Compare Distributions Among k Populations (Ch-Square Test)	8
Time to Event	Several Continuous or Dichotomous	Survival Analysis	14

CHAPTER 14: Introduction to Survival Analysis

14.1   Introduction

In Chapter 7 we presented estimation and statistical inference procedures for crude or unadjusted comparisons of two groups with respect to a continuous outcome variable. When the outcome variable in a prospective study or a clinical trial (i.e., a study in which the outcome is measured subsequent to treatment allocation) is continuous, the effect measure is the difference in means. The test of the effect of treatment is carried out using a two independent sample t test. To adjust for other variables when the outcome is continuous, multiple regression analysis can be used as was described in Chapter 11.

In Chapter 9 we presented estimation and statistical inference procedures for crude or unadjusted comparisons of two proportions and a technique for estimation and statistical inference adjusting for one categorical confounder. In Chapter 12 we presented logistic regression, which can be used to model the effect of independent variables on the risk of a dichotomous outcome. When the outcome variable in a prospective study or a clinical trial is dichotomous, the effect measure is the relative risk and the crude test of the effect of treatment is carried out using a chi-square test. If there are confounders to consider, the adjusted effect is assessed using the Mantel Haenszel chi-square statistic or logistic regression analysis.

In this section we give a brief overview of survival analysis which is used to compare groups with respect to a dichotomous outcome as is logistic regression, but the techniques also take into account different lengths of follow-up. Survival analysis techniques are used to estimate risks using all available data. These techniques are also used to estimate survival over time and to compare groups with respect to time to an event. Survival analysis techniques are beyond the scope of this book but we do present a brief heuristic explanation of the theory behind them.

14.2   Incomplete Follow-Up

In Chapters 9 and 12 we described procedures to compare proportions or risks between comparison groups. Some studies involve a long follow-up or observation period during which time each participant is measured for the occurrence or non-occurrence of the event of interest. When the follow-up or observation period is long, there are subjects who withdraw from the study before the end of the observation period. Some might move or simply decide that they no longer wish to participate. These subjects are often termed lost to follow-up. In addition, subjects may die during the study period. Subjects who are lost to follow-up or who die during the study period do not contribute to the overall counts of outcomes measured at the end of the study period, but their data should be included as much as possible in all analyses.

Example 14.1. Consider a very small study of five subjects in which the outcome of interest is stroke. Suppose that the study period is five years; that is, subjects are classified at the end of five years according to whether or not they had a stroke. Now suppose that among the five subjects in our study, two have strokes (after two and three years, respectively), one completes the five year period free of stroke, one subject dies after one year and one subject drops out of the study after four years.

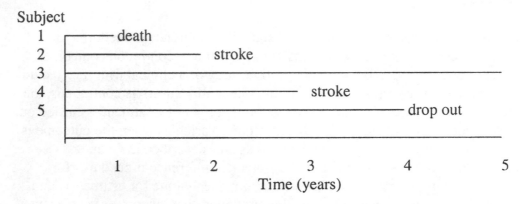

Based on these data, what can we say about the five-year risk of stroke? We could restrict our analysis to those whose five-year data are available, excluding the subject who died during the study period and the subject who dropped out. This would yield a five-year risk estimate of 2/3 = 67%. However, the person who dropped out survived for four of the five years free of stroke, and the person who died survived for one year free of stroke. Thus our estimate is probably too high. In fact, if we estimated one-year, two-year, three-year and four-year risks of stroke using this method, we would obtain estimates of 0% (0/5), 25% (1/4), 50% (2/4), and 50% (2/4), respectively. Then, with no additional strokes, our five-year estimate jumps to 67%!

Example 14.2. Suppose we sample a group of n=100 individuals from a population and follow them for two years. The goal is to estimate the probability of surviving for two years. Twenty people die each year, and we estimate the two-year survival probability as 0.60 (60/100).

Suppose that we then sample another group of n=100 from the same population and follow them for one year. Seventy-five people survive. How can we use this additional information to better estimate the two-year survival probability?

Recall the *conditional probability rule* from Chapter 4,

$$P(B \mid A) = \frac{P(A \text{ and } B)}{P(A)}.$$

Introduction to Survival Analysis

We can re-write this as P(A and B) = P(A)P(B|A) and apply it to the example above. Let A = survive year 1 and B = survive year 2. Then, using Sample 1 only,

$$P(A) = P(\text{survive year 1}) = 80/100 = 0.80$$

$$P(B|A)= P(\text{ survive year 2 given survival through year 1}) = 60/80 = 0.75.$$

$$P(A \text{ and } B) = P(\text{survive both years}) = P(A)P(B|A)= (0.80)\,(0.75) = 0.60$$

Note that this is the correct survival probability!

Now we can use the combined data from the two samples to estimate year 1 survival by combining the two samples:

$$P(A) = P(\text{survive year 1 (both samples combined)}) = \frac{(80+75)}{(100+100)} = 155/200 = 0.775$$

$$P(A \text{ and } B) = P(\text{survive both years}) = P(A)P(B|A)= (0.775)\,(0.75) = 0.58$$

Notice that P(A) is based on the combined sample and P(B|A) is based on sample 1 only. Thus our estimate of two-year survival is 0.60 if we only use sample 1 but is 0.58 if we include all available data. Survival analysis techniques allow us to do this.

14.3 Time to Event

Another application of survival analysis techniques is based on the date at which events occur, in addition to simply that they do or do not occur.

Example 14.3. Consider another very small study of five subjects in which the outcome of interest is stroke, also with a five-year study period. Suppose that in this study we do have complete data, and that two of the five had strokes during the study period. The estimated five-year risk of stroke is thus 2/5 = 40%. Consider the two graphs below.

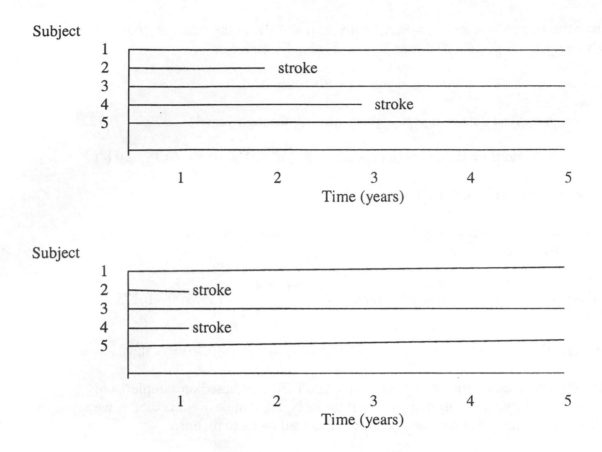

In both cases, there are two strokes among the five subjects. However, the strokes in the first graph occur much later than those in the second graph. Using a simple proportion to estimate risk yields does not allow us to distinguish these two scenarios. Survival analysis techniques take into account time to event in addition to the occurrence of the event.

14.4 Survival Analysis Techniques

Survival analysis methods are used when the time-to-event is important and/or when there are withdrawals or losses to follow-up. The probability of survival through a specified time is calculated using a) the probability of survival through the previous interval and b) the conditional probability of surviving the current interval given survival through the previous interval.

The Kaplan-Meier method creates a new interval each time there is an event; the life table method uses pre-specified intervals such as months or years. Cox regression analysis allows the inclusion of potential confounders in survival analysis models. We will not present either technique here, but do encourage the reader to explore them in more advanced texts.

Appendix A: Introduction to Statistical Computing Using SAS

A1.   Introduction to SAS

A *SAS program* is a list of commands in the SAS language. We use the SAS editor to create SAS programs and then execute them using the *submit* command. The components of a SAS program are outlined below. When a SAS program is executed, a *log* and *output* are produced. The program, log and output are described in detail below. Each of the three components, the program, the log, and the output can be printed and/or saved in files on your computer (or computer account).

SAS can be implemented on a variety of operating platforms (e.g., personal computer, mainframe). The operating platform does not dramatically change the structure and operation of the SAS program. There are also several operational modes in which SAS can be implemented. These include a mode in which the user enters SAS program code to perform a specific application as well as a menu-driven mode in which the user selects the desired statistical application from a list of options on a point-and-click basis. In the following, we illustrate the use of SAS in the former mode, in all cases specifying the SAS code to perform specific analyses.

A1.1.   Components of a SAS Program

Each basic SAS program is made up of three components:
- The **Data Step** which contains variable names and labels,
- The **Raw Data Section**, or the actual observations which are considered for analysis, and
- The **Procedure Section** which contains the call statements for the appropriate statistical analysis (e.g., summary statistics, two sample t test, analysis of variance). Several procedures can be invoked in this section to perform the desired analyses.

The following sample SAS program displays each component. The parts of the program presented in boldface represent parts that are specific to each application and can be changed by the user, as appropriate. The notes to the right of the SAS program (*in italics*) are for orientation and are not part of the program. In addition, the line spacing is only to accommodate the notes. In practice, the SAS program will be single-spaced. Indentation is used solely for readability and does not influence processing.

Sample SAS Program	Notes
data **one**;	*Beginning of the Data Step. SAS allows the user to choose a data set name (e.g., **one**).*
input **x**;	*Specification of variables. Here we consider a single variable which we name **x**.*
cards;	*Beginning of the Raw Data section.*
5	*Actual observations - one observation (x) per subject.*
6	
12	
run;	*End of the Raw Data section.*
proc print;	*SAS Procedure which prints the data.*
var **x**;	*Specification of the variable to be printed (**x**).*
run;	*End of the Procedure call.*
proc means;	*SAS Procedure to generate summary statistics (e.g., mean, standard deviation).*
var **x**;	*Specification of the analytic variable (**x**).*
run;	*End of the Procedure call.*

The dataset called **one** is a temporary SAS dataset and exists only for the duration of the program execution. It does not exist after your SAS session is over and you will not find it in your computer's directories or folders. It is possible to create permanent SAS datasets. At this point, our focus is on small, temporary datasets.

The statements that make up a SAS program are generally called the SAS code. The three components of SAS programs will be discussed in detail in this and subsequent chapters.

The sample SAS program produces the following output. The notes to the right of the SAS output (*in italics*) are for orientation and are not part of the output. In addition, the line spacing is only to accommodate the notes. In practice, the SAS output will be single-spaced.

Sample SAS Output	Notes

```
OBS    X
 1     5
 2     6
 3     12
```

Output from the Print Procedure:
SAS prints a column labeled "OBS"
denoting the observation number, followed
by each value of the variable x.

```
Analysis Variable: X
```

Output from the Means Procedure:
Summary statistics (See Chapter 3 for a complete
discussion).

N	Mean	Std Dev	Minimum	Maximum
3	7.666	3.7859	5.0000	12.0000

A1.2.   Overview of the SAS System

SAS uses three windows to display information:

- The **PROGRAM**, or **PGM** Window (also called the Editor or Advanced Editor Window) which contains the SAS program code,
- The **LOG** Window which displays the SAS execution statements, notes, error messages, etc., and
- The **OUTPUT**, or **OUT** Window which displays the results of the statistical analysis.

Throughout this book we present SAS programs and corresponding output. We do not illustrate the contents of the LOG Window. The information contained in the LOG Window is very useful and is used primarily to verify that the data are read correctly by SAS and that the procedure statements are entered and interpreted appropriately.

To view the contents of any window (i.e., PGM, LOG or OUTPUT) simply click on any part of that window. The box in the top corner of each window (i.e., PGM, LOG, or OUTPUT) is used to maximize or minimize that window. Clicking on the box will enlarge or shrink the current window display.

Program creation, editing and execution may be simplified by using the command line option within SAS. To move from one window to another, click on any part of that window or type PGM, LOG or OUTPUT in the command line at the top of any window followed by [enter] to move to the PROGRAM, LOG and OUTPUT windows, respectively.

The following table displays some of the more commonly used SAS commands, a description of their functions, and the windows in which the commands are used. All commands are issued in the command line. Either the arrow keys or the [Home] key, which moves the cursor directly to the command line, can be used to move to the command line.

On some platforms, SAS also displays a toolbox window where commands can be issued to perform specific functions. Commands that are issued at the command line (see table below) can also be issued in the appropriate space in the toolbox window. The toolbox also has some built in functions which appear as buttons in the toolbox window (e.g., submit, cut, copy, paste). Specific functions can be performed using the buttons.

SAS COMMAND	*Function*	*Windows*
submit	executes SAS program	PGM
recall	recalls last program executed into the PGM window	PGM
file filename (ext[1])	saves the contents of window into the file named "filename"	PGM (use ext=SAS) LOG (use ext=LOG) OUTPUT (use ext=LST)
inc filename	includes a copy of the file named "filename.sas" in the PGM window	PGM
pgm	moves to the PROGRAM window	LOG OUTPUT
log	moves to the LOG window	PGM OUTPUT
out	moves to the OUTPUT window	PGM LOG
help	displays on-line help	PGM LOG OUTPUT
clear	clears the contents of the window	PGM LOG OUTPUT
bye	exits SAS	PGM LOG OUTPUT

[1] The extension (ext) is a 3 character code used to distinguish between program files (extension 'SAS'), log files (extension 'LOG') and output files (extension 'LST'). SAS <u>automatically</u> appends these extensions when you save code in a particular window.

A1.3. Creating and Executing a SAS Program

The following is a step-by-step guide to create and execute a SAS program. This guide can be applied in any statistical computing application presented in this book.

1. Start SAS (Type SAS).
2. Click on the PGM window (or enter PGM from LOG or OUTPUT command lines).
3. Enter the program code (statements) by typing directly into the SAS Program window. The SAS Program editor has line numbers on the left hand margin. The program code is typed to the right of the line numbers. The column directly adjacent to the line numbers is reserved and cannot be used to enter data or program code. The arrow keys can be used to move around in the SAS Program editor to edit the program. Typing SAS commands into the SAS editor creates a *temporary* SAS program file which you will save as a *permanent* program file after checking that it works correctly.
4. Once the program code is complete, execute the program. Move the cursor to the command line in the PGM window (use arrow keys, or [Home] key). Enter <u>submit</u>.
5. Check the LOG for error messages and for documentation of desired procedure.
   If there are errors (in which case no output appears in the OUTPUT Window) or if the program did not execute correctly, do the following:
   - i) Return to the PGM window (click on PGM window or type PGM followed by [enter] in command line of LOG window). Recall the SAS program by typing <u>recall</u> followed by [enter] in the command line of the PGM window.
   - ii) Edit the program code to fix the errors and/or modify code to perform the desired analysis. Use the arrow keys to move around within the program window.
   - iii) Go to the LOG window (click on LOG window or enter LOG in command line of the PGM window). At the command line, type <u>clear</u> followed by [enter] to clear the contents of the window for the next program execution.
   - iv) Go to the OUTPUT window (click on OUTPUT window or enter OUT in command line of current window). At the command line, type <u>clear</u> followed by [enter] to clear the contents of the window for the next program execution.
   - v) Return to the PGM window (click on PGM window or type PGM followed by [enter] in command line of OUTPUT window).
   - vi) Return to Step 4 and continue the process.

If there are no errors, then continue to Step 6.

6.   Save the SAS PROGRAM code, the LOG and the SAS OUTPUT into files by performing the following steps:

         i)   Return to the PGM window (click on PGM window or enter PGM in command line of current window).

                Recall the SAS program by typing <u>recall</u> followed by [enter] in the command line.

                At the command line, type <u>file 'filename'</u> followed by [enter], where filename is the selected name (maximum 8 characters in length). The '.sas' extension is automatically appended to the filename.

         ii)   Go to the LOG window (click on LOG window or enter LOG in command line of current window).

                At the command line, type <u>file 'filename'</u> followed by [enter]. The '.log' extension is automatically appended to the filename.

         iii)   Go to the OUTPUT window (click on OUTPUT window or enter OUTPUT in the command line of current window). At the command line, type <u>file 'filename'</u> followed by [enter]. The '.lst' extension is automatically appended to the filename.

         **   You have now saved three files on your computer or computer account called filename.sas, filename.log and filename.lst.

7.   Exit SAS by typing <u>bye</u> followed by [enter] in the command line of the current window.

8.   Print hard copies of the 3 files (program, log and output) using the appropriate print command on your system. For example, <u>print filename.pgm [enter]</u>, <u>print filename.log [enter]</u>, and <u>print filename.out [enter]</u>

9.   Log off or sign off the system using the appropriate logout command.

A2.   The Data Step

      The Data Step is the first component of a SAS program and contains (among other items) the variable names, specifications and labels. We will illustrate a variety of programming statements that can be used in the Data Step.

      SAS, like many computing languages, offers a variety of options for inputting and manipulating data. SAS can read and analyze both numeric and alphanumeric, or character, data. We present several different techniques for inputting and manipulating data in the Data Step below. Sample data sets are provided to illustrate each technique.

A2.1. Inputting Data: Types of Data

In SAS, data are classified as either numeric, alphanumeric (or character), or as date variables (e.g., 01/15/1998). Numeric data include numbers such as 48, 4.8 and 0.48. SAS expects to read and analyze numeric data and no special considerations are necessary to input numeric data. Alphanumeric data, or character data, include letters and symbols (and possibly numbers). These include data elements such as male, female, M or F. To input character data, you must inform SAS that the data are character as opposed to numeric. To do this, simply follow the variable name in the input statement with a "$" symbol which denotes a character variable. The following data set includes age (in years) and gender (denoted M or F) recorded on each of four subjects:

```
20 F
32 M
24 F
31 M
```

The following SAS code will correctly input the data:

```
data one;
  input age gender $;
cards;
20 F
32 M
24 F
31 M
run;
proc print;
run;
```

The following output is produced by the Print procedure. Since no variables are specified (e.g., var age), SAS prints the values of all of the variables. Note that SAS inserts a column labeled "OBS" indicating the observation numbers:

```
OBS    AGE    GENDER
 1     20       F
 2     32       M
 3     24       F
 4     31       M
```

A2.2. Inputting Data: Types of Input

Two of the more common techniques for inputting data are: a) list, or free, format, and b) column, or fixed, format. (A third technique, formatted data, is described in Section A2.4. Advanced Data Input and Data Manipulation). To illustrate the different techniques, consider the following data set which includes 4 variables (listed in the columns below) recorded on 5 subjects (listed on each row):

```
10    12    11    1.4
22    12    99    4.3
15    12    64    3.1
12     5     3    8.9
99    45    66    0.5
```

a. List, or free, format. As long as the data are separated by at least one space, the following SAS code may be used to input the data. The data do not have to be lined up in columns, only separated by one or more spaces on each line. Notice that four variables, called $x1$, $x2$, $x3$ and $x4$, are input.

```
data one;
  input x1 x2 x3 x4;
cards;
10    12    11    1.4
22    12    99    4.3
15    12    64    3.1
12     5     3    8.9
99    45    66    0.5
run;
proc print;
run;
```

b. Column, or fixed, format. If the data set is rectangular (i.e., the observations are lined up in columns and rows), then the following SAS code is used to input the data, where the numbers following each variable name denote the columns in which the actual data elements are located in the Raw Data section.

```
data one;
  input x1 1-2 x2 4-5 x3 7-8 x4 10-12;
cards;
10    12    11    1.4
22    12    99    4.3
15    12    64    3.1
```

```
12      5      3      8.9
99     45     66      0.5
run;
proc print;
run;
```

Both programs shown above produce the following output:

```
OBS    X1     X2     X3     X4
 1     10     12     11     1.4
 2     22     12     99     4.3
 3     15     12     64     3.1
 4     12      5      3     8.9
 5     99     45     66     0.5
```

A2.3. Missing Values

Every effort should always be made to ensure that data are complete and accurate. However, even with intense effort on the part of investigators, it is not always possible to obtain values for every variable under study for each subject.

For example, suppose a study is conducted in which 7 variables are measured on each of 5 subjects: height (in inches), weight (in pounds), systolic blood pressure (in mmHg), diastolic blood pressure (in mmHg), gender (M or F), age (in years) and race (Asian, black, Hispanic, other, white). Suppose that one subject does not have his weight measured, while another does not report her race. We would not want to eliminate these subjects altogether since there are some variables collected on these subjects which are non-missing and can be analyzed.

SAS can input and maintain missing data values. Missing values of numeric variables (e.g., sbp) are denoted by a period ".", while missing values of alphanumeric, or character, variables are denoted by a blank space " ". The following data set includes 7 variables measured on 5 subjects and includes two missing values. Notice that the missing values are denoted with a period or a blank space for the numeric and alphanumeric variables, respectively (without quotation marks).

```
63 145 135 70 F  32 black
67 132 120 60 F  31 hispanic
60 117 130 80 F  33
68 187 140 60 M  26 white
72 145   . 55 M  29 white
```

The following SAS code inputs 7 variables using column formats:

```
data one;
  input height 1-2 weight 4-6 sbp 8-10 dbp 12-13 gender $ 15
      age 17-18 race $ 20-27;
cards;
63 145 135 70 F  32 black
67 132 120 60 F  31 hispanic
60 117 130 80 F  33
68 187 140 60 M 26 white
72 145  .  55 M 29 white
run;
proc print;
run;
```

The following output is produced by the Print procedure.

```
OBS HEIGHT WEIGHT SBP  DBP GENDER AGE RACE
1     63     145   135  70    F    32  black
2     67     132   120  60    F    31  hispanic
3     60     117   130  80    F    33
4     68     187   140  60    M    26  white
5     72      .    145  55    M    29  white
```

A2.4. Advanced Data Input and Data Manipulation*

**   * This section contains advanced statistical computing techniques. It is designed for more advanced users and can be omitted without disrupting the progression of material.***
This section covers inputting formatted data, the use of line pointers, and the drop and keep functions.

Inputting formatted data

If a data set is rectangular, the following SAS code may be used to input the data, where the "@" symbol "points to" the starting column location of each variable, and the number following each variable name denotes the variable format (i.e., 2.0 denotes a variable of length 2 with no decimal places, 3.1 denotes a variable of length 3 with 1 decimal place). Note, 2. can be used in place of 2.0.

```
data one;
   input @1 x1 2.0 @4 x2 2.0 @7 x3 2.0 @10 x4 3.1;
```

Notice that x1, x2 and x3 all have the same format (2.0). The following SAS code is equivalent to the above:

```
data one;
   input (x1 x2 x3) (2. 2. 2.) @10 x4 3.1;
```

The following is also equivalent, where the "+1" moves the column pointer to the right 1 column after inputting x1, x2 and x3:

```
data one;
   input (x1-x3) (2. +1) @10 x4 3.1;
```

SAS allows the user to mix input options (e.g., formatted input and list, or free, format):

```
data one;
   input (x1-x3) (2. +1) x4;
```

Inputting data: Line pointers

In addition to the column pointers, illustrated above, SAS allows the user to manipulate the rows, or lines in a data set. For example, suppose we record total serum cholesterol values on a sample of 5 subjects at three different points in time. The data set consists of three lines (rows) per person, with exam 1 values for total cholesterol on the first line, exam 2 values on the second line, and exam 3 values on the third line. The data are given below where the first entry on each line (row) is the exam number (i.e., 1, 2 or 3), the second entry is the subject identifier (sequential numbers) and the third entry is the observed total cholesterol value. A "." denotes a missing value.

1 1 180	_exam number=1, subject identifier=1, cholesterol=180_
2 1 160	_exam number=2, subject identifier=1, cholesterol=160_
3 1 177	_exam number=3, subject identifier=1, cholesterol=177_
1 2 240	_exam number=1, subject identifier=2, cholesterol=240_
2 2 220	
3 2 146	
1 3 .	
2 3 230	
3 3 210	
1 4 170	
2 4 200	
3 4 240	
1 5 180	
2 5 .	
3 5 190	_exam number=3, subject identifier=5, cholesterol=190_

Introduction to Statistical Computing

The data are input as follows where the "#" symbol denotes the line number in the data set. The numbers following the variable names denote the columns in which the data are stored. Notice that the exam number (i.e., 1, 2, 3), stored in column 1 on each line, is ignored (i.e., not input).

```
data one;
 input #1 id 3 tc1 5-7
        #2  tc2 5-7
        #3  tc3 5-7;
cards;
1 1 180
2 1 160
3 1 177
1 2 240
2 2 220
3 2 146
1 3 .
2 3 230
3 3 210
1 4 170
2 4 200
3 4 240
1 5 180
2 5 .
3 5 190
run;
proc print;
run;
```

The Print procedure generates the following output:

```
OBS     ID     TC1    TC2    TC3
1       1      180    160    177
2       2      240    220    146
3       3       .     230    210
4       4      170    200    240
5       5      180     .     190
```

The drop function

Suppose (for the same data set) we are only interested in the first two exams. The "drop" function can be used to remove variables from a data set. Note that the data are never deleted or removed permanently - a variable is simply omitted from a SAS data set:

Introduction to Statistical Computing

```
data one;
  input  #1 id 3 tc1 5-7
         #2 tc2 5-7
         #3 tc3 5-7;
  drop tc3;
cards;
1 1 180
2 1 160
3 1 177
1 2 240
2 2 220
3 2 146
1 3  .
2 3 230
3 3 210
1 4 170
2 4 200
3 4 240
1 5 180
2 5  .
3 5 190
run;
proc print;
run;
```

Or equivalently:

```
data one;
  input  #1 id 3 tc1 5-7
         #2 tc2 5-7
         #3 ;
cards;
1 1 180
2 1 160
3 1 177
1 2 240
2 2 220
3 2 146
1 3  .
2 3 230
```

```
3 3 210
1 4 170
2 4 200
3 4 240
1 5 180
2 5  .
3 5 190
run;
proc print;
run;
```

NOTE: The #3 indicates that there are three lines of data per subject and is necessary so that the correct observations are read from the appropriate lines in the data set.

Both programs shown above produce the following output:

```
OBS     ID      TC1     TC2
1       1       180     160
2       2       240     220
3       3        .      230
4       4       170     200
5       5       180      .
```

Suppose we only want the second exam:

```
data one;
  input #1 id 3 tc1 5-7
        #2 tc2 5-7
        #3;
  drop tc1;
cards;
1 1 180
2 1 160
3 1 177
1 2 240
2 2 220
3 2 146
1 3  .
2 3 230
3 3 210
1 4 170
```

```
2 4 200
3 4 240
1 5 180
2 5  .
3 5 190
run;
proc print;
run;
```

The following program is equivalent, and illustrates the use of line holders. The "@" symbol at the end of an input line holds the pointer at that line until the next input statement.

```
data one;
  input exam 1 @;
  if exam eq 2 then input id 3 tc2 5-7;
cards;
1 1 180
2 1 160
3 1 177
1 2 240
2 2 220
3 2 146
1 3  .
2 3 230
3 3 210
1 4 170
2 4 200
3 4 240
1 5 180
2 5  .
3 5 190
run;
proc print;
run;
```

Both programs shown above produce the following output:

```
OBS    ID    TC1
1      1     160
2      2     220
3      3     230
4      4     200
5      5      .
```

Now, suppose that only lines with non-missing data are in the data set. Again, suppose we are interested only in the exam 2 values. In this case we no longer have exactly 3 lines per person. Instead, the data is as follows:

1 1 180
2 1 160
3 1 177
1 2 240
2 2 220
3 2 146
2 3 230
3 3 210   *Notice that subject 3 has only two lines of data*
1 4 170
2 4 200
3 4 240
1 5 180
3 5 190

To input only the second exam value from a data set which does not contain three lines per person, we must use the line holder:

```
data one;
  input exam 1 @;
  if exam eq 2 then input id 3 tc2 5-7;
cards;
1 1 180
2 1 160
3 1 177
1 2 240
2 2 220
3 2 146
2 3 230
```

```
3 3 210
1 4 170
2 4 200
3 4 240
1 5 180
3 5 190
run;
proc print;
run;
```

The program shown above produces the following output:

```
OBS    ID    TC2
1      1     160
2      2     220
3      3     230
4      4     200
5      5      .
```

Suppose we record data (a continuous variable) on ten subjects and want descriptive statistics on that variable. The data could be entered in ten rows and SAS Proc Means (See Chapter 3) could be used to generate descriptive statistics:

data one;	*Data set name=one.*
input id x;	*Two variables are input - **id** and **x**.*
cards;	*Beginning of Raw Data Section.*
1 23	*id=1, x=23*
2 43	*id=2, x=43*
3 66	
4 31	
5 28	
6 73	
7 92	
8 19	
9 33	
10 55	
run;	
proc means;	*Procedure call to generate summary statistics.*
var x;	*Specification of analytic variable **x**.*
run;	

Introduction to Statistical Computing

The same can be achieved using line holders:

```
data one;
  input id x @ @;

cards;
1 23 2 43 3 66 4 31 5 28
6 73 7 92 8 19 9 33 10 55
run;

proc means;
  var x;
run;
```

NOTE: "@@" holds the line until all data have been read and then moves to the next line.

The following program is equivalent, only the id variable is created in the program instead of input:

```
data one;
  input x @ @;                    Input variable x.
  id=id+1;                        Create the variable id by adding 1 for
                                  each subject.

cards;
23 43 66 31 28 73 92 19 33 55
run;

proc means;
  var x;
run;
```

The three programs shown above produce the following output:

```
Analysis Variable: X

N      Mean        Std Dev      Minimum      Maximum
----------------------------------------------------------
10     46.300      25.2901      19.0000      92.0000
----------------------------------------------------------
```

Appendix B: Statistical Tables

I. Statistical Tables

II. SAS[1] Programs used to generate table entries

1: SAS Institute Inc. *SAS® User's Guide: Basics, Verson 6 Edition 8.* Cary, NC: SAS Institute Inc., 1985.

Statistical Tables

Table 1. Probabilities of the Binomial Distribution

Table entries represent $P(X=x)$ for n trials and $P(success) = p$
e.g., $P(X=4 \mid n=10, p=0.2) = 0.0881$

n	x	.10	.20	.30	.40	.50	.60	.70	.80	.90
2	0	0.8100	0.6400	0.4900	0.3600	0.2500	0.1600	0.0900	0.0400	0.0100
	1	0.1800	0.3200	0.4200	0.4800	0.5000	0.4800	0.4200	0.3200	0.1800
	2	0.0100	0.0400	0.0900	0.1600	0.2500	0.3600	0.4900	0.6400	0.8100
3	0	0.7290	0.5120	0.3430	0.2160	0.1250	0.0640	0.0270	0.0080	0.0010
	1	0.2430	0.3840	0.4410	0.4320	0.3750	0.2880	0.1890	0.0960	0.0270
	2	0.0270	0.0960	0.1890	0.2880	0.3750	0.4320	0.4410	0.3840	0.2430
	3	0.0010	0.0080	0.0270	0.0640	0.1250	0.2160	0.3430	0.5120	0.7290
4	0	0.6561	0.4096	0.2401	0.1296	0.0625	0.0256	0.0081	0.0016	0.0001
	1	0.2916	0.4096	0.4116	0.3456	0.2500	0.1536	0.0756	0.0256	0.0036
	2	0.0486	0.1536	0.2646	0.3456	0.3750	0.3456	0.2646	0.1536	0.0486
	3	0.0036	0.0256	0.0756	0.1536	0.2500	0.3456	0.4116	0.4096	0.2916
	4	0.0001	0.0016	0.0081	0.0256	0.0625	0.1296	0.2401	0.4096	0.6561
5	0	0.5905	0.3277	0.1681	0.0778	0.0313	0.0102	0.0024	0.0003	0.0000
	1	0.3280	0.4096	0.3601	0.2592	0.1562	0.0768	0.0284	0.0064	0.0005
	2	0.0729	0.2048	0.3087	0.3456	0.3125	0.2304	0.1323	0.0512	0.0081
	3	0.0081	0.0512	0.1323	0.2304	0.3125	0.3456	0.3087	0.2048	0.0729
	4	0.0005	0.0064	0.0283	0.0768	0.1563	0.2592	0.3601	0.4096	0.3281
	5	0.0000	0.0003	0.0024	0.0102	0.0313	0.0778	0.1681	0.3277	0.5905
6	0	0.5314	0.2621	0.1176	0.0467	0.0156	0.0041	0.0007	0.0001	0.0000
	1	0.3543	0.3932	0.3025	0.1866	0.0938	0.0369	0.0102	0.0015	0.0001
	2	0.0984	0.2458	0.3241	0.3110	0.2344	0.1382	0.0595	0.0154	0.0012
	3	0.0146	0.0819	0.1852	0.2765	0.3125	0.2765	0.1852	0.0819	0.0146
	4	0.0012	0.0154	0.0595	0.1382	0.2344	0.3110	0.3241	0.2458	0.0984
	5	0.0001	0.0015	0.0102	0.0369	0.0938	0.1866	0.3025	0.3932	0.3543
	6	0.0000	0.0001	0.0007	0.0041	0.0156	0.0467	0.1176	0.2621	0.5314
7	0	0.4783	0.2097	0.0824	0.0280	0.0078	0.0016	0.0002	0.0000	0.0000
	1	0.3720	0.3670	0.2471	0.1306	0.0547	0.0172	0.0036	0.0004	0.0000
	2	0.1240	0.2753	0.3177	0.2613	0.1641	0.0774	0.0250	0.0043	0.0002
	3	0.0230	0.1147	0.2269	0.2903	0.2734	0.1935	0.0972	0.0287	0.0026
	4	0.0026	0.0287	0.0972	0.1935	0.2734	0.2903	0.2269	0.1147	0.0230
	5	0.0002	0.0043	0.0250	0.0774	0.1641	0.2613	0.3177	0.2753	0.1240
	6	0.0000	0.0004	0.0036	0.0172	0.0547	0.1306	0.2471	0.3670	0.3720
	7	0.0000	0.0000	0.0002	0.0016	0.0078	0.0280	0.0824	0.2097	0.4783
8	0	0.4305	0.1678	0.0576	0.0168	0.0039	0.0007	0.0001	0.0000	0.0000
	1	0.3826	0.3355	0.1977	0.0896	0.0313	0.0079	0.0012	0.0001	0.0000
	2	0.1488	0.2936	0.2965	0.2090	0.1094	0.0413	0.0100	0.0011	0.0000
	3	0.0331	0.1468	0.2541	0.2787	0.2188	0.1239	0.0467	0.0092	0.0004
	4	0.0046	0.0459	0.1361	0.2322	0.2734	0.2322	0.1361	0.0459	0.0046
	5	0.0004	0.0092	0.0467	0.1239	0.2188	0.2787	0.2541	0.1468	0.0331
	6	0.0000	0.0011	0.0100	0.0413	0.1094	0.2090	0.2965	0.2936	0.1488
	7	0.0000	0.0001	0.0012	0.0079	0.0313	0.0896	0.1977	0.3355	0.3826
	8	0.0000	0.0000	0.0001	0.0007	0.0039	0.0168	0.0576	0.1678	0.4305

Statistical Tables

n	x	p .10	.20	.30	.40	.50	.60	.70	.80	.90
9	0	0.3874	0.1342	0.0404	0.0101	0.0020	0.0003	0.0000	0.0000	0.0000
	1	0.3874	0.3020	0.1556	0.0605	0.0176	0.0035	0.0004	0.0000	0.0000
	2	0.1722	0.3020	0.2668	0.1612	0.0703	0.0212	0.0039	0.0003	0.0000
	3	0.0446	0.1762	0.2668	0.2508	0.1641	0.0743	0.0210	0.0028	0.0001
	4	0.0074	0.0661	0.1715	0.2508	0.2461	0.1672	0.0735	0.0165	0.0008
	5	0.0008	0.0165	0.0735	0.1672	0.2461	0.2508	0.1715	0.0661	0.0074
	6	0.0001	0.0028	0.0210	0.0743	0.1641	0.2508	0.2668	0.1762	0.0446
	7	0.0000	0.0003	0.0039	0.0212	0.0703	0.1612	0.2668	0.3020	0.1722
	8	0.0000	0.0000	0.0004	0.0035	0.0176	0.0605	0.1556	0.3020	0.3874
	9	0.0000	0.0000	0.0000	0.0003	0.0020	0.0101	0.0404	0.1342	0.3874
10	0	0.3487	0.1074	0.0282	0.0060	0.0010	0.0001	0.0000	0.0000	0.0000
	1	0.3874	0.2684	0.1211	0.0403	0.0098	0.0016	0.0001	0.0000	0.0000
	2	0.1937	0.3020	0.2335	0.1209	0.0439	0.0106	0.0014	0.0001	0.0000
	3	0.0574	0.2013	0.2668	0.2150	0.1172	0.0425	0.0090	0.0008	0.0000
	4	0.0112	0.0881	0.2001	0.2508	0.2051	0.1115	0.0368	0.0055	0.0001
	5	0.0015	0.0264	0.1029	0.2007	0.2461	0.2007	0.1029	0.0264	0.0015
	6	0.0001	0.0055	0.0368	0.1115	0.2051	0.2508	0.2001	0.0881	0.0112
	7	0.0000	0.0008	0.0090	0.0425	0.1172	0.2150	0.2668	0.2013	0.0574
	8	0.0000	0.0001	0.0014	0.0106	0.0439	0.1209	0.2335	0.3020	0.1937
	9	0.0000	0.0000	0.0001	0.0016	0.0098	0.0403	0.1211	0.2684	0.3874
	10	0.0000	0.0000	0.0000	0.0001	0.0010	0.0060	0.0282	0.1074	0.3487
11	0	0.3138	0.0859	0.0198	0.0036	0.0005	0.0000	0.0000	0.0000	0.0000
	1	0.3835	0.2362	0.0932	0.0266	0.0054	0.0007	0.0000	0.0000	0.0000
	2	0.2131	0.2953	0.1998	0.0887	0.0269	0.0052	0.0005	0.0000	0.0000
	3	0.0710	0.2215	0.2568	0.1774	0.0806	0.0234	0.0037	0.0002	0.0000
	4	0.0158	0.1107	0.2201	0.2365	0.1611	0.0701	0.0173	0.0017	0.0000
	5	0.0025	0.0388	0.1321	0.2207	0.2256	0.1471	0.0566	0.0097	0.0003
	6	0.0003	0.0097	0.0566	0.1471	0.2256	0.2207	0.1321	0.0388	0.0025
	7	0.0000	0.0017	0.0173	0.0701	0.1611	0.2365	0.2201	0.1107	0.0158
	8	0.0000	0.0002	0.0037	0.0234	0.0806	0.1774	0.2568	0.2215	0.0710
	9	0.0000	0.0000	0.0005	0.0052	0.0269	0.0887	0.1998	0.2953	0.2131
	10	0.0000	0.0000	0.0000	0.0007	0.0054	0.0266	0.0932	0.2362	0.3835
	11	0.0000	0.0000	0.0000	0.0000	0.0005	0.0036	0.0198	0.0859	0.3138
12	0	0.2824	0.0687	0.0138	0.0022	0.0002	0.0000	0.0000	0.0000	0.0000
	1	0.3766	0.2062	0.0712	0.0174	0.0029	0.0003	0.0000	0.0000	0.0000
	2	0.2301	0.2835	0.1678	0.0639	0.0161	0.0025	0.0002	0.0000	0.0000
	3	0.0852	0.2362	0.2397	0.1419	0.0537	0.0125	0.0015	0.0001	0.0000
	4	0.0213	0.1329	0.2311	0.2128	0.1208	0.0420	0.0078	0.0005	0.0000
	5	0.0038	0.0532	0.1585	0.2270	0.1934	0.1009	0.0291	0.0033	0.0000
	6	0.0005	0.0155	0.0792	0.1766	0.2256	0.1766	0.0792	0.0155	0.0005
	7	0.0000	0.0033	0.0291	0.1009	0.1934	0.2270	0.1585	0.0532	0.0038
	8	0.0000	0.0005	0.0078	0.0420	0.1208	0.2128	0.2311	0.1329	0.0213
	9	0.0000	0.0001	0.0015	0.0125	0.0537	0.1419	0.2397	0.2362	0.0852
	10	0.0000	0.0000	0.0002	0.0025	0.0161	0.0639	0.1678	0.2835	0.2301
	11	0.0000	0.0000	0.0000	0.0003	0.0029	0.0174	0.0712	0.2062	0.3766
	12	0.0000	0.0000	0.0000	0.0000	0.0002	0.0022	0.0138	0.0687	0.2824

Statistical Tables

n	x	.10	.20	.30	.40	.50	.60	.70	.80	.90
13	0	0.2542	0.0550	0.0097	0.0013	0.0001	0.0000	0.0000	0.0000	0.0000
	1	0.3672	0.1787	0.0540	0.0113	0.0016	0.0001	0.0000	0.0000	0.0000
	2	0.2448	0.2680	0.1388	0.0453	0.0095	0.0012	0.0001	0.0000	0.0000
	3	0.0997	0.2457	0.2181	0.1107	0.0349	0.0065	0.0006	0.0000	0.0000
	4	0.0277	0.1535	0.2337	0.1845	0.0873	0.0243	0.0034	0.0001	0.0000
	5	0.0055	0.0691	0.1803	0.2214	0.1571	0.0656	0.0142	0.0011	0.0000
	6	0.0008	0.0230	0.1030	0.1968	0.2095	0.1312	0.0442	0.0058	0.0001
	7	0.0001	0.0058	0.0442	0.1312	0.2095	0.1968	0.1030	0.0230	0.0008
	8	0.0000	0.0011	0.0142	0.0656	0.1571	0.2214	0.1803	0.0691	0.0055
	9	0.0000	0.0001	0.0034	0.0243	0.0873	0.1845	0.2337	0.1535	0.0277
	10	0.0000	0.0000	0.0006	0.0065	0.0349	0.1107	0.2181	0.2457	0.0997
	11	0.0000	0.0000	0.0001	0.0012	0.0095	0.0453	0.1388	0.2680	0.2448
	12	0.0000	0.0000	0.0000	0.0001	0.0016	0.0113	0.0540	0.1787	0.3672
	13	0.0000	0.0000	0.0000	0.0000	0.0001	0.0013	0.0097	0.0550	0.2542
14	0	0.2288	0.0440	0.0068	0.0008	0.0001	0.0000	0.0000	0.0000	0.0000
	1	0.3559	0.1539	0.0407	0.0073	0.0009	0.0001	0.0000	0.0000	0.0000
	2	0.2570	0.2501	0.1134	0.0317	0.0056	0.0005	0.0000	0.0000	0.0000
	3	0.1142	0.2501	0.1943	0.0845	0.0222	0.0033	0.0002	0.0000	0.0000
	4	0.0349	0.1720	0.2290	0.1549	0.0611	0.0136	0.0014	0.0000	0.0000
	5	0.0078	0.0860	0.1963	0.2066	0.1222	0.0408	0.0066	0.0003	0.0000
	6	0.0013	0.0322	0.1262	0.2066	0.1833	0.0918	0.0232	0.0020	0.0000
	7	0.0002	0.0092	0.0618	0.1574	0.2095	0.1574	0.0618	0.0092	0.0002
	8	0.0000	0.0020	0.0232	0.0918	0.1833	0.2066	0.1262	0.0322	0.0013
	9	0.0000	0.0003	0.0066	0.0408	0.1222	0.2066	0.1963	0.0860	0.0078
	10	0.0000	0.0000	0.0014	0.0136	0.0611	0.1549	0.2290	0.1720	0.0349
	11	0.0000	0.0000	0.0002	0.0033	0.0222	0.0845	0.1943	0.2501	0.1142
	12	0.0000	0.0000	0.0000	0.0005	0.0056	0.0317	0.1134	0.2501	0.2570
	13	0.0000	0.0000	0.0000	0.0001	0.0009	0.0073	0.0407	0.1539	0.3559
	14	0.0000	0.0000	0.0000	0.0000	0.0001	0.0008	0.0068	0.0440	0.2288
15	0	0.2059	0.0352	0.0047	0.0005	0.0000	0.0000	0.0000	0.0000	0.0000
	1	0.3432	0.1319	0.0305	0.0047	0.0005	0.0000	0.0000	0.0000	0.0000
	2	0.2669	0.2309	0.0916	0.0219	0.0032	0.0003	0.0000	0.0000	0.0000
	3	0.1285	0.2501	0.1700	0.0634	0.0139	0.0016	0.0001	0.0000	0.0000
	4	0.0428	0.1876	0.2186	0.1268	0.0417	0.0074	0.0006	0.0000	0.0000
	5	0.0105	0.1032	0.2061	0.1859	0.0916	0.0245	0.0030	0.0001	0.0000
	6	0.0019	0.0430	0.1472	0.2066	0.1527	0.0612	0.0116	0.0007	0.0000
	7	0.0003	0.0138	0.0811	0.1771	0.1964	0.1181	0.0348	0.0035	0.0000
	8	0.0000	0.0035	0.0348	0.1181	0.1964	0.1771	0.0811	0.0138	0.0003
	9	0.0000	0.0007	0.0116	0.0612	0.1527	0.2066	0.1472	0.0430	0.0019
	10	0.0000	0.0001	0.0030	0.0245	0.0916	0.1859	0.2061	0.1032	0.0105
	11	0.0000	0.0000	0.0006	0.0074	0.0417	0.1268	0.2186	0.1876	0.0428
	12	0.0000	0.0000	0.0001	0.0016	0.0139	0.0634	0.1700	0.2501	0.1285
	13	0.0000	0.0000	0.0000	0.0003	0.0032	0.0219	0.0916	0.2309	0.2669
	14	0.0000	0.0000	0.0000	0.0000	0.0005	0.0047	0.0305	0.1319	0.3432
	15	0.0000	0.0000	0.0000	0.0000	0.0000	0.0005	0.0047	0.0352	0.2059

Statistical Tables

					p				
n	x	.10	.20	.30	.40	.50	.60	.70	.80	.90
16	0	0.1853	0.0281	0.0033	0.0003	0.0000	0.0000	0.0000	0.0000	0.0000
	1	0.3294	0.1126	0.0228	0.0030	0.0002	0.0000	0.0000	0.0000	0.0000
	2	0.2745	0.2111	0.0732	0.0150	0.0018	0.0001	0.0000	0.0000	0.0000
	3	0.1423	0.2463	0.1465	0.0468	0.0085	0.0008	0.0000	0.0000	0.0000
	4	0.0514	0.2001	0.2040	0.1014	0.0278	0.0040	0.0002	0.0000	0.0000
	5	0.0137	0.1201	0.2099	0.1623	0.0667	0.0142	0.0013	0.0000	0.0000
	6	0.0028	0.0550	0.1649	0.1983	0.1222	0.0392	0.0056	0.0002	0.0000
	7	0.0004	0.0197	0.1010	0.1889	0.1746	0.0840	0.0185	0.0012	0.0000
	8	0.0001	0.0055	0.0487	0.1417	0.1964	0.1417	0.0487	0.0055	0.0001
	9	0.0000	0.0012	0.0185	0.0840	0.1746	0.1889	0.1010	0.0197	0.0004
	10	0.0000	0.0002	0.0056	0.0392	0.1222	0.1983	0.1649	0.0550	0.0028
	11	0.0000	0.0000	0.0013	0.0142	0.0667	0.1623	0.2099	0.1201	0.0137
	12	0.0000	0.0000	0.0002	0.0040	0.0278	0.1014	0.2040	0.2001	0.0514
	13	0.0000	0.0000	0.0000	0.0008	0.0085	0.0468	0.1465	0.2463	0.1423
	14	0.0000	0.0000	0.0000	0.0001	0.0018	0.0150	0.0732	0.2111	0.2745
	15	0.0000	0.0000	0.0000	0.0000	0.0002	0.0030	0.0228	0.1126	0.3294
	16	0.0000	0.0000	0.0000	0.0000	0.0000	0.0003	0.0033	0.0281	0.1853
17	0	0.1668	0.0225	0.0023	0.0002	0.0000	0.0000	0.0000	0.0000	0.0000
	1	0.3150	0.0957	0.0169	0.0019	0.0001	0.0000	0.0000	0.0000	0.0000
	2	0.2800	0.1914	0.0581	0.0102	0.0010	0.0001	0.0000	0.0000	0.0000
	3	0.1556	0.2393	0.1245	0.0341	0.0052	0.0004	0.0000	0.0000	0.0000
	4	0.0605	0.2093	0.1868	0.0796	0.0182	0.0021	0.0001	0.0000	0.0000
	5	0.0175	0.1361	0.2081	0.1379	0.0472	0.0081	0.0006	0.0000	0.0000
	6	0.0039	0.0680	0.1784	0.1839	0.0944	0.0242	0.0026	0.0001	0.0000
	7	0.0007	0.0267	0.1201	0.1927	0.1484	0.0571	0.0095	0.0004	0.0000
	8	0.0001	0.0084	0.0644	0.1606	0.1855	0.1070	0.0276	0.0021	0.0000
	9	0.0000	0.0021	0.0276	0.1070	0.1855	0.1606	0.0644	0.0084	0.0001
	10	0.0000	0.0004	0.0095	0.0571	0.1484	0.1927	0.1201	0.0267	0.0007
	11	0.0000	0.0001	0.0026	0.0242	0.0944	0.1839	0.1784	0.0680	0.0039
	12	0.0000	0.0000	0.0006	0.0081	0.0472	0.1379	0.2081	0.1361	0.0175
	13	0.0000	0.0000	0.0001	0.0021	0.0182	0.0796	0.1868	0.2093	0.0605
	14	0.0000	0.0000	0.0000	0.0004	0.0052	0.0341	0.1245	0.2393	0.1556
	15	0.0000	0.0000	0.0000	0.0001	0.0010	0.0102	0.0581	0.1914	0.2800
	16	0.0000	0.0000	0.0000	0.0000	0.0001	0.0019	0.0169	0.0957	0.3150
	17	0.0000	0.0000	0.0000	0.0000	0.0000	0.0002	0.0023	0.0225	0.1668
18	0	0.1501	0.0180	0.0016	0.0001	0.0000	0.0000	0.0000	0.0000	0.0000
	1	0.3002	0.0811	0.0126	0.0012	0.0001	0.0000	0.0000	0.0000	0.0000
	2	0.2835	0.1723	0.0458	0.0069	0.0006	0.0000	0.0000	0.0000	0.0000
	3	0.1680	0.2297	0.1046	0.0246	0.0031	0.0002	0.0000	0.0000	0.0000
	4	0.0700	0.2153	0.1681	0.0614	0.0117	0.0011	0.0000	0.0000	0.0000
	5	0.0218	0.1507	0.2017	0.1146	0.0327	0.0045	0.0002	0.0000	0.0000
	6	0.0052	0.0816	0.1873	0.1655	0.0708	0.0145	0.0012	0.0000	0.0000
	7	0.0010	0.0350	0.1376	0.1892	0.1214	0.0374	0.0046	0.0001	0.0000
	8	0.0002	0.0120	0.0811	0.1734	0.1669	0.0771	0.0149	0.0008	0.0000
	9	0.0000	0.0033	0.0386	0.1284	0.1855	0.1284	0.0386	0.0033	0.0000
	10	0.0000	0.0008	0.0149	0.0771	0.1669	0.1734	0.0811	0.0120	0.0002
	11	0.0000	0.0001	0.0046	0.0374	0.1214	0.1892	0.1376	0.0350	0.0010
	12	0.0000	0.0000	0.0012	0.0145	0.0708	0.1655	0.1873	0.0816	0.0052
	13	0.0000	0.0000	0.0002	0.0045	0.0327	0.1146	0.2017	0.1507	0.0218
	14	0.0000	0.0000	0.0000	0.0011	0.0117	0.0614	0.1681	0.2153	0.0700
	15	0.0000	0.0000	0.0000	0.0002	0.0031	0.0246	0.1046	0.2297	0.1680
	16	0.0000	0.0000	0.0000	0.0000	0.0006	0.0069	0.0458	0.1723	0.2835
	17	0.0000	0.0000	0.0000	0.0000	0.0001	0.0012	0.0126	0.0811	0.3002
	18	0.0000	0.0000	0.0000	0.0000	0.0000	0.0001	0.0016	0.0180	0.1501

Statistical Tables

						p				
n	x	.10	.20	.30	.40	.50	.60	.70	.80	.90
19	0	0.1351	0.0144	0.0011	0.0001	0.0000	0.0000	0.0000	0.0000	0.0000
	1	0.2852	0.0685	0.0093	0.0008	0.0000	0.0000	0.0000	0.0000	0.0000
	2	0.2852	0.1540	0.0358	0.0046	0.0003	0.0000	0.0000	0.0000	0.0000
	3	0.1796	0.2182	0.0869	0.0175	0.0018	0.0001	0.0000	0.0000	0.0000
	4	0.0798	0.2182	0.1491	0.0467	0.0074	0.0005	0.0000	0.0000	0.0000
	5	0.0266	0.1636	0.1916	0.0933	0.0222	0.0024	0.0001	0.0000	0.0000
	6	0.0069	0.0955	0.1916	0.1451	0.0518	0.0085	0.0005	0.0000	0.0000
	7	0.0014	0.0443	0.1525	0.1797	0.0961	0.0237	0.0022	0.0000	0.0000
	8	0.0002	0.0166	0.0981	0.1797	0.1442	0.0532	0.0077	0.0003	0.0000
	9	0.0000	0.0051	0.0514	0.1464	0.1762	0.0976	0.0220	0.0013	0.0000
	10	0.0000	0.0013	0.0220	0.0976	0.1762	0.1464	0.0514	0.0051	0.0000
	11	0.0000	0.0003	0.0077	0.0532	0.1442	0.1797	0.0981	0.0166	0.0002
	12	0.0000	0.0000	0.0022	0.0237	0.0961	0.1797	0.1525	0.0443	0.0014
	13	0.0000	0.0000	0.0005	0.0085	0.0518	0.1451	0.1916	0.0955	0.0069
	14	0.0000	0.0000	0.0001	0.0024	0.0222	0.0933	0.1916	0.1636	0.0266
	15	0.0000	0.0000	0.0000	0.0005	0.0074	0.0467	0.1491	0.2182	0.0798
	16	0.0000	0.0000	0.0000	0.0001	0.0018	0.0175	0.0869	0.2182	0.1796
	17	0.0000	0.0000	0.0000	0.0000	0.0003	0.0046	0.0358	0.1540	0.2852
	18	0.0000	0.0000	0.0000	0.0000	0.0000	0.0008	0.0093	0.0685	0.2852
	19	0.0000	0.0000	0.0000	0.0000	0.0000	0.0001	0.0011	0.0144	0.1351
20	0	0.1216	0.0115	0.0008	0.0000	0.0000	0.0000	0.0000	0.0000	0.0000
	1	0.2702	0.0576	0.0068	0.0005	0.0000	0.0000	0.0000	0.0000	0.0000
	2	0.2852	0.1369	0.0278	0.0031	0.0002	0.0000	0.0000	0.0000	0.0000
	3	0.1901	0.2054	0.0716	0.0123	0.0011	0.0000	0.0000	0.0000	0.0000
	4	0.0898	0.2182	0.1304	0.0350	0.0046	0.0003	0.0000	0.0000	0.0000
	5	0.0319	0.1746	0.1789	0.0746	0.0148	0.0013	0.0000	0.0000	0.0000
	6	0.0089	0.1091	0.1916	0.1244	0.0370	0.0049	0.0002	0.0000	0.0000
	7	0.0020	0.0545	0.1643	0.1659	0.0739	0.0146	0.0010	0.0000	0.0000
	8	0.0004	0.0222	0.1144	0.1797	0.1201	0.0355	0.0039	0.0001	0.0000
	9	0.0001	0.0074	0.0654	0.1597	0.1602	0.0710	0.0120	0.0005	0.0000
	10	0.0000	0.0020	0.0308	0.1171	0.1762	0.1171	0.0308	0.0020	0.0000
	11	0.0000	0.0005	0.0120	0.0710	0.1602	0.1597	0.0654	0.0074	0.0001
	12	0.0000	0.0001	0.0039	0.0355	0.1201	0.1797	0.1144	0.0222	0.0004
	13	0.0000	0.0000	0.0010	0.0146	0.0739	0.1659	0.1643	0.0545	0.0020
	14	0.0000	0.0000	0.0002	0.0049	0.0370	0.1244	0.1916	0.1091	0.0089
	15	0.0000	0.0000	0.0000	0.0013	0.0148	0.0746	0.1789	0.1746	0.0319
	16	0.0000	0.0000	0.0000	0.0003	0.0046	0.0350	0.1304	0.2182	0.0898
	17	0.0000	0.0000	0.0000	0.0000	0.0011	0.0123	0.0716	0.2054	0.1901
	18	0.0000	0.0000	0.0000	0.0000	0.0002	0.0031	0.0278	0.1369	0.2852
	19	0.0000	0.0000	0.0000	0.0000	0.0000	0.0005	0.0068	0.0576	0.2702
	20	0.0000	0.0000	0.0000	0.0000	0.0000	0.0000	0.0008	0.0115	0.1216

Statistical Tables

n	x				p					
		.10	.20	.30	.40	.50	.60	.70	.80	.90
21	0	0.1094	0.0092	0.0006	0.0000	0.0000	0.0000	0.0000	0.0000	0.0000
	1	0.2553	0.0484	0.0050	0.0003	0.0000	0.0000	0.0000	0.0000	0.0000
	2	0.2837	0.1211	0.0215	0.0020	0.0001	0.0000	0.0000	0.0000	0.0000
	3	0.1996	0.1917	0.0585	0.0086	0.0006	0.0000	0.0000	0.0000	0.0000
	4	0.0998	0.2156	0.1128	0.0259	0.0029	0.0001	0.0000	0.0000	0.0000
	5	0.0377	0.1833	0.1643	0.0588	0.0097	0.0007	0.0000	0.0000	0.0000
	6	0.0112	0.1222	0.1878	0.1045	0.0259	0.0027	0.0001	0.0000	0.0000
	7	0.0027	0.0655	0.1725	0.1493	0.0554	0.0087	0.0005	0.0000	0.0000
	8	0.0005	0.0286	0.1294	0.1742	0.0970	0.0229	0.0019	0.0000	0.0000
	9	0.0001	0.0103	0.0801	0.1677	0.1402	0.0497	0.0063	0.0002	0.0000
	10	0.0000	0.0031	0.0412	0.1342	0.1682	0.0895	0.0176	0.0008	0.0000
	11	0.0000	0.0008	0.0176	0.0895	0.1682	0.1342	0.0412	0.0031	0.0000
	12	0.0000	0.0002	0.0063	0.0497	0.1402	0.1677	0.0801	0.0103	0.0001
	13	0.0000	0.0000	0.0019	0.0229	0.0970	0.1742	0.1294	0.0286	0.0005
	14	0.0000	0.0000	0.0005	0.0087	0.0554	0.1493	0.1725	0.0655	0.0027
	15	0.0000	0.0000	0.0001	0.0027	0.0259	0.1045	0.1878	0.1222	0.0112
	16	0.0000	0.0000	0.0000	0.0007	0.0097	0.0588	0.1643	0.1833	0.0377
	17	0.0000	0.0000	0.0000	0.0001	0.0029	0.0259	0.1128	0.2156	0.0998
	18	0.0000	0.0000	0.0000	0.0000	0.0006	0.0086	0.0585	0.1917	0.1996
	19	0.0000	0.0000	0.0000	0.0000	0.0001	0.0020	0.0215	0.1211	0.2837
	20	0.0000	0.0000	0.0000	0.0000	0.0000	0.0003	0.0050	0.0484	0.2553
	21	0.0000	0.0000	0.0000	0.0000	0.0000	0.0000	0.0006	0.0092	0.1094
22	0	0.0985	0.0074	0.0004	0.0000	0.0000	0.0000	0.0000	0.0000	0.0000
	1	0.2407	0.0406	0.0037	0.0002	0.0000	0.0000	0.0000	0.0000	0.0000
	2	0.2808	0.1065	0.0166	0.0014	0.0001	0.0000	0.0000	0.0000	0.0000
	3	0.2080	0.1775	0.0474	0.0060	0.0004	0.0000	0.0000	0.0000	0.0000
	4	0.1098	0.2108	0.0965	0.0190	0.0017	0.0001	0.0000	0.0000	0.0000
	5	0.0439	0.1898	0.1489	0.0456	0.0063	0.0004	0.0000	0.0000	0.0000
	6	0.0138	0.1344	0.1808	0.0862	0.0178	0.0015	0.0000	0.0000	0.0000
	7	0.0035	0.0768	0.1771	0.1314	0.0407	0.0051	0.0002	0.0000	0.0000
	8	0.0007	0.0360	0.1423	0.1642	0.0762	0.0144	0.0009	0.0000	0.0000
	9	0.0001	0.0140	0.0949	0.1703	0.1186	0.0336	0.0032	0.0001	0.0000
	10	0.0000	0.0046	0.0529	0.1476	0.1542	0.0656	0.0097	0.0003	0.0000
	11	0.0000	0.0012	0.0247	0.1073	0.1682	0.1073	0.0247	0.0012	0.0000
	12	0.0000	0.0003	0.0097	0.0656	0.1542	0.1476	0.0529	0.0046	0.0000
	13	0.0000	0.0001	0.0032	0.0336	0.1186	0.1703	0.0949	0.0140	0.0001
	14	0.0000	0.0000	0.0009	0.0144	0.0762	0.1642	0.1423	0.0360	0.0007
	15	0.0000	0.0000	0.0002	0.0051	0.0407	0.1314	0.1771	0.0768	0.0035
	16	0.0000	0.0000	0.0000	0.0015	0.0178	0.0862	0.1808	0.1344	0.0138
	17	0.0000	0.0000	0.0000	0.0004	0.0063	0.0456	0.1489	0.1898	0.0439
	18	0.0000	0.0000	0.0000	0.0001	0.0017	0.0190	0.0965	0.2108	0.1098
	19	0.0000	0.0000	0.0000	0.0000	0.0004	0.0060	0.0474	0.1775	0.2080
	20	0.0000	0.0000	0.0000	0.0000	0.0001	0.0014	0.0166	0.1065	0.2808
	21	0.0000	0.0000	0.0000	0.0000	0.0000	0.0002	0.0037	0.0406	0.2407
	22	0.0000	0.0000	0.0000	0.0000	0.0000	0.0000	0.0004	0.0074	0.0985

Statistical Tables

						p				
n	x	.10	.20	.30	.40	.50	.60	.70	.80	.90
23	0	0.0886	0.0059	0.0003	0.0000	0.0000	0.0000	0.0000	0.0000	0.0000
	1	0.2265	0.0339	0.0027	0.0001	0.0000	0.0000	0.0000	0.0000	0.0000
	2	0.2768	0.0933	0.0127	0.0009	0.0000	0.0000	0.0000	0.0000	0.0000
	3	0.2153	0.1633	0.0382	0.0041	0.0002	0.0000	0.0000	0.0000	0.0000
	4	0.1196	0.2042	0.0818	0.0138	0.0011	0.0000	0.0000	0.0000	0.0000
	5	0.0505	0.1940	0.1332	0.0350	0.0040	0.0002	0.0000	0.0000	0.0000
	6	0.0168	0.1455	0.1712	0.0700	0.0120	0.0008	0.0000	0.0000	0.0000
	7	0.0045	0.0883	0.1782	0.1133	0.0292	0.0029	0.0001	0.0000	0.0000
	8	0.0010	0.0442	0.1527	0.1511	0.0584	0.0088	0.0004	0.0000	0.0000
	9	0.0002	0.0184	0.1091	0.1679	0.0974	0.0221	0.0016	0.0000	0.0000
	10	0.0000	0.0064	0.0655	0.1567	0.1364	0.0464	0.0052	0.0001	0.0000
	11	0.0000	0.0019	0.0332	0.1234	0.1612	0.0823	0.0142	0.0005	0.0000
	12	0.0000	0.0005	0.0142	0.0823	0.1612	0.1234	0.0332	0.0019	0.0000
	13	0.0000	0.0001	0.0052	0.0464	0.1364	0.1567	0.0655	0.0064	0.0000
	14	0.0000	0.0000	0.0016	0.0221	0.0974	0.1679	0.1091	0.0184	0.0002
	15	0.0000	0.0000	0.0004	0.0088	0.0584	0.1511	0.1527	0.0442	0.0010
	16	0.0000	0.0000	0.0001	0.0029	0.0292	0.1133	0.1782	0.0883	0.0045
	17	0.0000	0.0000	0.0000	0.0008	0.0120	0.0700	0.1712	0.1455	0.0168
	18	0.0000	0.0000	0.0000	0.0002	0.0040	0.0350	0.1332	0.1940	0.0505
	19	0.0000	0.0000	0.0000	0.0000	0.0011	0.0138	0.0818	0.2042	0.1196
	20	0.0000	0.0000	0.0000	0.0000	0.0002	0.0041	0.0382	0.1633	0.2153
	21	0.0000	0.0000	0.0000	0.0000	0.0000	0.0009	0.0127	0.0933	0.2768
	22	0.0000	0.0000	0.0000	0.0000	0.0000	0.0001	0.0027	0.0339	0.2265
	23	0.0000	0.0000	0.0000	0.0000	0.0000	0.0000	0.0003	0.0059	0.0886
24	0	0.0798	0.0047	0.0002	0.0000	0.0000	0.0000	0.0000	0.0000	0.0000
	1	0.2127	0.0283	0.0020	0.0001	0.0000	0.0000	0.0000	0.0000	0.0000
	2	0.2718	0.0815	0.0097	0.0006	0.0000	0.0000	0.0000	0.0000	0.0000
	3	0.2215	0.1493	0.0305	0.0028	0.0001	0.0000	0.0000	0.0000	0.0000
	4	0.1292	0.1960	0.0687	0.0099	0.0006	0.0000	0.0000	0.0000	0.0000
	5	0.0574	0.1960	0.1177	0.0265	0.0025	0.0001	0.0000	0.0000	0.0000
	6	0.0202	0.1552	0.1598	0.0560	0.0080	0.0004	0.0000	0.0000	0.0000
	7	0.0058	0.0998	0.1761	0.0960	0.0206	0.0017	0.0000	0.0000	0.0000
	8	0.0014	0.0530	0.1604	0.1360	0.0438	0.0053	0.0002	0.0000	0.0000
	9	0.0003	0.0236	0.1222	0.1612	0.0779	0.0141	0.0008	0.0000	0.0000
	10	0.0000	0.0088	0.0785	0.1612	0.1169	0.0318	0.0026	0.0000	0.0000
	11	0.0000	0.0028	0.0428	0.1367	0.1488	0.0608	0.0079	0.0002	0.0000
	12	0.0000	0.0008	0.0199	0.0988	0.1612	0.0988	0.0199	0.0008	0.0000
	13	0.0000	0.0002	0.0079	0.0608	0.1488	0.1367	0.0428	0.0028	0.0000
	14	0.0000	0.0000	0.0026	0.0318	0.1169	0.1612	0.0785	0.0088	0.0000
	15	0.0000	0.0000	0.0008	0.0141	0.0779	0.1612	0.1222	0.0236	0.0003
	16	0.0000	0.0000	0.0002	0.0053	0.0438	0.1360	0.1604	0.0530	0.0014
	17	0.0000	0.0000	0.0000	0.0017	0.0206	0.0960	0.1761	0.0998	0.0058
	18	0.0000	0.0000	0.0000	0.0004	0.0080	0.0560	0.1598	0.1552	0.0202
	19	0.0000	0.0000	0.0000	0.0001	0.0025	0.0265	0.1177	0.1960	0.0574
	20	0.0000	0.0000	0.0000	0.0000	0.0006	0.0099	0.0687	0.1960	0.1292
	21	0.0000	0.0000	0.0000	0.0000	0.0001	0.0028	0.0305	0.1493	0.2215
	22	0.0000	0.0000	0.0000	0.0000	0.0000	0.0006	0.0097	0.0815	0.2718
	23	0.0000	0.0000	0.0000	0.0000	0.0000	0.0001	0.0020	0.0283	0.2127
	24	0.0000	0.0000	0.0000	0.0000	0.0000	0.0000	0.0002	0.0047	0.0798

Statistical Tables

n	x	.10	.20	.30	.40	p .50	.60	.70	.80	.90
25	0	0.0718	0.0038	0.0001	0.0000	0.0000	0.0000	0.0000	0.0000	0.0000
	1	0.1994	0.0236	0.0014	0.0000	0.0000	0.0000	0.0000	0.0000	0.0000
	2	0.2659	0.0708	0.0074	0.0004	0.0000	0.0000	0.0000	0.0000	0.0000
	3	0.2265	0.1358	0.0243	0.0019	0.0001	0.0000	0.0000	0.0000	0.0000
	4	0.1384	0.1867	0.0572	0.0071	0.0004	0.0000	0.0000	0.0000	0.0000
	5	0.0646	0.1960	0.1030	0.0199	0.0016	0.0000	0.0000	0.0000	0.0000
	6	0.0239	0.1633	0.1472	0.0442	0.0053	0.0002	0.0000	0.0000	0.0000
	7	0.0072	0.1108	0.1712	0.0800	0.0143	0.0009	0.0000	0.0000	0.0000
	8	0.0018	0.0623	0.1651	0.1200	0.0322	0.0031	0.0001	0.0000	0.0000
	9	0.0004	0.0294	0.1336	0.1511	0.0609	0.0088	0.0004	0.0000	0.0000
	10	0.0001	0.0118	0.0916	0.1612	0.0974	0.0212	0.0013	0.0000	0.0000
	11	0.0000	0.0040	0.0536	0.1465	0.1328	0.0434	0.0042	0.0001	0.0000
	12	0.0000	0.0012	0.0268	0.1140	0.1550	0.0760	0.0115	0.0003	0.0000
	13	0.0000	0.0003	0.0115	0.0760	0.1550	0.1140	0.0268	0.0012	0.0000
	14	0.0000	0.0001	0.0042	0.0434	0.1328	0.1465	0.0536	0.0040	0.0000
	15	0.0000	0.0000	0.0013	0.0212	0.0974	0.1612	0.0916	0.0118	0.0001
	16	0.0000	0.0000	0.0004	0.0088	0.0609	0.1511	0.1336	0.0294	0.0004
	17	0.0000	0.0000	0.0001	0.0031	0.0322	0.1200	0.1651	0.0623	0.0018
	18	0.0000	0.0000	0.0000	0.0009	0.0143	0.0800	0.1712	0.1108	0.0072
	19	0.0000	0.0000	0.0000	0.0002	0.0053	0.0442	0.1472	0.1633	0.0239
	20	0.0000	0.0000	0.0000	0.0000	0.0016	0.0199	0.1030	0.1960	0.0646
	21	0.0000	0.0000	0.0000	0.0000	0.0004	0.0071	0.0572	0.1867	0.1384
	22	0.0000	0.0000	0.0000	0.0000	0.0001	0.0019	0.0243	0.1358	0.2265
	23	0.0000	0.0000	0.0000	0.0000	0.0000	0.0004	0.0074	0.0708	0.2659
	24	0.0000	0.0000	0.0000	0.0000	0.0000	0.0000	0.0014	0.0236	0.1994
	25	0.0000	0.0000	0.0000	0.0000	0.0000	0.0000	0.0001	0.0038	0.0718

Statistical Tables

Table 2. Probabilities of the Standard Normal Distribution Z

Table entries represent P(Z < Z_i)

e.g., P(Z < -1.96) = 0.0250, P(Z < 1.96) = 0.9750

Z_i	.00	.01	.02	.03	.04	.05	.06	.07	.08	.09
-3.0	0.0013	0.0013	0.0013	0.0012	0.0012	0.0011	0.0011	0.0011	0.0010	0.0010
-2.9	0.0019	0.0018	0.0018	0.0017	0.0016	0.0016	0.0015	0.0015	0.0014	0.0014
-2.8	0.0026	0.0025	0.0024	0.0023	0.0023	0.0022	0.0021	0.0021	0.0020	0.0019
-2.7	0.0035	0.0034	0.0033	0.0032	0.0031	0.0030	0.0029	0.0028	0.0027	0.0026
-2.6	0.0047	0.0045	0.0044	0.0043	0.0041	0.0040	0.0039	0.0038	0.0037	0.0036
-2.5	0.0062	0.0060	0.0059	0.0057	0.0055	0.0054	0.0052	0.0051	0.0049	0.0048
-2.4	0.0082	0.0080	0.0078	0.0075	0.0073	0.0071	0.0069	0.0068	0.0066	0.0064
-2.3	0.0107	0.0104	0.0102	0.0099	0.0096	0.0094	0.0091	0.0089	0.0087	0.0084
-2.2	0.0139	0.0136	0.0132	0.0129	0.0125	0.0122	0.0119	0.0116	0.0113	0.0110
-2.1	0.0179	0.0174	0.0170	0.0166	0.0162	0.0158	0.0154	0.0150	0.0146	0.0143
-2.0	0.0228	0.0222	0.0217	0.0212	0.0207	0.0202	0.0197	0.0192	0.0188	0.0183
-1.9	0.0287	0.0281	0.0274	0.0268	0.0262	0.0256	0.0250	0.0244	0.0239	0.0233
-1.8	0.0359	0.0351	0.0344	0.0336	0.0329	0.0322	0.0314	0.0307	0.0301	0.0294
-1.7	0.0446	0.0436	0.0427	0.0418	0.0409	0.0401	0.0392	0.0384	0.0375	0.0367
-1.6	0.0548	0.0537	0.0526	0.0516	0.0505	0.0495	0.0485	0.0475	0.0465	0.0455
-1.5	0.0668	0.0655	0.0643	0.0630	0.0618	0.0606	0.0594	0.0582	0.0571	0.0559
-1.4	0.0808	0.0793	0.0778	0.0764	0.0749	0.0735	0.0721	0.0708	0.0694	0.0681
-1.3	0.0968	0.0951	0.0934	0.0918	0.0901	0.0885	0.0869	0.0853	0.0838	0.0823
-1.2	0.1151	0.1131	0.1112	0.1093	0.1075	0.1056	0.1038	0.1020	0.1003	0.0985
-1.1	0.1357	0.1335	0.1314	0.1292	0.1271	0.1251	0.1230	0.1210	0.1190	0.1170
-1.0	0.1587	0.1562	0.1539	0.1515	0.1492	0.1469	0.1446	0.1423	0.1401	0.1379
-0.9	0.1841	0.1814	0.1788	0.1762	0.1736	0.1711	0.1685	0.1660	0.1635	0.1611
-0.8	0.2119	0.2090	0.2061	0.2033	0.2005	0.1977	0.1949	0.1922	0.1894	0.1867
-0.7	0.2420	0.2389	0.2358	0.2327	0.2296	0.2266	0.2236	0.2206	0.2177	0.2148
-0.6	0.2743	0.2709	0.2676	0.2643	0.2611	0.2578	0.2546	0.2514	0.2483	0.2451
-0.5	0.3085	0.3050	0.3015	0.2981	0.2946	0.2912	0.2877	0.2843	0.2810	0.2776
-0.4	0.3446	0.3409	0.3372	0.3336	0.3300	0.3264	0.3228	0.3192	0.3156	0.3121
-0.3	0.3821	0.3783	0.3745	0.3707	0.3669	0.3632	0.3594	0.3557	0.3520	0.3483
-0.2	0.4207	0.4168	0.4129	0.4090	0.4052	0.4013	0.3974	0.3936	0.3897	0.3859
-0.1	0.4602	0.4562	0.4522	0.4483	0.4443	0.4404	0.4364	0.4325	0.4286	0.4247
-0.0	0.5000	0.4960	0.4920	0.4880	0.4840	0.4801	0.4761	0.4721	0.4681	0.4641

Statistical Tables

Table 2. Probabilities of the Standard Normal Distribution Z (continued)

Table entries represent $P(Z < Z_i)$

e.g., $P(Z < -1.96) = 0.0250$, $P(Z < 1.96) = 0.9750$

Z_i	.00	.01	.02	.03	.04	.05	.06	.07	.08	.09
0.0	0.5000	0.5040	0.5080	0.5120	0.5160	0.5199	0.5239	0.5279	0.5319	0.5359
0.1	0.5398	0.5438	0.5478	0.5517	0.5557	0.5596	0.5636	0.5675	0.5714	0.5753
0.2	0.5793	0.5832	0.5871	0.5910	0.5948	0.5987	0.6026	0.6064	0.6103	0.6141
0.3	0.6179	0.6217	0.6255	0.6293	0.6331	0.6368	0.6406	0.6443	0.6480	0.6517
0.4	0.6554	0.6591	0.6628	0.6664	0.6700	0.6736	0.6772	0.6808	0.6844	0.6879
0.5	0.6915	0.6950	0.6985	0.7019	0.7054	0.7088	0.7123	0.7157	0.7190	0.7224
0.6	0.7257	0.7291	0.7324	0.7357	0.7389	0.7422	0.7454	0.7486	0.7517	0.7549
0.7	0.7580	0.7611	0.7642	0.7673	0.7704	0.7734	0.7764	0.7794	0.7823	0.7852
0.8	0.7881	0.7910	0.7939	0.7967	0.7995	0.8023	0.8051	0.8078	0.8106	0.8133
0.9	0.8159	0.8186	0.8212	0.8238	0.8264	0.8289	0.8315	0.8340	0.8365	0.8389
1.0	0.8413	0.8438	0.8461	0.8485	0.8508	0.8531	0.8554	0.8577	0.8599	0.8621
1.1	0.8643	0.8665	0.8686	0.8708	0.8729	0.8749	0.8770	0.8790	0.8810	0.8830
1.2	0.8849	0.8869	0.8888	0.8907	0.8925	0.8944	0.8962	0.8980	0.8997	0.9015
1.3	0.9032	0.9049	0.9066	0.9082	0.9099	0.9115	0.9131	0.9147	0.9162	0.9177
1.4	0.9192	0.9207	0.9222	0.9236	0.9251	0.9265	0.9279	0.9292	0.9306	0.9319
1.5	0.9332	0.9345	0.9357	0.9370	0.9382	0.9394	0.9406	0.9418	0.9429	0.9441
1.6	0.9452	0.9463	0.9474	0.9484	0.9495	0.9505	0.9515	0.9525	0.9535	0.9545
1.7	0.9554	0.9564	0.9573	0.9582	0.9591	0.9599	0.9608	0.9616	0.9625	0.9633
1.8	0.9641	0.9649	0.9656	0.9664	0.9671	0.9678	0.9686	0.9693	0.9699	0.9706
1.9	0.9713	0.9719	0.9726	0.9732	0.9738	0.9744	0.9750	0.9756	0.9761	0.9767
2.0	0.9772	0.9778	0.9783	0.9788	0.9793	0.9798	0.9803	0.9808	0.9812	0.9817
2.1	0.9821	0.9826	0.9830	0.9834	0.9838	0.9842	0.9846	0.9850	0.9854	0.9857
2.2	0.9861	0.9864	0.9868	0.9871	0.9875	0.9878	0.9881	0.9884	0.9887	0.9890
2.3	0.9893	0.9896	0.9898	0.9901	0.9904	0.9906	0.9909	0.9911	0.9913	0.9916
2.4	0.9918	0.9920	0.9922	0.9925	0.9927	0.9929	0.9931	0.9932	0.9934	0.9936
2.5	0.9938	0.9940	0.9941	0.9943	0.9945	0.9946	0.9948	0.9949	0.9951	0.9952
2.6	0.9953	0.9955	0.9956	0.9957	0.9959	0.9960	0.9961	0.9962	0.9963	0.9964
2.7	0.9965	0.9966	0.9967	0.9968	0.9969	0.9970	0.9971	0.9972	0.9973	0.9974
2.8	0.9974	0.9975	0.9976	0.9977	0.9977	0.9978	0.9979	0.9979	0.9980	0.9981
2.9	0.9981	0.9982	0.9982	0.9983	0.9984	0.9984	0.9985	0.9985	0.9986	0.9986
3.0	0.9987	0.9987	0.9987	0.9988	0.9988	0.9989	0.9989	0.9989	0.9990	0.9990

Statistical Tables

Table 2A: Z Values for Confidence Intervals

Confidence Level	$Z_{1-\alpha/2}$	α (Total Tail Area)
99.99%	3.819	0.0001
99.9%	3.291	0.001
99%	2.576	0.01
95%	1.960	0.05
90%	1.645	0.10
80%	1.282	0.20

Table 2B   Z Values for Tests of Hypothesis

Lower Tailed Tests	α	Z_{α}	Decision Rule
$H_0: \mu = \mu_0$	.0001	-3.719	Reject H_0 if $Z \le Z_{\alpha}$
$H_1: \mu < \mu_0$	.001	-3.090	
	.005	-2.576	
	.010	-2.326	
	.025	-1.960	
	.050	-1.645	
	.100	-1.282	
Upper Tailed Tests	α	$Z_{1-\alpha}$	**Decision Rule**
$H_0: \mu = \mu_0$	.0001	3.719	Reject H_0 if $Z \ge Z_{1-\alpha}$
$H_1: \mu > \mu_0$	.001	3.090	
	.005	2.576	
	.010	2.326	
	.025	1.960	
	.050	1.645	
	.100	1.282	
Two Tailed Tests	α	$Z_{1-\alpha/2}$	**Decision Rule**
$H_0: \mu = \mu_0$	.0001	3.819	Reject H_0 if $Z \le -Z_{1-\alpha/2}$
$H_1: \mu \ne \mu_0$	.001	3.291	or if $Z \ge Z_{1-\alpha/2}$
	.010	2.576	
	.050	1.960	
	.100	1.645	
	.200	1.282	

Statistical Tables

Table 3. Critical Values of the t Distribution

Table entries represent values from t distribution with upper tail area equal to α.

e.g., $P(t_{df} > t) = \alpha$, $P(t_6 > 1.943) = 0.05$

Confidence Level CL[#]		80%	90%	95%	98%	99%
Two Sided α	α_2^*	.20	.10	.05	.02	.01
One Sided α	α	.10	.05	.025	.01	.005
df						
1		3.078	6.314	12.71	31.82	63.66
2		1.886	2.920	4.303	6.965	9.925
3		1.638	2.353	3.182	4.541	5.841
4		1.533	2.132	2.776	3.747	4.604
5		1.476	2.015	2.571	3.365	4.032
6		1.440	1.943	2.447	3.143	3.707
7		1.415	1.895	2.365	2.998	3.499
8		1.397	1.860	2.306	2.896	3.355
9		1.383	1.833	2.262	2.821	3.250
10		1.372	1.812	2.228	2.764	3.169
11		1.363	1.796	2.201	2.718	3.106
12		1.356	1.782	2.179	2.681	3.055
13		1.350	1.771	2.160	2.650	3.012
14		1.345	1.761	2.145	2.624	2.977
15		1.341	1.753	2.131	2.602	2.947
16		1.337	1.746	2.120	2.583	2.921
17		1.333	1.740	2.110	2.567	2.898
18		1.330	1.734	2.101	2.552	2.878
19		1.328	1.729	2.093	2.539	2.861
20		1.325	1.725	2.086	2.528	2.845
21		1.323	1.721	2.080	2.518	2.831
22		1.321	1.717	2.074	2.508	2.819
23		1.319	1.714	2.069	2.500	2.807
24		1.318	1.711	2.064	2.492	2.797
25		1.316	1.708	2.060	2.485	2.787
26		1.315	1.706	2.056	2.479	2.779
27		1.314	1.703	2.052	2.473	2.771
28		1.313	1.701	2.048	2.467	2.763
29		1.311	1.699	2.045	2.462	2.756
30		1.310	1.697	2.042	2.457	2.750
31		1.309	1.696	2.040	2.453	2.744
32		1.309	1.694	2.037	2.449	2.738
33		1.308	1.692	2.035	2.445	2.733
34		1.307	1.691	2.032	2.441	2.728
35		1.306	1.690	2.030	2.438	2.724 (continued)

Statistical Tables

Confidence Level CL[#]		80%	90%	95%	98%	99%
Two Sided α	α_2[*]	.20	.10	.05	.02	.01
One Sided α	α	.10	.05	.025	.01	.005
df						
36		1.306	1.688	2.028	2.434	2.719
37		1.305	1.687	2.026	2.431	2.715
38		1.304	1.686	2.024	2.429	2.712
39		1.304	1.685	2.023	2.426	2.708
40		1.303	1.684	2.021	2.423	2.704
41		1.303	1.683	2.020	2.421	2.701
42		1.302	1.682	2.018	2.418	2.698
43		1.302	1.681	2.017	2.416	2.695
44		1.301	1.680	2.015	2.414	2.692
45		1.301	1.679	2.014	2.412	2.690
46		1.300	1.679	2.013	2.410	2.687
47		1.300	1.678	2.012	2.408	2.685
48		1.299	1.677	2.011	2.407	2.682
49		1.299	1.677	2.010	2.405	2.680
50		1.299	1.676	2.009	2.403	2.678
51		1.298	1.675	2.008	2.402	2.676
52		1.298	1.675	2.007	2.400	2.674
53		1.298	1.674	2.006	2.399	2.672
54		1.297	1.674	2.005	2.397	2.670
55		1.297	1.673	2.004	2.396	2.668
56		1.297	1.673	2.003	2.395	2.667
57		1.297	1.672	2.002	2.394	2.665
58		1.296	1.672	2.002	2.392	2.663
59		1.296	1.671	2.001	2.391	2.662
60		1.296	1.671	2.000	2.390	2.660
61		1.296	1.670	2.000	2.389	2.659
62		1.295	1.670	1.999	2.388	2.657
63		1.295	1.669	1.998	2.387	2.656
64		1.295	1.669	1.998	2.386	2.655
65		1.295	1.669	1.997	2.385	2.654
66		1.295	1.668	1.997	2.384	2.652
67		1.294	1.668	1.996	2.383	2.651
68		1.294	1.668	1.995	2.382	2.650
69		1.294	1.667	1.995	2.382	2.649
70		1.294	1.667	1.994	2.381	2.648
71		1.294	1.667	1.994	2.380	2.647
72		1.293	1.666	1.993	2.379	2.646
73		1.293	1.666	1.993	2.379	2.645
74		1.293	1.666	1.993	2.378	2.644
75		1.293	1.665	1.992	2.377	2.643
∞		1.282	1.645	1.960	2.326	2.576

Statistical Tables

Table 4a. Critical Values of the F Distribution with Upper Tail Area = 0.05

$$P(F > F_{df1,df2}) = 0.05$$

$$e.g., P(F_{3,20} > 3.10) = 0.05$$

denominator df (df2)	1	2	3	4	5	numerator df (df1) 6	7	8	9	10	20	30	40	50
1	161.4	199.5	215.7	224.6	230.2	234.0	236.8	238.9	240.5	241.9	248.0	250.1	251.1	251.8
2	18.51	19.00	19.16	19.25	19.30	19.33	19.35	19.37	19.38	19.40	19.45	19.46	19.47	19.48
3	10.13	9.55	9.28	9.12	9.01	8.94	8.89	8.85	8.81	8.79	8.66	8.62	8.59	8.58
4	7.71	6.94	6.59	6.39	6.26	6.16	6.09	6.04	6.00	5.96	5.80	5.75	5.72	5.70
5	6.61	5.79	5.41	5.19	5.05	4.95	4.88	4.82	4.77	4.74	4.56	4.50	4.46	4.44
6	5.99	5.14	4.76	4.53	4.39	4.28	4.21	4.15	4.10	4.06	3.87	3.81	3.77	3.75
7	5.59	4.74	4.35	4.12	3.97	3.87	3.79	3.73	3.68	3.64	3.44	3.38	3.34	3.32
8	5.32	4.46	4.07	3.84	3.69	3.58	3.50	3.44	3.39	3.35	3.15	3.08	3.04	3.02
9	5.12	4.26	3.86	3.63	3.48	3.37	3.29	3.23	3.18	3.14	2.94	2.86	2.83	2.80
10	4.96	4.10	3.71	3.48	3.33	3.22	3.14	3.07	3.02	2.98	2.77	2.70	2.66	2.64
11	4.84	3.98	3.59	3.36	3.20	3.09	3.01	2.95	2.90	2.85	2.65	2.57	2.53	2.51
12	4.75	3.89	3.49	3.26	3.11	3.00	2.91	2.85	2.80	2.75	2.54	2.47	2.43	2.40
13	4.67	3.81	3.41	3.18	3.03	2.92	2.83	2.77	2.71	2.67	2.46	2.38	2.34	2.31
14	4.60	3.74	3.34	3.11	2.96	2.85	2.76	2.70	2.65	2.60	2.39	2.31	2.27	2.24
15	4.54	3.68	3.29	3.06	2.90	2.79	2.71	2.64	2.59	2.54	2.33	2.25	2.20	2.18
16	4.49	3.63	3.24	3.01	2.85	2.74	2.66	2.59	2.54	2.49	2.28	2.19	2.15	2.12
17	4.45	3.59	3.20	2.96	2.81	2.70	2.61	2.55	2.49	2.45	2.23	2.15	2.10	2.08
18	4.41	3.55	3.16	2.93	2.77	2.66	2.58	2.51	2.46	2.41	2.19	2.11	2.06	2.04
19	4.38	3.52	3.13	2.90	2.74	2.63	2.54	2.48	2.42	2.38	2.16	2.07	2.03	2.00
20	4.35	3.49	3.10	2.87	2.71	2.60	2.51	2.45	2.39	2.35	2.12	2.04	1.99	1.97
21	4.32	3.47	3.07	2.84	2.68	2.57	2.49	2.42	2.37	2.32	2.10	2.01	1.96	1.94
22	4.30	3.44	3.05	2.82	2.66	2.55	2.46	2.40	2.34	2.30	2.07	1.98	1.94	1.91
23	4.28	3.42	3.03	2.80	2.64	2.53	2.44	2.37	2.32	2.27	2.05	1.96	1.91	1.88
24	4.26	3.40	3.01	2.78	2.62	2.51	2.42	2.36	2.30	2.25	2.03	1.94	1.89	1.86
25	4.24	3.39	2.99	2.76	2.60	2.49	2.40	2.34	2.28	2.24	2.01	1.92	1.87	1.84
26	4.23	3.37	2.98	2.74	2.59	2.47	2.39	2.32	2.27	2.22	1.99	1.90	1.85	1.82
27	4.21	3.35	2.96	2.73	2.57	2.46	2.37	2.31	2.25	2.20	1.97	1.88	1.84	1.81
28	4.20	3.34	2.95	2.71	2.56	2.45	2.36	2.29	2.24	2.19	1.96	1.87	1.82	1.79
29	4.18	3.33	2.93	2.70	2.55	2.43	2.35	2.28	2.22	2.18	1.94	1.85	1.81	1.77
30	4.17	3.32	2.92	2.69	2.53	2.42	2.33	2.27	2.21	2.16	1.93	1.84	1.79	1.76
31	4.16	3.30	2.91	2.68	2.52	2.41	2.32	2.25	2.20	2.15	1.92	1.83	1.78	1.75
32	4.15	3.29	2.90	2.67	2.51	2.40	2.31	2.24	2.19	2.14	1.91	1.82	1.77	1.74
33	4.14	3.28	2.89	2.66	2.50	2.39	2.30	2.23	2.18	2.13	1.90	1.81	1.76	1.72
34	4.13	3.28	2.88	2.65	2.49	2.38	2.29	2.23	2.17	2.12	1.89	1.80	1.75	1.71
35	4.12	3.27	2.87	2.64	2.49	2.37	2.29	2.22	2.16	2.11	1.88	1.79	1.74	1.70
36	4.11	3.26	2.87	2.63	2.48	2.36	2.28	2.21	2.15	2.11	1.87	1.78	1.73	1.69
37	4.11	3.25	2.86	2.63	2.47	2.36	2.27	2.20	2.14	2.10	1.86	1.77	1.72	1.68
38	4.10	3.24	2.85	2.62	2.46	2.35	2.26	2.19	2.14	2.09	1.85	1.76	1.71	1.68
39	4.09	3.24	2.85	2.61	2.46	2.34	2.26	2.19	2.13	2.08	1.85	1.75	1.70	1.67
40	4.08	3.23	2.84	2.61	2.45	2.34	2.25	2.18	2.12	2.08	1.84	1.74	1.69	1.66
41	4.08	3.23	2.83	2.60	2.44	2.33	2.24	2.17	2.12	2.07	1.83	1.74	1.69	1.65
42	4.07	3.22	2.83	2.59	2.44	2.32	2.24	2.17	2.11	2.06	1.83	1.73	1.68	1.65
43	4.07	3.21	2.82	2.59	2.43	2.32	2.23	2.16	2.11	2.06	1.82	1.72	1.67	1.64
44	4.06	3.21	2.82	2.58	2.43	2.31	2.23	2.16	2.10	2.05	1.81	1.72	1.67	1.63
45	4.06	3.20	2.81	2.58	2.42	2.31	2.22	2.15	2.10	2.05	1.81	1.71	1.66	1.63
46	4.05	3.20	2.81	2.57	2.42	2.30	2.22	2.15	2.09	2.04	1.80	1.71	1.65	1.62
47	4.05	3.20	2.80	2.57	2.41	2.30	2.21	2.14	2.09	2.04	1.80	1.70	1.65	1.61
48	4.04	3.19	2.80	2.57	2.41	2.29	2.21	2.14	2.08	2.03	1.79	1.70	1.64	1.61
49	4.04	3.19	2.79	2.56	2.40	2.29	2.20	2.13	2.08	2.03	1.79	1.69	1.64	1.60
50	4.03	3.18	2.79	2.56	2.40	2.29	2.20	2.13	2.07	2.03	1.78	1.69	1.63	1.60
75	3.97	3.12	2.73	2.49	2.34	2.22	2.13	2.06	2.01	1.96	1.71	1.61	1.55	1.52
100	3.94	3.09	2.70	2.46	2.31	2.19	2.10	2.03	1.97	1.93	1.68	1.57	1.52	1.48
125	3.92	3.07	2.68	2.44	2.29	2.17	2.08	2.01	1.96	1.91	1.66	1.55	1.49	1.45
150	3.90	3.06	2.66	2.43	2.27	2.16	2.07	2.00	1.94	1.89	1.64	1.54	1.48	1.44
175	3.90	3.05	2.66	2.42	2.27	2.15	2.06	1.99	1.93	1.89	1.63	1.52	1.46	1.42
200	3.89	3.04	2.65	2.42	2.26	2.14	2.06	1.98	1.93	1.88	1.62	1.52	1.46	1.41

Statistical Tables

Table 4b. Critical Values of the F Distribution with Upper Tail Area = 0.025

$$P(F > F_{df1,df2}) = 0.025$$

$$e.g., P(F_{3,20} > 3.86) = 0.025$$

denominator df (df2)	\multicolumn{13}{c}{numerator df (df1)}												
	1	2	3	4	5	6	7	8	9	10	20	30	40	50
1	647.8	799.5	864.2	899.6	921.8	937.1	948.2	956.7	963.3	968.6	993.1	1001	1006	1008
2	38.51	39.00	39.17	39.25	39.30	39.33	39.36	39.37	39.39	39.40	39.45	39.46	39.47	39.48
3	17.44	16.04	15.44	15.10	14.88	14.73	14.62	14.54	14.47	14.42	14.17	14.08	14.04	14.01
4	12.22	10.65	9.98	9.60	9.36	9.20	9.07	8.98	8.90	8.84	8.56	8.46	8.41	8.38
5	10.01	8.43	7.76	7.39	7.15	6.98	6.85	6.76	6.68	6.62	6.33	6.23	6.18	6.14
6	8.81	7.26	6.60	6.23	5.99	5.82	5.70	5.60	5.52	5.46	5.17	5.07	5.01	4.98
7	8.07	6.54	5.89	5.52	5.29	5.12	4.99	4.90	4.82	4.76	4.47	4.36	4.31	4.28
8	7.57	6.06	5.42	5.05	4.82	4.65	4.53	4.43	4.36	4.30	4.00	3.89	3.84	3.81
9	7.21	5.71	5.08	4.72	4.48	4.32	4.20	4.10	4.03	3.96	3.67	3.56	3.51	3.47
10	6.94	5.46	4.83	4.47	4.24	4.07	3.95	3.85	3.78	3.72	3.42	3.31	3.26	3.22
11	6.72	5.26	4.63	4.28	4.04	3.88	3.76	3.66	3.59	3.53	3.23	3.12	3.06	3.03
12	6.55	5.10	4.47	4.12	3.89	3.73	3.61	3.51	3.44	3.37	3.07	2.96	2.91	2.87
13	6.41	4.97	4.35	4.00	3.77	3.60	3.48	3.39	3.31	3.25	2.95	2.84	2.78	2.74
14	6.30	4.86	4.24	3.89	3.66	3.50	3.38	3.29	3.21	3.15	2.84	2.73	2.67	2.64
15	6.20	4.77	4.15	3.80	3.58	3.41	3.29	3.20	3.12	3.06	2.76	2.64	2.59	2.55
16	6.12	4.69	4.08	3.73	3.50	3.34	3.22	3.12	3.05	2.99	2.68	2.57	2.51	2.47
17	6.04	4.62	4.01	3.66	3.44	3.28	3.16	3.06	2.98	2.92	2.62	2.50	2.44	2.41
18	5.98	4.56	3.95	3.61	3.38	3.22	3.10	3.01	2.93	2.87	2.56	2.44	2.38	2.35
19	5.92	4.51	3.90	3.56	3.33	3.17	3.05	2.96	2.88	2.82	2.51	2.39	2.33	2.30
20	5.87	4.46	3.86	3.51	3.29	3.13	3.01	2.91	2.84	2.77	2.46	2.35	2.29	2.25
21	5.83	4.42	3.82	3.48	3.25	3.09	2.97	2.87	2.80	2.73	2.42	2.31	2.25	2.21
22	5.79	4.38	3.78	3.44	3.22	3.05	2.93	2.84	2.76	2.70	2.39	2.27	2.21	2.17
23	5.75	4.35	3.75	3.41	3.18	3.02	2.90	2.81	2.73	2.67	2.36	2.24	2.18	2.14
24	5.72	4.32	3.72	3.38	3.15	2.99	2.87	2.78	2.70	2.64	2.33	2.21	2.15	2.11
25	5.69	4.29	3.69	3.35	3.13	2.97	2.85	2.75	2.68	2.61	2.30	2.18	2.12	2.08
26	5.66	4.27	3.67	3.33	3.10	2.94	2.82	2.73	2.65	2.59	2.28	2.16	2.09	2.05
27	5.63	4.24	3.65	3.31	3.08	2.92	2.80	2.71	2.63	2.57	2.25	2.13	2.07	2.03
28	5.61	4.22	3.63	3.29	3.06	2.90	2.78	2.69	2.61	2.55	2.23	2.11	2.05	2.01
29	5.59	4.20	3.61	3.27	3.04	2.88	2.76	2.67	2.59	2.53	2.21	2.09	2.03	1.99
30	5.57	4.18	3.59	3.25	3.03	2.87	2.75	2.65	2.57	2.51	2.20	2.07	2.01	1.97
31	5.55	4.16	3.57	3.23	3.01	2.85	2.73	2.64	2.56	2.50	2.18	2.06	1.99	1.95
32	5.53	4.15	3.56	3.22	3.00	2.84	2.71	2.62	2.54	2.48	2.16	2.04	1.98	1.93
33	5.51	4.13	3.54	3.20	2.98	2.82	2.70	2.61	2.53	2.47	2.15	2.03	1.96	1.92
34	5.50	4.12	3.53	3.19	2.97	2.81	2.69	2.59	2.52	2.45	2.13	2.01	1.95	1.90
35	5.48	4.11	3.52	3.18	2.96	2.80	2.68	2.58	2.50	2.44	2.12	2.00	1.93	1.89
36	5.47	4.09	3.50	3.17	2.94	2.78	2.66	2.57	2.49	2.43	2.11	1.99	1.92	1.88
37	5.46	4.08	3.49	3.16	2.93	2.77	2.65	2.56	2.48	2.42	2.10	1.97	1.91	1.87
38	5.45	4.07	3.48	3.15	2.92	2.76	2.64	2.55	2.47	2.41	2.09	1.96	1.90	1.85
39	5.43	4.06	3.47	3.14	2.91	2.75	2.63	2.54	2.46	2.40	2.08	1.95	1.89	1.84
40	5.42	4.05	3.46	3.13	2.90	2.74	2.62	2.53	2.45	2.39	2.07	1.94	1.88	1.83
41	5.41	4.04	3.45	3.12	2.89	2.74	2.62	2.52	2.44	2.38	2.06	1.93	1.87	1.82
42	5.40	4.03	3.45	3.11	2.89	2.73	2.61	2.51	2.43	2.37	2.05	1.92	1.86	1.81
43	5.39	4.02	3.44	3.10	2.88	2.72	2.60	2.50	2.43	2.36	2.04	1.92	1.85	1.80
44	5.39	4.02	3.43	3.09	2.87	2.71	2.59	2.50	2.42	2.36	2.03	1.91	1.84	1.80
45	5.38	4.01	3.42	3.09	2.86	2.70	2.58	2.49	2.41	2.35	2.03	1.90	1.83	1.79
46	5.37	4.00	3.42	3.08	2.86	2.70	2.58	2.48	2.41	2.34	2.02	1.89	1.82	1.78
47	5.36	3.99	3.41	3.07	2.85	2.69	2.57	2.48	2.40	2.33	2.01	1.89	1.82	1.77
48	5.35	3.99	3.40	3.07	2.84	2.69	2.56	2.47	2.39	2.33	2.01	1.88	1.81	1.77
49	5.35	3.98	3.40	3.06	2.84	2.68	2.56	2.46	2.39	2.32	2.00	1.87	1.80	1.76
50	5.34	3.97	3.39	3.05	2.83	2.67	2.55	2.46	2.38	2.32	1.99	1.87	1.80	1.75
75	5.23	3.88	3.30	2.96	2.74	2.58	2.46	2.37	2.29	2.22	1.90	1.76	1.69	1.65
100	5.18	3.83	3.25	2.92	2.70	2.54	2.42	2.32	2.24	2.18	1.85	1.71	1.64	1.59
125	5.15	3.80	3.22	2.89	2.67	2.51	2.39	2.30	2.22	2.15	1.82	1.68	1.61	1.56
150	5.13	3.78	3.20	2.87	2.65	2.49	2.37	2.28	2.20	2.13	1.80	1.67	1.59	1.54
175	5.11	3.77	3.19	2.86	2.64	2.48	2.36	2.27	2.19	2.12	1.79	1.65	1.57	1.52
200	5.10	3.76	3.18	2.85	2.63	2.47	2.35	2.26	2.18	2.11	1.78	1.64	1.56	1.51

Statistical Tables

Table 5. Critical Values of the Chi-Square Distribution

Table entries represent values from χ^2 distribution with upper tail area equal to α.
$$P(\chi^2 > \chi^2_{df}) = \alpha, \ \text{e.g.,} \ P(\chi^2_3 > 7.81) = 0.05$$

df	.10	.05	.025	.01	.005
1	2.71	3.84	5.02	6.63	7.88
2	4.61	5.99	7.38	9.21	10.60
3	6.25	7.81	9.35	11.34	12.84
4	7.78	9.49	11.14	13.28	14.86
5	9.24	11.07	12.83	15.09	16.75
6	10.64	12.59	14.45	16.81	18.55
7	12.02	14.07	16.01	18.48	20.28
8	13.36	15.51	17.53	20.09	21.95
9	14.68	16.92	19.02	21.67	23.59
10	15.99	18.31	20.48	23.21	25.19
11	17.28	19.68	21.92	24.72	26.76
12	18.55	21.03	23.34	26.22	28.30
13	19.81	22.36	24.74	27.69	29.82
14	21.06	23.68	26.12	29.14	31.32
15	22.31	25.00	27.49	30.58	32.80
16	23.54	26.30	28.85	32.00	34.27
17	24.77	27.59	30.19	33.41	35.72
18	25.99	28.87	31.53	34.81	37.16
19	27.20	30.14	32.85	36.19	38.58
20	28.41	31.41	34.17	37.57	40.00
21	29.62	32.67	35.48	38.93	41.40
22	30.81	33.92	36.78	40.29	42.80
23	32.01	35.17	38.08	41.64	44.18
24	33.20	36.42	39.36	42.98	45.56
25	34.38	37.65	40.65	44.31	46.93
26	35.56	38.89	41.92	45.64	48.29
27	36.74	40.11	43.19	46.96	49.64
28	37.92	41.34	44.46	48.28	50.99
29	39.09	42.56	45.72	49.59	52.34
30	40.26	43.77	46.98	50.89	53.67
40	51.81	55.76	59.34	63.69	66.77
50	63.17	67.50	71.42	76.15	79.49
60	74.40	79.08	83.30	88.38	91.95
70	85.53	90.53	95.02	100.4	104.2
80	96.58	101.9	106.6	112.3	116.3
90	107.6	113.1	118.1	124.1	128.3
100	118.5	124.3	129.6	135.8	140.2

Statistical Tables

Table 6. Critical Values of the Studentized Range Distribution, $\alpha=0.05$

K = Number of Treatments (Groups) being Compared

df_{Error} (df_{Within})	3	4	5	6	7	8	9	10
5	4.60	5.22	5.67	6.03	6.33	6.58	6.80	6.99
6	4.34	4.90	5.30	5.63	5.90	6.12	6.32	6.49
7	4.16	4.68	5.06	5.36	5.61	5.82	6.00	6.16
8	4.04	4.53	4.89	5.17	5.40	5.60	5.77	5.92
9	3.95	4.41	4.76	5.02	5.24	5.43	5.59	5.74
10	3.88	4.33	4.65	4.91	5.12	5.30	5.46	5.60
11	3.82	4.26	4.57	4.82	5.03	5.20	5.35	5.49
12	3.77	4.20	4.51	4.75	4.95	5.12	5.27	5.39
13	3.73	4.15	4.45	4.69	4.88	5.05	5.19	5.32
14	3.70	4.11	4.41	4.64	4.83	4.99	5.13	5.25
15	3.67	4.08	4.37	4.59	4.78	4.94	5.08	5.20
16	3.65	4.05	4.33	4.56	4.74	4.90	5.03	5.15
17	3.63	4.02	4.30	4.52	4.70	4.86	4.99	5.11
18	3.61	4.00	4.28	4.49	4.67	4.82	4.96	5.07
19	3.59	3.98	4.25	4.47	4.65	4.79	4.92	5.04
20	3.58	3.96	4.23	4.45	4.62	4.77	4.90	5.01
30	3.49	3.85	4.10	4.30	4.46	4.60	4.72	4.82
40	3.44	3.79	4.04	4.23	4.39	4.52	4.63	4.73
60	3.40	3.74	3.98	4.16	4.31	4.44	4.55	4.65
120	3.36	3.68	3.92	4.10	4.24	4.36	4.47	4.56
∞	3.31	3.63	3.86	4.03	4.17	4.29	4.39	4.47

Statistical Tables

II.   SAS Programs used to generate table entries

Table 1.   Probabilities of the Binomial Distribution

```
options ps=55 ls=80;
data in;
k=1;
file 'c:\sas\btable.out' noprint notitle;
do n=2 to 25 by 1;
x=0;
 do while (x le n);
  put +4 x 2.0 @@;
   do pi=0.1 to 0.9 by 0.10;
    if x=0 then p=probbnml(pi,n,x);
    else if x^=0 then p=probbnml(pi,n,x)-probbnml(pi,n,x-1);
    output;
    if mod(k,9) ^= 0 then put +2 p 6.4 @@;
    else if mod(k,9) = 0 then put +2 p 6.4 /;
    k+1;
   end;
   x+1;
 end;
end;
run;
```

Table 2.   Probabilities of the Standard Normal Distribution

```
options ps=55 ls=80;
data in;
k=1;
file 'c:\sas\ztable.out' noprint notitle;
do i=-3.09 to 3.09 by 0.01;
 p=probnorm(i);
 output;
 if mod(k,10) ^= 0 then put +2 p 6.4 @@;
 else if mod(k,10) = 0 then put +2 p 6.4 /;
 k+1;
end;
run;
```

Statistical Tables

Table 3.   Critical Values of the t Distribution

```
options ps=55 ls=80;
data in;
k=1;
file 'c:\sas\ttable.out' noprint notitle;
do df=1 to 75 by 1;
 do a=0.10,0.05,0.025,0.01,0.005;
  t=abs(tinv(a,df));
  output;
  if mod(k,5) ^= 0 then put +2 t 5.3 @@;
  else if mod(k,5) = 0 then put +2 t 5.3 /;
  k+1;
 end;
end;
run;
```

Table 4.   Critical Values of the F Distribution

```
options ps=55 ls=80;
data in;
k=1;
file 'c:\sas\ftable.out' noprint notitle;
do df2=1 to 49,50 to 200 by 25;
 do df1=1 to 9 by 1,10 to 50 by 10;
  f=finv(0.95,df1,df2);
  output;
  if mod(k,14) ^= 0 then put +2 f 5.2 @@;
  else if mod(k,14) = 0 then put +2 f 5.2 /;
  k+1;
 end;
end;
run;
```

Statistical Tables

Table 5. Critical Values of the χ^2 Distribution

```
options ps=55 ls=80;
data in;
k=1;
file 'c:\sas\ctable.out' noprint notitle;
do df=1 to 29,30 to 100 by 10;
 do a=0.10,0.05,0.025,0.01,0.005;
  c=cinv(1-a,df);
  output;
  if mod(k,5) ^= 0 then put +2 c 5.2 @@;
  else if mod(k,5) = 0 then put +2 c 5.2 /;
  k+1;
 end;
end;
run;
```